LABORS OF A MODERN HERCULES

LABORS OF A MODERN HERCULES

The Evolution of a Chemical Company

DAVIS DYER
AND
DAVID B. SICILIA

HARVARD BUSINESS SCHOOL PRESS
Boston, Massachusetts

Printed in the United States of America
94 93 92 91 90 5 4 3 2 1

Unless otherwise noted, all photographs were furnished by Hercules Incorporated. Page 26b, courtesy of the Hagley Museum and Library; page 186b, courtesy of Emile Progoff; page 280a, reprinted with permission from *Chemical Processing*; pages 26b, 66a, and 66e reproduced courtesy of the Institute of Makers of Explosives.

The paper used in this publication meets the requirements of the American National Standard for Permanence of Paper for Printed Library Materials Z39.49-1984.

Library of Congress Cataloging-in-Publication Data

Dyer, Davis.
Labors of a modern Hercules : the evolution of a chemical company / Davis Dyer and David B. Sicilia.
p. cm.
Includes bibliographical references and index.
ISBN 0-87584-227-5 (alk. paper)
1. Hercules Incorporated—History. 2. Chemical industry—United States—History. I. Sicilia, David B. II. Title.
HD9651.9.H47D94 1990 90-36847
338.7′66′00973—dc20 CIP

FOR JANICE AND LEILA

C O N T E N T S

8 Defense Contractor, 1940–1945 221

9 Joining the Petrochemicals Revolution, 1946–1954 253

10 Expansion and Metamorphosis, 1955–1961 291

11 Managing New Constraints, 1962–1973 337

12 Restructuring and Reorientation, 1974–1987 385

PREFACE

This book has been roughly two and a half years in the making, although, in a larger sense, it has roots that trace to the late 1970s. At that time, at the initiative of Richard B. Douglas, director of Public Relations at Hercules Incorporated, the company's PR Department launched a program to gather historical materials, conduct oral history interviews, and prepare a detailed chronology of key events in the company's past. At first, the goal of this work was to create an accurate record of Hercules' history to use as a reference tool. As time wore on, however, Douglas and other managers, including S. Raymond Clarke, senior vice president; Dr. Eugene D. Crittenden, Jr., vice president for administration and public affairs; and S. Maynard Turk, general counsel, hoped to expand the project and produce a book on the company. To these champions of the cause, Hercules' seventy-fifth anniversary, celebrated in 1988, provided an ideal occasion for commissioning a book on the company's history.

Late in 1986, outgoing chairman and CEO Alexander F. Giacco and his successor, David S. Hollingsworth, approved the project, and in February 1987, Hercules contracted with the Winthrop Group to write a history of the company's first seventy-five years. Although it was clear by then that a book could not possibly be completed before Hercules' formal birthday celebration in June 1988, the company was eager nonetheless to see its history handled in a scholarly fashion and willingly acceded to the timetables of professional historical research and academic publishing.

Hercules imposed no restrictions on us, save a general concern to avoid disclosing details of the company's current competitive position and trade secrets. We shared this concern: there is little point in a book of this sort—a history, after all—to airing the company's present-day strategy; nor, in any case, do we as historians feel confident in pronouncing on events and decisions whose ultimate significance remains to be seen.

Accordingly, Chapter 12, which chronicles the period between 1974 and 1987, treads more lightly than other chapters of the book. As it happens, these years have been unusually eventful. Although we did not scrutinize this period for the purposes of publication, we have gathered documents and records and conducted dozens of oral history interviews that bear on the period. These materials, deposited in an archive at Hercules, will perhaps aid future historians in understanding this turbulent era at the company.

Throughout this project, Hercules has provided an extraordinary degree of support to us, offering unrestricted access to vital company records and making available countless hours of employees' time for interviews. Most of the research for this book was conducted in Wilmington, Delaware, home of Hercules' modern headquarters building as well as its Research Center, Marketing Center, and Hall of Records. In addition, we made short (one- to three-day) research trips to current or formerly owned operations across the country. These sites included Kenvil (smokeless powder) and Parlin, New Jersey (nitrocellulose); Brunswick (naval stores); Savannah (paper chemicals); and Oxford, Georgia (synthetic fibers); Magna, Utah (rocket motors and graphite composites); Hopewell (chemical cotton and water-soluble polymers) and Radford, Virginia (military ordnance); Terre Haute, Indiana (polypropylene film); and Lake Charles, Louisiana (polypropylene resins; a plant now owned by HIMONT Incorporated, a subsidiary of the Italian chemical company, Montedison S.p.A.).[1]

During the project, we met periodically with an advisory committee consisting of present and former Hercules managers and a noted expert in business history. We are grateful for the advice, comments, and suggestions from this group, which included (in alphabetical order) Theodore M. Bednarski, Werner C. Brown, Eugene D. Crittenden, Jr., Richard B. Douglas, Carl Eurenius, John R. L. Johnson, Jr., Donald R. Kirtley, Carolyn R. Miller, Glenn Porter, and S. Maynard Turk.

Although Hercules was generous in support of the project, the organization, themes, interpretations, and conclusions of the book are solely our responsibility. The company did not provide any kind of subvention to Harvard Business School Press, which followed its customary review procedures before agreeing to publish the manuscript.

The subject of this book is the successive transformations of Hercules as it grew from a medium-sized maker of explosives created by a judicial decree in 1912 to its modern incarnation as a multibillion-dollar, diversified, multinational corporation. As such, this is a business history that focuses on the company's growth and change from the point of view of the leaders and decision makers responsible for guiding it into the future. In choosing this perspective, we were influenced by several factors.

First, we started with the conviction that the primary question to be asked of any large and older company is how it became what it is. The answer to this question, we believe, lies first of all in understanding and analyzing the opportunities the company faced, the decisions its leaders made, and the organization and policies that supported (or thwarted) its efforts. The best business histories, in our view, follow (at least implicitly) the strategy and structure paradigm developed by Alfred D. Chandler, Jr.[2]

There are, of course, other valid ways to approach a company's history: for example, focusing on group biography, technology, social relationships inside the company, or social and political contexts on the outside. Such approaches, we believe, work best when the outlines of a company's business history are well known or easily sketched. The particular case of Hercules—which had not published a detailed history of itself prior to this volume, which has an unusually intricate pattern of growth, and which operates in a highly complex industry—reinforced our conviction that the place to begin was with a high-level view of its strategy and structure.

Second, like all historians, we were influenced by the nature of our sources. In this respect, we were extremely fortunate. Hercules' history is well documented from the beginning. The company's origins are a matter of public record in the antitrust case against E.I. du Pont de Nemours & Company. Over the years, moreover, the Du Pont Company and the du Pont family donated voluminous records from the early twentieth century—including many items pertaining to the founding and early years of Hercules—to the Hagley Museum and Library in Wilmington, where they are accessible to scholars.

In contrast to most companies, Hercules began its independent life with experienced leadership and many management systems in place. From the outset, the company was organized into departments that reported periodically (in most cases monthly and also annually) to the board of directors until 1928, and thereafter until 1978 to the executive committee of the board. By a wonderful stroke of luck, we found a virtually complete run of these records covering the entire period. This massive resource—roughly a thousand pages of typed commentary, tables, and graphs for each year, the entire series housed in more than a hundred cartons—proved a gold mine: an extraordinarily rich vein of materials that illumines the growth and change of the company.

Other important sources survive, too: the Hercules company magazine, the *Mixer*, which runs from 1919 to the present; newsletters and magazines from the plants; advertisements and trade literature from the earliest days; published technical reports; detailed official histories of the wartime ordnance facilities; minutes of committee meetings; and many other significant items. To help flesh out these materials, we interviewed roughly two hundred present and former employees. About half of the interviews were in-depth, one-on-one discussions lasting several hours. We met several retirees whose memories stretched back to the company's earliest years (including one employee hired in 1914), but oral histories proved most valuable in shedding light on events from the 1930s forward. Most of the individuals interviewed held managerial or technical positions in the company. (A complete list of interviewees is included in Appendix I.)

In sum, we had a treasure trove of documents and records, supplemented by interviews, to support a business history of Hercules. The richness of the sources affords a rare opportunity to understand one company's growth and change in detail.

Although the focus of this book is Hercules itself, the company's story provides interesting insights into the growth of U.S. companies generally, and the dynamics of the chemical industry in particular. In many respects, Hercules' evolution is a paradigm of the development of the modern U.S. manufacturing corporation: growth through vertical integration and diversification; the move from a centralized, functional organization to a decentralized, multidivisional one; increasing emphasis on research and development; gradual penetration of foreign markets, beginning in Europe before moving to Asia and less developed countries (LDCs); increasing reliance on portfolio analysis to frame corporate strategies; formation of joint ventures and other kinds of relationships with foreign producers; and heavy use of information technology to streamline management and operations.

Hercules stood in the forefront of many of these developments in part because of the nature of competition in the chemical industry, and also because of the company's middling rank among chemical producers. Long neglected by academic researchers, the chemical industry has recently become the subject of several important books: by Arnold Thackray and his collaborators at the Beckman Center for the History of Chemistry, and by Joseph L. Bower, Sheldon Hochheiser, Peter H. Spitz, Robert Stobaugh, and David A. Hounshell and John Kenly Smith, Jr.[3] As these and other recent works show, the chemical industry is unusually dynamic: change is rapid, constant, and extensive (see Chapter 1). The chemical industry is more diversified, technology-dependent, capital-intensive, and global than any other major industry of the developed economies. As a result, it has led most other industries in the development of modern corporate strategies and management techniques. It is not too much to suggest that the chemical industry is to the twentieth-century corporation what the railroad was to its nineteenth-century forebear: the wellspring of management innovation.

Every successful competitor in the chemical industry has to be fast on its feet. Hercules, a medium-sized competitor (ranked eighth by sales among U.S.-based chemical producers in 1987), has had to be especially nimble. In contrast to larger companies such as Du Pont in the United States, ICI in the United Kingdom, or any of the great triumvirate in Germany—Bayer, BASF, and Hoechst—Hercules has seldom enjoyed a commanding position in major chemicals markets. As a result, it has been obliged to choose opportunities and weigh investments with unusual care and to regroup quickly in response to changing market conditions. In research, the company has generally been a fast follower, adapting and improving upon the inventions of others, a process innovator rather than a pioneer of new products. In production, Hercules has tended to avoid vertical integration (backward and especially forward) because its operations lacked sufficient scale. And in marketing, the company has focused on particular niches, providing its customers with a high level of technical support and service.

A fast-moving competitor in a fast-changing industry, Hercules has been in a constant state of transformation since its earliest days. At various times, Hercules has been the country's second-largest explosives company; number-one producer of nitrocellulose; largest source of chemical cotton; leading producer of wood naval stores and naval stores derivatives; principal supplier of chemicals to the paper industry; and largest or second-largest manufacturer of sodium carboxymethylcellulose (CMC—a versatile additive to foods and fluids), dimethyl terephthalate (DMT—an intermediate chemical for making polyester fiber), polypropylene, solid fuel for rockets, and graphite composites.

At its founding, Hercules defined itself as an explosives-maker. On its twenty-fifth anniversary, company publicity still mentioned explosives but emphasized achievements in cellulose products, naval stores, and paper chemicals. On its fiftieth, Hercules celebrated its position as "a well-diversified, major chemical manufacturing company" and highlighted its balanced sales to sixteen diverse industries. And on its seventy-fifth, it described itself as "a worldwide supplier of a broad line of natural and synthetic materials and products and related systems serving a wide range of industries, including the electronics, packaging, aerospace, food, synthetic fibers, automotive, graphic arts, adhesives, paper coatings, and personal-care industries."[4]

Hercules has come a long way. This book is intended to describe and explain how it happened.

II

The writing of a company history poses peculiar issues for authors. To begin with, the imperative to be comprehensive limits the amount of space that can be devoted to any one subject or theme, no matter how interesting and important. This is not a book about industrial marketing, although Hercules is an extremely effective practitioner, and a good book could be written about that topic using Hercules as a case study. Nor is it about process innovation or new product development, although these are historic strengths of the company. We touch on these themes, of course, and many others, too: a striking degree of loyalty to and affection for the company among its employees (reflecting in part the company's origins as an explosives-maker and its long-standing concern with safety); a preoccupation with achieving balanced growth; a high standard of analysis of business problems; periodic concern about the price and availability of raw materials; an aversion to making highly risky investments; a continual search for stable market niches; strong skills in chemical engineering; and rapid adoption and adaptation of new technologies.

All of these themes interplay, ebb, and flow across the history of Hercules, and testify to the complex, vital, and dynamic nature of that history. No single theme dominates throughout, however, and none by itself embodies or reflects the company's essential character. The closest to a dominant theme in the company involves the role of chemical engineering, a restless mindset, and skills that continually lead Hercules into new areas. This theme, however, must be joined with two others: the company's long-standing excellence in industrial marketing, a capability that helps identify new opportunites and establish secure market niches; and Hercules' generally conservative financial management, an outlook that determines the kinds of businesses in which it will compete.

Capturing the complexity of the corporation is one issue for historians. The problem of selection is another. Lest a corporate history become unduly encyclopedic, choices must be made about which themes and subjects to emphasize. In the case of a highly diversified company like Hercules—a maker of thousands of products, each with its own history of research and development, production, and distribution, each with its own challenges, setbacks, and victories, and each with its own fate in the marketplace—the problem of selection is particularly acute. In the account below, not every cellulose product, rosin derivative, materials breakthrough, rocket-motor innovation, market penetration, or domestic or foreign investment receives its due: inevitably, some important people and episodes receive scant attention, or are subsumed in representative stories.

Finally, the enterprise of a corporate history is open-ended: although a book must have a final chapter, the tale is not over. Healthy corporations continue to grow and change. This book is a biography of a living subject, a drama whose next act is unfolding even as this one closes. An uncompleted story, then, is one constraint on us as historians. A related issue is the indeterminate significance of recent events: as our narrative marches toward the present, we are transformed from historians into journalists and chroniclers. Our primary aim in discussing Hercules in the 1980s is to get the facts straight, so that readers and future historians will have a solid base for understanding the company during what appears to be a highly consequential period of its existence.

With these general issues in mind, let us turn to the organizing structure of the book.

III

In the 1950s, the history of Hercules appeared as a case study to illustrate a famous theory of how companies grow. Edith Penrose, a professor of economics at Johns Hopkins University, postulated in her book, *The Theory of the Growth of the Firm,* that the possibilities of growth for a given company are constrained by its accumulated resources, experience, and knowledge, as well as by the kinds of opportunities it investigates when it considers expansion. In an article written about Hercules from the vantage point of the early 1950s, Penrose saw that the company had been "imaginative, versatile, and venturesome in the introduction of new products," based on chemistry it understood, but also "cautious and conservative in entering new and alien fields of technology."[5] (Had she looked again a bit later, during Hercules' era of rapid expansion via acquisition and investment in petrochemicals, she might have reached a different conclusion.)

Penrose believed that the growth of a company is the result of a complex interaction between its resources and capabilities, its goals, and its market opportunities. One need not be so abstract to see that Hercules has, in fact, grown and changed at different rates in different periods for different reasons. The pace and direction of its growth have been determined by the company's management as well as by developments in the marketplace.

The chapter outline of this book attempts to capture this interplay of organization and environment in Hercules' history. Although the overall structure is chronological, the basic logic of the book reflects the notion that a company grows and changes in response to opportunities and threats. That is, in a given period, a company's performance and fate depend upon how well it deals with a set of primary challenges, those specific to its business and those generated by its political and social context.

When a company is young or small, for example, such challenges might include attracting investment capital, developing a new product or process, or opening or cultivating a market. As the company grows and matures, the challenges become different: satisfying investors' expectations for growth and profitability; responding to competitive threats; expanding the geographical scope; enlarging through vertical integration or diversification; forming structures and systems to institutionalize innovation; and developing formal planning and management systems (including personnel planning and executive succession) to sustain the "immortality" of the corporation.

Such challenges concern (or have concerned) every large and growing company. But companies also operate in a political and social milieu. For U.S. companies in the twentieth century, the context of business has changed dramatically through wars, depressions, and recessions. Long-term trends, such as escalated government intervention in the marketplace, changing demographic patterns, and the increasing interdependence of world markets, also affect the way a company conducts its business. Such "external" circumstances pose specific challenges: mobilization and demobilization, expansion and retrenchment, upgrading of skills in public affairs and marketing, and development of new strategies for global competition.

How well a company responds to the key business problems of a given period is the primary determinant of its survival and success. The organizing principle of this history of Hercules, then, is challenge and response. Each historical chapter (especially Chapters 3 through 12) deals with the company's growth and change in response to specific business challenges.

Chapter 1 introduces the company as it stood at a particular moment in time: 1987, Hercules' seventy-fifth year. This chapter is a descriptive and necessarily impressionistic portrait of a sophisticated, diversified, multinational corporation and includes a section on the dynamics of change in the modern chemical industry. Thus, Chapter 1 provides a sense of the modern company in its complexity.

Chapter 2 then flashes back to the other endpoint of the story, the conception of Hercules in 1911, when the U.S. Circuit Court for the District of Delaware ordered the dismemberment of the Du Pont explosives business. Here we discuss why and how Hercules was created and introduce the problems it was to confront on its first day of operation, January 1, 1913.

Chapter 3 picks up the story of Hercules as it faced its initial challenge to organize itself and develop a viable strategy for competing in the explosives industry. In its first two years, certain elements of the strategy became clear, as the company sought to broaden its product line into newer explosives such as smokeless powder and to compete on a nationwide basis. At that point, the outbreak of World War I brought a wholly new challenge: management of the vast expansion of business created by the war. Chapter 4 covers the World War I years and highlights how Hercules began to change from a mixer of explosives to a maker of chemicals. For instance, producing military smokeless powders and trinitrotoluene (TNT) required that the company master new chemical and engineering skills, a foundation on which it could build after the Armistice.

Chapter 5 deals with the postwar challenges: What would happen to the people and assets employed and the ideas and skills mastered during the war? How would the company develop a distinctive strategy of its own? The answers, after much debate, were an exploration of peacetime uses of nitrocellulose and entry into an entirely new area, the naval stores industry. Chapter 6 describes how Hercules slowly developed successful strategies for completing its new lines of business. As it eventually prospered in commercial nitrocellulose and naval stores, Hercules increasingly defined itself as a chemical company rather than an explosives-maker. This transformation was memorialized in 1928 when the company adopted a multidivisional structure that was to endure for fifty years.

In the dozen years after this reorganization—the period treated in Chapter 7—Hercules completed its transformation into a diversified chemical company. In the 1930s, Hercules established a modern, centralized R&D laboratory in Wilmington, acquired a major subsidiary in a new business (paper chemicals), and competed increasingly on the basis of its technical expertise in cellulose and rosin chemistry. The company's research in cellulosics, for example, led to promising new opportunities in protective coatings and plastics. At the same time, research in rosin and terpene chemistry resulted in many new products from naval stores, including synthetic resins and additives for a variety of markets. By the end of the period, the company's products and markets were highly specialized.

The outbreak of World War II posed the renewed challenge of scaling up as a military contractor. Chapter 8 tells this story and describes the new sciences and skills the company developed during the war: rocketry, cellulosic plastics, and insecticides derived from naval stores products.

The subject of Chapter 9 is Hercules' approach to the problems of postwar reconversion and the reestablishment of its base businesses. The company continued to develop new products from familiar chemistry, including toxaphene, a new insecticide, and CMC. At the same time, the company maintained its hand in chemical propulsion as a military contractor. This was a conservative growth strategy, however, and Hercules lagged behind other U.S. chemical producers in cultivating the rapidly expanding postwar market for petrochemicals. In the mid-1950s, to achieve faster growth, Hercules at last joined the petrochemicals revolution by working with European partners to develop intermediate chemicals and a new process for polymerizing ethylene and propylene.

Hercules' entry into petrochemicals proved to be more than a straightforward diversification: growth in this business, as Chapter 10 shows, required that the company make heavy investments and integrate forward into polyolefin fibers and films. At the same time—the late 1950s and early 1960s—the company used acquisitions to broaden its base in protective coatings, food additives, and chemical propulsion and to acquire technical know-how in related areas.

Chapter 11 depicts the recognizably modern Hercules as a diversified, multinational chemical company with operations in explosives, pine and paper chemicals, synthetics, pigments, polymers, fiber development, and cellulose and protein products. Sensing limits to growth in its traditional lines, Hercules pushed hard in the 1960s to develop opportunities in petrochemical plastics, including fibers and films. At the same time, the company redoubled its efforts to expand overseas through license agreements and joint ventures, as well as through direct investment. Finally, in these "go-go" years, Hercules embarked on a program of radical diversification into new areas through acquisitions and equity investments under the aegis of its New Enterprise Department.

In the years since the first oil crisis of 1973—the period covered in Chapter 12—Hercules has undergone a thorough and wrenching restructuring of its operations. The company abandoned most of the diversification efforts of the previous decade and divested many of its traditional businesses. In 1978, Hercules abolished the departmental structure that had served it for half a century in favor of a structure based on strategic business units. During this period, Hercules sought to reduce its dependence on commodity petrochemicals and dramatically refocused on value-added businesses such as aerospace, engineered polymers, and specialty chemicals. Chapter 12 concludes with a brief narrative of events at Hercules from 1987 through the end of 1989 and offers general observations on the company's history.

LABORS OF A MODERN HERCULES

CHAPTER 1

HERCULES AT SEVENTY-FIVE

In mid-June, the weather in Wilmington is contested terrain, a battleground of seasons fighting for supremacy. One day, it is cold and rainy, a reminder that winter has not quite given up; on the next day, it is midsummer, a foretaste of the long periods of steamy, hazy heat to come. Occasionally—tantalizingly—the gods relax and smile, and it is spring: warm, clear, dry, and breezy.

On the weekend of June 18–19, 1988, the gods were smiling on Wilmington. The source of their good humor, perhaps, was a gigantic birthday party for a favored child of Olympus. Ten thousand men, women, and children, organized in four shifts across the two days, gathered at the company-owned country club near Wilmington to celebrate. There was plenty of food: a ton and a half of ground beef, 16,000 hot dogs (which, had they been laid end to end, would have covered a mile and a half), 8,800 pieces of chicken, 3,000 pounds of baked beans, 8,000 cubic feet of potato chips and pretzels, 2,300 gallons of beverages, and 10,000 cupcakes. A local disc jockey played popular music, including "Happy Anniversary" and "We Are Family," while clowns, jugglers, and costumed characters worked their way through the crowds to the delight of the youngsters. Nearly 400 volunteer "ambassadors" and "diplomats"—identifiable by their red knit shirts and white pith helmets—served food, guided people to and from buses and parking lots, answered questions, and patrolled the grounds to pick up litter.

It was an extraordinary event, in some respects resembling a family reunion. People of all ages, small children to nonagenarians, mingled, formed into groups, and talked animatedly with friends and passersby. All of them were present to commemorate a special occasion: the guest of honor at this fete had marked a seventy-fifth birthday earlier in the year. This guest, however, was not a person, but a company: Hercules Incorporated.

THE GUEST OF HONOR

Seventy-five years is a long time in the life of any business, and it is well that Hercules Incorporated should have paused to mark the occasion.

To put some perspective on the achievement:

The survival rate of American corporations is a dismal statistic. More than 90 percent of new ventures fail in their first five years. To be sure, Hercules, created by an antitrust action against E.I. du Pont de Nemours & Company in 1912, was not a typical startup. It began with a thousand employees and millions of dollars in the bank. But even big companies frequently fail or disappear. *Forbes* magazine, which celebrated its seventieth birthday in 1987, recently published a list of the leading American corporations of 1917 with notes on what had happened to them since. Of the one hundred largest companies in the United States back then, only twenty-two are still around in recognizable form—a dropout rate of 78 percent.[1]

Consider another point: The judicial action that created Hercules seventy-five years ago also launched the Atlas Powder Company. This company was calling itself Atlas Chemical Industries in 1971 when it was bought by Imperial Chemical Industries (ICI) of Great Britain. The modern headquarters and laboratories Atlas built for itself in Wilmington in 1955 are now the home of ICI Americas.

Nor has Atlas been the only casualty of time and turnover among Hercules' rivals. In 1913, the six largest explosives companies in the United States were Du Pont, Hercules, Atlas, Aetna, Giant, and Grasselli. Today, only Hercules and Du Pont are still around, although neither is prominent in the commercial explosives business.

The Hercules of the late 1980s has journeyed far from its origins in the explosives business. In 1987, Hercules achieved sales of roughly $2.7 billion and reported an operating profit of $208 million. At the same time, the sale of its interest in a major joint venture—HIMONT Incorporated, a global producer of polypropylene plastic resins—resulted in a nonrecurring, after-tax gain of $1.15 billion. Hercules was operating facilities in more than eighty locations worldwide and employing nearly twenty-six thousand people. The company and its subsidiaries and affiliates were manufacturing approximately three thousand different products, ranging from highly specialized chemicals and additives to synthetic fibers and materials to solid-fuel motors for missiles and rockets.[2]

Hercules organized its businesses into three "operating companies": Hercules Aerospace Company (HAC), Hercules Specialty Chemicals Company (HSCC), and Hercules Engineered Polymers Company (HEPC).* In addition, the corporation owned and operated 50 percent of The Aqualon Group, a global joint venture that manufactured and marketed water-soluble polymers, and held equity positions ranging from 20 to 95 percent in three dozen other (primarily chemical) ventures around the world (see Exhibits 1.1 and 1.2).

* Since 1987, Hercules has changed its organization and mix of businesses. For an account of Hercules in 1989, see the epilogue of Chapter 12 below.

EXHIBIT 1.1 Hercules' Offices and Manufacturing Sites, 1987

HERCULES
SPECIALTY CHEMICALS
COMPANY
Amersfoort, The Netherlands
Barneveld, The Netherlands
Beringen, Belgium
Bremen, Federal Republic of Germany
Brunswick, Georgia
Burlington, New Jersey
Burlington, Ontario, Canada
Busnago, Italy
Chicopee, Massachusetts
Culver City, California
Franklin, Virginia
Gibbstown, New Jersey
Grenoble, France
Grossenbrode, Federal Republic of Germany
Harbor Beach, Michigan
Hattiesburg, Mississippi
Hopewell, Virginia
Hutchinson, Minnesota
Kalamazoo, Michigan
Lilla Edet, Sweden
Lille Skensved, Denmark
Louisiana, Missouri
Mexico City, Mexico
Middelburg, The Netherlands
Middletown, Delaware
Middletown, New York
Milwaukee, Wisconsin
Parlin, New Jersey
Pendlebury, England
Perivale, England
Portland, Oregon
St.-Jean, Quebec, Canada
Sandame, Sweden
São Paulo, Brazil
Savannah, Georgia
Sobernheim, Federal Republic of Germany
Tampere, Finland
Tarragona, Spain
Traun, Austria
Vero Beach, Florida
West Elizabeth, Pennsylvania
Zwijndrecht, The Netherlands

HERCULES ENGINEERED
POLYMERS COMPANY
Covington, Virginia
Deer Park, Texas
Iberville, Quebec, Canada
Oxford, Georgia
Terre Haute, Indiana
Varennes, Quebec, Canada
Winooski, Vermont

HERCULES AEROSPACE
COMPANY
Cedar Knolls, New Jersey
Chester, New Jersey
Clearwater, Florida
Hatfield, Pennsylvania
Kenvil, New Jersey
Magna, Utah
McGregor, Texas
Miami, Florida
Norwich, New York
Rocket Center, West Virginia
Vergennes, Vermont

GOVERNMENT-OWNED PLANTS
OPERATED BY HERCULES
AEROSPACE
Radford Army Ammunition Plant
Radford, Virginia
Sunflower Army Ammunition Plant
Lawrence, Kansas

SALES OFFICES
Agawam, Massachusetts
Akron, Ohio
Atlanta, Georgia
Beaverton, Oregon
Chicago, Illinois
Cincinnati, Ohio
Dallas, Texas
Detroit, Michigan
Green Bay, Wisconsin
Greenwich, Connecticut
Kalamazoo, Michigan
Los Angeles, California
Mobile, Alabama
Richmond, Virginia
San Francisco, California
Shreveport, Louisiana
Waterville, Maine
Wilmington, Delaware

Source: Hercules Incorporated, Annual Report for 1987.

EXHIBIT 1.2 Principal Hercules' Subsidiaries and Affiliated Companies, 1987

CONSOLIDATED,
WHOLLY OWNED SUBSIDIARIES
(Directly or Indirectly)

Austria
Patex Chemie GmbH, Traun

The Bahamas
Hercules International Trade Corporation Limited (HINTCO), Nassau

Belgium
Hercules Chemicals N.V., Beringen
S.A. Hercules Europe N.V., Brussels

Bermuda
Curtis Bay Insurance Co. Ltd., Hamilton

Brazil
Hercules do Brasil Produtos Quimicos Ltda., São Paulo

Canada
Hercules Canada Inc., Mississauga, Ontario

Denmark
A/S Københavns Pektinfabrik, Lille Skensved

England
Hercules Limited, London
PFW Ltd., Perivale
Zimmerman Hobbs Holdings Limited, Milton Keynes

Finland
Oy Hercofinn Ab, Helsinki

France
Hercules France S.A., Paris

Germany (Federal Republic of)
Hercules GmbH, Hamburg
Pomosin GmbH, Grossenbrode

Italy
Hercules Italia S.p.A., Milan

Japan
Hercules Far East K.K., Tokyo

Mexico
Quimica Hercules, S.A. de C.V., Mexico, D.F.

The Netherlands
Hercules B.V., Rijswijk
PFW (Nederland) B.V., Amersfoort

AFFILIATED COMPANIES
(Percentage Directly or Indirectly Owned by Hercules)

Australia
Australian Chemical Holdings Limited, Botany, New South Wales (63%)
Hercules Chemicals Investments Pty. Ltd., Melbourne (63%)

Belgium
Moplefan N.V., Brussels (50%)

Brazil
Polo Industria e Comercio, Ltda., São Paulo (49%)

Canada
Devron-Hercules Inc., North Vancouver, British Columbia (70%)
PPD Hercules Inc., Sherbrooke, Quebec (50%)

Chile
Algas Marinas S.A., Santiago (48%)

England
Moplefan (UK) Limited, Brantham (50%)

France
Aqualon France S.A. (50%)

Germany (Federal Republic of)
Abieta Chemie GmbH, Augsburg (50%)
Aqualon GmbH & Co. K.G., Düsseldorf (50%)

Indonesia
Hercules Mas Indonesia, Djakarta (49%)

Italy
Moplefan S.p.A., Milan (50%)

Japan
DIC-Hercules Chemicals Incorporated, Tokyo (50%)
Rika-Hercules Incorporated, Osaka (50%)
Sumika-Hercules Company, Ltd., Osaka (50%)

Mexico
Petrocel, S.A., Tampico (30%)
Taloquimia, S.A., Chihuahua (49%)

The Netherlands
Aqualon B.V., The Hague (50%)

EXHIBIT 1.2 (Continued)

CONSOLIDATED,
WHOLLY OWNED SUBSIDIARIES
(Directly or Indirectly)

Singapore (Republic of)
Hersean Pte Ltd., Singapore
Spain
Ceratonia S.A., Tarragona
Sweden
Hercules Kemiska Aktiebolag, Göteborg
United States
Champlain Cable Corporation, Wilmington, Delaware
Hercules Aerospace Display Systems, Inc., Hatfield, Pennsylvania
Hercules Credit, Inc., Wilmington, Delaware
Hercules Defense Electronics Systems, Inc., Clearwater, Florida
Hercules Trading Corporation, Wilmington, Delaware
Mica Corporation, Culver City, California
Simmonds Precision Products, Inc., Wilmington, Delaware
Virgin Islands
Hercules Overseas Corporation, St. Croix

AFFILIATED COMPANIES
(Percentage Directly or Indirectly Owned by Hercules)

New Zealand
A. C. Hatrick (N.Z.) Ltd., Auckland (63%)
Pakistan
Dawood Hercules Chemicals Limited, Lahore (40%)
Spain
Hercules Aerospace España, S.A., Madrid (90%)
Taiwan
Taiwan Hercules Chemicals Inc., Taipei (95%)
United States
Aeonic Systems, Inc., Billerica, Massachusetts (34%)
Aqualon Company, Wilmington, Delaware (50%)
BHC Laboratories, Inc., Warsaw, Indiana (50%)
EKC Technology, Inc., Hayward, California (35%)
Epicor Laboratories, Inc., Palo Alto, California (50%)
Intermarine, U.S.A., Savannah, Georgia (40%)
Texas Alkyls, Inc., Deer Park, Texas (50%)

Source: Hercules Incorporated, Annual Report for 1987.

Clearly, Hercules is a complex and diverse corporation. To begin to comprehend its dimensions, it is helpful to place the company in its contexts. Among large, U.S.-based industrial companies, for example, Hercules ranks in the middle: while big in absolute terms, it is by no means a corporate giant. Ranked by sales in 1987, Hercules was 150th of the *Fortune* 500 industrial corporations—about the same size as much better known companies such as Wang Laboratories, Apple Computer, and Zenith Electronics.[3] Ranked by market value (common stock price times the number of shares outstanding on March 18, 1988), Hercules stood at 215th among the *BusinessWeek* top 1000 American corporations, ahead of such prominent companies as Martin Marietta, General Dynamics, W.R. Grace, and Hershey Foods.

Within the U.S. chemical industry (where Hercules is placed by securities analysts), the company in its anniversary year was ranked eighth by both sales and assets, behind Du Pont, Dow, Monsanto, Union Carbide, PPG Industries,

American Cyanamid, and W.R. Grace, and just in front of Quantum Chemical and Rohm and Haas.[4] (This classification excludes the petrochemical subsidiaries of the major oil companies; had they been included, Hercules would rank lower, perhaps fifteenth.) Even in its native Wilmington, Hercules is only the third-largest local chemical company, with several others moving up fast.

A TOUR THROUGH THE COMPANY

These statistics help locate the company in its industry but they convey little of the scale, diversity, and complexity of the modern Hercules. Indeed, it is impossible to capture its essential character in a brief overview. Nonetheless, a tour of several representative operations provides a better sense of its products, processes, and people, as well as a sense of the distance the company has traveled.

Aerospace: The New and the Old

In Magna, Utah, high on the shoulder of the Oquirrh Mountains, sits Hercules' largest and most ambitious facility.[5] This is Bacchus Works, the heart of Hercules Aerospace Company, and the working home of roughly 4,300 employees. The Works—named after T.W. Bacchus, the company's vice president of operations at its founding in 1913—commands a fine view of the Great Salt Lake to the north, and eighteen miles across the valley to the east, Salt Lake City.

Here the new and the old in Hercules' history intermingle. In the fall of 1913, ground was broken for a dynamite plant, the first new operation built by Hercules as an independent company. When it opened in the spring of 1915, the plant represented an investment of about $1 million. Today, across the highway from the original plant, sits the largest and most recent major investment in Hercules' history: Bacchus West, a $150-million (first stage of appropriations), highly automated facility for casting propellants for large, solid-fuel motors.

A visit to Bacchus West produces eerie sensations, stemming in part from the inherent risks of proximity to explosive matter. (The plant, which was not fully operational when we visited, is in fact quite safe: visitors are not permitted anywhere near dangerous materials; in any case, explosives facilities have better safety records than most industrial plants.) But the eeriness also derives from the sheer scale of the facility and its extraordinarily high level of automation. The complex is scattered across more than four thousand acres on the mountainside and includes twenty-nine buildings. Yet, once up to speed, Bacchus West will be operated around the clock by just sixteen technicians.

Designed to process explosive materials, the facility is also designed to minimize and contain the dangers. Distinct operations are dispersed in separate, isolated locales and conducted by remote control from a concrete bunker that sits below a three-foot earthen roof. Thus, an accident in one building will not affect operations in another.* The rocket fuel ingredients—aluminum and ammonium

*Indeed, in March 1989, a fire and resulting "overpressure" in the mixing building at Bacchus West resulted in considerable damage to property, but no injuries or loss of life.

Built as a gamble on the commercialization of space, Bacchus West is a highly automated, state-of-the-art plant to make large solid-fuel rocket motors.

In the late 1980s, Hercules spent $30 million to modernize one of its oldest and most profitable businesses: naval stores products.
At left: Workers at Brunswick, Georgia, install a continuous rosin extractor to replace the batch process invented by Homer Yaryan.

Peering through the bubble: Chairman and CEO David S. Hollingsworth surveys one of two film-forming technologies on display at Hercules' plant in Terre Haute, Indiana.

Headquarters since 1983, Hercules Plaza delights architecture critics and, more important, the people who work there.

perchlorate and HMX (a handy but obscure abbreviation for cyclotetramethylene tetranitrotriamine)—arrive at the facility in trucks and are immediately sorted into batches and stored in a massive, electronically controlled inventory area. When needed, they are carried in 5,000-pound bins from one location to another by automated guided vehicles (AGVs), robotically operated trucks that lumber along slowly and faithfully on their rounds under the watchful eyes of video cameras.

The heart of Bacchus West is the mix building, where explosive ingredients are blended into propellants. On the building's fifth level, the AGVs dump the bins into hoppers and the ingredients filter down into a mammoth mixmaster. As the bowl rotates, planetary blades in the center move in opposite directions, working the material constantly from top to bottom. As much as 28,000 pounds of propellant can be mixed in a single batch (more than four times the amount handled by conventional rocket-fuel mixers). This process resembles nothing so much as the mixing of bread dough or cake batter—though on a scale hard to fathom. Indeed, to test the equipment after its installation, employees mixed and baked a 425-pound loaf of bread.

From the mix building, the "loaded" mixing bowl is hauled by a mix bowl transporter (MBT), a huge robotic vehicle capable of bearing sixty thousand pounds of cargo. (To help humanize these behemoths, employees have christened the plant's two MBTs as "Lulu" and "FeFe.") The MBT moves its cargo through a cleaning operation, and then to the cast/cure building. At this point, the propellant is cast into a huge pit more than twenty-one feet across and over fifty feet deep. After a specified time, the propellant—now a rocket motor—is said to be "cured." (When ready, cast propellant is a rubbery substance that resembles in color the inside of a ripe avocado.)

In the final stage of production, cranes lift the motor onto a heavy-duty vehicle, which transports it to the finishing area. There it is mated to its case, nozzle, igniter, instrumentation, and other hardware and readied for pickup by the customer. In time, the motor may help propel a Delta II or a Titan IV rocket on its way into space.

The futuristic operations of Bacchus West are only a part of the dazzling technology on display at Bacchus Works. On these grounds, or at facilities nearby, other wonders are manufactured: carbon fiber and "prepreg" (a woven fabric or tape of carbon fibers that is impregnated with polyester or epoxy resins for use in composite parts or structures); filament-wound, composite rocket-motor cases, huge canisters as long as 400 feet and up to 120 feet in diameter for Titan IV boosters; motors for smaller rockets and ballistic missiles; and "3-D carbon-carbon," an incredibly light and tough material made from carbon fibers woven in three dimensions and "densified" (heated under pressure). The toughest manmade material, 3-D carbon-carbon is used to make high-performance parts for aerospace applications, such as the exit cones of rocket nozzles.

One other aspect of life at Bacchus deserves mention. As part of the U.S.–Soviet Intermediate Nuclear Forces (INF) Treaty of 1987, teams of Russian observers are posted on the grounds of Plant 1, a dispersed facility for making solid-fuel propellants for ballistic missiles. From a fenced area that overlooks part of the plant, the Soviets have the Kafkaesque assignment of watching for movements of rocket motors (for Pershing II intercontinental ballistic missiles [ICBMs]) no longer in production—a curious aspect of modern geopolitics.

Amid such up-to-date concerns, signs of an older Hercules are also present at Bacchus Works. Plant 1 is situated on land formerly occupied by the original dynamite lines. The company has made nitroglycerin—the explosive ingredient of dynamite, and now an element of rocket fuels—on this land since 1915.

The original plant was opened to serve a new copper mine owned and operated by Kennecott Copper Company, up the road in Bingham Canyon. Since 1906, five billion tons of rock and minerals have been removed from Bingham Canyon, the bulk of it blasted and reduced to rubble by Hercules dynamite. Once a mountain, the mine is now the largest manmade hole in the world, more than a mile deep and nearly two miles across.

During its first forty-five years, operations at Bacchus Works surged and declined with the fortunes of the mine, a business whose performance reflected the peaks and valleys of the U.S. economy. In its first decade, the Works saw both extremes: during World War I, it ran flat out; in the postwar recession, business was so bad that it closed for more than a year.

Bacchus Works began to change character in the mid-1950s. At that time, Hercules invested heavily in modern, automated equipment for making nitroglycerin. Soon thereafter, however, cheaper explosives came on the market and the mine stopped using dynamite. In 1958, Hercules was left with a state-of-the-art explosives facility, plenty of land, and virtually no business in the region. This misfortune was only momentary, however. By then, another explosives-related business was booming: chemical propulsion.

In 1958, Hercules had already accumulated nearly fifteen years of experience with solid-fuel rockets. During World War II, the U.S. government, recognizing the company's expertise in double-based powders (containing both nitroglycerin and nitrocellulose), contracted with Hercules to manage six federal ordnance plants. Double-based propellants that produced high-muzzle velocities in rifles and cannons proved to be suitable for small rockets as well. At several of the ordnance facilities—Radford, Virginia; Sunflower, Kansas; and Baraboo, Wisconsin—Hercules pioneered "large-grain" sticks of double-based powders for bazookas and other early rockets. After the war, Hercules kept its hand in chemical propulsion at the U.S. Navy's Allegany Ballistics Laboratory (ABL) near Cumberland, Maryland, where researchers developed small, tactical rocket motors for the U.S. military.

The launching of the Russian satellite *Sputnik* in October 1957 spurred a major buildup of the U.S. arsenal in response. As its dynamite business in Utah wound down, therefore, Hercules seized the opportunity to retool Bacchus Works to make rocket motors. The company soon won development contracts for the third stage on the Minuteman ICBM. From there, Hercules increased its participation (sometimes via joint venture) in other major programs, including the Polaris and Poseidon submarine-launched ballistic missiles (SLBMs) for the U.S. Navy. By the early 1960s, Hercules had become a significant factor in the aerospace industry. Since then, continued investments at Bacchus, ABL, and other locations have made the company one of the leading defense contractors in the country. Hercules rocket motors power missiles across the defense triad: land-based ICBMs, sea-based Trident missiles, and many air-based tactical rockets.

Hercules' involvement in the aerospace industry extends beyond making propellants and rocket motors. Over time, the company's strategy has been to increase its level of participation in major programs. In 1958, for example, the company acquired Young Development Laboratories of Rocky Hill, New Jersey, which had pioneered an exciting new technology for making lightweight cases for rocket motors. By winding canisters out of fiberglass filaments and treating them with resins and epoxies, Young Development produced cases significantly stronger and lighter than those made from steel. These properties were particularly valuable on the upper stages of missiles, where weight is at a premium: less weight in the rocket motor means a greater payload. Hercules' ability to make cases from composite materials helped the company win its breakthrough contract to produce the second stage on the Minuteman ICBM.

The strategy of increasing market share on major programs has proved successful. Over time, Hercules has developed new propellants and cases, as well as energetic "binders" to mate the two. Work on advanced materials and composites, which began in earnest in the late 1960s, has resulted in cases and exit cones made from graphite composites. Hercules' composite materials also appeared on structural parts for advanced aircraft and other high-performance vehicles in which strength-to-weight ratios are critical to success. In the 1980s, Hercules acquired two companies that rounded out its role in the aerospace industry: Simmonds Precision Products (1983) brought in instrumentation and electronics, and Sperry Microwave (1986) broadened Hercules' role in providing advanced communications.

In recent years, Hercules has targeted applications of its technology in commercializing space. The Bacchus West investment, for example, was made partly to fulfill contracts for the Trident II missile, but also on the possibility that Hercules would become a second source for the booster rockets that help launch the space shuttle. Although the *Challenger* disaster in January 1986 thwarted this plan, Bacchus West did prove an ideal facility in which to build boosters for alternatives to the shuttle, such as the Delta II and Titan IV expendable launch vehicles. At the close of its seventy-fifth year, Hercules was positioning itself to bid on futuristic programs such as Pegasus,* an economical, air-launched rocket-propelled vehicle designed to boost small payloads into earth orbit.

In 1987, Hercules Aerospace Company (HAC) reported total sales of $1.007 billion and earnings of $74 million. HAC has grown rapidly in the past decade and boasts a very strong position in solid-fuel rocket motors and a proprietary lead in graphite composites, a highly promising material of the future. In addition to the quality of its products, HAC's former president, Edward J. Sheehy, credits Hercules' emphasis on improving process technology for much of its success in aerospace. "Manufacturing technology has made incredible strides," he says, pointing to the company's ability to make larger, more

*On its maiden flight in April 1990, Pegasus successfully boosted a small satellite into orbit.

powerful, and more reliable rocket motors from one product generation to the next. To Sheehy, the building of Bacchus West prepares the way for Hercules to remain a leading aerospace contractor well into the 1990s.[6]

Bacchus West offers "a good platform for growth in the aerospace business," agrees Hercules' chairman and CEO David S. Hollingsworth. He is also frank about the risks: the company has made a major commitment of time and resources in the expectation that aerospace will "take off like a skyrocket between 1990 and 1995."[7] Other issues must be sorted out in the future as well: Will the strategy of vertical integration, as represented by the acquisitions of Simmonds and Sperry Microwave, pan out? With aerospace now so central a part of Hercules, can the company continue to attract investors, given tough competition and tight margins in the industry?

Specialty Chemicals: The Historical Core

Along the southeastern coast of Georgia lies the small city of Brunswick.[8] Since colonial times, the local economy has been dominated by the area's most visible resource: *Pinus palustris,* the Southern white pine. The most valuable property of these tall, slender trees was not the wood itself—although it makes serviceable lumber—but the sticky matter that it contains. In the seventeenth and eighteenth centuries, the British navy prized the gooey, resinous pitch of the Southern white pine as a waterproofing agent for sealing the hulls of wooden ships. Since that time, the tree has been the primary feedstock of the naval stores industry, in which the primary products have been rosin, turpentine, and (at a later date) pine oil.

For centuries, naval stores products were collected by hand, laboriously, from live trees. Each winter, workers—usually independent contractors—trooped into the forest and hung buckets below slashes they cut into the trees. Weeks later, the workers revisited the trees to collect the buckets, whose contents could be processed into rosin and turpentine. In the early part of this century, several factors converged to change this age-old method of cultivation. First was the widespread belief that the pine forests were being depleted. Second, the U.S. Department of Agriculture was eager to help farmers throughout the South open new land for cultivation by clearing it of pine stumps. And third, breakthroughs in mechanical engineering led some entrepreneurs to believe that stumps could be harvested and processed to make naval stores products—that is, that modern techniques of management and production could industrialize the naval stores industry.

One such entrepreneur was Homer T. Yaryan, a salty engineer who in 1909 built a stump-processing plant in Gulfport, Mississippi, and another in Brunswick two years later. The business was plagued by equipment problems and poor financing, however, and in 1919, the Yaryan Rosin & Turpentine Company declared bankruptcy. A year later, a court-appointed receiver sold Yaryan's plants to Hercules, which has remained in the business ever since.

Hercules' entry into naval stores was its first major diversification outside of explosives. The company eventually became a leading manufacturer of naval stores products, but not by following the expected path. Although Hercules overcame the mechanical difficulties that had discouraged Yaryan, two serious problems remained: first, wood naval stores products (derived from stumps) proved less attractive to consumers than gum naval stores products (derived from live trees)—wood rosin came in a single dark grade suitable for only a few uses, while wood turpentine carried an unpleasant and pungent odor—and second, the gum naval stores industry experienced something of a renaissance because, surprisingly, second-generation pine trees matured quickly.

In these circumstances, Hercules regrouped and, over a decade, developed a fresh approach to competing in the industry. Manufacturing process improvements helped lower costs and allowed the production of new, lighter grades of rosin. At the same time, the company used its marketing clout with the hardware trade to create national distribution of Hercules-brand turpentine and pine oil. Hercules also cultivated the market for these products in Europe: N.V. Hercules, organized in the Netherlands in 1925 to sell naval stores products in Europe, was the company's first significant foreign-based subsidiary.

Finally, and most important, the company explored the basic chemistry of naval stores commodities to isolate and control their constituents. Through Hercules' research, rosin became the source of products used in papermaking, paints and varnishes, adhesives, asphalt emulsions, and many other products; turpentine was fractionated into terpene mixtures, which served as solvents for oils, waxes, and resins, and as ingredients of disinfectants, cleaners, insecticides, textile chemicals, and many other products; and specialized uses in protective coatings, adhesives, and textile processing were even found for pine oil (a mixture of terpene alcohols).

By the mid-1930s, Hercules realized its long-sought dream of becoming a dominant producer in the naval stores industry. Along the way, the company helped transform the industry by a process of chemicalization as well as of industrialization. Henceforth, Hercules competed by offering controlled and standardized products, manufactured in bulk and at relatively low cost, and by providing technical support to its customers in many industries around the world. In short, Hercules transformed its naval stores commodities into specialty chemicals.

In 1987, Hercules' plant at Brunswick was the scene of much activity. The facility is large, sitting on 350 acres, and employs about 700 people. Set between palm trees, the administration buildings and laboratory area, built in the 1930s, have a tropical look, as though part of a World War II movie set in the Pacific. The grounds here are grassy and green, though elsewhere around the facility they have the muddy and sterile feel of chemical plants anywhere. The smell of burnt pine pervades the plant; it is a strong odor, but not unpleasant.

Since Homer Yaryan's time, the basic operation of the plant has changed little, although Hercules has, of course, succeeded in lowering costs, improving yields, raising quality, and increasing the range of products that emerge from the plant. In the traditional process, stumps were delivered to the plant by rail or truck. (At any given moment, about one hundred thousand tons of stumps are sitting in the plant.) In the first step, huge cranes lifted the stumps onto a rising conveyor belt, which carried them to the top of a multistory building where they were dropped into a trough and sprayed to remove dirt, sand, and other impurities. From there, the washed stumps were dumped into a mammoth hopper. Inside, a huge rotating cylinder (called a "hog") ground and shredded the stumps into pencil-sized chips, which were then conveyed to the extraction area and poured into giant, five-story tanks, each holding fourteen tons of material. The chips were then boiled in a solvent to remove the resinous material.

At the end of this step, the liquid was drawn off and the spent chips were "pulled" from the tanks by hand—a hot, dirty, and arduous task—to be dried and burned as fuel. The liquid was piped to a refining area to remove the solvent and process the remaining material into valuable products. About two-thirds of the resinous matter was crude rosin that was treated and refined in the Pexite® rosin area through subsequent steps into resins of many types and grades; the other third, still in liquid form, was distilled into products such as pine oil and turpentine, or processed into terpene resins.

In the late 1980s, however, Hercules was developing a wholly new approach to processing stumps. A $30-million modernization plan (completed in 1989) was aimed at combining, automating, and making continuous the steps of milling, extracting, and refining. The new approach is based on a process that has proved successful in the soybean industry. Hercules' transformation of a process that is workable for small, pliable matter such as soybeans into one that handles large, tough, and irregular items such as stumps will involve a major engineering effort. If Hercules succeeds, however, it will result over time in substantial cost savings and perhaps will help carry its naval stores business into the twenty-first century.

The outlines of Hercules' approach to naval stores—continuous improvement of manufacturing operations, research to upgrade products, and marketing to the needs of specific groups of customers—carry over into other HSCC businesses. At the same time as it appeared in naval stores, for example, the strategy of specialization emerged in another line: nitrocellulose and other cellulose products for the protective coatings, artificial fibers, and plastics industries. Indeed, a period from the late 1930s through the early 1950s marked the zenith of pure research at Hercules: chemists such as Emil Ott, Harold Spurlin, Robert Cairns, and many others developed new processes and products from rosin and cellulose and obtained new information on the composition, properties, and reactions of these materials.

Later, Hercules developed from its own research, or used acquisitions to provide, specialty chemicals for other industries: rosin size and other products for the paper industry; natural gums and other additives and flavorings for the food processing industry; *para*-xylene and terephthalates for the synthetic fibers and plastics industries; water-soluble polymers for coatings, food ingredients, mud drilling, and many other applications; hydrocarbon resins for adhesives and printing inks; emulsifiers, tackifiers, vulcanizing agents, and other products for rubber and other elastomers; flocculants, defoamers, and other chemicals for treating water; photopolymer plates and printing systems for graphic artists and newspapers; and a miscellaneous collection of products—magnetic oxides, toners, and laminate boards—for the recording, photocopying, and computer industries.

The portfolio of such businesses has churned over time: product lines have been emphasized and then downplayed, added and then dropped. At the close of 1987, HSCC consisted of five groups: Food and Fragrance Ingredients, Paper Chemicals, Coatings and Additives, Resins, and Electronic and Printing Products.

In 1987, with annual sales surpassing $1.056 billion and profits of $116 million, HSCC was the largest and most profitable of Hercules' operating companies. As they looked toward the future, leaders of the corporation and the unit shared similar hopes and concerns: "The most obvious opportunity," says Hollingsworth (who spent most of his career in specialty chemicals businesses), "is in the fragrance and food ingredients business, which is a very, very good business." Thomas L. Gossage, a former Monsanto executive who was recruited to become president of HSCC in 1988, agrees, calling fragrance and food ingredients "the most exciting business opportunity today for the chemical part of Hercules, and if not the most important growth opportunity for the corporation as a whole, then certainly it is among the top two or three." To justify this optimism, Gossage cites rapidly expanding markets in Europe and South America, as well as the strong consumer trend toward natural foods in the United States, and eventually elsewhere.

As for concerns, both leaders sound the same refrain: apart from fragrance and food ingredients, where will HSCC find growth? Its major subunits—paper chemicals and resins—show signs of maturity. Growth will continue to come, but share point by share point. The principal task facing the unit, therefore, involves "an analysis of the assets inside the chemical company, and the positioning of them to do what they can do best." In other words, HSCC is attempting to blend the subunits into a portfolio of assets: some will be managed for cash, others for margins, and still others nurtured for growth.[9]

Long the mainstay of Hercules and the source of its essential character, the chemical businesses now pose challenges of their own: Are they mature? Can they grow? Where will new opportunities lie?

Engineered Polymers: From Resins to Materials

The city of Terre Haute lies at the crossroads of two national highways in southwestern Indiana.[10] One leads west from Indianapolis toward St. Louis; the other flows north, connecting the industrial counties of the southern part of the state with the great metropolitan area around Chicago. This is prime farm country and, of course, basketball territory. Terre Haute is the home of Indiana State University, where the great basketball star Larry Bird played college ball in the late 1970s.

A few miles to the north of the city, the visitor encounters the familiar white water tower embellished in red with the Hercules logo. This landmark presides over a facility that makes polypropylene film, a product used primarily in packaging cigarettes and snack foods. For such products, packaging in paper or cellophane is diminishing because modern thermoplastics such as polypropylene offer superior properties: strength, light weight, uniformity, and moisture protection.

Although Terre Haute is a young plant in comparison with Bacchus or Brunswick, it nonetheless offers visitors an interesting industrial archaeology. Construction on the plant started in 1967, when Hercules sought to augment its rapidly growing but cash-consuming polypropylene resins business by developing downstream applications such as fibers and film. In 1967, Hercules had already made film commercially for several years at its plant in Covington, Virginia. By that time, business had developed well enough that Hercules decided to increase its capacity by a factor of more than two.

At the time, both Covington and Terre Haute formed film using the tubular or bubble process, which makes film by blowing tubes of hot polypropylene into enormous bubbles four to six feet in diameter and three or four stories high. The tubular process, then, is a vertical process: on the top floor of the manufacturing area, tubes of molten polypropylene are extruded. As they emerge, hot air is blown into the tubes from above, making them into free-floating bubbles that then become transparent as they stretch and are pulled by gravity and machinery from below. When they arrive at the ground, the bubbles are sheared in half and the pieces are wound onto big spools, which are then slit into various widths, depending upon the application. Once a bubble materializes, things move swiftly: the bubbles form at a rate of 100 meters per minute. To the untrained eye, the end products resemble (on a much larger scale) the plastic wraps that consumers use to keep food fresh in a refrigerator. (That product is made from polyvinylidene chloride.)

During the early 1970s, Hercules added more and larger bubbles to the film-manufacturing process at the Terre Haute facility. By the end of the decade, under the leadership of Fred L. Buckner, general manager of the film business, polypropylene film had dislodged cellophane in many packaging applications. To substitute successfully for glassine paper, the next market targeted by Buckner, Hercules had to develop films with new properties, such as opacity and printability. These considerations led the company to license an alternative technology, the tenter process, which stretches and pulls sheets of molten polypropylene into heavier gauged, heat-sealable films. The new process was faster (by as much as 500 percent) and more highly automated than the tubular; it also proved capable of producing films with a wider variety of properties, including opacity and heat-sealability. As a result, Hercules was able to capitalize on fast-growing new markets, such as that for labels on two-liter beverage bottles.

Although Hercules' strategy in the film business involves step-by-step substitution of its products for existing products in specific packaging markets, the company conceives of its business more broadly as the marketing of properties. Under Paul Mayfield, Werner C. Brown, and Alexander F. Giacco, who successively led the Polymers Department from the 1960s through the early 1970s, Hercules promoted the importance of "adding value" to customers, moving beyond vending film by the foot for the lowest price to selling more abstract notions such as coverage, softness, strength-to-weight ratios, density, and light-

and moisture-barrier properties. The shift has proved liberating: no longer positioned (and circumscribed) as a manufacturer of polypropylene film, Hercules now also supplies materials based on polypropylene, such as multilayered films and metallized wraps, for a variety of applications.

This approach extends to Hercules' other major downstream polypropylene business, synthetic fibers, as well. At Covington, Virginia, in the early 1960s, the company had started making polyolefin fibers for carpets under the brand name Herculon. At the time, carpets made from nylon and other synthetic fibers were rapidly becoming the floor covering of choice for American homeowners. Hercules eventually built a strong business in carpets, although not in the expected fashion.[11] It turned out that Herculon is not as resilient as nylon, and hence makes less satisfactory deep-pile carpets for residences; on the other hand, when densely woven into short-filament commercial carpets, polypropylene's superior stain resistance and colorfastness come to the fore. As a result, by the late 1960s, Herculon became a leading fiber in commercial carpets, and the company added a new carpet fibers plant in Oxford, Georgia.

From that base, Hercules sought other applications for its fibers, such as indoor-outdoor carpets and furniture upholstery. (By far the biggest market for synthetic fibers—textiles for apparel—was closed to Hercules because of a critical feature of polypropylene: it melts at low temperatures and thus is unsuitable for garments that must be ironed or machine-dried.) Progress in those areas was slow and steady until, in one of the first actions of Giacco's presidency, the company made a surprising and audacious move: it committed itself to building a second huge plant at Oxford to make an untested product, fine-denier polypropylene staple fibers (short, squiggly, extremely thin fibers that resemble cotton or wool and can be gathered into bales).

The motive behind this investment was, again, the concept of properties. Giacco reasoned that polypropylene staple fibers, if made in sufficient volume, might compete successfully with cotton and polyester in certain applications.[12] Among these, the company quickly identified disposable diapers as a product of significant potential: polypropylene staple fiber has the look and soft feel of cotton, but it is cheaper to process than other natural and synthetic fibers and has a critical advantage in properties as well: superior "wickability" (that is, polypropylene fiber wicks moisture away from the surface and keeps it there). The gamble paid off: in the 1980s, Hercules' polypropylene staple fibers were used to make cover sheets for most disposable diapers sold in the United States.

Hercules Engineered Polymers Company, the youngest and smallest of Hercules' operating groups, is essentially in the business of tailoring materials derived from polypropylene for specific applications. In 1987, HEPC's sales totaled $399 million and the company reported earnings of $42 million. Although film and fibers accounted for more than 90 percent of HEPC's sales that year, the unit was also engaged in developing new materials (including some not derived from polypropylene) and new applications from old ones. The most promising of the new materials, says HEPC's former president James Knox, is METTON® liquid-molding resin. Knox expects sales of METTON,® a lightweight, extremely tough plastic for use in large structures such as automobile body panels and bumpers, to surpass $200 million by the early 1990s. Among the new applications of film and fiber that Knox singles out are FreshHold® freshness protector, a controlled-atmosphere package that can prolong the freshness of produce by as much as a factor of three, and staple tow, a substitute for cellulose acetate in cigarette filters.[13]

Looking ahead, the leaders of Hercules and HEPC see a future full of opportunities and challenges. According to Buckner, now president and chief operating officer of Hercules, the long years of investment in building the fibers and film businesses will pay off handsomely in the 1990s. "During our first twenty years in the business," he recalls, "it was pure grief. None of those ventures made any money because they were expanding fast, consuming lots of cash, developing products, and entering new markets." By the mid-1980s, however, "that foundation of hard work gives our products broad recognition and high value." He sees a similar pattern unfolding in the development of METTON,® which he expects to become a major contributor to the company into the next century.[14]

As for challenges, 1987 produced a major one: the sale of Hercules' 50 percent interest in HIMONT Incorporated, a global producer of polyolefin resins and the source of HEPC's raw materials, raises critical questions for the future. As a result of this step, which netted Hercules $1.5 billion in cash, "the Engineered Polymers Company was forever impacted," says Hollingsworth.[15] Now that Hercules is no longer vertically integrated, how will it fare as a downstream producer of engineered products? Can it defend its position against oil companies and potential new competitors such as HIMONT? Can HEPC sustain growth in fibers and film while it waits for the market for METTON® to develop?

Home Base

It is one of the most distinctive buildings on the city's skyline.[16] The last in a line of tall structures that march east from the Christina River to the Brandywine—a distance of fourteen blocks—it appears, when viewed from afar, to be a kind of sentinel, poised to defend the city from invaders from the north or west.[17] On closer approach, the cool, glassy exterior seems less imposing as it reflects sun, sky, and images of buildings nearby. On entering, one feels welcome: the huge, 200-foot-high, 90-foot-square atrium admits ample light. Greenery and a cascading waterfall amplify the sensation of being outdoors. This is both a public and a private space, suitable for large meetings but equally appropriate for conversations among a few.

Above, after a ride on an escalator and transfer to an elevator, are a dozen floors divided into geographical quadrants to help identify locations. The offices and meeting rooms on each floor are mostly open. Flexible partitioning allows space to be changed, depending on the needs and requirements of office workers. Hallways and pathways between sections of the building occasionally open up into areas for meetings or private discussions.

This is Hercules Plaza, from which the far-flung and disparate activities of

Hercules Incorporated are directed. Known variously as "the Plaza" or "the Home Office," the building was completed in 1983 to the delight of architecture critics and, more important, most Hercules employees. The building is one of the most distinctive in Wilmington, and an unusually stunning structure for a company that traditionally maintains a low profile. Therein lies a tale.

For most of its history, Hercules has lived in the shadow of its giant forebear Du Pont, by far the biggest U.S.-based chemical company and Delaware's largest private employer. Created from Du Pont's assets in 1912, Hercules operated out of Du Pont's own headquarters building until 1921, when it moved across the street from the Hotel Du Pont to occupy the upper floors of the Delaware Trust Building. For the next six decades, the company remained in the bank building, faithfully renewing its lease at five-year intervals and occasionally expanding to new floors. In 1959, Hercules helped fund and occupy a twenty-two-story tower built between the wings of the Delaware Trust Building.

In the late 1970s, several factors converged to persuade Hercules to seek an alternative headquarters location. To begin with, these were tough years for the company: in the post–oil shock world, Hercules was forced to restructure its businesses to reduce dependency on petrochemical feedstocks. At the same time, under President Giacco, the company reorganized in such a way as to abolish traditional channels of reporting and alter flows of information. At the Home Office, everything was in flux: the company needed better control of its inventories and costs, and it needed to reevaluate its basic communications systems. Giacco, an adroit manager of symbols, realized that moving to a new and modern building might accomplish several objectives at once: reinforce the company's new strategy and structure; highlight a fresh corporate image among such stakeholders as employees (including potential recruits), competitors, the trade press, and the financial community; and help move Hercules out from under the shadow of Du Pont.

Other factors were at work, too. Hercules' drive for greater efficiency appeared to clash with the relatively high cost of operating in Wilmington. The high cost of living in the area was also an issue: the company worried that it might not be able to attract and hold key people without paying significantly higher salaries and benefits than its competitors. And then there was the condition of Wilmington itself: the city, like most of urban America, suffered a decline in the 1960s and 1970s. Could it be reborn? Was Hercules obliged to help?

The impending expiration of Hercules' lease in the Delaware Trust Building provided occasion for all these issues to be addressed. In the late 1970s, Giacco let it be known that the company was contemplating a move, then waited to gauge the local reaction. The company was rumored to be heading to Texas, to Pennsylvania, and (most plausibly) to a site near the Research Center in the suburbs of Wilmington, where Hercules owned more than 1,000 acres.

Almost immediately, forces in the state and the city began to mobilize. In 1979, the Delaware General Assembly voted to reduce personal income taxes. Soon thereafter, Giacco announced that the company would remain in the state, citing the company's roots in Delaware and implying that Hercules wished to avoid the heavy financial and personal burden of a major relocation. After Giacco's announcement, William McLaughlin, the mayor of Wilmington, designated a liaison to work with Hercules and the state on finding a suitable location in the city; he also negotiated financial incentives for the company to stay, agreed to build a park between the building and the Brandywine River to the north, and helped to solve such mundane matters as the anticipated parking problems. These efforts paid off, and in the spring of 1980, the Hercules board approved the initial appropriation for a new headquarters building to be situated on land at the northwest end of downtown Wilmington, with an unobstructed view of Fletcher Brown State Park and the Brandywine. (The site includes property once owned by Henry Thouron, president of Hercules from 1961 to 1970.)

Hercules Plaza cost $88 million to build, although, through the arcana of modern tax laws and tricks of the trade in commercial development, the company's outlay was negligible. The building was planned, designed, developed, and built with careful attention to detail.[18] In the early 1980s, a project team consisting of employees John Greer, Edward Lacy, and Elizabeth Ronat met regularly with consultants, bankers, public officials, architects, developers, and builders and served as a conduit between these outsiders and senior corporate management. As a result of their work and a high degree of cooperation among the principal actors, Hercules Plaza achieved a remarkable dual success: an award-winning design and a creative use of architecture to support not only the company's strategy but its ideal way of functioning.

One of the most striking aspects of the building is the ubiquity of information technology: in 1987, a satellite dish adorned the roof; inside, computers, word-processing equipment, fax machines, and other electronic marvels were everywhere; meeting rooms were configured for tele- and videoconferencing. The aggressive use of this technology, like the building itself, has roots in the corporate restructuring that began in the late 1970s. Along with the need to reposition the company's portfolio of businesses and the need to boost employee morale were imperatives to improve white-collar productivity and forge tighter links between headquarters and the company's distant operating units.[19] To help meet those needs, Hercules committed heavily to information technology, believing that its advanced office systems would significantly improve corporate performance.

Hercules Plaza illustrates a growing trend in corporate America: an effort to achieve a congruence of environment, ergonomics, and business objectives. At a more personal level, the Plaza's user-friendly aspects—the atrium, well-lit and open offices, ample meeting spaces, and abundant information technology—appear to have realized the goal of raising morale: employees enjoy working in the building. The Plaza has also succeeded in making a statement about Hercules and its attitudes about its past and future: the building aims northwest, away from Du Pont and toward the horizon.

Herculites

To suggest that Hercules consists primarily of products and places, to attempt to deduce its shape and character from its buildings and its sales volume, is, of course, to form an incomplete sketch. The company consists of people, and it is people who have made Hercules what it now is.

On our travels, we met hundreds of Hercules employees and retirees and spoke with many of them in depth. From these interviews, a composite picture of these "Herculites" (as they sometimes call themselves) emerges. As in the rest of the U.S. corporate world, sociologically speaking, white American males are predominant. Although the company professes to make good-faith efforts to comply with equal employment and opportunity guidelines and features increasing numbers of women, minorities, and foreign nationals in managerial positions, the nature of the labor market for professional and technical people slows the pace of change.

Moving beyond these considerations, it seemed to us that Herculites are noteworthy in other respects: articulateness; friendliness (everyone, including the chairman, is addressed on a first-name basis); a high degree of technical proficiency; a concern for observing procedure; and (as the opening scene of this chapter suggests) loyalty toward and affection for their employer, "Uncle Herc." At Hercules, everybody seems to know everybody else.

None of this is meant to say that Hercules seeks and rewards conformists, or that its corporate culture is overbearing. On the other hand, that such people are found in such numbers, and that such warm feelings generally arise toward the company, are circumstances that can be explained at least partly by the company's history. Hercules' origins as an explosives-maker, for example, account for at least some of the corporation's benevolent paternalism. From the very beginning, the company placed a premium on safety that extends to the present day. An employer that looks out for the well-being of its employees can expect to be repaid in kind.

The company's employment policies and practices also reflect aspects of the era in which it was founded. In the 1910s and 1920s, many progressive employers or welfare capitalists manifested a broad, general concern for their employees. At Hercules, this attitude took the form of open communications and sponsorship of social events outside of work. The *Mixer,* the company magazine, started in 1919 and continues today. In its early years, the *Mixer* served to reinforce a family spirit among employees, reporting on life events such as weddings, births, and deaths, as well as less momentous matters such as vacation plans and trips to the theater. In the 1920s and 1930s, the company supported the formation of sports leagues and annual picnics—traditions that persist today in many locations—in which senior executives as well as hourly workers have always participated. The company built its country club—open to Hercules employees only—on ground adjacent to its Experiment Station in 1934.

This friendly spirit lingers on today in such areas as the Plaza cafeteria—which consists of a smallish room furnished with big tables, a setting that encourages people of disparate levels and backgrounds to sit together—and the Hercules Guest House in Wilmington, a revamped private house where short-term visitors are made to feel at home, literally, by husband-and-wife caretakers Raphael and Angela Rossetti.

The feeling of family also stems from generally long records of service employment. Employees join and tend to stay. To cite the most prominent example, chairman and CEO Hollingsworth hired on in 1948. Among the senior management team, only one—Tom Gossage, head of HSCC—is an outside hire.* Pensioners also remain close to the company, retiring near company operations and showing up at the country club or at periodic gatherings such as picnics. At Kenvil, a large group of retirees calling themselves "the Hercules Has-Beens" meets weekly for lunch.

Finally, the feeling of family is literally true in many instances. Hercules' second president, Charles Higgins, was the son-in-law of the first. Family dynasties are commonplace in both managerial and hourly ranks. To cite but two examples, Dick Best, the Kenvil plant manager, is the son of retiree Paul Best, whose long career at Hercules centered on cellulose products and water-soluble polymers; at Brunswick, octogenarian Jim Grant is the first of three generations of his family to tackle the tough job of pulling spent wood chips.

In sum, Herculites work together, play together, and occasionally even live together. Throughout most of the company's history, employee morale has appeared high. Even in the Great Depression, when substantial layoffs were necessary, the company first attempted to save jobs by working short weeks. In more recent times, the wrenching restructuring of the past decade took its toll; most Herculites we interviewed remarked that things are not what they used to be. Nonetheless, when asked what single message he would most like to communicate to employees, Chairman Hollingsworth thought a moment, then replied:

> I want the people of Hercules to have confidence that the best of Hercules is very important to us. We're going to try to maintain and preserve it. Hercules is a good place to work, and we're going to see that it will continue to be a good place to work in the future. We grow and change. We move out in new directions. But we're not going to lose sight of what we are and where we've been. Our heritage and culture: these are constants that our employees can count on.[20]

HERCULES AND THE DYNAMICS OF CHANGE

The flip side of continuity is change. And the constancy of change pervades the industry and environment in which Hercules has grown up and now competes.

The pace of change is particularly rapid in modern business. Markets dry up. Buyer preferences shift. Important customers fall on hard times. New discoveries make current products or processes obsolete. A new product thought to be an adjunct to existing lines becomes a major business on its own. New leaders develop new strategies. Some competitors grow more powerful, others fall by

*In 1989, Hercules added another ousider: Richard Schwartz as president of HAC.

the wayside. New entrants or suppliers or buyers arrive to rewrite the rules of the game. Substitutes suddenly dislodge once popular products. Governments, or labor unions, or consumer advocates, force changes in management and organization.

All of this has happened more than once at Hercules. Yet it would be misleading in the extreme to imply that its transformations are the result of a one-way flow from the economic environment to the company. Hercules is not merely a survivor. It has also initiated important developments in the chemical industry. It pioneered significant advances in the uses of explosives and propellants, the products and processes of the naval stores industry, and the chemistry of cellulose, polyolefins, and advanced materials. Nor has innovation at the company been confined to the laboratory or the plant. Among American companies, for example, Hercules has been a leading participant in worldwide joint ventures and coalitions, as well as an early devotee of advanced communications systems. The company has shown a consistent ability to identify and meet the key challenges of each era in its history. Over the long haul, Hercules has generally achieved above-average financial performance and has grown at faster rates than the U.S. chemical industry or the U.S. economy as a whole. Between 1913 and 1987, Hercules grew at a compound annual growth rate of 8.15 percent, the chemical industry at 7.17 percent, and the U.S. economy at 6.51 percent (see Exhibits 1.3 and 1.4).

The peculiar dynamics of competition in the chemical industry largely account for the pace and extent of change in Hercules' history. The industry has a number of exacting characteristics that demand adaptability from its participants.[21] To summarize the most important:

Size and complexity. The chemical industry is one of the world's largest and most complex. In 1987, in the United States, the "value of shipments" (the total value of all products and services sold) in the chemicals and allied products industry stood at $210 billion, a total well ahead of such prominent industries as motor vehicles, metal fabrication, electric and electronic equipment, and printing and publishing.[22] Elsewhere, the chemical industry is even larger. Only five of the fifteen largest chemical companies in the world are based in the United States. The remaining ten are to be found in Europe, where three of the top four (BASF, Bayer, and Hoechst) are based in West Germany.

The operations and activities of the industry are also extremely complex. This stems in part from the highly technical and scientific nature of chemical processes. But it also reflects history and economics: most large manufacturers supply thousands of basic, intermediate, and finished products; to manage these businesses, companies tend to be decentralized, customers for many of these products are likely to include competitors, and most producers are linked through a variety of mechanisms—license agreements, joint ventures, global partnerships—with a half-dozen or more competitors.

Commodities and specialties. The industry is usually classified into two camps: production of bulk or commodity chemicals in very high volumes for sale to a large number of customers; and manufacture of differentiated or specialty chemicals for a narrower range of customers in particular market segments.

EXHIBIT 1.3 Growth Rates: Hercules Compared with the U.S. Chemical Industry, 1913–1987

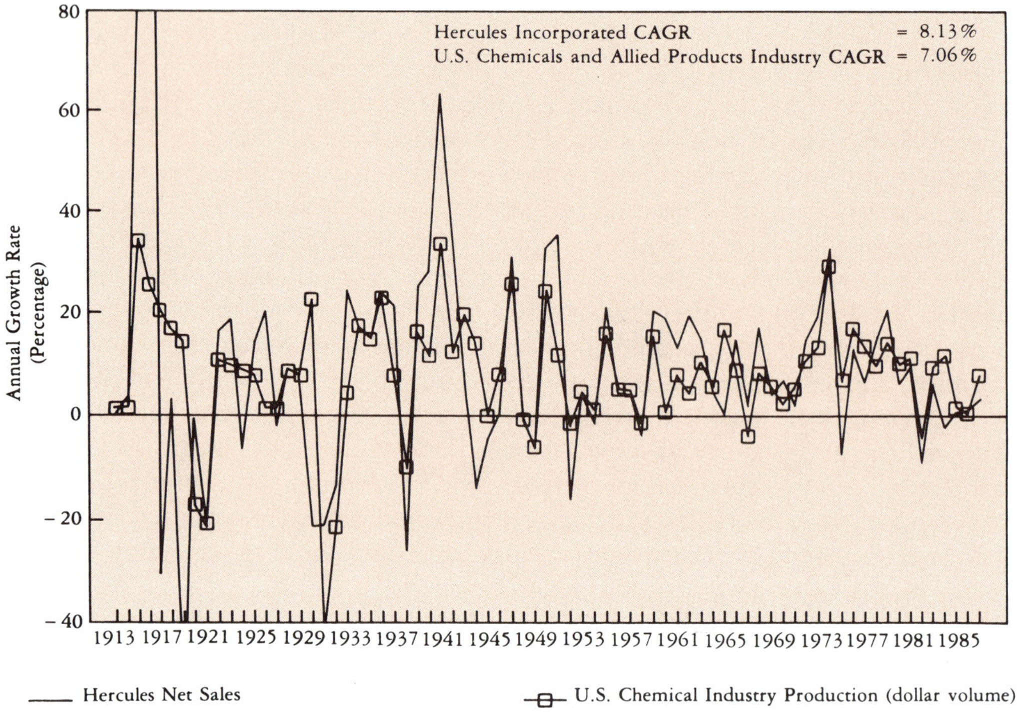

Source: Hercules Powder Company, Annual Reports 1913–1965; Hercules Incorporated, Annual Reports 1966–1987; Edward H. Hempel, *The Economics of Chemical Industries* (London: John Wiley & Sons, 1939), p. 4; Jules Backman, *The Economics of the Chemical Industry* (Washington, D.C.: Chemical Manufacturers Association, 1970), p. 325; Chemical Manufacturers Association, Government Relations and Economics Department.

Competition in commodity chemicals tends to be based on low prices and normally involves only the largest producers; in specialty chemicals, competition is based on many factors—for example, performance, branding, or service—and engages many smaller manufacturers. In general, supplies of commodity chemicals seek to transform their products into specialty items, while manufacturers of specialty chemicals are concerned about preventing their products from becoming commodities.

Capital intensity. A modern petrochemical plant can cost from $100 million to $500 million or more. Even modest operations represent tens of millions of dollars in investment. As a result, competitors think long and hard about the magnitude and timing of investments. Such concerns are particularly acute at medium-sized competitors such as Hercules, where major investments could involve substantial risk.

Compared with its role in most other mass-production industries, labor in the chemical industry tends to be a less significant element of the cost structure.[23] Modern chemical plants can be operated successfully by a handful of people. As a

EXHIBIT 1.4 Growth Rates: Hercules Compared with U.S. GNP, 1913–1987

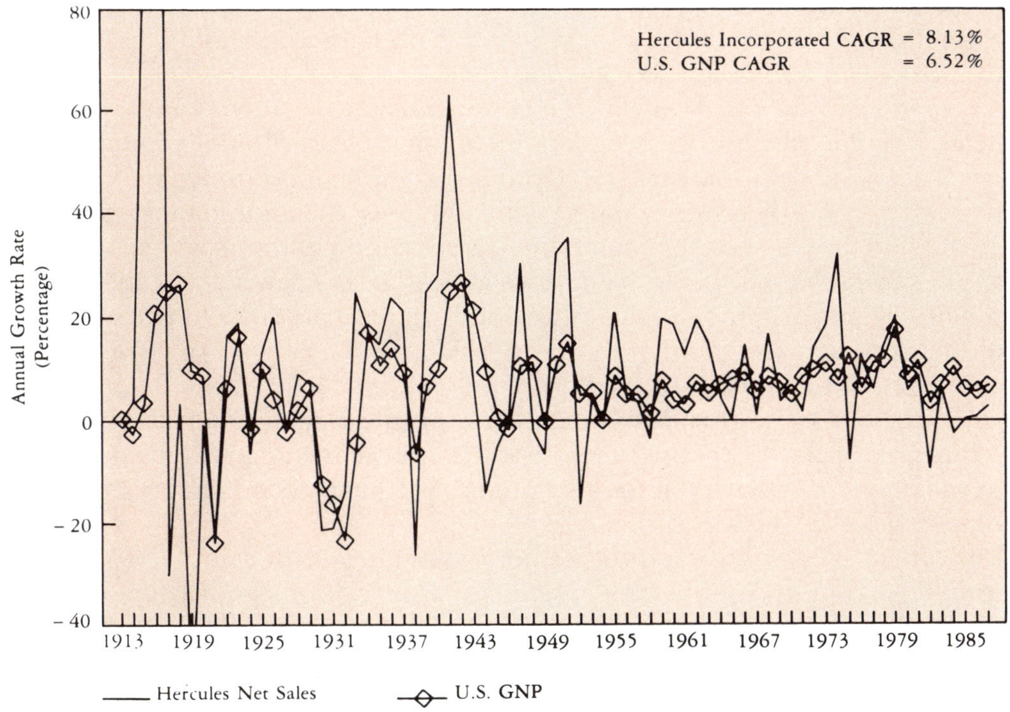

Source: Hercules Powder Company, Annual Reports, 1913–1965; Hercules Incorporated, Annual Reports, 1966–1987; U.S. Department of Commerce, Bureau of Economic Analysis.

result, employers occupy a relatively powerful bargaining position, and protracted strikes occur infrequently. (On the other hand, the need to keep plants up and running encourages employers to pay premium wages and salaries compared with other manufacturing industries—another factor that encourages generally harmonious industrial relations.) Hercules has been no exception: although some of its U.S. plants are unionized, the company has generally avoided strikes or other serious industrial disputes.

Continuous processes. Chemical plants are designed to operate continuously, twenty-four hours a day. This factor has two consequences: first, competitors place a premium on process improvements—indeed, the industry has been extremely aggressive in introducing new process technologies; second, competitors also place a premium on engineering skills as a source of process innovations and improvements. As a result, the industry is one of the most productive in the Western economies, over the long term significantly outperforming other manufacturing industries in productivity measures.

Economies of scale. Competitors gain an advantage when increasing the output of a single facility results in lower unit costs.[24] This is true of virtually all chemical processes: bigger plants mean cheaper products. Consider a famous example: the impact of increasing scale on the early oil industry. About 1870, a typical oil refinery was capable of producing about 900 barrels of oil per week at a cost of 6¢ each; within two years, with the advent of new production techniques, refineries could produce 1,000 barrels per day at a unit cost of 3¢ per barrel. John D. Rockefeller's insight into this phenomenon was the first step in the accumulation of a fabulous fortune.[25]

Vertical integration. Chemical companies tend to occupy a middle position in an economic chain that consists of several links: extraction or production of raw materials; one or more stages of processing or manufacturing; and distribution and retailing of finished products. In most cases, chemical companies are not involved in the production of raw materials at one end of the chain, nor do they, at the other end, sell brand-name consumer goods. Typically, chemical companies buy raw materials, modify them, and sell their products to fabricators or formulators that use them in products sold subsequently to other industrial customers and eventually to the public.

The cost of raw materials often represents the biggest element in chemical companies' operating cost structure. For this reason, the largest companies tend to be integrated backward into raw materials production, and most competitors constantly assess this option.

In most cases, chemical companies do not integrate forward, lest they find themselves competing with their customers. Chemical companies have occasionally been victimized by forward integration of their suppliers. The most obvious example is petrochemicals: since the 1930s, and particularly after World War II, most large oil companies have become major producers of petrochemicals and compete directly with many chemical companies.

Economies of scope. Competitors gain another sort of advantage when they can use a single facility to produce or distribute more than one product—that is, they achieve savings when they can spread the costs of production or distribution across more than one product line.[26] In the chemical industry, economies of scope are perhaps less significant than economies of scale, but they nonetheless produce powerful effects. These economies occur more often in distribution than in production. As a result, chemical companies tend to grow by diversifying into new product lines. Since the 1920s, chemical companies have ranked among the most diversified in the international economy.

R&D intensity. In the United States, the chemical industry employs about 5 percent of all industrial workers, but more than 10 percent of all scientists and engineers. Chemical companies also produce a disproportionate number of patents: in the 1960s, for example, the industry accounted for more than 20 percent of patents issued in the United States. R&D expenditures as a percentage of sales average between 3 and 4 percent, a total higher than for all but two other industries in the United States: electrical products and aerospace, both of which—in contrast to the chemical industry—enjoy significant research contributions from federal funds.[27] Since the early part of this century, and since the 1930s especially, investment in chemical R&D has been a major source of the industry's growth and innovation.

Product substitution. It has been estimated that between 85 and 90 percent of the chemical industry's growth comes from substitution for existing products or materials; the industry has not created wholly new markets, as, for example, Xerox, Polaroid, and Apple Computer have done. Substitution takes many forms: replacement of natural products, materials produced by other industries, or other chemical products. Although most new chemical products have long development cycles—often ten to fifteen years—substitution can take place rapidly once a superior product is introduced. To protect their investments, therefore, chemical manufacturers constantly endeavor to differentiate their products in various ways: by developing high-performance products; by protecting process and product innovations with patents; by selling under brand names; by incorporating products into a larger family or system of products; and by providing customers with a high level of technical support and service.[28]

Environmental regulation. Since the late 1960s, the chemical industry has been a particular focus of environmental concern and regulation. For chemical companies, the problem is twofold. First, many chemical processes involve toxic materials and by-products. Chemical companies are therefore required to take extraordinary and often expensive precautions to protect employees and control pollution. Second, chemical companies are required to test new products and periodically retest existing products for toxicological effects. Such tests typically involve many technical personnel and sophisticated equipment. As a result, environmental regulation has become a significant element of production cost, a circumstance that tends to favor large-scale producers and industry concentration.

Patterns of growth. Competitors in the chemical industry generally have grown in similar ways.[29] At the level of a single product line (or business unit), management's central challenge is to ensure that assets are fully employed and that continuous processes are not interrupted. To keep a plant running flat out all the time, it is paramount to maintain a steady supply of raw materials. If the nature of competition so warrants and a company can afford it, integrating backward may be attempted. At the same time, the company will attempt to develop superior capabilities in marketing and distribution to pull products through its plant. It may integrate forward to the next stage of production, although (as noted earlier) this is a relatively rare occurrence in the chemical industry and an option generally open only to the largest companies.

Growth via a single product line and vertical integration normally approaches limits imposed by the extent of demand, the nature of competition, or public policies such as antitrust. When that happens, companies seek growth through diversification, which occurs occasionally through internal development of new product lines but more often (and more often successfully) through acquisition. In the chemical industry, the pressures to diversify are formidable: to obtain growth, offset product line cyclicity, realize economies of scope, and find sources of corporate renewal.

Oligopolistic competition. The characteristics just listed tend to reward bigness and concentration in the chemical industry. As early as 1947, more than two-thirds of the output of the U.S. chemical industry was controlled by a handful of big companies.[30] Most large companies, then, are still significant competitors. Since World War II, however, two factors have introduced change in the industry's basic structure: the entry of the oil companies into petrochemicals; and more recently, the globalization of competition (see below).

Political salience. The chemical industry's sheer size and scale tend to elevate its political importance in developed nations. Indeed, in most of these countries, it is a significant positive contributor to the international trade balance. Partly as a result, the developed world boasts many "national champions" in the industry: in addition to the formidable West German companies, major European competitors include ICI in Great Britain, Rhône-Poulenc in France, Montedison in Italy, Ciba-Geigy in Switzerland, and Solvay in Belgium. And the Japanese chemical industry is coming on strong: it is estimated that two or three Japanese companies could break into the ranks of the global industry's top twenty by the early 1990s.[31]

Even among less developed nations, the economic salience of the industry evokes considerable interest. Many LDCs view the industry as an engine of development and seek to nurture native chemical producers. The phenomenon is most pronounced in the oil-producing nations, where governments encourage local manufacture of petrochemicals.

Globalization. Since World War II, and especially since the early 1970s, international competition in the chemical industry has intensified dramatically. Many factors (in addition to those just cited) account for this change: declining costs of transportation; major advances in communications; international trade and monetary policies; domestic industrial policies; and the long-term trend of world markets to become homogeneous.

This long list of chemical industry characteristics is rehearsed to make a point: the industry is hard on its participants, requiring a high degree of adaptability and creativity. It is no accident that, since the early part of this century, chemical companies have been the source of many significant innovations in modern business management. To cite a few examples: improved accounting practices for vertical integration; the strategy of diversification; management of decentralized operations; development of return-on-investment (ROI) criteria; rapid adoption of automation and automated process control; innovations in industrial marketing; and early formation of joint ventures and global partnerships.

In short, the industry's constant is change. This elemental fact holds especially powerful implications for medium-sized competitors such as Hercules. The company has been obliged to adapt to major swings of fortune: its history has been particularly protean, its future seldom predictable. To have maintained generally above-average performance in the industry for so long, and to have managed the ups and downs of its principal product lines, is testimony to its resilience and resourcefulness.

That Hercules would eventually build strong businesses in aerospace, specialty chemicals, and engineered polymers and materials was never predestined. The unpredictable elements in the company's nature date from the start—in the peculiar circumstances of an antitrust suit that took place eight decades ago.

At its campus-like research center near Wilmington, Hercules concentrates on development of new products as well as improvement of existing lines.

Birth of a trademark—and a corporate name: The California Powder Works competed against rival Giant Powder Company by recalling mythology: Hercules slaying the giant.

Breaking up is hard to do: Late in 1912, the Du Pont lawyers who handled the antitrust case and negotiated the birth of Hercules and Atlas assembled for a final time over dinner. Second from the left facing the camera behind the table in the foreground is Gould G. Rheuby, Hercules' first general counsel and later a company director.

CHAPTER 2

Birth of a Company

The history of Hercules Powder Company opens in a corner room on the seventh floor of the Du Pont Building on an autumn afternoon in 1912. Although the new company's certificate of incorporation was recorded on October 19, the formal affairs of the company did not commence until the twenty-ninth. On that day, at 2:00 P.M., three men gathered to launch a new company in the presence of several witnesses. The three incorporators were Judge J.P. Laffey, Du Pont's general counsel, who had been intimately involved in the antitrust suit; John J. Raskob, assistant treasurer of Du Pont and Pierre du Pont's closest business associate; and Russell H. Dunham, formerly Du Pont's comptroller and now the president-designate of the new company.[1]

The incorporators' first acts were to approve the company's bylaws and to reassign its initial capital stock, which consisted of twenty shares totaling $2,000 subscribed by the three incorporators.[2] Dunham, who had subscribed eight shares, transferred three to James T. Skelly, retaining five for himself. Laffey, who had subscribed six shares, distributed three each to George H. Markell and Frederick W. Stark. Raskob, who had also subscribed six shares, split his evenly between T.W. Bacchus and Clifford D. Prickett. At that point, the new shareholders present among the witnesses took their places in the meeting, constituted themselves as the board of directors, and assumed responsibility for the business of the Hercules Powder Company.[3]

The formalities on that October afternoon culminated a long chain of events that gave definition and shape to Hercules. The events involved many participants, but the key actors were the U.S. Department of Justice, the U.S. Circuit Court for the Third District of Delaware, more than forty defendants—most of them affiliated with E.I. du Pont de Nemours & Company of Wilmington, Delaware—in a criminal antitrust case, and the newly designated officers of Hercules itself.

On June 13, 1912, the district court had decreed the general terms under which Du Pont, by far the largest explosives-maker in the United States, would be broken up. The court ordered that roughly half of Du Pont's capacity to make explosives be divided between two new companies (one of which would become the Hercules Powder Company) and specified how the new companies would be financed, the properties each would own, and the general terms under which they would compete.[4]

The court's final decree terminated a federal antitrust suit launched against Du Pont in 1907. That suit had gone to trial in 1908 and was concluded on June 21, 1911, when the court found that twenty-eight of the defendants had violated the Sherman Antitrust Act. The court ordered the defendants and the government to prepare plans for the dissolution of Du Pont's dominant position in the explosives industry.[5] The year between that order and the final decree was the period in which the future Hercules Powder Company took shape in negotiations between Du Pont executives and government lawyers.

Of course, the verdict against Du Pont has a history. Here the tale must be unraveled into its separate strands: the origins and development of the explosives industry in America, the Du Pont company's rise to dominance in the industry, and the government's specific charges against Du Pont. This chapter examines each of the strands before returning to the conclusion of the trial, the negotiations leading to the final decree, and the period of its implementation before the birth of the Hercules Powder Company.

The business of explosives

Consider for a moment the world into which Hercules was born. In December 1911, the Nobel Prize in chemistry (an award established in memory of the inventor of dynamite) went to Marie Curie for the discovery of two new elements, radium and polonium. The following spring, the *Titanic* sailed to its doom, carrying with it the lives of 1,595 people. That summer, at the Olympics in Stockholm, Jim Thorpe won two gold medals (for the pentathlon and the decathlon), which he was later forced to return when his amateur standing was called into question.

In politics, 1912 was an election year in the United States. A few weeks after the incorporation of Hercules, Woodrow Wilson, a Democrat, won a four-way fight for the presidency. He defeated not only the Republican incumbent, William Howard Taft, but also former president Theodore Roosevelt, who had broken ranks with Taft and had lost a "rough and riotous" contest for the Republican nomination. Roosevelt entered the general election at the head of the progressive Bull Moose party and finished second in the voting, ahead of Taft.[6] The fourth candidate in the race, socialist Eugene V. Debs, pulled nearly a million votes—6 percent of the total. Many issues divided the electorate: A continuation of the tariff, or free trade? Suffrage for women? A federal income tax? Federal regulation of big business?

The last question was very much at the forefront of public concern, as well as a major issue in the election. Teddy Roosevelt had earned, in the public's eye, an image as a "trustbuster." And it is true that during his presidency (1901–1909) the Department of Justice had launched more than fifty antitrust investigations, including those against John D. Rockefeller's Standard Oil Corporation, James Buchanan Duke's American Tobacco Company, and the "powder trust" of E.I. du Pont de Nemours & Company. But during the four years of the Taft administration, ninety antitrust investigations were initiated, many at the instigation of U.S. Attorney General James B. Wickersham, whom critics described as "the most radical man who has held that office for years," and as a man possessed by "a fanatical love of action."[7] It was under Taft that, in the span of four remarkable weeks in May and June 1911, the Supreme Court upheld the dissolutions of Standard Oil and American Tobacco and the federal district court ordered the dismembering of Du Pont.

The public's interest in big business was not all focused on the trusts. The economy was booming. The gross national product (GNP) of the United States had doubled between 1900 and 1910, rising to $35.3 billion. At the same time, the population of the country had grown by 25 percent, reaching 92.4 million in 1910. Virtually all the population growth was in the cities, where in the decade after 1910 Americans began to outnumber those on farmland for the first time.[8] Industry was feeding much of the GNP growth. In Detroit, Henry Ford was marketing the Model T, though the moving assembly line that would make him famous was more than a year away from operation. At the four-year-old General Motors, Charles Kettering had just invented the automatic self-starter. The automobile revolution was on its way, although when people thought of big business, they still thought of the railroads, or steel, or perhaps J.P. Morgan's financial empire, which became the focus of a celebrated congressional investigation at about the time Hercules was born.

Although it might not have sprung to anyone's mind in 1912, America's industrial might would have been inconceivable without the largely unheralded contributions of the explosives industry. The steel industry's growth would have been stunted without explosives to blast iron ore free. The coal and oil industries would not have developed as rapidly unless explosives had made those energy sources accessible. Railroads could not have crossed the continent without explosives to clear the way. Perhaps then we can forgive the hyperbole of T.W. Bacchus, vice president and general manager of the Hercules Powder Company at its birth, who would write of "the super-force—dynamite," as "the greatest boon that has ever been created for overcoming the obstacles in Nature, and the masterpiece of the chemist's art."[9]

The versatility of explosives in commercial as well as military applications helped to create a business whose significance was disproportionate to its size. Between 1869 and 1909, the total value of explosives products sold in the United States grew roughly tenfold, from $4.2 million to $40.1 million. Over the same period, total employment in the industry grew from about 1,000 to over 7,000. Total capital invested in the industry rose from $4.1 million to $50.2 million.[10]

This was impressive growth, although industries such as railroads, iron and steel, tobacco, food products, and petroleum-refining were much larger and were growing at a much faster clip. To put some perspective on the explosives industry's position in the economy, the chemical industry, of which explosives was only a part, was medium-sized. Total capital investment in chemicals and "allied substances" in 1909 was about one-third that of textiles or food and kindred products, and about half that of machinery or iron and steel. If the chemical industry was in the middle of the economic picture, explosives ranked somewhere in the middle of the middle. The value of explosives products sold in 1910 was less than that of general chemicals, patent medicines, paints and varnishes, or soap; about the same as that of fertilizers; and larger than that of druggist preparations, cottonseed products, linseed oil, toilet preparations, grease and tallow, or salt.[11]

At the dawn of the twentieth century, the explosives industry in the United States was passing through a period of sweeping change. Its products and processes, markets, and competitive dynamics were all affected. The transformation of the industry had important implications for the future of the principal explosives-makers, including the Hercules Powder Company.

A Chemical Revolution in Explosives

Modern science draws distinctions between *explosives*—materials that, when "properly initiated," produce large volumes of hot gases at exceedingly rapid rates, measured in kilometers per second—and *propellants,* materials that, as they burn, also decompose rapidly into hot gases, albeit at controlled, predetermined rates usually measured in centimeters per second. Modern science further distinguishes between *primary explosives* or *initiators,* which detonate when subjected to shock or heat, and *secondary* or *high explosives,* which detonate under the influence of the shock of a primary explosive.[12]

At the start of the twentieth century, these distinctions were not so clear, although they were becoming apparent as the explosives industry's major products evolved. Before the late nineteenth century, explosives and black powder had been synonymous (although black powder is now classified as a propellant).[13] The manufacture of black powder—which is a mixture, not a chemical compound—was a straightforward process that had changed little over the previous six centuries: the raw materials were pulverized and mixed together in proper proportions, usually 75 percent saltpetre (potassium nitrate, a substance that Europeans and Americans normally imported from India), 10 percent sulphur, and 15 percent charcoal. These ingredients were mixed in small batches to ensure safety. Later, the mixture was rolled or pressed into cakes, which were then broken apart into small pieces or grains to be polished, dried, and blended.[14]

The properties of manufactured black powders varied with the purity and proportions of the raw materials, the size of the finished grains, and the density of packing for use. In the Crimean War (1853–1856), when saltpetre from India was hard to come by, Lammot du Pont, a junior partner of E.I. du Pont de

Nemours & Company, developed a substitute powder using sodium nitrate imported from South America. This formulation, called "B" blasting powder, carried slightly greater explosive power and was more economical to produce.[15]

Throughout its long history, black powder served a wide range of uses, military and civilian, big jobs and small. In general, finer grained powders were used for guns and rifles and coarser grained powders were used for military cannons or commercial blasting applications, such as mining and quarrying. By the late nineteenth century, however, the development of more sophisticated explosives and propellants based on chemical compounds was destroying black powder's primacy.

The most common of the new compounds was nitroglycerin, a very powerful, very sensitive chemical that was very dangerous to handle. The explosive properties of nitroglycerin and a basic process for making it had been identified in the 1840s, although the extreme instability of the compound restricted its commercial development. In 1867, the Swede Alfred Nobel found a practical way to handle nitroglycerin when he developed the first dynamite. Nobel discovered that the compound could be impregnated in sticks of kieselguhr (a kind of clay), which could be handled with relative safety and detonated with a primary explosive.

The advantages of dynamite over black powder were manifold. To begin with, dynamite is a far more powerful and violent explosive. It decomposes into gases in a few millionths of a second—about a thousand times faster than black powder.[16] Dynamite can be tailored for a variety of conditions and applications and can be mixed with other substances to control the rate and force of detonation. Black power is simply black powder; the only variations possible—in the formula, the size and other characteristics of the grains, and the density of packing—make relatively little difference to the nature of the explosion. Dynamite can be manufactured as a water-resistant gelatin (also in stick form) for blasting in wet areas; black powder cannot be used reliably in wet conditions. Ammonium nitrate can be substituted for some of the nitroglycerin to make a product called ammonia dynamite, which can be used in cold weather; again, black powder is unreliable in such conditions.

The new high explosives had other advantages as well. Because it requires a separate explosive to detonate it, dynamite is safer to transport, store, and handle than black powder. In addition, ammonia dynamite and its derivative products detonate without flames. That feature is important in underground mines, where black powder explosions often touched off deadly fires. The new explosives could also be formulated to limit the release of poisonous gases—a serious problem in enclosed areas such as underground mines, where black powder explosions left deadly traces of carbon monoxide behind. Indeed, after 1910, the U.S. Bureau of Mines classified some types of the new high explosives as "permissible" for use in underground coal mines and actively promoted them.[17]

Dynamite was manufactured in several stages. The first step was to make nitroglycerin, which was done by introducing glycerin into a mixture of nitric and sulfuric acids. Next, the nitroglycerin was incorporated into a "dope," which would make it safe to handle. At first, Nobel had used an inert dope (kieselguhr), which had no effect on the characteristics of the explosion. By the 1870s, however, using oxidants such as sodium nitrate or ammonium nitrate, explosives-makers (including Nobel) had formulated active dopes that contributed to the explosion, as well as different absorbents such as wood pulp and flour. The exact formula for the dope depended on the intended use of the dynamite, and differences in the formula allowed dynamites to carry different brand names. For instance, adding nitrocellulose to the nitroglycerin, or using special absorbents, produced the water-resistant gelatin form of dynamite. The final step in the manufacture was to pack the incorporated material in paper shells that were soaked in paraffin to prevent leaking. The dynamite was then ready for storage and shipping.[18]

In the late nineteenth century, technological change was also eroding another traditional market for black powder: as a propellant in firearms and cannons. Once again, the new product, smokeless powder, was a chemical compound, this one based on nitrocellulose. The ability of nitrocellulose powders to burn without smoke was an obvious attraction, but the greater advantages were the higher muzzle velocities and controllable burning rates possible with the new product. A projectile fired with black powder, for example, might achieve an initial velocity of seventeen hundred feet per second. Under the same conditions but fired with smokeless powder, the projectile could reach an initial velocity of three thousand feet per second.[19]

The production of smokeless powder involved a more sophisticated process than either black powder or dynamite. The first step was the nitration of cellulose (usually purified cotton linters—the short fuzzy fibers that remain attached to the seed after ginning). That delicate operation consisted of reacting the raw cellulose with large amounts of nitric and sulfuric acids, the latter acting as a catalyst. The resulting compound—nitrocellulose—was then washed, boiled, and reduced to pulp before it was colloided with a mixture of ether and alcohol. The colloided substance was stirred and kneaded vigorously, then extruded into long, spaghetti-like strands. Finally, the strands were cut into grains that could be coated and polished for packing.

The properties of smokeless powders could differ significantly depending on the nitrogen content of the nitrocellulose (a higher percentage of nitrogen rendered greater explosive power), the addition of other ingredients (including nitroglycerin, which enhanced the powder's brisance, or power), and the size of the grains (which could vary considerably, from those suitable for small weapons to those required by large naval guns).

The differences between black powder and the new chemical explosives had important implications for competitors in the industry. Plants that made dynamite and smokeless powders were much larger and more sophisticated than black powder mills. The level of investment required to compete successfully was much higher for dynamite-makers. Given the advantages of large-scale operations, few small companies or individual proprietorships could hope to stay in the business for long.

Production of the new explosives, moreover, placed a premium on scientific research and analysis and on technical engineering. The nitration of cellulose or glycerin utilized enormous quantities of nitric and sulfuric acids. To make nitrocellulose, for instance, required at least thirty-three parts of mixed acid for each part of cellulose, and the manufacture of nitric acid itself involved large amounts of sulfuric acid. Nitration of glycerin used a smaller proportion of mixed acid, but so much nitroglycerin was required to satisfy the demand for dynamite that it too consumed large quantities of acids in absolute terms.[20] Although these chemicals could be bought, such huge amounts were needed that most producers developed a capacity to manufacture and recover acids, a complicated, capital-intensive process. Along the way, they developed and mastered new skills in engineering, construction, and chemical analysis that went far beyond those required to make black powder.

The Marketplace for Explosives

As the products of the explosives industry were in transition at the turn of the century, so were its markets. Demand for explosives divided into two broad categories, commercial and military. Commercial powders ranged from sporting firearms to civilian blasting applications in mining and construction; military powders were designed for specific firearms and weapons such as rifles and cannons.

The market for commercial explosives was vastly larger than that for military explosives—even in wartime. During the four years of the Civil War, for example, munitions suppliers for the North produced 9 million tons of black powder—in contrast to the yearly production of roughly 25 million pounds of black powder in the country just before the war.[21] The next conflict that engaged the United States, the Spanish-American War (1899–1902), was fought primarily at sea, and it consumed proportionately fewer explosives. The U.S. Navy shot about 2.2 million pounds of brown prismatic powder, a special formulation designed for large guns, during the three years of the war. In 1902, Du Pont alone made nearly three times that amount of commercial black blasting powder. By that time, the U.S. Navy had begun to use smokeless powders in significant quantities, which nevertheless paled in comparison to commercial volumes. In the last year of the war, the total amount of smokeless military powder sold by all U.S. producers was 1.86 million pounds; in the same year, nearly 11 million pounds of black sporting and smokeless sporting powders were sold in the commercial market.[22]

At the turn of the century, the total size of the commercial market was large and growing. In 1902, producers sold 148.6 million pounds of black blasting powder, 114.6 million pounds of high explosives, 9.2 million pounds of black sporting powder, and 1.8 million pounds of smokeless sporting powder. A decade later, sales stood at 193 million pounds of black blasting powder, 248.3 million pounds of high explosives, 7.3 million pounds of black sporting powder, and 4.8 million pounds of smokeless sporting powder. These figures translate into growth rates of 30 percent for black blasting powder, 117 percent for high explosives, and 167 percent for smokeless sporting powder, and a decline of 21 percent for black sporting powder.[23]

These numbers reveal two important points about the commercial market. First, production of blasting explosives (black blasting powder and high explosives) was far greater—twenty-five to thirty times greater—than production of sporting powders. Nearly fifty black blasting powder mills and two dozen dynamite plants operated in the United States in 1902, but only nine black sporting powder mills and two smokeless sporting powder mills were active. Second, the substitution of the new chemical explosives for black powder in both blasting and sporting applications was proceeding rapidly.

The market for blasting explosives divided into three generic applications: mining and quarrying, construction, and agriculture (in descending order of importance). Coal operators consumed the largest share of explosives, but mining of iron ore, copper, nickel, lead, and the precious metals and quarrying of limestone and other crushed stones consumed significant amounts of explosives as well. In the early twentieth century, most of the mining and quarrying industries were growing at a healthy rate and employing great quantities of explosives along the way.

The construction market was expanding rapidly as well. The two decades spanning the turn of the century witnessed massive investment in large-scale public works—canals, roads, railroads, aqueducts, harbors, dams, subways, tunnels, and bridges—which required significant amounts of explosives. To cite but a few examples: in 1900, the railroads were still laying an average of four thousand miles of track per year, much of it in the West, where the rocky terrain demanded enormous amounts of explosives to clear the way. The Catskill Aqueduct, which helped supply water to the booming population of New York City, used 17 million pounds of dynamite and 700,000 pounds of black blasting powder from 1907 to 1922. The granddaddy project of them all was the Panama Canal, built over the period 1904–1914. Nearly 60 million pounds of high explosives were used to move more than 200 million cubic yards of earth during the construction.[24]

In both the mining and construction industries, competition was becoming concentrated in a few hands as a smaller number of larger customers accounted for an increasing portion of explosives sales. Producers sold directly to the big coal operators, mining companies, quarries, and construction firms through purchasing offices in major cities such as New York, Chicago, St. Louis, and San Francisco. These buyers were sophisticated customers, and the technical support provided by explosives-makers was relatively less important than other sales criteria: price, reputation, reliability, and delivery terms.

The agricultural market for explosives was very different. To begin with, it was much smaller. In the early part of the century, farmers bought between ten and twenty million pounds of dynamite per year to dig drainage ditches, drain swamps, and remove rocks and stumps. Such uses accounted for between 5 and 10 percent of total explosives production in the country. The agricultural market was served by the general hardware trade, and the technical support furnished by explosives-makers was extremely important.[25]

For most blasting applications, powerful and versatile high explosives were supplanting black powder. Given the advantages of the new products, one may wonder why any customers still used black powder. In part, they did so because black powder had advantages over high explosives for some applications. Some coal operators, for example, preferred black powder because they wanted the coal to come apart in large lumps; high explosives tended to pulverize the coal. Besides, black powder had the virtue of being familiar to many customers: Why learn to use a different, potentially dangerous product when black powder was so well understood and got the job done? As Clifford T. "Pat" Butler, a forty-one-year veteran dynamite-maker at Hercules, puts it, for many customers failing to switch "was a case of not learning to not use black powder."[26]

The shift from black powder to new chemical explosives was also under way in sporting markets. The cartridge and loading companies—Winchester, Remington, and the owners of other great brands—were the principal customers for sporting powder. These companies, like the large mining and construction firms, normally bought in bulk through centralized purchasing offices. A small replacement market (less than 10 percent of sales) also existed for individual hunters and sportsmen who bought brand-name powder at hardware stores. In the early twentieth century, smokeless sporting powder was becoming increasingly popular for both rifle and shotgun ammunition because of its superior power and ballistic properties.

The armed forces were also actively investigating new chemical explosives. The military market diverged from the commercial market in significant ways. Although the military sometimes used black powder and dynamite for demolition and for construction and engineering purposes, it tended to value different characteristics in explosives from those prized by commercial buyers. In particular, the armed forces were interested in propellants that would deliver projectiles at great velocities and with great force without destroying the gun or cannon that launched them. The brown prismatic powder that the U.S. Navy had used in the Spanish-American War had significant ballistic advantages over conventional powders, even though it was worthless as a commercial explosive. Similarly, the high-powered smokeless powders increasingly preferred by military leaders in Europe and the United States were inappropriate for all but a tiny segment of the sporting market.[27]

The military was also considering new explosives as the contents of projectiles. Shells and bombs were sometimes filled with black powder, although such weapons were dangerous to use and were decreasingly effective against modern armor and fortifications. Nitroglycerin-based explosives carried enough destructive power but were also so sensitive that they tended to detonate with the shock of firing. And the search was still on for a suitable high explosive to fill shells. Military scientists around the world were experimenting with many compounds. Some researchers developed new, less sensitive forms of dynamite and nitrocellulose, but most work was concentrated on nitrated derivatives of aromatic compounds made from coal tar. Among these, picric acid (and picrates) and trinitrotoluene (TNT) were the most promising. After both compounds were battle-tested in the Russo-Japanese War (1904–1905), the European militaries generally adopted TNT, which was easier to manufacture, less sensitive, and less likely than picric acid to react with metal shell casings. The U.S.

military also experimented with both compounds and adopted TNT as its standard after 1910. Before that, production of picric acid and TNT in the United States had been negligible.[28]

In sum, the military sought characteristics in explosives—muzzle velocity, sheer destructive power, and insensitivity to shock—that were increasingly specialized and tailored to the needs of modern warfare. On the one hand, the military might underwrite innovations and new product development. But on the other hand, explosives made for the military would not necessarily or readily find commercial uses. And again, the market for military explosives in the United States before World War I was small and tightly controlled. In 1900, only four commercial suppliers produced military smokeless powders, and after 1903, just one remained in the business—Du Pont, which had acquired the others. The U.S. military also made smokeless powders itself at ordnance plants in Picatinny, New Jersey, and Indian Head, Maryland.

The Powder Trust

Between 1902 and 1912, a single interest—Du Pont—dominated the explosives industry in the United States. The company and its principal subsidiaries and affiliated companies directly controlled well over two-thirds of total output. In 1907, for example, its sales amounted to 64 percent of the market for "B" blasting powder, 72 percent of saltpetre black blasting powder, 72 percent of dynamite, 73 percent of black sporting powder, 64 percent of smokeless sporting powder, and 100 percent of military ordnance powder not made by the government itself. In the same year, the company earned nearly $4 million on sales of $31.7 million, making it one of the most profitable businesses in the country.[29]

Although Du Pont had been a prominent player in the explosives industry since its founding in 1802, its dominance dated from the period after the Civil War. The company acquired its eventual position in the industry in three stages. First, from the 1870s on, Du Pont and the industry's second-largest producer, the Laflin & Rand Powder Company (L&R) of New York, helped organize the black powder business through the Gunpowder Trade Association (GTA). Second, starting in the 1880s, Du Pont and L&R jointly supervised the development of dynamite in the eastern part of the United States. Their control was institutionalized in 1895 with the formation of the Eastern Dynamite Company, in which Du Point and L&R were the leading investors. Finally, in the third stage, the ownership of Du Pont passed to three younger members of the family in 1902. The young du Ponts, assisted by some of the most talented executives in American industry, brought the best part of the explosives business under a single corporate umbrella by acquiring L&R, major interests in the leading West Coast explosives-makers, and the last independent producer of smokeless powder.[30]

We shall dwell on this history for a moment, because in it can be found the roots of Hercules, as well as the assets it acquired at its birth.

Stage I: The Gunpowder Trade Association

Before the Civil War, competition in explosives (that is, black powder) had been confined within distinct regions. In part, localized competition reflected the obvious risks in transporting powder over great distances. But it also reflected the preferences of customers—chiefly mine operators—who wished to have suppliers nearby so as to avoid the cost and risk of carrying explosives inventory and to benefit from the powdermakers' technical support. As a result, most explosives companies were small and located in pockets of mining activity around the country: near the coal fields of the eastern and southern states, and near the metal ore mines of the upper Midwest and the mountain states. Even a relatively large company such as Du Pont, which had sales agents around the country, confined production to one or two specific regions.

The Civil War and its aftermath altered the traditional dynamics of competition. The war had diverted the capacity of the larger and more successful powdermakers to military production and allowed many new companies to organize to serve commercial markets. Most of the new entrants were small mills, but a few, including the California Powder Works (CPW) of Santa Cruz, California, were well capitalized and became formidable competitors later on. After the war, when the military producers reentered the commercial business, prices began to fall rapidly. In 1865, a keg of powder sold in the East for $4.25 and in the West for $7; a year later, the prices fell to $1.25 and $2.75, respectively. The situation was made worse by the government's efforts to sell surplus stocks of powder accumulated during the war—an effort that continued through the 1880s.[31]

Explosives-makers responded to the new competitive conditions in the same way that producers of oil, steel, sugar, tobacco, whiskey, glass, and many other products did at the time: they tried to control competition. Their efforts proceeded in several stages. First, larger companies tried to buy all or part of smaller rivals—perhaps secretly—to control their behavior. Next, if that option was not available, companies attempted to negotiate price and output agreements with their competitors. If neither alternative proved effective, then it was competition to the death.

In the explosives industry, all three strategies were used. Companies gobbled up other companies. In 1869, the Laflin Powder Company and the Smith & Rand Powder Company, both of which were combinations of other companies and partnerships, joined to form L&R. About the same time, the Du Pont company also began a long spree of growth via acquisition.[32]

In 1872, representatives of the leading eastern manufacturers (including the four largest companies in the East, Du Pont, L&R, the Hazard Powder Company, and the Oriental Powder Company) established the Gunpowder Trade Association. The group was empowered to set prices for black powder for the territory east of the Rocky Mountains. Three years later, the GTA extended its reach by agreeing on prices with the largest western producer, the CPW of Santa Cruz. Still later, the association set production quotas, supervised marketing agreements among its members, and tried to inhibit the development of the new high explosives and smokeless powders—products that had been pioneered by outsiders.

As historians Alfred D. Chandler, Jr., and Stephen S. Salsbury have remarked, the GTA "was certainly one of longest-lived, most stable, and effective" industry cartels, primarily because of the interlocking pattern of ownership in the industry and the close cooperation between the two largest companies, Du Pont and L&R.[33] Both companies held equity positions—sometimes surreptitiously—in other member companies and affiliates. Close ties between Du Pont and L&R also reflected long-time personal and professional associations. Laflins and du Ponts, for example, had shared price information as early as 1853. The two companies routinely bought powder from each other and repackaged it for sale under their own brand names after plant explosions or when the icing of rivers blocked shipments from northern mills.[34] The relationship grew especially close after 1873, when Solomon Turck became president of L&R. When the GTA founded the Gunpowder Export Company in 1876 to extend its reach into Canada, the new company's first president was Turck, who was supported by Vice President Lammot du Pont of the Du Pont company. Du Pont and L&R also jointly organized several new powder companies.

The modern reader may wonder about the legality of such actions. Although some states had laws against combinations in restraint of competition inside their borders, until 1890 no federal statute dealt with the problem of combinations whose activities crossed state lines. Even the passage of the Sherman Antitrust Act in that year did not immediately change the situation. The statute proved difficult to enforce and even more difficult to interpret. It did not stand up to a definitive constitutional test until the late 1890s, and it was not used to dissolve a huge combination until 1904.[35]

Stage II: The Eastern Dynamite Company

Cooperation among the leading producers of black powder extended to their attitudes toward the new chemical explosives. In the 1880s, after trying for a decade to thwart the development of dynamite, Du Pont and L&R collaborated to produce high explosives in the East. This is a complicated story, and to appreciate it fully we must backtrack a moment.

In 1868, Alfred Nobel, the Swedish inventor of dynamite, had sold the U.S. patent rights to the product to the Giant Powder Company of San Francisco. Dynamite, which still used kieselguhr as the dope, caught on quickly. Giant expanded to the East in 1870 by establishing a subsidiary, the Atlantic Giant Powder Company, and building a dynamite plant a year later at Kenvil, New Jersey. When other companies tried to make and sell dynamite, Giant promptly challenged them with infringement of Nobel's patents. The CPW tried vainly to compete with Giant by mixing nitroglycerin and mealed black powder in a formulation called "Black Hercules." The trade name was apparently suggested by General William S. Rosecrans, the Civil War officer and a friend of Joseph W. Willard, a CPW salesman. Rosecrans pointed out that the mythological Hercules was renowned not only for his strength but also for slaying giants.[36]

Black Hercules proved to be a poor product, but the CPW continued to work on nitroglycerin explosives. In 1874, Willard received a patent for dynamites using an active base made from magnesium carbonate, potassium nitrate, potassium chlorate, and sugar. The CPW marketed these dynamites under the trade name "Hercules"; they were manufactured at plants near Pinole, California (a site subsequently known as Hercules, California), and (after 1877) in Cleveland, Ohio. Giant, meanwhile, filed suit to enjoin the CPW and other explosives-makers from producing the dynamites, claiming that even dynamites with active dopes were covered by the Nobel patents. In 1880, however, the Supreme Court rejected that argument and ruled that the original patents did not apply to the new formulations. The way was then clear for competitors nationwide to make the new high explosives.[37]

This episode is important, not only because it marks the birth of the Hercules brand name in explosives, but also because it led the eastern powder-makers to enter the dynamite business. Indeed, with the arrival of Atlantic Giant in New Jersey, and the CPW's Hercules plant in Cleveland, Du Pont and L&R had little choice but to respond. Neither company built its own dynamite plants. Rather, they chose to invest together in new companies or to acquire existing ones. The first of these ventures, in 1880, was the founding of the Repauno Chemical Company, whose capital was supplied largely by Du Pont and L&R. A year later, the same investors purchased the Cleveland plant from the CPW, named it the Hercules Powder Company, and established it as a subsidiary of Repauno. At the same time, they also bought into Atlantic Giant and soon reorganized it on their own terms as the Atlantic Powder Company.[38]

Their cooperation in high explosives became more formal in 1895, when members of the du Pont family and executives of both the Du Pont company and L&R founded the Eastern Dynamite Company. This enterprise was a holding company that owned, among other properties, Repauno, Hercules, and a stake in the Atlantic Powder Company. The Eastern Dynamite Company, like the GTA before it, set price and production terms and immediately expanded by acquiring all or part of many other dynamite-makers. It also negotiated marketing agreements with other companies. In 1897, for example, Eastern Dynamite joined with the California companies and several smaller manufacturers to sign an agreement with the leading French, English, and German suppliers to allocate markets around the world.[39]

Stage III: The "Big Company" at Du Pont

In the next several years, the explosives industry passed through its third period of consolidation. The pivotal event was the sale in 1902 of controlling interest in the Du Pont company to three younger members of the family: cousins Alfred I., T. Coleman, and Pierre S. du Pont. The particulars of what happened next form one of the great case studies in the development of modern management in the United States.[40]

In brief, as Chandler writes, the young cousins "discarded the policy of horizontal combination for one of administrative centralization and vertical integration." They believed the old strategy of controlling competition by cutting prices and buying stock in rivals' companies was inefficient because it "meant that the leading firms . . . often had to purchase unplanned and unwanted plant capacity that was rarely located in the place best suited to meet market and supply conditions and rarely equipped with the most modern facilities."[41] The cousins also believed that the traditional arrangements created enormous administrative waste and inefficiency, since the activities of the constituent companies were, at best, loosely coordinated, and, at worst, the source of needless conflict and duplication of effort. In short, the cousins believed that the business would be better run—and would make more money for its shareowners—as a single, integrated enterprise, what they called the "Big Company."

To carry out this strategy, the cousins had first to obtain full control of many companies that had been affiliated with Du Pont, the GTA, or the Eastern Dynamite Company. In October 1902, in a brilliant stroke, Du Pont acquired L&R, whose owners were approaching retirement. The purchase gave Du Pont not only more plants and facilities (including the nation's second-largest smokeless powder plant) and valuable brand names, but also complete control of the Eastern Dynamite Company. Soon thereafter, Du Pont acquired several smaller explosives-makers outright. It also purchased controlling interest in the CPW and bought into several other West Coast producers. Finally, it arranged to purchase the last independent producer of military smokeless powders in the United States, the International Smokeless Powder and Chemical Company.

By the end of 1903, most of these acquisitions were completed and Du Pont's leaders turned their attention to reining in the company's many properties and improving their management. Responsibility for the various plants was split among three operating departments, Black Powder, High Explosives, and Smokeless Powder. Modern techniques of cost accounting were established throughout the company under the leadership of Russell H. Dunham, Du Pont's comptroller. The sales force was also reorganized: salaried managers and employees replaced commission agents across the nation.

These actions made Du Pont a leaner, better managed, and more competitive enterprise and brought it into closer conformity with the antitrust law. At that point, Du Pont made a fateful decision. The executive committee split as to how far to proceed with the consolidation and how quickly to move.[42] One group, led by Pierre du Pont and Arthur J. Moxham, wished to eliminate the old trade agreements and subordinate companies entirely, and as soon as possible. Pierre argued for taking such action not only because he thought it would lead to greater efficiency, but also because the company's lawyers had advised him that the old ways were "absolutely illegal," given recent interpretations of the Sherman Antitrust Act. On the other side, however, executive committee members J. Amory Haskell (formerly president of L&R) and Hamilton M. Barksdale (formerly president of the Eastern Dynamite Company) saw benefits in proceeding with deliberate speed and maintaining at least some of the old subsidiaries and holding companies. The benefits included keeping successful brand names and identities in the marketplace, limiting liability for accidents, reducing exposure to labor organizing, and retaining the "great money value" of existing arrangements.

In the end, the two sides compromised. Over the next several years, Du Pont began to liquidate most of the old subsidiary companies, and it resigned from the GTA in 1904. However, it left some subsidiaries intact, at least as sales organizations or corporate shells, including L&R, the Hazard Powder Company, and the Eastern Dynamite Company.

THE DEMOLITION OF THE POWDER TRUST

The reign of the Big Company lasted but a short time. In 1906, Robert S. Waddell, a disgruntled former employee whose efforts to start a powder company in the Midwest had failed, launched a campaign to persuade Congress and the Department of Justice to investigate Du Pont's position in the explosives industry.[43] Waddell's initial charges focused on Du Pont's monopoly position in military smokeless powders, but he also gave the U.S. attorney general a sheaf of documents that amply illustrated the inner workings of the GTA. On July 31, 1907, armed with reams of evidence about price-fixing and predatory pricing, secret takeovers, dummy corporations, cross-ownership of competing firms, and other damning items, Attorney General William H. Moody filed a criminal antitrust suit against forty-three defendants, including the owners and top executives of Du Pont, the company itself, its principal holding companies and unconsolidated subsidiaries, and several unaffiliated companies that had been members of the GTA.

The suit went to trial before the U.S. Circuit Court for the Third District of Delaware in September 1908. It dragged on for almost three years, and the arguments and counterarguments, testimony, and exhibits filled thousands of pages of published text. The government contended that the defendants had "entered into and are now parties to an agreement, combination, or conspiracy in the shipment and sale of gunpowder and other high explosives . . . to control, regulate, and monopolize or attempt to monopolize said trade and commerce." The government cited the actions of the GTA and the Eastern Dynamite Company in evidence and argued that the reorganized Du Pont after 1902 was merely an attempt to carry on the old practices in a new guise. The government specifically pointed to Du Pont's delay in resigning from the GTA and the continuing existence of subsidiaries such as L&R, Hazard, and the Eastern Dynamite Company as proof that only the form, not the substance, of the illegal combination had changed.[44]

Du Pont defended itself by claiming that it was a fundamentally different company after 1902, and that it had not violated federal law. Instead, the company argued, its "object was solely to extend [its] business" by legitimate means, and its acquisitions were intended not to monopolize the industry, but to assure "well-located and well-equipped plants." As for the charge that it had been slow to resign from the GTA, Du Pont could only muster the weak argument that the purpose of the GTA had not been to deter new competitors, or to fix prices, but "to prevent 'unreasonable competition.' "[45]

The district court reached its verdict on June 21, 1911. Charges against fifteen of the defendants, who were found not to have been parties to illegal agreements, were dismissed. As for the other twenty-eight defendants—including the Du Pont company, its officers, and its principal holding companies and subsidiaries—the court agreed with the government's contention that they had acted, and were acting, in combination to restrain trade, and that they had violated Sections 1 and 2 of the Sherman Antitrust Act. The court pointed to the illegal practices of the GTA and found that the reorganized Du Pont company of 1902 was merely "the successor of the combination" that preceded it. The court enjoined the defendants from continuing the combination and gave them and the government lawyers four months to prepare a plan for its dissolution.[46]

Establishing Hercules

Deadlines for implementation of court orders are easily broken in major legal actions. In the case of *United States* v. *Du Pont,* the plan of dissolution required a year to gestate from the initial bargaining positions of the government and the company into the court's final decree. On June 13, 1912, the district court approved the plan negotiated by the two parties and thereby established the general structure of the explosives industry of the future.[47]

The final decree ordered Du Pont to dissolve the holding companies and subsidiaries that had lingered after 1902. The heart of the decree, however, concerned the terms under which Du Pont's explosives operations would be divided between itself and two new companies. The court specified the facilities that Du Pont would retain, as well as the plants and mills that would be divested (see Exhibits 2.1–2.3). Du Pont would keep 58 percent of its capacity to make dynamite, 50 percent of its capacity to make black powder, and all of its capacity to make smokeless powders, whether sporting or military. The new companies would be of different size and character, the larger being roughly half again the size of the smaller. The court did not specify names for the new companies but suggested that they might be identified as Laflin & Rand and the Eastern Dynamite Company. Here, for a moment, we will call them Company A and Company B.

The larger company, Company A (which would become the Hercules Powder Company), would have dynamite plants in New Jersey, Michigan, and California; black blasting powder mills (including three formerly owned by L&R) in New York, Pennsylvania, Ohio, Wisconsin, Kansas, and California; and black sporting powder mills (including one formerly owned by L&R) in New York and Connecticut. The court also required Du Pont to build for Company A a new plant with "a capacity sufficient to manufacture 950,000 pounds per annum of smokeless sporting powder." The new facility, which would make "the brands now or heretofore owned by the Laflin & Rand Powder Company," would be situated at Company A's dynamite plant in New Jersey, or at "some other suitable Eastern point."

EXHIBIT 2.1 Properties of Company A Decreed by the Court (Hercules Powder Company)

PLANT	*ORIGINS*
HIGH EXPLOSIVES	
Kenvil, New Jersey (cap.: 15 million lbs.) (1911 pdn.: 10,039,699)	Built in 1871 by Atlantic Giant Company. Du Pont and L&R acquired first stake in 1880. Acquired outright by Eastern Dynamite Company in 1895 and passed to Du Pont in 1902.
Marquette, Michigan (cap.: 9 million lbs.) (1911 pdn.: zero)	Idle in 1912. Built in 1881 by Lake Superior Powder Company; Lake Superior had been acquired by Du Pont, Hazard, and L&R in 1876 and passed to Du Pont in 1902. Largely dismantled in 1910 when Du Pont built Senter plant.
Hercules, California (cap.: 40 million lbs.) (1911 pdn.: 31,140,369)	Built in 1879 by CPW. (Du Pont acquired initial stake in CPW in 1869, and control in 1903.)
BLACK BLASTING POWDER	
Rosendale, New York (cap.: 210,800 kegs) (1911 pdn.: 58,858)	Idle during most of 1912. Built in 1855 by Smith & Rand. Passed to L&R in 1869, and to Du Pont in 1902.
Pleasant Prairie, Wisconsin (cap.: 263,500 kegs) (1911 pdn.: 115,063)	Built in 1899 by L&R. Passed to Du Pont in 1902.
Columbus, Kansas (cap.: 447,950 kegs) (1911 pdn.: 385,441)	Built in 1889 by L&R. Passed to Du Pont in 1902.
Ringtown, Pennsylvania (cap.: 250,355 kegs) (1911 pdn.: 234,825)	Ferndale mills: described by court as two mills, but really only one. Built by Titman Powder Company in the 1880s. Remodeled extensively in 1896. Acquired by Du Pont about 1903.
Youngstown, Ohio (cap.: 243,738 kegs) (1911 pdn.: 216, 712)	Built by Ohio Powder Company in 1881. Ohio Powder acquired by Du Pont, Hazard, and L&R in 1886. Control passed to Du Pont in 1902.
Santa Cruz, California (cap.: 737,800 kegs) (1911 pdn.: 414,425)	Built by CPW in 1863. Du Pont acquired initial stake in 1869 and completed control in 1903.
BLACK SPORTING POWDER	
Hazardville, Connecticut (cap.: 92,225 kegs) (1911 pdn.: 64,975)	Built by Hazard Powder Company in 1835. Acquired by Du Pont secretly in 1876.

EXHIBIT 2.1 (Continued)

PLANT	*ORIGINS*
Schaghticoke, New York (cap.: 105,400 kegs) (1911 pdn.: 78,644)	Built in 1812 by private interests at government request during the War of 1812. Control passed through various hands before acquisition by Schaghticoke Powder Company in 1858, by L&R in 1872, and by Du Pont in 1902.

Note: cap. = capacity; pdn. = production.

COMPANY A TOTALS	*CAPACITY*	*PRODUCTION (1911)*	*UTILIZATION*
High Explosives	64,000,000	41,180,068	64%
Black Blasting Powder	2,254,143	1,425,324	63
Black Sporting Powder	197,625	143,619	73
Smokeless Sporting Powder[a]	950,000	0	—
Smokeless Military Powder	—	—	—

Note: Capacity and production totals for high explosives and smokeless powders given in pounds; for black powders in kegs.

a. Final decree required Du Pont to build a facility with this capacity for Company A.

COMPANY A	*CAPACITY SHARE[b] (1911)*	*MARKET SHARE[c] (1911)*
High Explosives	16%	17%
Black Blasting Powder	18	18

b. Capacity allocated by the court divided by total U.S. capacity in 1911.

c. Total production of plants allocated by the court divided by total U.S. *sales* in 1911.

Sources: Final Decree, 4–5; *Defendants' Additional Profits,* Exhibit A, 111–116; Van Gelder and Schlatter, *Explosives Industry,* 480–497 (Kenvil); 655–658 (Marquette); 497–518 (Hercules); 98–99 (Rosendale); 230 (Pleasant Prairie); 239 (Columbus); 156–157 (Ferndale); 137–138n. (Youngstown); and 283–286 (Santa Cruz).

The smaller entity, Company B (which would become the Atlas Powder Company), would receive dynamite plants in New Jersey, Missouri, Michigan, and California, and its black blasting powder mills would number two in Pennsylvania and one each in Tennessee, Illinois, and Kansas. Company B was to receive no black sporting powder mills, nor any capacity to make smokeless powders.

The final decree also specified how the new companies would pay for the assets they would receive. Company A's initial capitalization would be $13 million, and Company B's $6 million. They would each issue securities in these

EXHIBIT 2.2 Properties of Company B Decreed by the Court (Atlas Powder Company)

PLANT	ORIGINS
HIGH EXPLOSIVES	
Hopatcong, New Jersey (cap.: 11 million lbs.) (1911 pdn.: 8,378,835)	Forcite plant. Built by American Forcite Powder Company in 1883. Acquired by Eastern Dynamite Company in 1899. Control passed to Du Pont in 1902.
Webb City, Missouri (cap.: 6 million lbs.) (1911 pdn.: zero)	Atlas plant. Built by Du Pont in 1912.
Point Isabel, California (cap.: 12 million lbs.) (1911 pdn.: zero)	Vigorite plant. Idle in 1912. Built in 1900 by Stauffer Chemical Company to make brands acquired from Vigorite Powder Company. Acquired by Du Pont in 1903. Equipment transferred to Hercules, California, in 1911.
Hancock, Michigan (cap.: 18 million lbs.) (1911 pdn.: 6,469,554)	Senter plant. Built by Du Pont in 1909 to replace Marquette plant.
BLACK BLASTING POWDER	
Riker, Pennsylvania (cap.: 158,000 kegs) (1911 pdn.: zero)	Idle in 1912. Possibly acquired by Du Pont through L&R in 1902.
Shenandoah, Pennsylvania (cap.: 105,400 kegs) (1911 pdn.: 24,354)	Idle during much of 1912. Date of construction unknown. Sold by Shenandoah Powder Company to Du Pont in 1902.
Ooltewah, Tennessee (cap.: 316,200 kegs) (1911 pdn.: 143,378)	Built by Chattanooga Powder Company in 1890. Partial ownership acquired by Du Pont and L&R in 1895. Passed to Du Pont in 1902.
Belleville, Illinois (cap.: 303,025 kegs) (1911 pdn.: 278,297)	Built in 1892 by Phoenix Powder Company. Acquired by Du Pont, L&R, and other interests in 1896. Passed to Du Pont in 1902.
Pittsburg, Kansas (cap.: 263,500 kegs) (1911 pdn.: 223,250)	Construction announced in 1901 by Pennsylvania & Kansas Powder Company, a front for Du Pont and L&R. Mill not actually built until 1903, when Pennsylvania & Kansas was fully owned by Du Pont.

Note: cap. = capacity; pdn. = production. At the time of the dissolution, Atlas Powder Company was awarded an additional black powder mill at Patterson, Oklahoma (cap.: 316,200; 1911 pdn.: 174,256), which had been built by Du Pont in 1907. See Van Gelder and Schlatter, *Explosives Industry,* 211, 309.

EXHIBIT 2.2 (Continued)

COMPANY B TOTALS	*CAPACITY*	*PRODUCTION (1911)*	*UTILIZATION*
High Explosives	47,000,000[a]	14,848,389	32%
Black Blasting Powder	1,146,125	669,279	58[b]
Black Sporting Powder	—	—	—
Smokeless Sporting Powder	—	—	—
Smokeless Military Powder	—	—	—

Note: Capacity and production totals for high explosives and smokeless powders given in pounds; for black powders in kegs.

a. Capacity figure includes Atlas plant built in 1912 but not operated that year.

b. Figures do not include Patterson, Oklahoma, plant deeded to Atlas after the final decree.

COMPANY B	*CAPACITY SHARE*[c] *(1911)*	*MARKET SHARE*[d] *(1911)*
High Explosives	12%	6%
Black Blasting Powder	15	9

c. Capacity allocated by the court divided by total U.S. capacity in 1911.

d. Total production of plants allocated by the court divided by total U.S. *sales* in 1911.

Source: Final Decree, 4–5; *Defendants' Additional Proofs,* Exhibit A, 111–116; Van Gelder and Schlatter, *Explosives Industry,* 453–465 (Forcite); 469–470 (Atlas); 640–646 (Vigorite); 466–469 (Senter); 219 (presidency of John L. Riker at L&R, 1895–1900); 1089 (Shenandoah); 141–144 (Ooltewah and Belleville); 240–241 (Pittsburg).

amounts, of which half would be common stock and half bonds bearing an interest rate of 6 percent. The interest would be payable only if the new companies made sufficient earnings. The common stock and half of the bonds would be distributed to Du Pont's stockholders, and the remaining half of the bonds would be retained by the Du Pont company in its treasury.

The court also ordered Du Pont to furnish the new companies with sufficient working capital, cash, and facilities "to efficiently carry on" their business, and also to transfer to them "so far as practicable, a fair proportion" of the explosives business Du Pont had under contract. The decree required Du Pont to provide the new companies with access to its sales records and its purchasing, research, and engineering departments. Access would continue for up to five years, and "upon some reasonable terms as to the cost thereof to the two corporations." Finally, the court gave Du Pont six months to implement the terms and conditions of the decree.

The final decree settled many questions, but it also raised others: How was the decision reached? Why this particular division of the business? Why the

EXHIBIT 2.3 Properties Retained by Du Pont under the Final Decree

	CAPACITY (LBS.)	*PRODUCTION (1911)*
HIGH EXPLOSIVES		
1. Ashburn, Missouri	15,000,000	12,499,226
2. Barksdale, Wisconsin	30,000,000	20,611,070
3. Du Pont, Washington	24,000,000	19,978,290
4. Emporium, Pennsylvania	12,000,000	7,398,039
5. Hartford City, Indiana	6,000,000	0
6. Louviers, Colorado	15,000,000	11,588,730
7. Gibbstown, New Jersey	45,000,000	36,747,515
8. Lewisburg, Alabama	6,000,000	0
BLACK BLASTING POWDER *(kegs)*		
1. Augusta, Colorado	230,559	0
2. Connable, Alabama	316,220	125,359
3. Olephant Furnace, Pennsylvania	421,600	322,173
4. Mooar, Iowa	1,159,400	1,111,385
5. Nemours, West Virginia	421,600	376,165
6. Patterson, Oklahoma[a]	316,200	174,256
7. Wilpen, Minnesota	316,200	19,698
BLACK SPORTING POWDER *(kegs)*		
1. Brandywine, Delaware	92,225	16,440
2. Wayne, New Jersey	144,925	140,838
SMOKELESS SPORTING POWDER *(lbs.)*		
1. Carneys Point, New Jersey	2,467,500	2,291,555
2. Haskell, New Jersey	2,873,700	1,552,736
SMOKELESS MILITARY POWDER *(lbs.)*		
1. Carneys Point, New Jersey	3,750,000	2,273,118
2. Haskell, New Jersey	2,400,000	1,848,686
3. Parlin, New Jersey	2,250,000	1,607,082

a. Patterson mill transferred to Atlas Powder Company after the final decree.

DU PONT TOTALS	*CAPACITY*	*PRODUCTION (1911)*	*UTILIZATION*
High Explosives	153,000,000	108,822,870	71%
Black Blasting Powder	3,181,779	2,129,036	67
Black Sporting Powder	237,150	157,278	66
Smokeless Sporting Powder	5,341,200	3,844,291	72
Smokeless Military Powder	8,400,000	5,728,886	68

Note: Capacity and production totals for high explosives and smokeless powders given in pounds; for black powders in kegs.

EXHIBIT 2.3 (Continued)

DU PONT	*CAPACITY SHARE[b] (1911)*	*MARKET SHARE[c] (1911)*
High Explosives	38%	44%
Black Blasting Powder	25	28

b. Capacity allocated by the court divided by total U.S. capacity in 1911.

c. Total production of plants allocated by the court divided by total U.S. *sales* in 1911.

Source: Defendants' Additional Proofs, Exhibit A, 111–116.

unequal positions of the two new companies? Why these particular assets? Why these capital structures? And why these terms requiring Du Pont's continuing support? The answers to those questions—the basic definition of Hercules—are contained in the dialogue that took place between government lawyers and Du Pont executives in the year that followed the interlocutory decree.[48] The dialogue was full of misunderstandings and false steps on both sides, and neither side developed a coherent position until the very end. Indeed, viewed at this distance, the whole episode has a comical aspect to it. (See Exhibit 2.4 for a chronology of the negotiations between Du Pont and the Justice Department.)

To begin with, the du Ponts were divided as to how to proceed. Their first instinct was to either appeal the decision or ask for a rehearing. But they couldn't—recourse to the Supreme Court was not an option for the simple reason that an interlocutory decree cannot be appealed. The company would have to wait until the district court issued a final decree—which it would not do until the company submitted a plan for dissolution that the court would approve. Next, the company considered asking the court to reopen the case. Du Pont's lawyers, however, "were unanimous in saying that [the company] could not consistently ask for a rehearing . . . with any chance of success," and they warned "that such a step might unduly antagonize the court."[49]

The impossibility of an appeal and the impracticality of a rehearing left Du Pont with two alternatives: find a way to evade the decree, or find a way to comply that would limit the damage. Coleman du Pont, a prominent figure in the Republican party, tried the first course, contacting his allies in Congress and the Taft administration to see whether the decision could be modified or reversed. Even when his efforts got him nowhere, Coleman persisted, but with a clumsiness and lack of result that proved embarrassing and perhaps counterproductive.[50] Du Pont's main strategy, developed by Pierre du Pont and the company's lawyers, centered on finding a way to comply with the decree while minimizing the damage done to the company.

The federal attorneys also disagreed among themselves about what settlement would be appropriate. Of Attorney General Wickersham's three assistants on the case, one (James Scarlet) was prepared to offer a very mild solution,

EXHIBIT 2.4 Creation of the Hercules Powder Company Negotiations between Du Pont and the U.S. Department of Justice, June 1911–May 1912

Context

June 21, 1911 U.S. District Court finds 28 defendants, including Du Pont and its senior officers guilty of violating the Sherman Antitrust Act. Defendants and Justice Department lawyers to agree on a plan of dissolution.

Date	*Du Pont*	*Department of Justice*
June 1911	Offers to dissolve holding companies and subsidiaries not consolidated after 1902.	Offer rejected.
August	In addition to former offer, proposes to divest one smokeless military powder plant.	Offer rejected.
September	In addition to former offers, agrees to divest former L&R properties, including 4 black blasting power and 1 black sporting powder mills.	Offer rejected.
October	Proposal rejected.	Asst. A.G. Glasgow proposes creation of 2 to 4 new companies from Du Pont assets, including all properties formerly owned by L&R or by Du Pont and L&R together. The new companies would oversee 6 black blasting powder and 1 black sporting powder mills and include capability to make high explosives.
	Offers to capitalize divested L&R company at $5 million. New L&R company to have capacity to make high explosives.	Offer rejected.
November	Proposal rejected.	Asst. A.G. Roadstrum proposes divestiture of 3 high explosives plants plus 4 former L&R black blasting powder and 2 black sporting powder mills, and 1 smokeless military powder plant. Book value of divested assets at $6.3 million.

EXHIBIT 2.4 (Continued)

Date	*Du Pont*	*Department of Justice*
December	Offers to divest 3 high explosives plants, 4 former L&R black blasting powder mills, 1 former L&R black sporting powder mill, and 1 smokeless sporting powder mill. New company to be capitalized at $12 million.	Offer rejected.
February 1912	Unable to pursue unauthorized discussions with Scarlet.	Asst. A.G. Scarlet prepared to accept dissolution of unconsolidated holding companies and subsidiaries, plus divestiture of 1 smokeless military powder plant.
March	Offers to divest 4 high explosives plants, 7 black blasting powder mills, 1 smokeless military powder mill, and 1 smokeless sporting powder mill.	Offer rejected. A.G. Wickersham threatens to place Du Pont in hands of court-appointed receiver. Proposes creation of 2 new companies from divested assets, including 42 percent of Du Pont's capacity to make high explosives and 50 percent of its capacity to make black blasting powder. Divested assets to include 7 high explosives plants, 12 black blasting powder mills, 2 black sporting powder mills, and capability of 1 smokeless sporting powder mill. Value of divested assets set at $20 million.
May	Agrees to government's terms.	
Conclusion		
June 12, 1912	U.S. District Court's final decree specifies creation of two new companies from assets divested by Du Pont.	

Source: Hagley Museum and Library: Coleman du Pont's presidential papers, papers of E.I. du Pont de Nemours & Co., series II, part 3, boxes 131 and 132, accession 472; Papers of Pierre S. du Pont, accession 616; Papers of Hamilton M. Barksdale, papers, EIDPDN, series II, part 2, boxes 1006 and 1007, accession 518; Chandler and Salsbury, *Pierre S. du Pont and the Making of the Modern Corporation,* 277–90.

another (William A. Glasgow) favored a drastic dismemberment, and the third (Victor Roadstrum) took a middle position. The attorney general himself, whose attention to the case was sporadic and erratic, took a line that was more extreme than that of even his most zealous assistant. The problem was partly a lack of precedent. The dissolutions of Standard Oil and American Tobacco were being negotiated simultaneously, but there were no close parallels among the cases. More serious was the federal attorneys' lack of knowledge and information about the explosives industry in general, and about Du Pont's operations in particular. Rather than initiate proposals themselves, they were obliged, most of the time, to react to Du Pont's plans for the dissolution. The plans that the government attorneys did offer tended to betray astonishing ignorance.

Finally, there was a mix-up about exactly what the court really wanted. In the closing passages of the interlocutory decree, the court had said that it needed more information before it could frame a final decree, and that, on the appointed day, it would hear the two sides "as to the nature of . . . any plan" for the dissolution that they might submit. The court evidently believed that the two parties would get together and negotiate terms for the breakup. That was also Wickersham's view. The du Ponts, however, changed their minds at several points about what the court meant. They began by talking with the government, but after several months of frustrating discussions and disagreements, they took a new tack. They interpreted the court's language to mean that the company could present its own plan, and that the court would then reconcile that plan with the suggestions forwarded by the government attorneys.[51] As a result, Du Pont's relations with the government attorneys became further strained.

In brief, the negotiations unfolded in the following sequence. Two weeks after the verdict was in, Du Pont offered to dissolve immediately its holding companies and subsidiaries that had not already been liquidated. That offer had already been rejected by the Department of Justice as inadequate in 1910, when Du Pont had made a futile effort to settle the case, and it was summarily dismissed again in the summer of 1911.[52] The next offer came in August: the company would not only liquidate the holding companies and subsidiaries, but also end its monopoly in smokeless military powders by divesting the plant acquired from the International Smokeless Powder and Chemical Company at the end of 1903.[53]

When that suggestion failed to impress the government, the company considered giving a little more. It was now prepared to divest the remaining assets of the old L&R and began to define the financial arrangements of such a deal. According to Pierre du Pont, the new company, to be named Laflin & Rand, would pay for the assets by issuing a fourth of their value in stock, to be distributed to Du Pont's stockholders, with the remaining three-quarters covered by bonds, which would be held in Du Pont's treasury. The net effect of the proposal for Du Pont would have been the loss of four black blasting powder plants, one black sporting powder plant, and a small drop in its capital surplus. Because the old L&R had not itself made dynamite but had participated in the business only through its share of the Eastern Dynamite Company, the new

L&R would not have the capacity to make high explosives. If the government were to accept this proposal, Du Pont would retain all of its dynamite plants.[54]

At that point, the government made its first counterproposal. The plan, which was presented by Assistant Attorney General Glasgow, a hard-liner who probably had not yet consulted his boss, called for Du Pont to divest the remaining assets of L&R and other properties that Du Pont and L&R had jointly acquired over the years. The properties would be distributed between at least two, and perhaps four, new companies, according to the following terms. First, Glasgow accepted Du Pont's offer to divest L&R's black blasting powder mills and create a new corporation. He then proposed, however, that Du Pont set up a second new business to manage the facilities that had been acquired jointly by Du Pont and L&R. The second corporation would own two additional black blasting powder mills.

Third, Glasgow argued that Du Pont should "segregate into another corporation, which may be the same as No. 1 [the first spinoff]," a dynamite capacity equivalent to L&R's stock holdings in the Eastern Dynamite Company—roughly 18 percent of Du Pont's capacity in 1911. Finally, Glasgow proposed that Du Pont should "segregate into another corporation, which may be the same as No. 2 [the second spinoff], the proportion of dynamite business represented by the stockholdings of the jointly owned companies in the Eastern Dynamite Company." The government's ignorance was beginning to show. Pierre du Pont observed that Glasgow was "not aware that no such stockholdings existed," and added, sardonically, "I believe, therefore, this segregation may be omitted."[55]

The government proposal would have required Du Pont to divest 32.4 million pounds of its capacity (18 percent) in high explosives (which Pierre thought could be achieved by spinning off three small plants or one big one), six black blasting powder plants, and one black sporting powder facility. Pierre was beginning to despair of a negotiated solution, and he instructed his lawyers to draft a plan to be presented for the court to reconcile with whatever the government proposed. Pierre's plan renewed the offer already made to reestablish L&R—this time with some dynamite capacity—but he made only one other concession. He was prepared to change the terms of the financing of the new L&R; he estimated that it could be capitalized at $5 million, of which half would be distributed to Du Pont's shareowners, and half retained in Du Pont's treasury.[56]

Within a few weeks, it was apparent that the court was not willing to accept the latest Du Pont offer. Pierre by then believed that "we cannot expect fair play from the Government or the Courts; that they have determined to make a showing with our company and that facts in support of our righteous position will count for but little." According to Pierre, Assistant Attorney General Roadstrum wanted the divestiture of fifty million pounds of Du Pont's capacity (28 percent) in high explosives in three specified plants, plus the L&R black blasting powder plants, the International smokeless powder plant, and two black sporting powder plants. The book value of the divested assets would be $6.3 million.[57]

Discussions between Du Pont and the government lawyers resumed at the end of November, and the upshot was more concessions by the company. In mid-December, Du Pont offered to relinquish the three dynamite plants specified by Roadstrum to the reconstituted L&R. In addition, the new L&R would get four black blasting powder mills and one black sporting powder mill formerly owned or jointly owned by the old L&R. The package also included an unnamed plant for the manufacture of smokeless sporting powder. L&R would pay for the properties by issuing half the amount in stock, and half in bonds to remain in Du Pont's treasury. Du Pont would also supply "a sufficient amount of working capital to insure the stability" of the new company. The total capitalization was creeping up to $12 million, as Pierre toted up the book value of the plants, the working capital to get the new company started, and the value of its brand names and goodwill.[58]

The plan, which Du Pont was confident enough to have printed in galley proofs, fell prey to a difference of opinion among the attorney general's men. What settlement would emerge depended on which government lawyer the company had talked to last. Roadstrum seemed to like the most recent plan, although there were ominous signs that Glasgow and Wickersham did not.[59] In February, Coleman du Pont briefly saw some hope in discussions with yet another assistant attorney general, James Scarlet. Scarlet was evidently prepared to recommend a settlement along the lines Du Pont had suggested the previous August: dissolution of the holding companies and corporate shells, and divestiture of the plant acquired from the International Smokeless Powder and Chemical Company. Just when the talks seemed to be getting somewhere, however, Scarlet abruptly disappeared and would not return his messages. It later turned out that he had taken a leave from the Taft administration—without telling his superiors—to work for the insurgent campaign of Theodore Roosevelt.[60]

By early March, Glasgow appeared to be back in charge of the government's position, and Du Pont had now increased its offer to include four dynamite plants and seven black blasting powder mills, in addition to the International smokeless powder mill and perhaps one other smokeless sporting powder mill. The offer, however, did not include a black sporting powder mill.[61]

The plan was finally presented to the court on March 4, 1912, on what turned out to be an amazing occasion. According to Pierre du Pont, the government lawyers rejected Du Pont's offer out of hand, saying "they thought a receivership was the proper decree." In other words, the government was now asking the court to appoint a third party to take over Du Pont's assets and design the breakup. At that point, the parties withdrew to the presiding judge's chambers, where the discussion continued along the same lines. As Pierre recounted it, the judge

> was quite plain in his statement that . . . the court could not make a plan of dissolution, nor administrate; that its only alternative was receivership, or a very drastic decree (meaning, I suppose, injunction against doing Interstate business) and that this form of a decree must be handed down unless the two parties came together in a "consent"

> decree. This statement was, at best, equivalent to saying that unless we agreed with the Government attorneys, we should receive an extremely harsh form of punishment. However, "in chambers" the court went so far as to say they believed, in a general way, the duPont company was entitled to a general position in the trade equivalent to that enjoyed by it in 1902.

Next followed an extraordinary scene, as again described by Pierre:

> After the above occurrences one of our Executive Committee [Alfred I. du Pont] had opportunity to meet President Taft, Attorney General Wickersham and Mr. Glasgow, his assistant, coming into the meeting later on. At this meeting the Attorney General used very violent language, claiming that the operation of our property should not be left with us any longer; that the court should appoint a receiver and continue the receiver in charge even though we were to appeal to the Supreme Court. Though the positive statement was not made, it was very strongly intimated that the court had already agreed on this course.[62]

What accounts for the government's suddenly tough stance? All indications prior to the meeting—at least those indications evident to Coleman and Pierre du Pont—were that the company and the government were approaching agreement.[63] There appear to be at least three explanations for the government's new position. First, and most important, the attorney general, the most aggressive trustbuster ever appointed to that office, had become directly involved in the case. Wickersham was doubtless irritated by Du Pont's efforts to divide and conquer the government attorneys, as well as by Coleman du Pont's persistent lobbying of Congress and the administration for help in getting around the court decision. Wickersham's talk of receivership may have been intended to scare Coleman off that tactic, and to convince the other senior Du Pont executives that the government wanted a serious plan, right away.[64]

Second, the attorney general had at last, apparently, developed a definite standard that he would apply to the dissolution. He would look beyond the terms previously considered—the requirement that Du Pont either divest specific properties it had acquired or taken control of after 1902 or divest specific percentages of production capacity added since then. Relying on the standards behind those terms would not prevent Du Pont from continuing to dominate the industry. Rather, Wickersham sought to ensure that Du Pont would not retain a majority of the nation's production capacity in high explosives and black powder. In other words, Du Pont would have to give up substantially more than had been discussed to date.[65] Wickersham was partly following the precedent of the breakup of American Tobacco at the end of 1911. In that case, no single entity after the dissolution controlled more than 37 percent of output of a major product line.[66]

Finally, there was the timing of the situation. President Taft faced reelection in 1912, and Theodore Roosevelt had just challenged him for the Republican nomination. The primaries were under way, and the campaign was being waged at least partly on the candidates' policies toward big business. Taft had positioned himself as tougher than Roosevelt on this issue, so it was no moment to appear soft.

After the March 4 meetings, things moved swiftly. The du Ponts had little choice but to wait for the details of Wickersham's plan. After several meetings, the outlines became clear. As Pierre du Pont described it,

> We are to form two new companies; to these companies will be turned over about 42% of the dynamite business and about 50% of the black blasting powder business. One of the companies will be provided with a new factory to be built for the purpose of manufacturing smokeless sporting powder. . . . They have conceded that we should keep the Government smokeless powder business intact, owing to the insistent demands of both the Army and Navy Department officials that it is best for the Government—which it is. It is for this reason that we are obliged to provide a smokeless sporting powder factory for one of the new companies, as our present factories are so closely interwoven with the operations of the military powder manufacture that they cannot be separated.
>
> All this segregation of property will call for the separation of about $20,000,000 of our assets, which are to be paid for by the receiving companies through the issue of $10,000,000 [in] 6% bonds and $10,000,000 of common stock; the bonds will remain in the Treasury of the parent company and the stock will be distributed as a dividend to the common stockholders, or possibly sold to them at an attractive figure.[67]

At a meeting between Roadstrum, Glasgow, Pierre and Alfred du Pont, and one of the Du Pont lawyers on March 17, the broad structural issues of the dissolution were essentially decided. The total business to be divested would be divided between two new companies, which would not be equal in size or capability. The larger would own three dynamite plants, seven black blasting powder mills, and two black sporting powder mills, and Du Pont would furnish it with the capacity to make smokeless sporting powders. The smaller company would own four dynamite plants and five black blasting powder mills but would have no capacity at all to make black or smokeless sporting powder.[68]

Subsequent talks centered on details of the support services Du Pont would extend to the new companies in their infancy, as well as on details of their financing. In April, Glasgow wanted Du Pont to provide the new companies with access to its Trade Bureau, where information about trends among customers and competitors was kept, and to the benefits of Du Pont's nitrate of soda operations in Chile, which had just been completed. He also sought to change the terms of the financing of the new companies to ensure that interest on their bonded debt would not burden them unnecessarily.

On each point, Du Pont either accepted the suggestions or proposed acceptable alternatives. As a result, the new companies would be obligated to pay interest on the bonded debt only if they made sufficient earnings. They would also have access to Du Pont's Trade Bureau and its purchasing department, "upon some reasonable division of the expense thereof," for up to five years. The purchasing provision was offered as a substitute for the plan to involve the new companies in the Chilean nitrate operation.[69] Du Pont also agreed to provide access to its facilities for "experimentation [and] development of the art and scientific research" for five years. The company thereby succeeded in retaining its R&D capabilities intact. By early May, the negotiations were virtually complete, and the court approved the terms in its final decree of June 13.[70]

In deciding which facilities would be assigned to each new company, the court ratified the agreement made in the meeting between Du Pont and the government lawyers on March 17. The government lawyers had apparently been guided by several principles. First, they obviously believed that more than one new company should be established to ensure competition. Glasgow's initial plan in October 1911 had called for the creation of between two and four new companies.[71] In that belief, the government may have been influenced by the settlement of the American Tobacco case in November 1911. As President Taft described it, American Tobacco would be split "between two or more companies with a division of the prominent brands in the same . . . products, so as to make competition not only possible, but necessary."[72]

As to why the government divided the production capacity shares so unevenly between the two new companies, we can only offer conjectures. It appears that the government was at least partly influenced by Du Pont's many plans and proposals. Company A bears a marked resemblance to the reconstituted L&R that Du Pont had been talking about for months, and more than half of the plants assigned to Company B had not been mentioned before the March 17 meeting.

It would seem, then, that something like the following happened: after the confrontations of March 4, Du Pont was apparently ready to accept virtually any solution to escape receivership. With Wickersham now actively involved in the settlement, the government lawyers proposed that in addition to divesting L&R, Du Pont would have to spin off still more assets to a second company. The size of that company would be determined by the percentage of production shares left to Du Pont after the assets of the new L&R (Company A) were subtracted. As proposed by Du Pont in early March, Company A would have roughly 17 percent of the nation's capacity in high explosives and black blasting powder. With only Company A spun off, Du Pont would have retained 51 percent of the nation's capacity to make these products. The government apparently decided that Du Pont's share should be substantially less, and that a second company should therefore be created (see Exhibit 2.5 for percentage allocations). At the meeting on March 17, the government lawyers and the du Ponts worked out the definition of Company B.[73]

A second principle—besides percentage of capacity—that apparently guided the government lawyers was geographic balance in both black powder

EXHIBIT 2.5 Comparison of Capacity and Market Shares of Du Pont, Company A, and Company B

	CAPACITY SHARE[a]		MARKET SHARE[b]	
COMPANY	DYNAMITE	BLASTING POWDER	DYNAMITE	BLASTING POWDER
Du Pont	38%	25%	44%	28%
Company A	16	18	17	18
Company B	12	15	6	9

a. Capacity allocated by the court divided by total U.S. capacity in 1911.

b. Total production of plants allocated by the court divided by total U.S. *sales* in 1911.

Source: Defendants' Additional Proofs, Exhibit A, 111–116.

EXHIBIT 2.6 High Explosives Plants

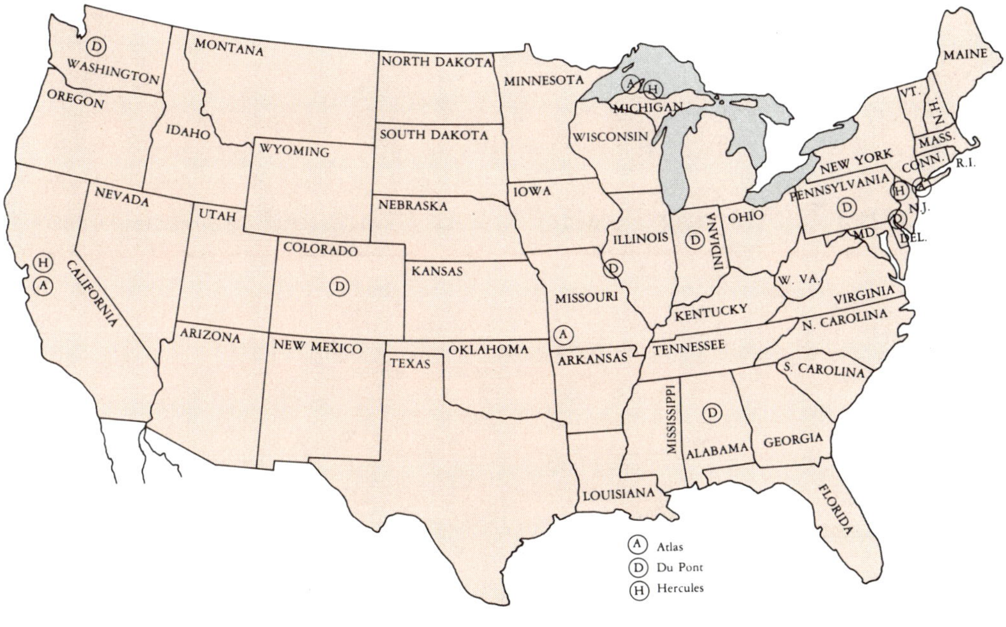

and high explosives. Both new companies had, at least on paper, facilities spread across the country (see Exhibits 2.6–2.8). Where several plants were clustered in a single location—in northwestern New Jersey, western Pennsylvania, western Kansas, the upper peninsula of Michigan, eastern Kansas, and the San Francisco Bay area—the government assigned them to different competitors. This principle was used to round out Company B: it was given a dynamite plant in the East and several powder mills in the Pennsylvania coal district. Concern about geo-

EXHIBIT 2.7 Black Blasting Powder Mills

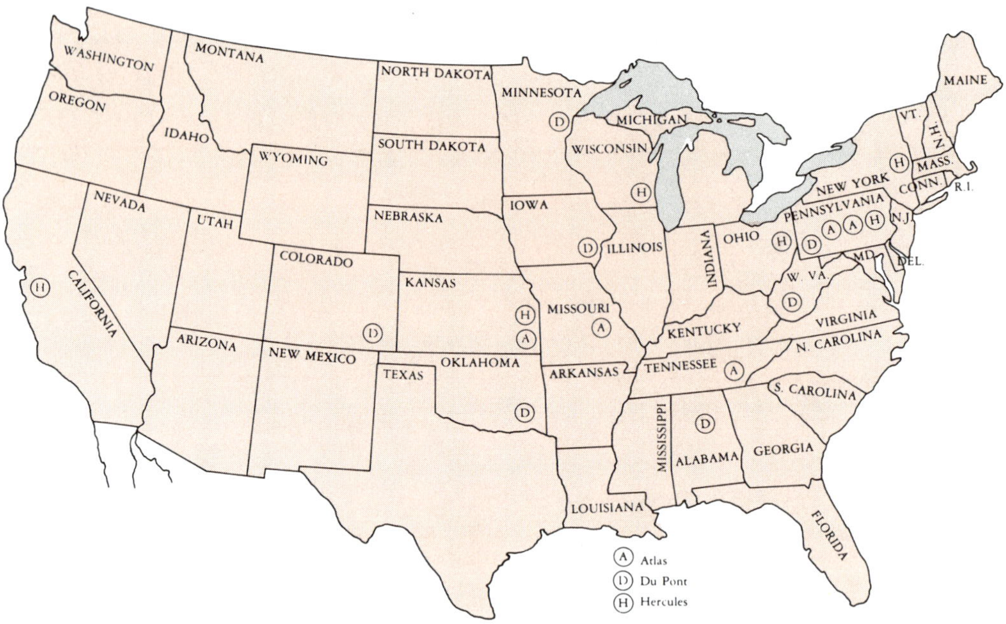

EXHIBIT 2.8 Sporting Powder Plants (Black and Smokeless)

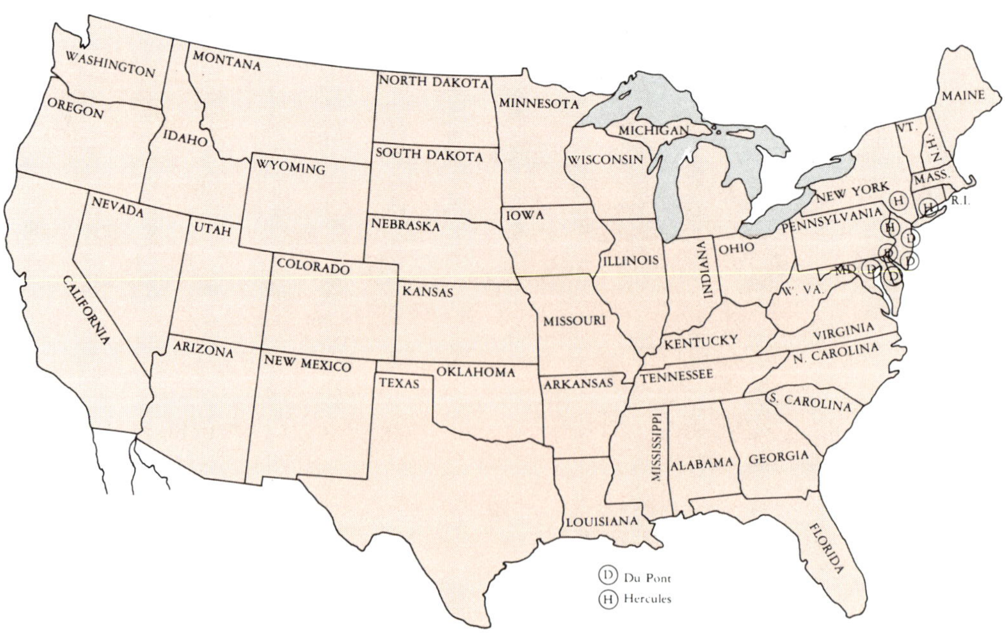

graphic balance appears also to have influenced the government's thinking about Company B, which had no foundation such as the old L&R at its core. On paper, Company B had nationwide market coverage in high explosives, although it had no black powder capacity in the West.

A third principle that apparently shaped the settlement was the importance of brand names. Company A essentially would be a merger between the old L&R and the CPW, with some smaller properties thrown in. In making this decision, the lawyers were doubtless influenced by Du Pont's suggestion, repeated from November 1911 to March 1912, that L&R be revived and spun off. Company A was assigned three of the four remaining L&R powder mills and its remaining sporting powder mill. To give Company A capacity in high explosives and black powder on the West Coast, the decree awarded it the Hercules dynamite plant and the Santa Cruz powder mill that had been built by the CPW, as well as the CPW's brand names. The brand names in sporting and smokeless powders given to Company A also came from L&R and the CPW. Company B had no comparable corporate identity at its core, but it was awarded several important brand names in high explosives: Vigorite, Forcite, and Atlas.

If these principles cast light on why the government made the decisions it did, the settlement is nonetheless full of ironies and anomalies. To put it mildly, the government made some curious if not bizarre judgments. In the first place, the government had been agitated in the beginning over Du Pont's 100 percent ownership of the military smokeless powder capacity not operated by the government, yet, ironically, the final decree left Du Pont's position as a military supplier untouched. Although Du Pont had been prepared to divest at least one smokeless powder plant in its offers from November to March, its capacity was kept intact at the specific requests of the chief ordnance officers of the U.S. Army and Navy. The court agreed with the military, but this aspect of the settlement remains a strange and conspicuous exception to the logic behind the rest of the decree.[74]

Second, the government appears not to have been aware that several of the facilities it assigned to the new companies were worthless, or nearly so. Company A's dynamite works at Marquette "was nothing but a plant site with water power," its equipment having been transferred to Company B's Senter plant two years earlier. The situation was the reverse on the West Coast: Company B's Vigorite plant was nothing but a shell whose equipment had been shipped to Company A's Hercules plant in 1911.[75] Thus, Company A had no real high explosives capacity in the Midwest, and Company B had none on the West Coast. Du Pont had been considering closing one of the black sporting powder mills (Hazardville) assigned to Company A.[76] One Company A powder mill (Rosendale) had been operating at minimal capacity for years, and both of Company B's powder mills in Pennsylvania, which represented about 20 percent of its theoretical capacity, were also mothballed.[77] Company B, in fact, would start out in a very weak position: it had no high explosives or black powder capacity on the West Coast, and its three functioning black powder mills were located in areas where competition was likely to be extremely tough because of the proximity of other mills.

The government was clearly handicapped by a lack of information about the facilities and by Du Pont's unwillingness to help, except in a general way. In April 1912, Glasgow had specifically requested performance figures on the plants, admitting that "we are absolutely in the dark as to whether, heretofore, the plants which these two companies will take over under the plan, have been enabled to make any net earnings in their operation." Both Pierre du Pont and Hamilton Barksdale opposed disclosing these figures; Pierre offered to "give every assurance that all the plants turned over can and have manufactured at a profit."[78] The government also made the strange decision to focus on operating capacity, rather than actual output, even though they had information on both. A glance at Du Pont's exhibits for the March 4 court appearance would have shown the government lawyers that several of the facilities were not operating, or were merely limping along, in 1911.[79]

Third, in making its allocations of business, the government ignored not only actual production figures for the plants but also their asset value. It was true that, on paper, Du Pont lost half its trade in black powder—a declining and only moderately profitable business—and nearly half its capacity in high explosives. However, the value of the assets Du Pont lost (about $20 million) represented only about one-third of the value of the assets it retained (about $60 million). The financing of the new companies—they would purchase their assets from Du Pont by paying half ($10 million) in stock and half ($10 million) in income bonds—favored Du Pont even more. Because Du Pont could retain half the income bonds in its treasury, the net effect of the transaction would be to reduce its assets by only $15 million (the stock portion of the payment plus half of the income bonds). "In other words," wrote Pierre du Pont, "[the company's financial position] will be reduced to approximately its condition in the year 1906"—the year its antitrust troubles began. Small wonder, then, that Pierre worried in April 1912, "I am fearful that the Government may try to cause further trouble when they find how little injury has been wrought by the plan so far agreed upon."[80]

Fourth, the government appears to have ignored the issue of vertical integration. At its eastern dynamite plant, for example, Company A had no capacity to make acids, a key element of manufacturing cost. Although both companies were awarded high explosives business, neither was granted capacity to make initiators, blasting caps, fuses, and other supplies. For these and many other significant items, the new companies would have to rely on Du Pont's purchasing department. Nor were the new companies given capabilities to perform their own research and engineering services—functions that were also critical to their future success in high explosives.

Finally, the government severely underestimated the market for commercial smokeless powders, which was a small business in 1912 but growing rapidly. It would soon overtake conventional black sporting powders. Du Pont retained its entire capacity (three functioning plants), while Company A received a new plant whose design and range of capacities would be determined by its principal competitor—Du Pont. Company B, of course, received no capacity at all.

In sum, the legal process and court decree that created the two new competitors in the explosives industry also burdened them in significant ways: outmoded plants, uneven market coverage, heavy reliance on purchased inputs, and limited access to the latest technologies. From the moment of their birth, the two companies would each face formidable strategic challenges.

Name, Location, and Leadership

Although it solved basic structural issues in the explosives industry, the final decree was not the last word on the organization of the two new companies. The court left many administrative issues unresolved: for example, what the new companies would be named, where they would be headquartered, who would manage them, and how they would be set up. Most of these matters were settled between May and October 1912 when the new companies filed for incorporation.

The most important decisions were made in late May by Du Pont's finance committee, which consisted of Alfred, Coleman, and Pierre du Pont. The first issue was the corporate names. The du Ponts considered—and the court was prepared to allow—calling them L&R and Eastern Dynamite. The finance committee, however, supported the recommendation of Du Pont's Sales Department that the two new companies be named after the principal brands of dynamite assigned to them. The larger (Company A) would be called Hercules Powder Company, and the smaller (Company B), Atlas Powder Company. Pierre recommended that the new companies be located in Wilmington as a matter of convenience. Not only did the decree require Du Pont to share the facilities of its purchasing, research, and engineering departments and the records of its Trade Bureau, but also "a great part of the administrative officers of the new companies will be recruited from those who are already residents of Wilmington."[81]

"The most important question" remaining, wrote Pierre, was "the selection of the two men to assume charge of the affairs of the two new companies." The final decree stipulated only that the leaders of the new companies could not be directors or officers of Du Pont.[82] Du Pont's problem, then, was to find executives who would be capable of making the new companies viable, without losing irreplaceable talent itself. It also had to avoid selecting people the court or the Justice Department might suspect of maintaining excessively close ties with it. Pierre's concern over the viability of the new companies, from this perspective, was quite genuine. After all, the securities of the companies would represent a substantial investment for Du Pont's stockholders, himself included. In addition, Du Pont management had no desire to remain under the continuing scrutiny of government antitrust lawyers.[83]

On May 25—three weeks before the court issued the final decree—Pierre suggested to the finance committee that "we select two men to be the General Managers of the companies and, that being done, we consult with them in the selection of an efficient staff." To speed the process, he compiled a list of all Du Pont employees who earned a salary of $5,000 or more and added the "names of lesser men who might be considered available." The du Ponts apparently discussed each name on the list and created a short list of twenty names. The list was then shortened to six names, and the du Ponts cast separate votes by each in a kind of proportional representation system. The leading candidates were Russell H. Dunham, Du Pont's comptroller; Frank Turner, another top financial man; three men from the Sales Department, Charles A. Patterson, William Coyne, and Dale Bumstead; and Ferdinand Lammot Belin, Pierre du Pont's brother-in-law. The voting was unanimous: the man who would be offered the leadership of the Hercules Powder Company was Russell H. Dunham.[84]

So the name would be Hercules Powder, the location Wilmington, and the leader Dunham. Once Dunham accepted the position, he began negotiating with his former employer for the services of others who would help him. By early October, "the five or six principal men to take charge of the new properties" had been selected.[85] The new senior officers of Hercules included T.W. Bacchus, James T. Skelly, Clifford D. Prickett, Frederick W. Stark, and George H. Markell. It was an impressive cast.

Dunham himself had been Du Pont's auditor and comptroller for the decade since that company's reorganization. One of the first new appointments in 1902, he had previously worked with Coleman du Pont as assistant comptroller of the Lorain Steel Company and later, on his own, as comptroller of Bethlehem Steel. At the turn of the century, both companies were leading practitioners of modern techniques of cost accounting and financial controls, and Dunham played an important role in shaping their policies. Indeed, at both companies, he worked with the noted efficiency expert and architect of "scientific management," Frederick Winslow Taylor.

At Du Pont, Dunham was responsible for adapting these management techniques to the explosives industry and improving on them. He possessed a sharp mind and an eagle eye for detail. In the explosives industry, he wrote,

> where a fraction of a cent a pound in some minor operation runs into thousands of dollars in the total, it is of little value to know that a loss has occurred without some means of determining where it occurred. It is for this reason that it is of great importance that most particular care be exercised to see that every expenditure, no matter how insignificant in itself, be distributed into the correct account.

Dunham's history of preparing accounts "with the greatest care and the closest possible approach to absolute accuracy" earned him many significant assignments at Du Pont. He was not only a functional specialist. He spent several years, for example, helping to consolidate the California companies into the Big Company. Later he helped reorganize the company's sales force, working to install incentives based on profitability measures. He also served as a director of Du Pont's operation north of the border, the Canadian Explosives Company, Ltd., and he served on several top management committees. By the time he assumed the presidency of Hercules, he was well prepared for general management.[86]

Bacchus, a fifty-year-old Englishman, was the senior member of the crew. He also boasted the most impressive credentials as an explosives-maker. He had come to America in 1887 already familiar with the selling of explosives, having been an agent of a hardware company that handled powder and dynamite. After working in several sales jobs in the United States, he joined the Cleveland office of the old Hercules Powder Company in 1893 and quickly rose through the ranks of several explosives companies in which Du Pont had interests. By 1903, he had become general superintendent of the Repauno dynamite plant—probably the most advanced facility in the industry—and, by 1912, general superintendent of high explosives for the company.[87]

As for the gregarious Skelly, the author of a brief biography wrote that he "started making friends in Nashville, Tenn., on February 23, 1877," his date of birth. Skelly had started in the explosives industry at the age of fifteen as an office boy at L&R. From there he moved into various positions in the hardware trade before returning to L&R as a salesman for Hercules Dynamite and L&R Powders. When that company merged with Du Pont in 1903, Skelly took charge of L&R business in the new general sales office. By 1912, Skelly was director not only of sporting powder and saltpetre black powder sales, but also of the Advertising Department at Du Pont.[88]

Prickett hailed from a long line of powdermen. His father and grandfather had each managed the sporting powder mill at Hazardville, Connecticut, which was built by the Hazard Powder Company, subsequently acquired by Du Pont, and transferred to Hercules at the end of 1912. Young Clifford also started at Hazardville, rising to be general superintendent of the plant before moving to Du Pont as assistant general manager for sporting powder. On the eve of the dissolution, he was assistant general manager of Du Pont's black powder mills.[89]

Stark had started out when the job of powder salesman involved no small dangers. Besides meeting with potential customers, the job included the responsibilities of keeping magazines and making deliveries. In those days, as the *Mixer* put it, "when you took an order for from fifty pounds to a ton, you went to the magazine and got it, put it in the buggy, and delivered it, making several trips if necessary." At the time of the Du Pont breakup, Stark had spent eight years as manager of the company's important sales office in Hazleton, Pennsylvania, in the heart of the anthracite coal district. He was the only Hercules executive who had testified in the Du Pont antitrust trial, where he disclosed something of his sales philosophy. Rather than compete purely on price, he sought to provide "quality and service" and to emphasize "safety and . . . serving the customers with what is best suited for their requirements."[90]

At age twenty-seven, Markell, the new secretary-treasurer, was perhaps the most brilliant of the officers. Markell had begun his career eight years earlier as a clerk in Accounts Receivable at Du Pont. There he evidently caught the eye of Dunham, who advanced his young protégé through a series of positions, culminating in that of assistant comptroller, before he followed his mentor to Hercules.[91]

Conclusion

The men who assumed control of the Hercules Powder Company in October 1912 were experienced and capable executives. Indeed, except for Markell, they had been executive officers in the administrative revolution that had transformed Du Pont from a loose federation to a tightly organized, well-managed company after 1902. The new leaders of Hercules were well prepared to operate the new business.

That fall they faced abundant challenges. Although Hercules would start out with ample cash and, at least on paper, production facilities spread across the country, it also started out with many basic issues to be resolved. The company would have to be organized, responsibilities defined, additional personnel recruited. Policies about appropriations, financing, employment, compensation, pensions, and a dozen other matters would have to be established. Above all, a coherent strategy for competing in the explosives industry would have to be worked out. The plants Hercules inherited were uneven in capacity and quality. They may have looked fine to government lawyers and the federal court, but in reality, the company would begin without good coverage in the marketplace. Nor could it offer a full line of explosives products until the new smokeless powder line at the Kenvil plant was completed.

The dissolution of the Big Company was, moreover, the breakup of a family. The infant Hercules would have to handle parental and sibling rivalry with Du Pont and Atlas. How would it establish its own identity? What sort of company would it be?

CHAPTER 3

STRUCTURE AND STRATEGY, 1912–1914

The unusual circumstances of Hercules Powder Company's birth created unusual challenges for its managers. Unlike most new companies, this one did not start out as a small entrepreneurial venture. Rather, Hercules sprang forth as a mature company in a mature industry. It started with money in the bank and facilities, sales offices, and salesmen distributed across the country. It inherited some well-known brand names and a fine reputation with customers. Its executives, among the most talented in the industry, started with intimate knowledge of the cost positions and strategies of their principal competitors.

This is not to say that Hercules faced smooth sailing from the start. To the contrary, many formidable problems lay in wait for the new company. The plants and mills it inherited from Du Pont had been designated without serious investigation of their capabilities and performance records. Those assets had to be evaluated and coordinated in ways that made sense for Hercules. Several important details of the final decree were yet to be implemented, including the full assignment of accounts to the new company and the construction of its smokeless powder facility. Although some management systems and policies could be borrowed or adapted from Du Pont, if the new company was to become more than a clone of its forebear, it had to establish its own ways of doing business.

Beyond all this, Hercules had to adapt to a new competitive environment. Friends and colleagues had suddenly become rivals. Although Hercules inherited a sales force and enjoyed a solid reputation, much confusion lingered in customers' minds as to the impact of the final decree. The confusion was bound to work initially to Du Pont's advantage. Thus, Hercules had to make fundamental choices about its business—sales strategy, product line, geographic scope, and internal policies and systems—while simultaneously creating an independent image in the marketplace and, for the first few years, continuing to rely on Du Pont for some essential staff services.

In short, as it started out, Hercules found itself in a peculiar situation. Most companies begin with at least an implicit strategy before they create an organizational structure to help carry out their plans.[1] At Hercules, however, the normal sequence was reversed. As it began operations, therefore, the company faced an unprecedented challenge: to reorder its assets and develop a strategy that would guide it into the future.

Getting started

On November 6, 1912, Hercules' board of directors held its first lengthy meeting. Russell Dunham, who had turned forty-two the day before, was elected president; T.W. Bacchus, vice president and general manager, "with full managerial control of [the company's] manufacturing operations"; James Skelly, vice president and director of sales, "with full managerial control of its selling operations"; and George Markell, secretary-treasurer.[2] The board did not meet again officially until the day after Christmas, when it authorized its full capitalization at $13 million to purchase the Du Pont assets specified by the court. The Hercules Powder Company opened its doors for business to the public on January 1, 1913.

These events were the formal milestones in the startup of the new company. At the same time, however, Hercules' officers were in constant communication with their former colleagues at Du Pont and were attending to the last details of the final decree. By the time the board constituted itself, the implementation of the decree was well under way. Several important matters, however, were still being negotiated late in 1912. The most important were recruitment of personnel for the Hercules Home Office and the assignment of contracts and business to the new company. Both matters remained topics of discussion well into 1913.[3]

By the summer of 1914, the Home Office staff numbered 138: fourteen in administration (senior management and clerical support), forty-two in sales and advertising, eleven in operations, forty-nine in accounting, fifteen in purchasing (who, presumably, were working closely with Du Pont's purchasing managers), and others in miscellaneous positions.[4] The corresponding numbers for late 1912 were surely much lower, although the company began processing accounts and recording business information almost immediately.[5] It is likely, therefore, that personnel in functions such as accounting, ordering, record-keeping, and filing were transferred from Du Pont in groups after the naming of the senior executives in early October. As Dunham himself had done, the other executives presumably identified personnel they wanted to hire or were presented by Du Pont with personnel options. Senior-level managers and key technical personnel were either requested by Hercules or identified by Du Pont as available before they were hired.[6]

The second lingering detail in the implementation of the final decree was the assignment of contracts and business to Hercules. According to Pierre du

Hercules' first management team revolved around President Russell H. Dunham (top center). Clockwise from top right: G.G. Rheuby, J.T. Skelly, Norman Rood, Fred Stark, C.D. Prickett, T.W. Bacchus, and George Markell.

For obvious reasons, explosives were stored and moved in small lots. Teamster Harry C. Scheitrumpf poses in front of the company's magazine at Hazleton, Pennsylvania, in 1915.

The young company sought to expand swiftly by acquiring a dynamite plant near Joplin, Missouri, and by building another in Bacchus, Utah. Above: The acid production area at Bacchus during the early years.

Because dynamite plants were located far from centers of population, explosives-makers often built settlements nearby for employees. Above: Bacchus Village in the 1920s.

Hercules' purchase of the Union Powder Company in 1915 brought capacity to make nitrocellulose. Following the war, Union's plant in Parlin eventually became one of Hercules' most valuable assets.

During World War I, Hercules acquired new — and surprising — skills in chemistry and engineering. Among its most creative acts was harvesting kelp to make acetone and potash. Above: Mechanical harvesting. Below: Pumping macerated kelp into barges for shipment to the Chula Vista plant.

At its smokeless powder plant in Kenvil, New Jersey, the company began independent research in 1915. Two years later, Hercules built its first central laboratory on the grounds.

Filling shells: At Hercules, California, during the war, the company operated the world's largest TNT works.

In 1918, Hercules contracted to manage construction and operation of a huge government arsenal at Nitro, West Virginia. Unfinished at the Armistice, Nitro included more than 700 buildings and housing for more than 10,000 employees.

Pont, this was "a very difficult and tedious task" that involved determining "the natural trade of the new companies in order that magazines and sales offices could be furnished them to their best advantage." Pierre believed that the assignment work was virtually complete by early October 1912, but negotiations between the two companies continued for several more months.[7] The principle used in assigning trade was to transfer a volume of accounts in each region that corresponded to Hercules' local production capacity. Du Pont was generally scrupulous in observing the spirit of the final decree and, by the end of the year, had provided Hercules with sales offices in Hazleton, Pennsylvania; San Francisco, California; Salt Lake City, Utah; and Pittsburg, Kansas, as well as with accounts in those cities and Chicago.[8] When Hercules began its operations in January 1913, two additional offices—in Chicago and Pittsburgh, Pennsylvania—were still in the process of being organized.

ASSETS AND CAPABILITIES

Hercules was organized around functional lines of responsibility in a structure patterned after Du Pont's (see Exhibit 3.1). Reporting to Dunham were the general managers of the Operating Department (Bacchus) and the Sales Department (Skelly), as well as the treasurer (Markell). The Operating Department was divided into separate units for high explosives, which Bacchus attended to, and for black powder and sporting powders, the responsibility of Prickett.[9] In the Sales Department, Skelly supervised the company's six offices across the country. Markell's small office included a comptroller and several assistants.

Hercules probably had another department as well. Although there is no formal mention of a Legal Department before September 1914, the company employed at least two lawyers, Judge G.G. Rheuby and Robert H. Richards, in its first year.[10] Other staff functions, such as research, engineering, and purchasing, were provided under contract by Du Pont as required by the final decree.[11]

In December 1912, shortly before the start of business, the board developed a number of general policies for Hercules. Guidelines for appropriations requests, for example, set maximum amounts that could be approved by various levels of authority. Requests for $5,000 or less could be authorized by department heads. The president could approve requests between $5,000 and $10,000, and sums above that required the stamp of the board. Forming a general policy on safety and the related matter of benefits for accident victims was extremely important in the explosives industry. The company focused on preventive measures, for obvious reasons, but also outlined a general procedure for handling the claims of accident victims.[12]

Hercules' compensation, incentive, and pension plans, which were carried over from Du Pont, were progressive policies for an American business at that time. Employees were eligible for several sorts of bonuses. Stock could be awarded for inventions or conspicuous service to the company, and department heads could recommend stock bonuses for employees with at least two years'

EXHIBIT 3.1 Hercules Organization, 1913

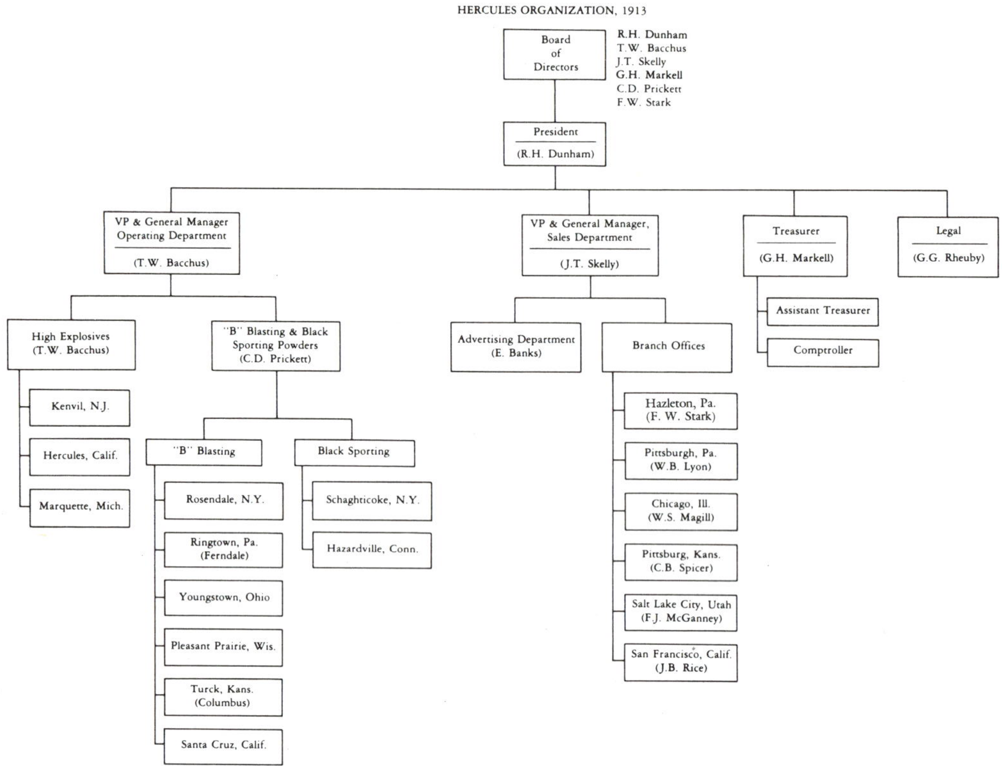

service. At the plant level, wage increases were based on both merit and seniority. The company's pension plan covered employees with at least fifteen years of service. Pensions were based on an employee's length of service (including service with Du Pont) and monthly salary averaged over the final ten years.[13]

Hercules' board of directors actively governed the company from the start. Regular meetings were set for the first Tuesday of each month, but as the company got under way, the board tended to meet more frequently. Certain matters routinely required board consideration—eligibility of employees for pensions, determination of dividends, settlements with accident victims and their heirs, real estate transfers, and so forth.

The board occupied most of its time, however, with the concerns of the major departments. Starting in March 1913, the board received detailed monthly reports from the Operating Department on production and costs of high explosives and powders; from the Sales Department on prices and volumes of each product at each branch office, as well as on general business trends by region; and from the treasurer on receipts and profits in aggregate and by product for the period.

These documents reveal a careful, probing, hardworking management.

The initial reports were probably modeled on those of Du Pont, though the Hercules board requested many changes and modifications in the early months. For example, the Sales Department was instructed to include the particulars of business gained and lost from competitors (a virtually irrelevant consideration at the predivestiture Du Pont), current and historical information about cost of sales at each of the branch offices, three- and six-month sales forecasts, and additional information on the costs of freight.[14] The Operating Department reports gradually became heavily annotated with cost data for each input, historical comparisons, and commentary swelling to sixty or seventy pages each month.

From time to time, the board also commissioned extra reports on special subjects, which ranged from inquiries into the high cost of foreign fuse to an explanation for high energy costs at one of the black powder mills, to cost estimates for adding capacity to make smokeless powder.[15]

Together, the board minutes and departmental reports give a composite picture of the new company's business. Three broad goals stand out: a drive to achieve a stronger national presence in high explosives; the desire to expand manufacturing capacity in smokeless powders; and a determination to put the company on a sound financial footing. But before Hercules could develop these plans, it had first to evaluate its assets.

Taking Stock

The final decree awarded Hercules Powder Company three lines of business: high explosives, black blasting powder, and black and smokeless sporting powders. Of these, high explosives—various grades of dynamite, blasting caps, and supplies—made up the fastest growing segment and produced the greatest profits. In Hercules' first month, high explosives amounted to 69 percent of operating income. Next in importance came black blasting powder (chiefly "B" blasting powder made from Chilean nitrates rather than saltpetre) at 28 percent; the sporting powders accounted for only 3 percent.[16]

As described in Chapter 2, Hercules' original operations consisted of high explosives plants in Kenvil, New Jersey; Marquette, Michigan; and Hercules, California; black powder mills in Rosendale, New York; Ringtown, Pennsylvania (also known as the Ferndale mill); Youngstown, Ohio; Pleasant Prairie, Wisconsin; Turck, Kansas (also known as the Columbus mill); and Santa Cruz, California; and black sporting powder mills in Schaghticoke, New York, and Hazardville, Connecticut. Hercules started without a smokeless powder facility, although the court had ordered Du Pont to "transfer to or furnish" it with a plant to make the brands once owned by Laflin & Rand (L&R). The court specified that the plant ought "to be located at Kenville [sic], New Jersey, or some other suitable Eastern point," with enough capacity to produce 950,000 pounds per year.[17]

The transfer of assets presented Hercules with several serious problems. First, some of the facilities had been mothballed for years. The dynamite plant in Marquette, Michigan, for example, had been essentially dismantled in 1910.

Moreover, because it was situated dangerously close to a growing settlement, it was unlikely ever to be restarted. The black powder mill at Rosendale, one of the company's older facilities, dating to about 1835, had also been shut for several years owing to lack of business. One of Hercules' first acts was to order the dismantling and sale of the mill.[18]

Second, among the functioning plants, several were marginal operations that would require close management attention to bring them into line. Du Pont had considered closing the antiquated black sporting powder mill at Hazardville in 1912, and perhaps would have done so if it had not been for the final decree.[19] Among the black blasting powder mills, Youngstown and Santa Cruz were replete with problems. Operating costs at Youngstown ran exceptionally high because of an unfavorable contract with a local utility. This matter was straightened out in time, but the situation at Santa Cruz, home of the old California Powder Works (CPW), the Hercules brand name, and the new company's largest black blasting powder mill by a factor of two, was more worrisome.[20] Like the dynamite plant at Marquette, the Santa Cruz mill was located perilously close to a rapidly growing community. The area, moreover, was prone to forest fires.[21] The mill had been idle for long stretches, and sales forecasts for the region were dismal. In addition, when the mill was open, it tended to be expensive to operate, chiefly because of high labor costs and electricity rates, as well as excess capacity.[22]

Third and most serious, Hercules' high explosives plants were poorly situated for the company to compete nationally. The company had large facilities on both coasts, but no plant to serve business between California and New Jersey. The Kenvil facility, moreover, lacked a nitric acid plant. Although Du Pont had agreed to build such a plant for Hercules, for the time being the company was obliged to purchase nitric acid from Atlas.

Complicating matters still further, less than two weeks after Hercules began operations, a violent explosion at Hazardville effectively destroyed the sporting powder mill, killing two employees and severely injuring a third. Prickett, who hailed from Hazardville, filed a gloomy, albeit vivid, report to the directors describing what happened. The explosion began in the press mill at about 1:30 P.M. on January 14. The two men inside the press mill were killed instantly, most of their remains being found on the bank of a canal about 400 feet away. A third man, "driving a transportation wagon within 250 feet of the building . . . was thrown from the wagon to the ground and severely bruised and shocked." Three wheel mills, 325, 475, and 590 feet away, respectively, also exploded, touched off by burning debris from the press mill. A barricade of chestnut poles twenty feet tall and sunk four or five feet in the ground, erected near the press mill the previous August, was

> shattered to long splinters and thrown distances as great as 350 feet, [and some poles] were broken into sections of 5 to 6 feet long and hurled to distances as great as 750 feet. Two poles were thrown whole into the air at a sufficient height that when they returned to earth 150 feet distant they stuck in a nearly vertical position and penetrated into the ground 5 to 6 feet.

Fifty-pound shards of barricade were strewn across a public road, and in the nearby town, windows, ceilings, and plaster walls cracked in many houses and "rather serious damage" was done to stained-glass windows in three local churches. The destruction of the Hazardville mill was so complete that Hercules "decided that for the immediate future no steps will be taken to rebuild this plant."[23]

The company had barely recovered from this tragedy when another struck. As again recounted by Prickett, at the Columbus mill in Turck, Kansas, which supplied about one-third of all black blasting powder made by the company, on February 18 at "about 10:30 P.M., Assistant Superintendent Durkee . . . was shot down in cold blood by one of the employees and died within a half hour." Ardon Pattyson, a mill employee who doubled as a night watchman, had been "somewhat intoxicated in the afternoon," and Durkee had ordered him not to report for work that night. "This action apparently angered the man to an unreasonable extent," Prickett reported, and sparked the fatal shooting. Hercules' board donated $3,000—more than a year's salary for the deceased superintendent—to his wife as partial compensation for her loss.[24]

At the end of its first two months, then, the operating situation at Hercules was considerably less promising than the court had projected. Indeed, the situation appeared grim. Hercules had no facility from which to supply the profitable and growing dynamite business in the Midwest, had idle and threatened operations in black blasting powder, had had half its capacity in sporting powders demolished, and had no capacity at all to manufacture smokeless powder. Hercules' product strategy, therefore, was to expand its high explosives business to ensure nationwide coverage, consolidate its black blasting powder operations, and develop the capability to make smokeless powders.

Becoming a National Company

In the explosives industry of 1913, high explosives—various grades of dynamite, blasting caps, and supplies—composed the fastest growing segment of the industry and generated the most profits. In 1913, dynamite sales amounted to 62 percent of Hercules' total receipts and contributed 65 percent of its net income. Those totals should be contrasted with the 19 percent of sales and 12 percent of net income accounted for by black blasting powder, and the 8 percent of sales and 17 percent of net income by sporting powders.[25]

In view of such statistics, it is not surprising that the board devoted attention to increasing its trade in high explosives. In 1913, the company took two actions specifically to improve its competitive position in dynamite. It started construction of a new plant to serve the mining trade in the Rocky Mountains, and it began prospecting for ways to add capacity in the Midwest.

Dynamite was a good business across the country, but especially so in the West, where the land was rocky and rich in minerals, railroads were still being built, and development was booming. In Hercules' first few months of operations, sales of high explosives from the San Francisco office averaged between one and a half and two million pounds per month, more than twice the sales of its next busiest office, in Salt Lake City.[26] Together, the two offices absorbed about 85 percent of the rated capacity of the company's huge dynamite plant at Hercules, California. If the company's sales were to grow in the region, a means had to be found of increasing output. With this situation in mind, as well as the difficulties of transporting high explosives from the Bay Area to the Rockies, Hercules executives decided to build a new plant to serve the mining trade. As early as February 1913, the company was considering putting the facility in Utah, although it may also have investigated sites in Arizona.[27]

In the spring of 1913, the company acquired twenty-three hundred acres on a mountainside about twenty miles west of Salt Lake City. The property, which Hercules christened "Bacchus" in honor of its vice president and general manager, was located several miles below Bingham Canyon, home of the Utah Mining Company (a forerunner of Kennecott Copper) and site of an extraordinarily rich vein of copper. Construction of the new dynamite plant, which would have an annual capacity of twelve million pounds, started in September 1913 and was substantially complete by the end of the following year. The building of the Bacchus plant, incidentally, helped the company solve another pressing problem: what to do with the venerable, underutilized black blasting powder mill at Santa Cruz. In December 1913, the board decided to close the mill and appropriated funds to transfer the equipment and personnel to new and smaller powder lines at Bacchus and Hercules, California.[28]

The lack of a facility to make dynamite in the Midwest proved a more difficult problem to solve. The company had trouble retaining assigned business near the coal mines in central South areas and the metal ore mines of the Great Lakes, and the problem only grew worse as time wore on. With the advent of competition in the industry, dynamite prices fell and shipping costs ate into shrinking margins. Both Du Pont and Atlas were aggressive in lowering prices. Because "our nearest Dynamite Plant [is] at Kenvil, N.J.," lamented Skelly, "we are hardly in an advantageous position to make an effort for any of the large business, especially when competitors are willing to take it at such a low figure."[29] Nor was it a simple matter to ship explosives over long distances. Then, as now, states and communities were concerned about boxcars loaded with powder and dynamite passing through their jurisdictions.

The need for a dynamite factory "West of Pennsylvania, East of Kansas, and North of the Ohio river" was a frequent refrain in Skelly's early sales reports.[30] On June 3, the board directed President Dunham to investigate the possible acquisition of facilities in the Midwest. Dunham, in turn, invited the board to a general discussion of the issue on August 6. In addition to the usual argument that Hercules could not compete with its rivals in the region, board members pointed out that the company's black blasting powder salesmen in the territory were handicapped because they could not offer a full line of products. "There are a number of customers to whom it is difficult to dispose of Blasting Powder unless Dynamite also can be supplied, and it is uneconomical to have salesmen selling Blasting Powder alone."[31]

Given the need for a plant in the Midwest, the next question was how to come by it. Build or buy? The board's decision to acquire was simple. Studies of the territory had shown that there was already excess capacity.

> [It] was, therefore, agreed . . . that if an existing Dynamite Plant or Plants could be acquired along with the business that the plant or plants may now be enjoying at a reasonable price, that such acquisition would be preferable . . . provided, of course, that the purchase of such plant or plants be not in restraint of trade and is honestly in harmony with the Sherman Law and the Decree of the Court under which this company was created.[32]

The latter point was a major concern. Hercules' purchase of a plant from either Du Pont or Atlas—in the unlikely event that either company would sell—would violate the spirit if not the letter of the final decree. That left only two acquisition candidates in the region, the Independent Powder Company of Joplin, Missouri, and the Aetna Powder Company with dynamite plants at Aetna, Indiana, and Thebes, Illinois.[33] Of these alternatives, Aetna possessed better locations but had the disadvantage of operating black blasting powder mills in the Midwest that would necessarily be part of any sale. This, in turn, raised additional questions about antitrust law—and production capacity—since Hercules already had three powder mills in the region. The board consulted company lawyers Rheuby and Richards, who argued that the purchase of either Independent or Aetna would not violate antitrust law as long as the contract did not prohibit the seller from reentering the business. The lawyers also "particularly recommended that in a purchase of this sort, it was advisable to purchase the assets of a corporation rather than the corporation itself, but that it was not material whether the consideration be cash, notes or preferred stock."[34]

The company eventually heeded this advice. Dunham pursued negotiations with both acquisition candidates throughout the fall and winter of 1913–1914 until Norman Rood, president of the Independent Powder Company, agreed to sell.[35] How Hercules went about its first acquisition provides some interesting glimpses into the young company's operations.

As the negotiations became serious, Vice President Bacchus visited the Independent dynamite plant in Missouri and made his own valuation of its assets. He liked what he saw. The plant had a rated capacity of about twelve million pounds per year, compared with forty-two million per year at Hercules, California, and fifteen million at Kenvil.[36] While the plant buildings at Joplin were "of a differing design and standard than the buildings which we would build if we were building a plant," he allowed that "they are of such a design and standard that we would be justified in using same should they come into our possession, and we are not prepared to say that these buildings are not of as good construction as our own for their several purposes." The plant's machinery particularly impressed Bacchus. The gelatin mixing machine "was a revelation . . . and, in my opinion, is the best Gelatin Mixing Machine in existence today . . ." and "far surpasses" anything at Hercules. He also admired the gelatin cartridge machine, which, he believed, was also superior to Hercules' equipment, confessing, "I hate to say this inasmuch as I was the inventor of the machine which both ourselves and the duPont people now use." Bacchus concluded his report by speculating that the Joplin facility "would be a pretty efficient plant to operate," and that even better results might ensue if the plant's management, which "seems to be very good . . . had the opportunity of putting into effect knowledge which we now have and practices which we now follow at our other plants." In particular, he suggested, if the deal were to be made, Hercules should add a nitric acid plant, one or two Hall cartridge machines, and some improvements to other machinery.[37]

Dunham's letter to the board proposing the acquisition noted that Bacchus put a higher valuation on Independent's permanent assets than had the seller itself.[38] Dunham also pointed out that the business had been profitable, averaging better than $90,000 per year between 1909 and 1912. He believed that after the additional investments proposed by Bacchus and others were made, profits of the plant should average between $125,000 and $150,000 per year.[39] Accordingly, the board agreed unanimously to acquire Independent's "property and assets of every kind and character" (but not its capital stock) for the price of $800,000. The sum reflected Independent's own assessment of its properties at $720,164, plus the value of "certain trademarks, brands, formulae, and secret processes." Thus, by early 1914, Hercules had its dynamite plant in the Midwest.[40]

Over the next several months, the board appropriated funds for a nitric acid plant, a new shell house, and a new packing house at Joplin.[41] Poor business conditions prevented a real test of the plant's capabilities in 1914, and operating costs were initially high until the nitric acid plant came onstream in November. By year's end, however, Hercules had sunk just over $126,000 into improvements at Joplin, and operating costs were almost as low as those at Hercules, California, and Kenvil.[42]

Of greater long-term significance than the added capacity Joplin represented was the management it brought to Hercules. Norman Rood stayed on to direct dynamite sales in the Midwest and eventually rose to become a vice president and director of Hercules. After World War I, Rood presided over the demobilization of the company's operations and guided its first efforts to diversify outside the explosives business. The inventor of the machinery that so impressed Bacchus, Herbert Talley, was instrumental in expanding Hercules' operations during the war; he later became chief engineer of the company when it set up its own Engineering Department in 1917. The Independent Powder Company's chief chemist, Leavitt N. Bent, remained at Hercules for many years, later directing the company's naval stores operations in the 1920s and 1930s and serving as a vice president and director.[43]

Starting the Smokeless Powder Line

Hercules inherited some popular brands from L&R in the antitrust settlement: E.C. and Infallible smokeless shotgun powders; Flag in Wreath, Lightning, Sharpshooter, W.A. 30 Cal., Unique, 1908 Bear, and 1908 Stag smokeless rifle powders; Bullseye smokeless revolver powder; and Orange Extra and New York black sporting powders.[44] These brands were familiar to the leading loading companies such as Winchester and Remington, as well as to sportsmen, and many of them were used at national rifle tournaments and trap-shooting contests.[45] Hercules also made an unbranded black sporting powder (sometimes called "A" blasting powder, which contained saltpetre), which it sold to loading companies, to other powder companies (especially Du Pont), and to the general hardware trade.

The sporting powder business was small but very lucrative.[46] As we saw in Chapter 2, the business was in the throes of change at the birth of Hercules. Black sporting powders were being phased out rapidly in favor of single-based smokeless powders made from "guncotton" (a kind of nitrocellulose) or double-based powders that combined nitrocellulose and nitroglycerin. The L&R brands the court had allotted to Hercules were of the double-based variety. At the start of 1913, Hercules purchased these powders from Du Pont and merely repackaged them for sale.

In compliance with the final decree, Du Pont had started work on the smokeless line at Kenvil in October 1912.[47] Hercules planned to supply its own nitroglycerin for the project from its dynamite factory, but a key issue was whether to build a guncotton plant at the facility, given the small volumes Hercules expected to produce. The final decree was silent on the point, and Du Pont Vice President Hamilton M. Barksdale figured that if his company could "supply the new company with Nitro-Cotton of proper quality at a price so low as to make the necessary capital investment by the latter in a Nitro-Cotton plant unwise, I don't think any possible criticism would attach." The plan evidently worked, and at the end of 1912 Hercules agreed to buy guncotton from Du Pont for cost plus 15 percent rather than make its own.[48]

Construction of the new facility proceeded slowly because Hercules made several changes in the design. In addition, Bacchus was concerned about finishing a new nitric acid plant at Kenvil, a project that absorbed his attention for most of 1913. The board and the Sales Department, however, were anxious not only to get the smokeless line operating but also to expand it. In February 1913, Hercules engaged noted ballistics expert E.A.W. Everitt to advise on the project. By spring, with sales of black sporting powders falling off, the Sales Department was urging the company to install additional capacity for a broader line of products, including the semismokeless E.C. shotgun powder. The board endorsed this suggestion "because of the serious disadvantages resulting from purchasing this powder from our principal competitor." At the same time, Hercules hired Bernhard Troxler, who had been in charge of Du Pont's ballistics laboratory at Haskell, New Jersey, as assistant superintendent for smokeless powder operations.[49]

Construction delays persisted, however, and production of the new powders at Kenvil did not begin until February 1914. Right away there was a disaster. A fire in the smokeless packing house on February 12 took four lives. "The cause of the accident still remains a mystery," reported Bacchus six weeks later, adding, somewhat awkwardly and obtusely, "We have made certain speculations as to the probable reasons, and have from these speculative theories evolved certain safeguards to guard against some things which might be amongst the number that probably caused the accident."[50]

Despite this setback, smokeless powder production began to build smoothly, and by the end of the year, Hercules was making all of its major brands and slowly phasing out purchases from Du Pont. In November, Bacchus reported the startup of a new dehydrating press at Kenvil that "puts us in a position to manufacture upwards of 9000 pounds of Smokeless powder per day." Pleased with this progress, the board appropriated additional funds to add more capacity.[51] By the end of the year, Bacchus was generally satisfied with the operations at Kenvil, pointing out that, as far as he could tell, Hercules' costs compared favorably with Du Pont's except for one item—guncotton. The price of this product included, of course, Du Pont's markup, a source of continuing irritation to Bacchus.[52]

Financial Strategy

The accounting value of the assets Hercules acquired from Du Pont at the end of 1912 amounted to $13 million. The assets included permanent investments such as plants and equipment, valued at $5.4 million; working capital, such as cash, sales offices, magazines, inventories, and some transportation equipment, totaling $4.3 million (of which more than half was cash); and intangible properties such as brand names and goodwill, amounting to $3.3 million. In exchange for these assets, as mandated by the final decree, Hercules issued to Du Pont's shareowners 65,000 shares of its authorized capitalization of 100,000 shares of common stock at a par value of $100 per share and $6.5 million in the form of gold bonds redeemable in ten years at 6 percent interest. Half of the bonds were distributed to Du Pont's shareowners, and half were retained by the Du Pont company itself.[53]

This transaction raised two interesting issues for the company: What would it do with all that cash? And was this an appropriate capital structure? On the first point, Hercules started with a high cash balance—more than $2 million—because accounts receivable were not included in the working capital transferred from Du Pont on January 1, 1913.[54] This cash balance, according to Markell, was nearly one million dollars more than Hercules needed to operate its business. And Markell expected the sum to grow by about $50,000 a month in operating profit as the company got under way. "I feel it is incumbent on us," he urged, "to find some profitable employment for this excess." He proposed several options, including simply depositing the money in the bank to earn interest. That course, however, struck Markell as merely postponing the problem: "We, of course, do not wish to enter the banking business. . . . Therefore, [we] must endeavor to find permanent profitable employment for this surplus, either by enlargement or improvement of our present manufacturing facilities, or by the building and acquiring of others."[55]

Markell continued to prod the directors throughout the spring of 1913 to find appropriate uses for the cash. On March 26, the company parked $500,000 with the Liberty National Bank of New York as approved stock exchange collateral.[56] Hercules also bought $110,000 worth of Pennsylvania Railroad stock in

June.[57] The real change came in the fall, when the board appropriated nearly $900,000 to build the Bacchus facility and to add black powder capacity there and at Hercules, California.[58]

As for the capital structure of the new company, the board acted to retire the debt as quickly as possible and, early in 1914, to increase its capital stock to $20 million. We can only speculate on why it took these actions, since surviving records are silent as to the company's motives. In the first place, the one-to-one ratio of debt to equity must have seemed too high to the board of directors. Not only did the ratio exceed the norms of industry at the time and raise concerns about the impact of an economic downturn, but it was also much higher than Du Pont's, which had hovered at about 27 percent for the preceding five years.[59] In practical terms, the one-to-one ratio at Hercules meant that the company would have to set aside a substantial portion of its earnings to pay off interest on its loans. Indeed, that is exactly what Markell chose to do, believing that "it was proper to set aside an amount sufficient to take care of this interest as the entire earnings for the year 1913 are, in a sense, mortgaged to the extent of 6 percent on the bonds outstanding."[60]

In the fall of 1913, Markell submitted a special report to the board on the company's capital structure. He evidently recommended canceling at least some of the debt by an issue of preferred stock. In December, the board reached a consensus that "it was to the best interests of the Company to retire the Income Bonds by the issue of some permanent form of security, preferably a preferred stock" with an attractive annual dividend. The board decided to increase the company's total capitalization through the issue of 100,000 shares of preferred stock (par value $100 each), which would pay an annual dividend of 7 percent. The initial plan was to use 65,000 of these new shares to retire the $6.5 million debt. This plan was authorized later at a special meeting of stockholders on January 22.[61]

Exchanges of the new stock for the old bonds took place gradually during 1914. By year end, $5.4 million of the total debt had been retired, and Hercules had established a new capital structure that substantially reduced its exposure to risk and gave it greater flexibility to manage its own growth.

Important elements of Hercules' financial strategy were the board's policies toward growth and investment. As the restructuring of its finances reveals, the company expected to pay for its growth out of retained earnings and, perhaps, from the sale of additional stock, rather than by borrowing. To guide its decisions about future investments—allocations to existing operations as well as new construction—the board depended not only on the detailed accounts that Dunham and Markell had devised but also on a key aspect of Dunham's philosophy. In the early twentieth century, most American companies evaluated their performance by looking at net earnings from year to year and perhaps relating them to costs. At the companies where Dunham learned his trade, however, a different standard was being developed: return on total investment. As he himself put it, "It is . . . often overlooked that the true test of whether the profit is too great or too small is the rate of return on the money invested in the business and not the per cent. of profit on the cost."[62]

Thus, Hercules tracked its permanent investment carefully and sought a return of 15 percent when allocating resources. This standard probably influenced the board's thinking about the closing of the Santa Cruz powder mill, construction of the Bacchus dynamite plant, and acquisition of the Joplin plant, for example. Return-on-investment (ROI) criteria also guided choices about whether to make or buy certain key inputs. In 1912, Dunham had used an ROI calculation in deciding to purchase guncotton from Du Pont rather than build a plant at Kenvil.[63]

Asserting independence

Hercules' relationship to Du Pont was admittedly awkward. On the one hand, at the start of 1913, every Hercules employee had worked at Du Pont—some for many years. Friendships, and at least some professional associations, were presumably expected to survive judicial separation. The new company was even housed in the Du Pont building, where, for the next few years, it would rely on the Du Pont company to provide essential services and information. On the other hand, Hercules needed to make its own way in the world if it was to be successful. The contrary pulls of cooperation and competition with Du Pont proved difficult to live with, and as time wore on, Hercules asserted its independence.

Arrangements for access to Du Pont's facilities and pesonnel varied. In general, Du Pont was supposed to serve Hercules and Atlas on the same terms that it served itself. For R&D, payment was to be made at the end of each calendar year on a pro rata basis. The total value of research services supplied by Du Pont was apportioned among the three companies based on the ratio of total sales of each to total sales of all three. Project-related expenses, such as those for chemical analysis or engineering services, were charged as used.[64]

As for purchasing, the contract called for Hercules to buy raw materials that Du Pont itself purchased at cost, plus a commission that varied by product. For example, the commission for nitrate of soda was 1.25 percent, and for glycerin 1.125 percent. As these commodities were bought in bulk, Du Pont required Hercules to pay cash in advance or to post collateral. The commission for other purchased commodities, such as caps and fuses, was 0.085 percent. Finally, for products that Du Pont manufactured, such as guncotton and some acids, Du Pont was to charge cost plus 15 percent.[65]

The arrangements worked well for the most part. Hercules relied heavily on Du Pont's Chemical Department and received a steady stream of research reports on its key product lines.[66] The Du Pont Engineering Department managed the construction of Hercules' new dynamite plant at Bacchus and also built the nitric acid plant and smokeless powder line at Kenvil to Hercules' specifications.[67]

But there was also, inevitably, tension in Hercules' relationship with Du Pont, heightened in part by Hercules' familiarity with Du Pont's policies and costs. Early in 1913, for example, Dunham reacted to the first bill from Du Pont's Chemical Department by claiming it was "higher than we had any idea that it would amount to." Dunham asked for a detailed breakdown of charges and added, "We are inclined to imagine that there may be some overhead charges included in this that from our point of view should not be."[68]

Hercules also frequently asked for clarifications of Du Pont's purchasing policies. The stakes for the new company were high. Because it was not vertically integrated, Hercules had to rely heavily on Du Pont, not only for raw materials but also for some manufactured goods. Indeed, purchased inputs, or "ingredient cost," accounted for 40 percent of the total manufactured cost of dynamite, and more for other explosives. Purchases of nitrate of soda alone represented about half the total ingredient cost.[69] Once again, Hercules asked for breakdowns of overhead charges and monitored prices of nitrate, glycerin, and guncotton closely.[70]

The tension in Hercules' relationship with Du Pont was most keenly felt on the front lines among the sales force. In part, the Hercules salesmen resented Du Pont's actions in stocking up its customers in December 1912, as well as what appeared to be its deliberate slowness in assigning trade to the new company. Hercules' Chicago office, for example, did not get its full assignment of contracts until September 1913, and only then after the board looked into the matter.[71] Hercules salesmen were also frustrated by Du Pont's entrenched advantages. "We have an uphill game" in sporting powders, noted Skelly, "owing to the strong hold the duPont brand has on the trade."[72]

The new company's struggle to differentiate itself from Du Pont was full of annoyances. In the summer of 1914, after many delays, Hercules completed its first general product catalog, only to withdraw it suddenly. The problem, according to Skelly: "The greater part of it corresponds practically word for word with the du Pont catalogue. If this book had been issued at the first of last year as was originally intended, this similarity would not have been so bad, but we do not feel that it would be the proper thing to put it out now."[73]

Such concerns led Skelly to propose a radical step. Hercules, he believed, ought to move out of Wilmington. He had urged that course in his initial sales report of March 1913, and he repeated it often.[74] By the summer, the board ordered him to prepare a report listing his reasons for a move and discussing the likely impact it might have on business. Late in the year, he finally complied. "Practically all of the managers and field men have called attention to the appearance and the linking with du Pont because of our being located in Wilmington," he wrote. "Some claim that this is a disadvantage and that reference is frequently made to it by customers and by competitors." Skelly went on to advocate moving to New York, where many large customers were headquartered, or to another location—"most any of the large cities would be better than Wilmington." He believed that Hercules could set up independent departments for research, purchasing, and engineering within six months.[75]

The board received Skelly's report politely and filed it for later discussion. There is no record of what his colleagues thought of the proposal, and the matter soon disappeared in the press of other business; it would resurface, however, from time to time in later years. We can speculate that the board members were sympathetic to Skelly's problems, but that they believed, nonetheless, in the value of keeping close to Du Pont's services and its pool of management talent, as well as to their own homes. Although Hercules chose to stay in Wilmington, the frictions between it and Du Pont were symptomatic of the intensifying rivalry in the newly restructured explosives industry.

Early returns

The final decree had launched Hercules into troubled waters as the U.S. economy entered what looked to be the worst recession since 1893. The first real competition in the explosives industry since the 1870s only made its situation worse. In 1913, signs of intense rivalry—excess capacity, declining demand, and falling prices—were soon apparent across the country. Consumption of explosives in the United States was tied closely to the fortunes of the coal operators, the mining trade, and the construction industry, all of which faced hard times at the moment of Hercules' birth. Dynamite sales, which had grown by 20 percent between 1909 and 1912, would rise slightly in 1913, only to drop 10 percent the following year. The situation was even worse in black blasting powder: sales had been slumping for a decade as customers switched to dynamite and permissible explosives. Total consumption of black powder fell gradually from 1909 to 1912 and plunged 11 percent more in the next two years.[76]

These trends were not fully evident when Hercules mapped out its sales strategy early in 1913. The company allocated nearly $60,000 for its first advertising budget. Following Du Pont's example, Hercules divided its promotions into four categories: general, sporting, industrial, and agricultural. The first was a miscellaneous grouping for "electrotypes, framed pictures, imprints . . . memorandum books with inserts . . . blotters, greetings cards," and other items intended to get the company's name across to its customers. Sporting advertisements were aimed at hunters and gun enthusiasts. Hercules planned to advertise moderately in magazines and trade journals and to publish booklets giving advice and tips on shooting. The company also planned some giveaways, including gun club trophies, tournament score sheets, signs, and posters.

The industrial category consisted of booklets and catalogs for the branch offices to use in their sales efforts with the mining and construction industries. The agricultural promotions included advertisements in newspapers and journals catering to farmers who used explosives to clear land. Hercules' first advertising manager, Edward Banks, was particularly proud of a proposed "Farmer's Hand Book," which he expected "will be a publication which I know will reflect great credit upon the Company. It was written especially for us by the man best qualified to do so and will be standard."[77]

Although advertising may have helped with name awareness, Hercules leaned on its technical sales force to develop and maintain business. The company's top salesmen, Skelly and Stark, were amply experienced in the hands-on approach. Stark's practice was to "study each customer's requirements and if we find they are not using what we consider is best for their requirements, we suggest something better, and, if necessary, send a man to demonstrate something better for their use, or, at least, develop whether or not they are using what is best for them."[78] In 1913, in addition to its normal branch office personnel, Hercules' sales force included seven "shooter salesmen" who specialized in rifle and shotgun demonstrations and two sales demonstrators who roamed across the country giving lectures and shooting explosives.[79]

Despite its preparations, Hercules faced heated competition in every market. The battle for black blasting powder sales was especially fierce. As early as the spring of 1913, W.S. Magill, manager of the Chicago branch office, complained of "decidedly unpleasant competition" for black blasting powder. He reported that "one of [his] competitors is resorting to unfair means in the endeavor to secure business, and that all of [his] competitors are cutting prices." The Pittsburg, Kansas, office, which accounted for about one-third of all black blasting powder sales, was also hard hit. In July, Hercules lost its biggest customer, the Central Coal & Coke Company, to the Excelsior Powder Company. At Hazleton, Stark claimed that sales for July were the smallest monthly total in his experience, partly because the Pennsylvania mines were suffering a recession and partly because Hercules' efforts to get an extra nickel per keg "enabled competitors who did not make the advance to make inroads on our business." At Salt Lake City, Branch Manager F.J. McGanney reported a July loss of four thousand kegs of business to Du Pont and seventeen hundred kegs to another company.[80]

The depressing conditions continued into the winter, when the normal seasonal drop-off seemed that much worse. Nor was there much improvement in 1914. "Blasting powder business has gone all to pieces, and we will not even reach half of the amount we should sell monthly of this commodity," Skelly predicted in January.[81] He was wrong, but not by much. In September, he reported dropping prices 5¢ per keg in the Midwest. "It is unfortunate that competitive conditions brought this about, as the net we were receiving at the higher price was not any too satisfactory on the small volume we enjoy."[82] Sales at every office save Chicago were down for the year. At year end, sales were down 20 percent from 1913, which itself had been a dismal year.[83]

Hercules fared a bit better in the nationwide contest for dynamite sales after the acquisition of the Joplin plant. The company increased its total sales of dynamite from forty-four million to forty-six million pounds, and its market share from 15.3 percent to 17.8 percent.[84] Nonetheless, the market was troubled, especially in the West. In the summer of 1913, J.B. Rice, head of the important sales office in San Francisco, complained that "our sales are all off because the business is not here this year . . . mining consumption is about the same as last year; but our losses are due to the lack of construction work." McGanney reported similar news from Salt Lake City.[85] By the next summer, Skelly's outlook was gloomy; the depression in construction had extended to the mines.

> There is no question but that we are feeling the general business depression which has had its effect in the explosives industry. . . . At this time, when everyone is apparently feeling the depression, all seem to be more active in soliciting business, attacking others' steady trade, [and] quoting extreme and sometimes ridiculous prices on open business with the hope of obtaining volume regardless of profit.[86]

The new Bacchus plant was ready to start operations in November 1914 but delayed opening for five months owing to lack of business.

Competition was also keen in sporting powders. As time wore on, the need for Hercules to produce a broad range of its own smokeless powders grew obvious. In 1914, Du Pont was aggressively pushing new nitrocellulose powders to the loading companies. Skelly reported that Hercules' Lightning brand was particularly vulnerable, since the switch from nitroglycerin to nitrocellulose powders would allow comparable performance with 10 percent fewer grains and "would effect quite a savings to the loader." He continued:

> From recent developments and from what has been said to us by some of our ammunition customers, Nitro Cellulose Powders will soon displace Nitro Glycerin Powders, as they have features, especially with respect to erosion, which make them more desirable. We also know that practically every Government with one exception has adopted Nitro Cellulose Powders, and we believe with the loading companies knowing of the advantages, they will, as soon as proper Nitro Cellulose Powders are developed, begin to show preference for them as against Nitro Glycerin Powders.

"When this condition arrives," Skelly worried, "we will be left high and dry with an expensive plant and organization without business to support them."[87]

Conclusion

Despite the troubles of the explosives industry, Hercules was able to make earnings of $1.4 million on sales of $7.6 million in 1913 and achieved similar results the next year.[88] Hercules easily met its obligations to bondholders and even inaugurated an employee stock ownership plan. The company paid out its first dividend of $1.50 per share (an annual rate of 6 percent) on schedule at the close of the third quarter of 1913.[89] Hercules continued to pay dividends at this level until the end of 1914, when, on Markell's recommendation, the annual rate was raised to 8 percent.[90] The company also made its dividend payments on its new issue of preferred stock in 1914.

In light of declining demand and escalating competitive rivalry, this was an impressive performance. Hercules certainly benefited from relatively low fixed costs, as well as modest overhead. The addition of high explosives capacity in the

Midwest and West also helped. The young company's record was also testimony to the quality of its management. Indeed, Hercules had accomplished a lot in its first two years. Key personnel were working together smoothly and productively, and essential management policies and systems were in place. The company was accumulating and tracking detailed information about its business that contributed to prudent decisions. Hercules had also acquired enough capital to finance growth in the explosives business once demand picked up. With the acquisition of the Joplin plant and the building of Bacchus, the company was well positioned to compete nationally in high explosives. The decline of the black blasting powder business was certainly a long-term concern; Hercules' managers, however, were not only well aware of the situation but confident of their ability to compensate by increasing sales of dynamite. Finally, the company was rapidly acquiring the capacity to manufacture smokeless powders, a product that suddenly seemed to have enormous potential as the guns of August—the outbreak of war in Europe—were heard across the Atlantic.

CHAPTER 4

War and Transformation, 1915–1918

At the close of 1914, Hercules' sales totaled $7.9 million, and it employed about 1,500 people. Two years later, sales surpassed $60 million, and Hercules employed nearly 10,000 people. The catalyst for this amazing growth, of course, was the First World War. The conflict, besides stimulating Hercules' sales, also sparked its initial—and fundamental—transformation from a powdermaker to a chemical company.

Among its many far-reaching effects, World War I transformed the U.S. chemical industry from a relatively small, dispersed, and unsophisticated group of companies into a large, concentrated, and highly technical industry.

The Great War was the first modern technological war; it was fought using advanced weapons based on research at the frontiers of scientific knowledge. Chemical companies that supplied munitions were forced to respond to challenges that were wholly new: rapid development of new products, intensive research on new aspects of chemistry, snythesis of basic chemicals once imported from Europe, massive increases in production and capacity, and management of operations on a scale much vaster than had ever existed in peacetime.

At Hercules, the specific challenges posed by the Great War involved adapting the company's smokeless sporting powder operations to make military powders, and developing the capacity to make the new "disruptive" explosives such as trinitrotoluene (TNT). Production of some military explosives, moreover, required Hercules to pioneer methods for producing key intermediate ingredients and solvents. These challenges raised yet another: the urgent need to extend the company's expertise in research and development and in plant design and engineering. Hercules' efforts to develop these capabilities were complicated by its reliance on Du Pont for R&D, engineering, and other services under the terms of the final decree. The United States' entry into the war in April 1917 complicated matters still further by involving the company in the administration of a mammoth federal ordnance facility.

Finally, scaling up to wartime production levels raised significant administrative issues for Hercules: Where would it find the scientists, managers, and employees to operate this vastly increased business? And how would the company deal with the substantial increases in cash flow and profits brought about by the war?

Modern warfare

The range and extent of the Great War were also without precedent, as was its technological nature: the airplane, the submarine, the tank, poison gases, the machine gun, and artillery featuring automatic breech-loading all made their full-scale debuts between 1914 and 1918.[1] Compared with their modern descendants, these were primitive weapons, but they proved deadly nonetheless. Eight million people perished during World War I, and twenty million more were wounded. (In comparison, the American Civil War, which had been one of the bloodiest sustained conflicts in history, resulted in 620,000 casualties.) And the World War I figures do not include the victims of famine, disease, and privation. The greatest casualties were in Germany and Russia, though millions of French, Austro-Hungarians, and British also perished in the conflict. The United States, which was directly engaged in the war for nineteen months, lost 115,000 soldiers.

The scale and savagery of World War I were unprecedented. During its four years and four months, armies from virtually every developed nation and their colonial dependents participated in protracted land battles on four major fronts—Belgium and France, Northern Italy, the Balkans and Turkey, and Central Europe and Russia. The war also engaged navies in conflicts around the world, and its ripple effects were felt in revolutions in Mexico, Ireland, and Russia, in clashes of imperial powers in Africa and China, and in the collapse of centuries-old empires in Austria-Hungary and Turkey.

Within two months of the outbreak of fighting in August 1914, the general shape of what was to come became clear. In the East, the Germans delivered a series of crushing blows to the Russians in Central Europe and began their long, slow march into the Ukraine and White Russia. Over the next several years, the Russians could only delay the advance until, in 1917, the czarist regime collapsed and the revolutionary government negotiated a separate peace with the Germans.

In the West, the first weeks of the war saw a rapid German advance toward Paris, but in September, the French dug in near the Marne River and stopped the invasion. But the Germans also dug in and could not be pushed back. From then on, the two sides struggled ineffectually to dislodge one another. Along the Western Front, the conflict would be a defensive struggle, a battle for position, a war of attrition. Coils of barbed wire appeared in front of both trenches, and strategically placed machine guns, protected by earth barriers and sandbags,

EXHIBIT 4.1

Rates of Artillery Fire per Gun per Day in Recent Wars

War	*Army*	*Approximate rounds per gun per day*
1851–1856. Crimean	British and French*	5
1859...... Italian	Austrian	0.3
1861–1865. Civil	Union	4
1866...... Austro-Prussian	Austrian	2.2
	Prussian	0.8
1870–1871. Franco-Prussian	German**	1.1
1905–1905. Russo-Japanese	Russian	4
1912–1913. Balkan	Bulgarian	7
World War I		
September, 1914	French**	8
Jan. 1–Oct. 1, 1918	Italian**	8
Jan. 1–Nov. 11, 1918	United States**	30
Jan. 1–Nov. 11, 1918	French**	34
Jan. 1–Nov. 11, 1918	British**	35

*Siege of Sebastopol. **Field gun ammunition only.

Past Wars Compared with One Month of World War I

War	*Army*	*Rounds expended during war*
1859...... Italian	Austrian	15,328
1861–1965. Civil	Union	5,000,000
1866...... Austro-Prussian	Austrian	96,472
	Prussian	36,199
1870–1871. Franco-Prussian	German	817,000
1905–1905. Russo-Japanese	Russian	954,000
1912–1913. Balkan	Bulgarian	700,000
1918...... World War I	British and French	12,710,000 (In one month)

Expenditures for One Year, Civil and World War I

1864∿..... Civil	Union	1,950,000
1918∿*.... World War I	United States	8,100,000
1918∿*.... World War I	British	71,445,000
1918∿*.... World War I	French	81,070,000

∿·Average, year ended November 10, 1918 ∿ ∿ Year ended June 30, 1864 ∿· Year ended November 10, 1918

Source: Crowell, *America's Munitions,* 28–29.

swept the no-man's-land between the opposing sides with monotonous, mechanical efficiency.

This sort of warfare consumed gigantic, almost unimaginable quantities of ammunition. Offensives were mounted behind intensive barrages of artillery (although they seldom proved effective against the opponents' precautions), and the machine guns repelled charges with devastating effect. The British attempt to break through along the Somme River in 1916, for instance, was preceded by a barrage of heavy artillery that lasted for a week, nonstop, twenty-four hours a day. It is estimated that one and a half million artillery shells, weighing forty-two million pounds, were launched in a single week in the battle of the Somme alone.[2] Those figures do not include the staggering amounts of propellant used to launch the shells, nor the answering fire by the Germans (see Exhibit 4.1).

Ironically, such bombardments were largely ineffectual. The problem lay with the shells themselves, which had heavy metal jackets to protect the contents from the shock of firing, and which also transported metal pieces designed to explode into shrapnel on contact. Because of the shell's weight, only a small charge of black powder or high explosives could be delivered. Of the forty-two million pounds of shells launched by the British at the battle of the Somme, only 1.8 million pounds represented explosive payload. The week-long barrage, therefore, had little effect on the German troops, who had dug in below the flying shrapnel. As British troopers climbed "over the top" at the Somme and raced for the enemy trenches, they were raked by German machine guns, which had survived the shelling mostly intact. On the first day of the infantry battle (July 1, 1916), the British sustained 60,000 casualties, including 21,000 deaths, "most in the first hour of the attack, perhaps the first minutes."

The battle of the Somme raged on for four and a half months, during which the British managed to advance the trench line about three miles along a 6-mile front. In all, the British lost 420,000 men, the French 200,000, and the Germans a number roughly equal to the Allied total.[3] Like so many battles of the war—the Marne, Ypres, Verdun, Passchendaele, the Argonne, Gallipoli, and many others—the Somme was an inconclusive but deadly struggle that involved extravagant expenditures of ammunition.

Many of the military innovations of the war were designed to overcome the defensive advantages of the trenches. The Germans first used poison gas at the first battle of Ypres in April 1915. Its use permitted a momentary breakthrough, but in the long run poison gas only made trench warfare more difficult, since both sides had to conduct battle wearing gas masks. The Germans also began dropping bombs from zeppelins in 1916. Because it did not have to withstand the shock of firing, a bomb could carry a much greater payload of high explosives than an artillery shell. The state of aircraft power and reliability, however, limited the impact of aerial bombardment until the very end of the war. The tank made its first appearance late in 1916 and would later spearhead the Allied offensives that finally broke the stalemate on the Western Front in 1917 and 1918. By then, the Allies had received a fresh infusion of troops from the United States, an event that changed the course of the war and made its outcome clear.

The Chemists' War

A war in which the combatants lobbed artillery shells at one another relentlessly, and in which the two sides relied increasingly on mechanical methods of firing, was bound to be a bonanza for suppliers of propellant and shell fillings. Indeed, it was.

The combatants in World War I took—and reaped—full measure of the chemical revolution in explosives. In the last nineteen months of the war—a time when the fighting was winding down—the Allies alone fired 1.3 billion pounds of nitrocellulose powders and exploded more than 1.8 billion pounds of disruptives (picric acid [trinitrophenol] and TNT).[4] Even before the United States entered the war, while it was ostensibly neutral, the country was acquiring its reputation as "the arsenal of democracy." Between 1914 and 1917, U.S. explosives-makers greatly expanded their productive capacity, supplying enormous quantities of munitions to England, France, and Russia. During these years, for example, investments in plant and equipment nearly tripled at Du Pont, and doubled at Hercules and Atlas. At the same time, gross revenues rose more than fivefold at Atlas (from $5.1 million to $27.5 million) and Hercules (from $7.9 million to $44.1 million), and more than tenfold at Du Pont (from $25.2 million to $269.8 million).[5] U.S. investment banks, meanwhile, handled more than $5 billion worth of transactions on behalf of the Allies, a volume of business that helped transform the United States from a debtor to a creditor nation and transferred the center of world finance from London to New York.[6]

The tremendous wealth accumulated by munitions suppliers during World War I attracted the attention of politicians and historians alike.[7] Although the war's impact on corporate finance and the American economy was far-reaching and fundamental, its effect on American technology was perhaps greater: the war transformed chemistry and the chemical industry. The transformation was twofold. To begin with, the war stretched the explosives companies' abilities to manufacture nitrocellulose and nitro-aromatic compounds to an unimagined extent. A critical problem was the shortage of nitric acid. As early as 1916, the Allies' demand for nitrates—the essential raw material for making nitric acid—began to exceed supplies from the Chilean mines. Munitions suppliers were forced to find new sources of nitric acid.

The most obvious solution to the problem involved synthesis of ammonia, from which nitric acid could be derived. That course, however, required levels of investment and chemical engineering expertise well beyond those typical of the prewar industry. To make synthetic ammonia from the air (the Haber process), for example, engineers had to work with large quantities of liquid air (production of which was a tricky, energy-intensive process) as a source of pure nitrogen, and then combine the gas with pure hydrogen under the pressure of 200 atmospheres and at a temperature of 1,022 degrees Fahrenheit.[8] Other ways of producing nitric acid from the nitrogen in air, such as the Birkeland-Eyde arc process, or the cyanamide process, were similarly complex. To operate any of them on a commercial scale required huge amounts of electricity and heroic feats of engineering.

The second major impact of the war on American chemistry involved the synthesis of other raw materials for explosives, industry, and agriculture. Because they could no longer rely on imports from Europe, American chemical companies were forced to find new methods for making potash (potassium oxide) and coal-tar derivatives such as phenol and toluol. Potash was an essential chemical in fertilizers, but it was also used in some explosives as a substitute for saltpetre or sodium nitrate. Before the war, the United States had imported virtually all of its potash from Germany. Between 1914 and 1918, American companies built nearly 130 plants to produce potash from a variety of unlikely sources: brine, waste water from beet-sugar refineries, cement-flue dust, seaweed, wood ashes, and such minerals as alunite, greensand, and feldspar. The potash plants ranged from simple factories that concentrated salts from briny lake water to sophisticated chemical plants, such as Hercules' kelp-processing operation near San Diego.[9]

The Germans had also built up a near monopoly of coal-tar derivatives before the war. Phenol and toluol were important ingredients of dyes and pharmaceuticals, among many other products. After 1914, phenol and toluol were much in demand to make picric acid and TNT as well as commercial products. When the German supplies became subject to blockade, however, American manufacturers were forced to develop substitutes. As made conventionally, phenol and toluol are by-products of the transformation of coal to coke, a feedstock in the production of raw steel. Typically, a ton of coking coal yields about twenty-one pounds of light oils, which includes three pounds of toluol and one and a half pounds of phenol.[10] Early in the war, the chemical companies obtained those products from the steel industry's coking ovens, and the steel companies themselves built several thousand more by-product ovens during the war.[11] As the conflict wore on, however, the demand for toluol overwhelmed traditional sources of supply. The demand was met partly by emergency measures, such as stripping toluol from illuminating gas at public utilities, but wartime conditions also called forth extraordinary feats of innovation and creativity in organic chemistry and engineering.[12]

Such projects as fixation of ammonia and synthesis of nitric acid, potash, and coal-tar derivatives posed great challenges to the American chemical industry. The demands of wartime production placed a premium on research and engineering skills and required much larger levels of investment than ever before. Between 1914 and 1919, total capital invested in the American chemical industry tripled, with the fastest growth occurring among manufacturers of acids, potassium, plastics, and coal-tar products.[13] These years saw revolutionary changes take place inside many companies, including Du Pont, General Chemical (the nucleus of today's Allied-Signal), American Cyanamid, Monsanto, Union Carbide, Koppers, and Atlas, as well as Hercules. At the same time, the distance—measured in both sales and technological capability—between these companies and their smaller rivals widened appreciably. It is not too much to say that World War I created the modern American chemical industry—an industry in which Hercules would play an important role over the next seven decades.

But we are getting ahead of the story. The challenges and changes lay in the future when, in the late summer of 1914, Hercules pondered what, if any, role it would play in the global conflict.

Supplying the Allies

In 1914, at the outbreak of the war, Hercules seemed unlikely to become a major supplier of munitions. The United States would not become directly involved in the fighting for nearly three more years, and the American military had an assured supply of military explosives from its own ordnance works and exclusive contracts with Du Pont. Hercules' capabilities in commercial explosives, moreover, matched poorly with military requirements. Armies and navies around the world had little use for dynamite and black powder. Nor were Hercules' brands of smokeless powder attractive to the military. Most of them were tailored for shotguns and small arms and could not be adapted for larger weapons; nor could they generate the extraordinarily high-muzzle velocities sought by the military. Hercules' powders, moreover, were double-based, containing both nitrocellulose and nitroglycerin, whereas most militaries preferred single-based nitrocellulose powders. Such single-based propellants, often called "pyrocellulose" or simply "pyro" powders, sacrificed some power, but they were less corrosive on gun barrels than powders containing nitroglycerin.

To become a major munitions supplier, therefore, Hercules needed to upgrade and extend significantly its capabilities. At the urging of the Sales Department, the company had already begun to take steps in this direction at the end of 1914. As we saw in Chapter 3, Vice President Skelly had been worried about the growing popularity of pyrocellulose sporting powders and had urged the company to make them. The board asked Vice President Bacchus to study the feasibility of producing the powders at Kenvil. He reported back in January 1915 with a recommendation to proceed as quickly as possible. The board agreed and appropriated $125,000 to build the new pyrocellulose plant.[14]

It was the first of many expansions at Kenvil in 1915. Within a matter of weeks, Hercules became a major supplier of propellant, negotiating a series of contracts to supply cordite M.D., a special type of double-based smokeless powder, to the British military for use in field artillery. The development was partly a stroke of luck, or at least a coincidence. By early February, the British had already come to the realization that the war would not end quickly and that huge amounts of propellants and explosives would be needed. Indeed, at that moment, the administration of Prime Minister Herbert Asquith—the last Liberal government in British history—was about to collapse, divided over the doomed efforts to outflank the Germans at Gallipoli and embarrassed by mounting evidence that domestic munitions factories could not nearly supply requirements for shells and propellants on the Western Front.[15]

The British began to seek other sources of powder. The War Office looked first to suppliers within the Empire; it placed a large order for cordite with the Canadian Explosives Company early in 1915. As the war settled into its dreadful stalemate, the British also looked to the United States, where, through the agency of J.P. Morgan & Company, they were put in touch with Hercules, a leading supplier of double-based smokeless powders.[16]

Hercules' first contract, dated February 6, 1915, called for the company to supply two million pounds of cordite at a price of 97.5¢ per pound. Two weeks later, the company negotiated a second contract for another two million pounds at 98¢ per pound. The orders were to be delivered in installments by the end of 1915. Hercules insisted that it be paid a substantial sum up front for the cordite produced to cover its investment and to protect it in the event of cancellation of the order. The arrangements called for Hercules to receive half of the money down and then to receive the balance in one or two installments, the last being on delivery of the final batch of powder. At the same time, Hercules was required to post a substantial bond to guarantee its performance. For the initial contract, the bond was for $250,000.[17]

Although Hercules was an experienced maker of double-based powders, contracting to supply cordite was nonetheless audacious, since the powder had never been produced in the United States. Cordite takes its name from its appearance after the mixture is squirted through a die: it emerges in strands resembling cords or strings. The strands are then cut into various lengths, depending on the application. In 1915, the British sought cordite for three-inch field artillery cannons, whose specifications called for the powder to come in strands about ten inches long and weighing four grams each.

Cordite differs from other double-based powders in that it contains the most highly nitrated (and therefore most explosive) form of guncotton (nitrocellulose). Because this form tends to be difficult to dissolve, most other double-based powders used a nitrocellulose that contained less nitrogen (pyrocellulose) and was soluble in ether alcohol, a common, inexpensive chemical. In the 1880s, British chemists Sir Frederick Abel and James Dewar had discovered that the highly nitrated guncotton could be dissolved in acetone, and that the resulting solution could then be incorporated with nitroglycerin. Abel and Dewar had also added mineral jelly to the mixture with the hope of offsetting the tendency of nitroglycerin powders to corrode metal gun barrels. The mineral jelly failed to do so—it was consumed completely in the explosion—but it greatly improved the stability of the powder and its resistance to varying climatic conditions.

These were important considerations for the British, who adopted cordite for official use in 1889. Its power and stability made it especially effective as a propellant for British naval guns because it bore up well under the temperature and humidity extremes of the far-flung Empire. After the Boer War (1899–1902), in which it was found that cordite still tended to corrode gun barrels at a rapid rate, the formula was altered by increasing the proportion of guncotton to nitroglycerin. The resulting product, modified cordite, or cordite M.D., sacrificed some power for improved gun barrel life. The new powder also consumed more acetone than the original formula—which was a significant disadvantage in most countries, acetone being rare and expensive.[18]

For Hercules, then, the manufacture of cordite M.D. involved more than simply adapting existing production methods. In particular, the company had to learn new ways to handle and treat the guncotton, incorporate the dough, press and cut the powder, and blend it for packing. Before starting the cordite line, T.W. Bacchus and smokeless powder experts E.A.W. Everitt and Bernhard Troxler visited the Beloeil plant of the Canadian Explosives Company to study the process. On their return, they supervised modifications of the Kenvil plant. The $125,000 that had been appropriated for the pyrocellulose plant was consumed by the end of February. Over the next two months, the company poured another $200,000 into construction at Kenvil.[19]

In modifying the process for making cordite, Hercules engineers made significant improvements that resulted in substantial savings of time and expense. It was found, for example, that the guncotton could be dehydrated more easily if its solubility was kept near the high limit (12 percent) of the official specifications. In addition, Troxler made changes to the pressing and cutting operations that raised output by a factor of five. For example, he noticed the tendency of the cords to become entangled as they squirted through the dies. To remedy the problem, he put a revolving table under the dies, so that

> slow-running strands would coil up near the center of the tables, faster ones near the outside. When the press charge was run out, the table with the powder was transferred to another spindle and revolved in the opposite direction to take off the strands. These were placed on a cutting table, about 16 feet long, where they were cut into the desired lengths.

Troxler later added a moving conveyor belt below the dies to draw away the extruded cords without tangling them. Besides reducing the amount of scrap, the belt also allowed the use of a 100-hole rather than a 40-hole die and more than doubled production at that stage.[20]

The first batch of cordite emerged from Kenvil at the end of February 1915. The company shipped 36,000 pounds of finished powder in March, 164,000 pounds in April, and 200,000 pounds in May. The initial cordite contract was completed by October—a remarkable achievement by any measure.[21]

The Union Acquisition

Although Hercules concentrated on cordite production in 1915, it did not abandon its work on pyrocellulose powders. The company managed small sales of the powders to the loading companies for rifle ammunition, and the business grew slowly in the face of intense competition.[22] Hercules' growth in smokeless powders of any type, however, was severely constrained by lack of capacity to make nitrocellulose. The company was still dependent on Du Pont to supply this critical ingredient, which came at a premium price that ate into Hercules' margins on finished powder.

In the fall of 1915, Hercules spotted an opportunity to ensure its own supply of nitrocellulose and lessen its dependence on Du Pont. Once again, the company benefited from a stroke of luck. Union Powder Corporation of Richmond, Virginia, which owned a smokeless powder plant under construction near Sayreville, New Jersey, was running out of money to complete the unit. The plant had been thrown up hastily, but it possessed one obvious attraction for Hercules: facilities for processing and nitrating raw cotton.

The story begins in the spring of 1915, when Canadian Car & Foundry Company negotiated two contracts to supply more than three million pounds of pyro cannon powder to the Russians for 98¢ per pound.[23] Canadian Car & Foundry had engaged Thomas A. Gillespie, head of a New York construction engineering company, to make the shells and had also given him authority to subcontract for the propellant. Gillespie had arranged to purchase mixed acid and raw cotton from suppliers near Philadelpia and planned personally to oversee the construction of the powder operations near Sayreville (a site subsequently known as Parlin).

Although the Russians had put down a deposit of $350,000 on the order, the contracts required Gillespie to post a bond of $350,000 to guarantee delivery. Strapped for capital, Gillespie apparently persuaded several businessmen in Richmond, Virginia, to pledge $350,000 to get the operation started. Thus was the Union Powder Corporation born. At the same time, Gillespie engaged Hugo Schlatter, an experienced maker of guncotton who had been chief chemist at the Naval Powder Factory, as the plant superintendent. Schlatter, in turn, hired powdermakers from wherever he could find them. Several employees, including Joseph Marx, followed him from the Naval Powder Factory, and others were recruited from Du Pont, which operated a smokeless powder works nearby. The new chemist at the plant, Charles A. Higgins, was a young Englishman who had arrived in the United States in July and would later become president of Hercules; M.G. Milliken, a young engineer who helped survey the property, would later become a director.

Union's contracts called for the acids and the cotton to arrive at Parlin in June and for the first finished powder to be shipped in October 1915. The supplies appeared on time, but delays in the delivery of key equipment held back production. Although the first batch of powder was made in August, it could not be produced in sufficient volume to meet the delivery schedule. As a result, the company fell behind on its payments to suppliers and scrambled desperately to stay afloat. Union's board of directors managed to postpone the initial delivery for a month and issued a plea to the original investors to pay up their subscriptions in full. A few weeks later, on November 15, help arrived from a new quarter in the form of a $100,000 infusion of capital from Hercules Powder Company.

With this investment, Hercules assumed control of the operation at Parlin; it cemented the relationship a year later by acquiring the assets and business of Union Powder Corporation for $550,000. Because the plant had been erected under severe budgetary pressures in 1915, Hercules found it necessary to make substantial changes in the operation. In 1916, the company spent more than $400,000 on improvements, building entirely new nitrating and solvent apparatus and more than doubling the capacity to process raw cotton.[24] By early 1916, Hercules at last had its own capacity to make nitrocellulose, which it incorporated into the smokeless powder made at both Parlin and Kenvil.

Hercules' new capacity at Parlin allowed it to expand its military business as well. Shortly after the company took over the plant, it signed an agreement with the Swedish-Russo-Asiatic Company of Stockholm to supply an additional two and a half million pounds of pyro cannon powder to the Russians over the next year.[25]

Inside the Shells

Hercules took its first order for shell fillings on May 25, 1915, from Col. N. Golejewski, the Russian military attaché in Washington, who contracted for 4.1

million pounds of TNT at $1.00 per pound.[26] Although Hercules had never made military disruptive explosives, it had worked with nitrated forms of toluol as an ingredient of certain grades of dynamite designed for use in low-temperature conditions. Indeed, at the end of 1914, the company sold surplus stocks of triton (a commercial grade of TNT) from Hercules, California, and Joplin to Du Pont for conversion into military explosives.[27]

The process for making TNT was similar to that for making other chemical explosives involving nitration. As the name TNT suggests, the conversion of toluol into TNT involves nitration in three stages. In the first two, which normally take place in the same reaction, pure toluol is mixed with sulfuric and nitric acids to produce first mononitrotoluol, and then dinitrotoluol (DNT). That product is then separated out and, in the third stage, added to a stronger acid solution to form TNT. Once the nitration is complete, the TNT is washed, purified by melting and remelting into crystals, and granulated for boxing and packing.[28]

Hercules' involvement in producing TNT was apparently sparked early in 1915 by a broker claiming to represent the Russian government. His inquiry led the company to make direct contact with Colonel Golejewski, who verified his government's interest. According to the *Mixer,* the first TNT contract required Hercules to post a bond because the Russians knew that the company "had never manufactured this explosive before. . . [and] were in some doubt as to our ability to do so."[29] Hercules began to gear up to make TNT early in May by requesting permission for chemists H.G. Hawthorne from Hercules, California, and George M. Norman from the Home Office to visit Du Pont's production facilities. Although Du Pont was not obligated to provide such support under the final decree, it had already done so "under a misapprehension." It seems that in response to an inquiry from Hercules about commercial triton, Du Pont had supplied "a full written description of [its] present manufacture . . . including the acid mixtures" for military TNT. With nothing further to lose, therefore, Du Pont allowed the Hercules chemists to visit its plants.[30]

The company chose to make TNT at Hercules, California, which possessed ample nitric acid capacity, modest facilities for making commercial triton, and three thousand acres of rolling hills acting as natural barricades. Construction of the military TNT line began in the spring of 1915, and the first delivery of one hundred thousand pounds was made in August. During the summer and early fall, Hercules contracted through J.P. Morgan & Co. to make an additional thirteen million pounds for the British and the Russians to be delivered over the following year.[31]

In addition to TNT, Hercules supplied other shell fillings for the Allies. By late 1915, toluol was in scarce supply, and the British and French began diluting TNT with ammonium nitrate to make amatol, a disruptive explosive roughly equivalent in force to TNT. The sudden popularity of amatol created an opportunity for Hercules to supply ammonium nitrate, produced at its dynamite plants.[32]

At the same time, the company increased its production of black powder for detonators and bursting charges in artillery shells. For such applications, black powder was preferred to smokeless powder on high explosives because it was less likely to cause damage to gun barrels in the event of a misfire. During the war, the company retooled the Hercules Works and its other black blasting powder mills at Ferndale and Youngstown to produce gunpowder for these purposes.[33] The first large purchases were made on behalf of the Russians early in 1916, including an order for five million pounds to be delivered during the year.[34]

THE SOLVENT SOLUTION

In the spring of 1915, Hercules sharply escalated its commitment to cordite production. The company was prepared to make four million pounds of the powder at Kenvil, and with incremental investments, it could easily produce much more. On May 25, the company contracted with the British to supply an additional six million pounds. At about the same time, Hercules approached the British with an offer to deliver an additional twenty-four million pounds by the end of 1916, at a price of 98¢ per pound. The British, however, seemed "strangely cool toward this proposition," although they obviously needed the propellant. The reason for their coolness, it turned out, was that they were extremely worried about the availability of acetone: they were concerned that Hercules' efforts to produce cordite in such volume would make an already scarce chemical prohibitively expensive to obtain. By the summer of 1915, demand for acetone already exceeded the total world supply two years earlier, and the last thing British buyers wanted was another competitor for this critical solvent.[35]

To counter the problem, Hercules offered in July not only to deliver the cordite but also to supply its own acetone. The company, moreover, promised to supply the British with an additional amount of the solvent equal to its own consumption. To get started, Hercules borrowed 900,000 pounds of acetone from the British, but the company agreed to repay this amount plus an additional 7.2 million pounds by the end of 1916. The acetone would be made from "sources not at that time available."[36]

For Hercules, scaling up to make cordite and TNT was challenging, but straightforward. Producing acetone, however, was a different matter altogether. As conventionally made, acetone was a product of the dry distillation of acetate of lime, which was, in turn, a product of the dry distillation of hard wood. The problem with the process, in 1915, was one of efficiency and scale. It took 80–100 tons of wood to yield a single ton of acetone.[37] To produce the volumes Hercules needed would have involved processing between 320,000 and 400,000 tons of wood, the vast bulk of which, in the form of charcoal, would have created massive logistical, inventory, and disposal problems for the company.

As negotiations for the big cordite contract proceeded in June and July, Hercules scientists worked with the Du Pont company's Chemical Department to find other ways to make acetone.[38] The most promising route involved the distillation of acetone from acetic acid (vinegar). Makers of table vinegars fermented from wine, apples, molasses, and other materials occasionally produced stronger solutions for industrial purposes. Similarly, grain distillers produced acetic acid as a by-product of the fermentation of wood grains. In early July, Hercules went so far as to commission Du Pont researchers and the Badger Company, an engineering construction firm, to develop plans for an experimental vinegar plant at Kenvil. At the same time, Hercules inquired about prices and production capabilities of several vinegar and alcohol producers, including the Fleischmann Company of New York City, a well-known producer of yeast.

"Of all the methods considered," wrote George Markell later, "the vinegar process seemed the only certain and satisfactory method and we definitely decided to undertake the manufacture of acetone through this process."[39] Markell was referring to the "quick vinegar" process that had been used to make industrial vinegar in small batches for nearly a century. The process involved fermenting wood shavings in a tank (in the United States, the most common wood was beech). Then, according to an account in the *Mixer*:

> the shavings are placed in these [tanks] with a small amount of vinegar to provide the bacteria which produce acetic acid. Alcohol is then admitted and acetic is continually formed through the bacteriological action set up by the priming charge of vinegar.[40]

At that point, acetone could be separated from acetic acid by several methods, such as neutralizing the acid with lime, or employing a distillation process.[41]

The volumes of acetone required by Hercules in 1915 were much greater than had ever been achieved using the quick vinegar process. A critical question, therefore, was whether the process, which required careful control of bacteriological action, could work on such a large scale. Observers expressed "grave doubts," "people in the vinegar business" made "predictions of failure," and "experts who were consulted stated that it could not be done."[42]

Nonetheless, Hercules set out on its own to make acetone from vinegar. In addition to making plans for an experimental vinegar plant at Kenvil, the company arranged to lease a distillery to make the alcohol and negotiated contracts for molasses as a possible raw material.

Curtis Bay

Just as these efforts were being launched, however, a new opportunity arose. As later recalled by George Markell, the U.S. Industrial Alcohol Company approached Hercules "through certain mutual banking friends in New York, and made us such an attractive offer that, after some negotiations, we contracted with them to supply us with all of the acetone that we would require in excess of that already contracted for . . . with the right of renewal for four succeeding years." The offer provided that U.S. Industrial Alcohol, through a subsidiary, would build and operate a plant to manufacture acetic acid and acetone on Curtis Bay, near Baltimore. The subsidiary, Curtis Bay Chemical Company, was confident enough of its abilities that it gave Hercules the right to take over the operation and to purchase alcohol "at an extremely low price" should the acetone fail to be delivered on time.[43]

The agreement with U.S. Industrial Alcohol was dated August 20, 1915.[44] Hercules contributed $500,000 to help start up the Curtis Bay operation, which cost "considerably over $1,000,000" for the vinegar plant alone. U.S. Industrial Alcohol erected an immense alcohol distillery nearby at its own expense.[45] Hercules also sent a team of engineers and chemists to work on the project. The actual facility was huge, occupying fifty-seven acres and employing 250 people. The plant included more than a thousand vinegar generators or tanks, most of which were ten feet in diameter, eighteen feet high, and packed with a mixture of beech shavings and vinegar. The mixture had to be maintained at a constant temperature, which the changeable Chesapeake Bay weather made troublesome. At first, chemists were able to improve the fermentation by adding beer to the alcohol mixture. According to the *Mixer,* however, the bacteria, known as vinegar bugs, posed difficult problems:

> As warm weather drew on they seemed to suffer from spring fever and later from heat prostration. At the first sign of weariness, the obliging chemists increased the beer ration. This had the desired effect, but not for long. The bugs got too much beer, with the inevitable result, and malt extract had to be substituted before they could again be induced to produce a satisfactory amount of vinegar.[46]

"The process is simple enough from a chemical stand point," acknowledged Carl F. Bierbauer, Hercules' technical representative at the plant, "but we ought to have an animal trainer to make the bugs behave."[47] Other contemporaries noted (less colorfully) that "many difficulties were encountered before the plant was in successful operation, as the experience of vinegar makers with small generators could not be applied directly to the large units."[48]

By the spring of 1916, the situation was becoming serious. Because of the acetone shortage, Hercules had fallen short a million pounds per month in deliveries of cordite. The company started shipments of the solvent itself at the end of May—five months behind schedule. At that time, the Hercules board appointed a committee consisting of Dunham, Bacchus, and Skelly to investigate the situation at Curtis Bay.[49] Their report does not survive, although the company opened negotiations with the British to extend the delivery schedules for both cordite and acetone. The Curtis Bay operation continued to struggle, however, and by the end of 1916, the company managed to deliver only 2.4 million pounds of acetone—less than one-third of the amount originally promised—and was two months behind on cordite production. At that time, the second of the two large acetone contracts (for 24 million pounds to be delivered in 1916) was canceled.[50]

Chemicals from Kelp

Fortunately, Hercules had not placed its entire bet on the venture with U.S. Industrial Alcohol. During 1915, it continued to investigate making acetone from various sources, including beer slop, cider vinegar, molasses, glycerin foots, alcohol, wood pulp, and kelp salts.[51] By the late summer of 1915, the company settled on kelp as its feedstock. According to a contemporary Hercules publication, "in reading about acetic acid in an old encyclopedia," George Markell "ran across a paragraph discussing kelp, and it occurred to him that we might harvest this seaweed to make acetone."[52]

Markell's inspiration may have come from other sources as well. The U.S. Department of Agriculture (USDA) and several companies had established experimental plants to make potash from seaweed in 1915.[53] (Seaweed had been used for fertilizer and as a source of potash in Scotland and Japan for centuries.) A USDA survey of the kelp beds off the Pacific coast from California to Alaska had promised "inexhaustible supplies of potash" if ways could be found to harvest the kelp and produce the chemical on a commercial scale.[54]

Under Markell's prodding, Hercules pushed forward on both harvesting and processing kelp. Markell himself left Wilmington for the West Coast in late summer. At the end of September or in early October, he sent $30,000 to George Simmons, owner of the Simmons Hardware Company in St. Louis, "for the purpose of developing a kelp harvesting machine and for the carrying on of experiments for the recovery of acetic acid from kelp." Simmons had already begun work on his own plant to recover potash from kelp. Two weeks later, Markell wrote to T.W. Bacchus from San Francisco describing his activities on the Coast. He enclosed a clipping from a local newspaper that reported Hercules' plans to investigate making potash from kelp and was pleased to note that he had succeeded in keeping the reporter focused on potash "so that he didn't even ask if we expected to get anything else." Markell also remained optimistic that the operation would be economical, basing his assessment on research being done at Hercules, California.[55]

Hercules pushed ahead with its investigations, and also with designs for the kelp harvester. A review of existing patents and equipment revealed that no current marine harvesting technology would be capable of producing the volumes of kelp that Hercules needed.[56] In November and December 1915, Markell and Herbert Talley (the engineer who had joined Hercules with its acquisition of Independent Powder Company in 1914) visited California to inspect the Simmons harvester. Despite its builder's claims, the vessel worked well only for short bursts of time; it proved incapable of sustained performance hour after hour. In January, Hercules gave up on Simmons, and Talley began to work with a marine designer at the Pusey & Jones Company in San Diego. Cecil Weaver was "looked upon as a very able designer and one who would not be unduly bound by precedent when an entirely new field was under consideration."[57]

The design eventually chosen called for a vessel with several unusual features. The harvester would have a tunnel running lengthwise through the center of the hull, where the dredging machinery, "consisting of two revolving cutters," would shear the tops of the kelp plants about six feet below the surface. Once cut, the seaweed would be chopped up, carried up from the water by a mechanical conveyor, and deposited in a storage bin on the upper deck of the ship.[58] Because plans called for the harvesters to operate around the clock, Hercules also designed barges to carry the macerated kelp to the processing plant onshore.

As soon as the design was tested and approved, Hercules proceeded rapidly to build its fleet of six harvesters and sixteen barges. At the same time, the company was working on the design of a plant to process the kelp.[59] The process was straightforward. At the dock, cranes lifted the chopped seaweed from barges and poured it into storage tanks, where it fermented under controlled conditions for a period of ten days to two weeks. The resulting liquor was filtered and evaporated to produce crystals of crude salts, including fatty acids, calcium acetate, iodine, and potassium chloride. The salts were then heated in retorts to produce acetone, while potash was recovered by leaching and crystallization.[60]

Hercules made a huge commitment to the kelp operation. The company retained the Charles C. Morse Company of San Francisco to build the plant. Construction began in February 1916 and proceeded throughout the year. The finished plant in Chula Vista (near San Diego) covered about thirty acres, employed more than a thousand people, and represented more than $5 million in investment.[61] Under the leadership of General Manager Leavitt Bent, the plant produced its first acetone in January 1917—too late to satisfy the original contracts with the British, but soon enough to allow Hercules to complete its cordite contracts and to terminate its agreement with U.S. Industrial Alcohol. In March, the company managed to negotiate a new contract with the British to supply nearly one million pounds of the solvent during the next three months. By the end of May 1917, Hercules had delivered 857,920 pounds against the contract, which was deemed completed.[62]

Fortunes of War

Scaling up for wartime production at Hercules was replete with challenges: to expand capabilities to make existing explosives; to make entirely new ones; to develop new production processes and management skills; and to branch into new areas of chemistry by making solvents and intermediate products. In meeting these challenges, the company showed an ability to respond quickly to opportunities and problems. The company obviously had good instincts about opportunity: its leaders recognized that the war afforded a rare chance to enlarge its skills and augment its technical knowledge with minimal financial exposure. But Hercules also showed a willingness to assume risks if the potential rewards justified them. Given the state of its technical knowledge and abilities, the Chula Vista venture was a bold stroke, its success testimony to the company's engineering resourcefulness. Indeed, throughout the war, Hercules showed a remarkable ability to solve immediate technical and administrative problems: improving the cordite process and raising output; scaling up rapidly to make smokeless powders and TNT; undertaking the refurbishing of the Parlin plant on the fly; and exploring unconventional technologies at Curtis Bay and Chula Vista. The company's healthy financial condition, of course, helped smooth the way.

Hercules was well rewarded for its efforts during the war. In 1915, total revenues were twice those of 1914, and net income nearly five times as high. The next year was even better: revenues soared to $63.4 million (eight times the level of 1914), and net income reached $16.7 million (nearly twelve times the level of 1914). Cordite and TNT accounted for the vast majority of the profits (see Exhibits 4.2 and 4.3).

The tremendous influx of cash sparked a round of discussions at Hercules about how it should be spent. The issue surfaced simultaneously with the first cordite contract in February 1915. Although the company decided to fund the expansion of Kenvil (and later the Hercules Works) out of advance payments, Markell still forecast a surplus of $2.5 million by the end of the year. "It is, of course, too early now to determine upon the disposition of this increase in cash," he admitted, but he went on to suggest three options: putting the cash in a reserve fund; paying out a special dividend; and applying the funds to reduce the company's capital obligations, specifically to retire the 6 percent income bonds still outstanding from the company's formation.[63] The company chose the last route and, by the end of 1915, no longer carried any debt on its balance sheet.

Yet money continued to pour in during 1915 and 1916. It was used in various ways: to acquire Union Powder Corporation at the end of 1915; to invest in the acetone operations at Curtis Bay and Chula Vista; to purchase and upgrade a small black powder mill in Marlow, Tennessee, in March 1916; and to pay special bonuses of 20 percent to every employee. At the same time, the board doubled its dividend payments (from 2 to 4 percent per quarter) on common stock in the third and fourth quarters of 1915, doubled the rate again for the first quarter of 1916, and added 5 percent more in the next two quarters. At the end of 1916, the company paid out a special dividend of 53 percent, most of it in the form of Anglo-French bonds, which the cash-starved Allies had begun to use to finance their purchases of munitions.[64]

Surplus cash became much less of a "problem" after 1916 because the munitions business was never again as lucrative. Hercules' total revenues dropped to $44.1 million in 1917 and climbed only modestly to $45.6 million during the last year of the war. By then, profits had fallen to $2.3 million—well above prewar levels, though far below the wartime peak. The decline of military business was in part the result of losing key customers. Sales to Russia, for example, ceased at the end of 1916 as the government teetered and collapsed into bankruptcy and revolution. At the same time, the British were managing to supply more of their own needs for propellants and solvents. They achieved partial independence by developing a substitute smokeless powder at the Royal Arsenal in Woolwich. The new powder, cordite R.D.B. (Research Department B), was ballistically equivalent to cordite M.D. but used ether alcohol rather than acetone as the solvent.[65]

A remarkable discovery by a young Jewish chemist, Dr. Chaim Weizmann, also made the acetone shortage less urgent in Britain. In pursuing research on synthetic rubber at the University of Manchester in 1914, Weizmann stumbled onto a bacillus that converted starch into acetone and butyl alcohol. His finding became the basis of an experiment in producing the solvent on a large scale at a nearby gin factory. After months of trials and tests, Weizmann succeeded in

EXHIBIT 4.2 Hercules' Commercial vs. Military Business, 1915–1918 (Dollars in Thousands)

	Total		*Commercial Business*				*Military Business*			
Year	*Sales*	*Profit*	*Sales*	*Percentage*	*Profit*	*Percentage*	*Sales*	*Percentage*	*Profit*	*Percentage*
1915	$15,510	$4,745	$8,914	57.5%	$1,608	33.9%	$6,596	42.5%	$3,137	66.1%
1916	62,787	16,070	13,215	21.0	2,589	16.1	49,572	79.0	13,481	83.9
1917	43,867	6,656	16,598	37.8	3,280	49.3	27,269	62.2	3,376	50.7
1918	45,298	2,855	17,571	38.8	1,875	65.7	27,727	61.2	980	34.3

Source: Treasurer's Reports for December, 1915–1918.

EXHIBIT 4.3 Analysis of Hercules' Military Business, 1915–1918 (Dollars in Thousands)

	1915		*1916*		*1917*		*1918*	
Product	*Sales*	*Profit*	*Sales*	*Profit*	*Sales*	*Profit*	*Sales*	*Profit*
Cordite	$4,909	$2,162	$24,437	$6,052	$11,146	$1,477	$0	$0
TNT	1,098	587	17,190	7,260	5,955	2,513	14,526	2,347
Pyro powder	0	0	6,042	(9)	8,788	(325)	10,043	(842)
Ammonium nitrate	254	56	151	51	(5)	(7)	261	79
Bursting charge	0	0	1,208	198	205	(31)	0	0
Acetone	0	0	520	(80)	778	(292)	238	(32)
Other	335	332	24	9	402	31	2,659	(572)
Total	6,596	3,137	49,572	13,481	27,269	3,376	27,727	980

Source: Treasurer's Reports for December, 1915–1918.

fermenting one hundred tons of maize and chestnut shells into eleven tons of acetone, thereby proving the process to be not only feasible but also economical. In 1916, the British Admiralty built a plant based on the Weizmann process and produced ninety thousand gallons of acetone the following year.[66] By 1917, the success of cordite R.D.B., and the availability of cordite and acetone from the United States and Canada, limited demand for the solvent produced by Weizmann's process.

As a result of these developments, as well as its own delays in delivering acetone, Hercules was able to sell only small amounts of the solvent after completion of the initial contracts. The company partially offset the decline of the acetone business by finding markets for other products made at Chula Vista. Potash from the plant was sold as fertilizer and ground into black powder for bursting charges. Butyl propionate and other chemicals were sold as industrial solvents. Ketones became ingredients of lacquers and dopes, particularly to coat fabrics on the wings and fuselages of airplanes.

Another problem after 1916 was falling prices for finished powders. At the beginning of the war, most powders and explosives sold for roughly $1.00 per pound, a price that reflected the customer's willingness to pay for the supplier's capacity investments. Once Hercules' new plants were complete, however, it could no longer expect high prices. The final cordite contract, for example, was negotiated at the end of 1916. It called for Hercules to deliver twelve million pounds at a price of 60¢ per pound by the middle of 1917. The price of pyro cannon powder fell to 50¢ per pound in 1916, and TNT began to be traded for 43¢ per pound.[67]

Still another problem was rising prices for key raw materials. Before assuming control of Parlin, for example, Hercules relied exclusively on Du Pont for guncotton. The initial orders, placed at the time of the first cordite contracts, continued the practice of paying for materials at cost plus 15 percent. As Hercules' military business grew, however, Du Point added a 6 percent amortization charge, which it then raised to 8.3 percent. By early 1916, Hercules was paying 55¢ per pound of guncotton, more than twice what it had paid before the war.[68] The toluol situation was even worse. Toluol had traded for 20–30¢ per gallon before the war but soared to $6.00 per gallon early in 1917. The situation improved once the United States entered the war because the government stabilized the price at $1.50 per gallon. Nonetheless, TNT became markedly less profitable during the last eighteen months of the conflict.[69]

Finally, Hercules' net income became subject to heavy federal taxes. In 1916, the company paid $2.3 million in compliance with a special "munition manufacturer's tax." Although that tax was repealed in 1917, Hercules was still required to pay in that year an "excess profits tax" that totaled nearly $1 million.[70]

Hercules was able to offset some of the decline in its military business by increasing its commercial sales in 1916 and 1917 and by finding new customers for its ordnance products. In September 1916, the British arranged to buy six million pounds of pyro cannon powder on behalf of the Russians, an order that was subsequently increased to a total of twenty million pounds. In the spring of

1917, the company negotiated the first of several contracts to supply TNT for the French government.[71] After 1916, however, most new military sales were made to the U.S. government.

HERCULES AND THE AMERICAN WAR EFFORT

When the United States declared war on the Central Powers on April 6, 1917, the American military had already begun stockpiling ammunition in anticipation of war. Indeed, Hercules had contracted to supply smokeless powder and TNT to the U.S. military the previous December.[72] With the declaration of war, additional orders poured in: one million pounds of TNT and 778,000 pounds of smokeless powder in May; thirteen million pounds of TNT in June; fifteen million pounds of TNT in July; six and a half million pounds of TNT in September; eighteen million pounds of smokeless powder in November; thirty million pounds of smokeless powder in December; and a whopping forty-two million pounds of TNT in March 1918.[73]

The contracts required Hercules to make substantial new investments at Kenvil and the Hercules Works. At Kenvil, the cordite lines were retooled to make the pyro powders required by the U.S. military, and two TNT units were built. At the Hercules Works, six new TNT units were added to the five already in operation. By the end of the war, Hercules was able to produce TNT at a rate of seven million pounds per month.[74]

Nitro

Despite the efforts of Hercules and other munitions suppliers, the United States soon encountered the same problem that had plagued the Allies in 1914: lack of capacity for making propellants and shell fillings. The output of most commercial plants was already dedicated to the needs of the Allies. The existing federal ordnance plants at Indian Head, Maryland, and Picatinny, New Jersey, moreover, were incapable of meeting the demands of modern warfare. Accordingly, in February, the chief of ordnance for the U.S. Army opened negotiations with Du Pont to build and operate at least two new smokeless powder plants. They would be huge facilities, capable of producing at least five hundred thousand pounds of smokeless powder per day in round-the-clock shifts. Sites were chosen near Nashville, Tennessee, and Charleston, West Virginia. The government was prepared to invest $50 million in each plant and hoped to be making powder by the end of 1917.[75]

It was not to be. Construction of the plants was delayed for nearly a year as Congress acted slowly to approve the funds and the War Department debated the wisdom, as well as the propriety, of entrusting both projects to Du Pont. At the close of 1917, Secretary of War Newton D. Baker decided to divide responsibilities for the new plants. Du Pont would proceed to build and operate the Tennessee plant (subsequently known as Old Hickory), but the West Virginia plant would be supervised by a new federal appointee, Daniel C. Jackling, a former Kennecott Copper executive. In January 1918, Jackling engaged Thompson-Starrett Company of New York to build the West Virginia plant, which was immediately christened "Nitro." Construction began at once. The plan was to build five huge smokeless powder lines, each with a capacity of 125,000 pounds per day. The first line was to be ready by July 1; the additional units were to come onstream at a rate of one per month until all were finished.[76]

To operate Nitro, Jackling called on Hercules, a company he had worked with for several years when he was purchasing dynamite from Bacchus for Kennecott's mine in Bingham Canyon, Utah. The agreement, signed on May 9, 1918, specified that the new ordnance plant would produce 101 million pounds of pyro cannon powder over the next year and a half. The choice of Hercules to be the plant's operator reflected not only Jackling's knowledge of the company, but also its position as one of only four private smokeless powdermakers in the country, and the only one besides Du Pont with broad experience in making double-based powders. The financial arrangements called for the company to be compensated on a cost-plus basis; the government also advanced a considerable sum of money—more than $14 million by the end of the war—to cover purchases of supplies, payroll, and operating expenses.[77]

Leavitt Bent moved from Chula Vista to become general manager of the Nitro plant early in June, assisted by John S. Shaw from Parlin and C.C. Hoopes from the Home Office. Because construction had fallen behind schedule, they and other Hercules employees worked closely with the Thompson-Starrett engineers and the military to finish the first line. As described by a company representative in the summer of 1919:

> Taken as a whole, we found that the design and layout of the plant were admirable. But there were many changes which had to be made in connection with certain parts of it. Most of these were of a minor character, but were absolutely necessary to safe and efficient manufacture. Many of them were discovered in time to make changes before equipment was actually installed, while others were only disclosed by trial of the equipment, and cases of the latter class caused serious delays.[78]

The task of construction was immense. The manufacturing complex alone featured 737 buildings, including multiunit plants for making sulfuric and nitric acids and purifying cotton. There were separate refrigerating units for each of the five lines, and the single ice plant could make 100 tons per day. The facilities also included many machine shops and buildings for storage, box-making, and shipping of finished powder. An electric power plant provided electricity for the entire operation, including a railway, and the water system was capable of pumping sixty million gallons a day, enough for a small city.

Nitro was, in fact, a small city. To house the 10,000–12,000 workers needed once the plant was fully operational, as well as their dependents, preparations were made to accommodate a population of more than 20,000. By November 1918, Nitro boasted restaurants, hotels, clubhouses, recreation centers, a bank, a YMCA and a YWCA, four churches, a 24-room schoolhouse "for white children," "a large school for colored children," a 450-bed hospital, a fire station, and a police station complete with a jail. Plans were made for "six moving picture theatres," which "probably would have been running, had not the epidemic of influenza temporarily put a ban on public gatherings." Four large blocks of stores were scattered around the city; each included "a grocery, meat market, dry-goods store, delicatessen, shoe store, barber shop and tailor." The civic center featured a department store "said to be the largest in West Virginia." Running through the settlement were paved streets and sidewalks and a modern sewer system.[79]

All this activity was, inevitably, attended by havoc. In the months after construction started, 20,000 laborers poured into Nitro, "equipped with every known form of machinery for scraping and gouging and scarifying the earth's surface." By late spring, according to the *Mixer*'s picturesque account, "not a bit of sod, not a blade of grass, was left in this stretch of valley":

> The whole surface of the earth was turned into a soft red mulch that shot up like a fine spray at the lightest footstep, was ceaselessly churned about by the constant movement of horses, machines, and pedestrians, and was kept suspended in the air by the softest breath of wind. In a gale it filled the atmosphere like a dense fog. Nothing was proof against it. . . . In dry weather you ate dust, you drank dust, you breathed dust. You were covered with dust by day, and at night after you had shaken the dust off your pillow it settled gently down upon you as you slept.
>
> If you arrived at Nitro during a dry spell, your first prayer was for rain. When your prayer was answered you understood the folly of asking for material blessings, and from then on you prayed only for strength to endure your tribulations.[80]

The presence of so much activity in such conditions; the competing demands of Hercules, Thompson-Starrett, the military, and several federal agencies, not to mention the voluntary associations that took root quickly at Nitro; and the great influenza epidemic of 1918 that swept through the town—all this added up to substantial delays in completing the works. By Armistice Day, the plant was about 60 percent completed, and the town about 90 percent. Only two powder lines were operating. The first had started on September 14, the second on October 23. When the word came down from Washington to close the facility at the end of November, Nitro was producing 175,000 pounds of finished powder per day and had reached a total of 4.5 million pounds. So far as is known, none of it was actually fired during the war.[81]

Completing the Company

At the close of 1917, Hercules observed an important milestone: the end of its five-year arrangement to obtain key support services from Du Pont under the terms of the final decree. By that time, Hercules had already taken steps to establish its own capabilities in purchasing, engineering, and R&D.

Hercules had set up a purchasing function in the Treasurer's Department at the Home Office in 1914, with R.B. McKinney (formerly of Du Pont) in charge. At that time—and until the expiration of the final decree—Hercules relied upon Du Pont chiefly for bulk purchases of key raw materials, such as the sodium nitrate and glycerin used to make commercial explosives. To purchase most other ingredients, as well as miscellaneous supplies and equipment, Hercules employed a buyer and a staff of clerks to check and process orders from the plants. Although those purchases were made on its own account, Hercules nonetheless paid a commission to Du Pont as required by the final decree.[82]

During the war, Hercules' purchasing requirements grew enormously. The company's addition of more staff to cope with the increased work load combined with its takeover of Union in the fall of 1915 raised the issue of whether Hercules' military business fell under the purview of the final decree. In December 1915, Pierre du Pont and Hamilton Barksdale recommended that Du Pont's Purchasing Department cease its work in connection with Hercules' military production. Thereafter, Hercules assumed full responsibility for purchasing supplies for Parlin, and later for the Chula Vista kelp operation as well.[83]

On January 1, 1918, when Hercules formally established its Purchasing Department under McKinney, it merely upgraded the status and expanded the scope of an existing organization. The department had two principal divisions: the Works Supply Division, which monitored and processed orders from the plants, and the buyers, who were assigned actual purchasing responsibility.[84]

The development of capabilities in research and engineering was a more formidable challenge. Once again, Hercules' decision to supply military powders was the stimulus to action. The company relied on Du Pont engineers to construct the initial cordite plant at Kenvil, but Hercules representatives supervised construction closely and (as we have seen) suggested several modifications as the work proceeded. In December 1915, however, Du Pont decided that it would no longer provide direct support to Hercules' military business, although, at Hercules' request, Du Pont chemists continued their research on potential sources of acetone until January 1916.[85]

Hercules was thus left to fend for itself in dealing with the technical problems arising from the war business. Since the commercial plants continued to operate during the war, the company had few technical personnel to spare. Although Hercules had stationed a chief chemist, George Norman, at the Home Office in 1913, he was a department of one until Carl Bierbauer was hired in December 1915.[86] Bierbauer was sent almost immediately to the Curtis Bay acetone plant as the company's technical representative.

Hercules began a search for new talent and promoted or hired many young people for important technical assignments. Bernhard Troxler, who had engineered improvements in cordite manufacturing at Kenvil, was thirty-one in 1915. That summer, the company hired a young Purdue graduate, Anson B. Nixon, and assigned him the task of chemical research on cordite. Nixon was just twenty-five when he reported to work. His counterpart at Parlin, Charles Higgins, was twenty-seven. Leavitt Bent was twenty-nine when he took over as superintendent of the Chula Vista plant, and thirty-one when he supervised at Nitro. Bent was two years older than his chief assistant at Nitro, John Shaw. Herbert Talley, the engineer who designed the kelp harvesters and supervised the Chula Vista plant's construction, was a venerable thirty-six in 1916, the same age as George Norman and Carl Bierbauer. In short, a generation of young men in their twenties and thirties—Leon W. Babcock, V.R. Croswell, Albert R. "Slick" Ely, Bill Hunt, M.G. Milliken, C.A. Murphy, Luke Sperry, Ernest Symmes, and many more—garnered significant technical experience at Kenvil, Parlin, Hercules Works, Chula Vista, and Nitro during the war.

At the end of 1917, these men formed the nucleus of Hercules' new Engineering and Chemical departments. Talley became chief engineer, assisted by Murphy, whose wartime training included work on the smokeless powder lines at Kenvil and Parlin. Together, Talley and Murphy presided over a staff of ten engineers whose responsibilities included design and development of plants, safety engineering, fire prevention, and civil engineering.[87]

Norman remained Hercules' chief chemist. The Home Office Chemical Department consisted of six chemists and four support staff, including Frances Fairbanks, a librarian. In addition, several other chemists traveled on special assignments to the company's plants. The department was responsible for supporting operations at the plants, and its staff was organized according to the company's product lines—smokeless powder, acids, high explosives, and other miscellaneous products, such as solvents and commercial grades of nitrocellulose.[88]

The Chemical Department also bore responsibility for centralized R&D. During the war, a laboratory at Kenvil directed by D.M. Jackman, the plant chemist, became the center of Hercules' first original experimental research. That work probably started in the late summer or early fall of 1915 with A.B. Nixon's studies of the chemistry of cordite and acetone substitutes. Between November 1915 and October 1917, the Kenvil laboratory issued more than ninety reports on various subjects, most of which involved the chemistry or manufacture of smokeless powder.[89]

Sometime during this period—perhaps as early as 1915—Hercules decided to expand the Kenvil laboratory into a central research facility for the military business. According to Norman's later recollection (in 1929), Hercules had found in 1915 that the research supplied by Du Pont, "owing to the pressure of war work, was not satisfactory." In anticipation of the final decree's eventual expiration, Norman recalled,

> it was deemed advisable that we prepare to handle the work with our own organization. Kenvil being our nearest plant with plenty of avail-

> able ground, a suitable spot was selected and a laboratory built. Later (1917) all smokeless ballistic work was transferred to the Station, [and] a new and larger building [was] built to handle the increased amount of work.[90]

Hercules' dissatisfaction with Du Pont's research service coincided with Du Pont's belief that it was not obligated under the final decree to support Hercules' work on military powders.[91] During the summer of 1915, Hercules' board appropriated funds to construct a new laboratory at Kenvil, a "two story fireproof structure of brick and concrete" that was apparently completed by the following spring. This structure (which today is the administration building at the Kenvil plant) is often called "the original Experimental Station." However, it was actually called either "Laboratory No. 2" or the "Kenvil Experimental Station" until the summer of 1917.

At that point, Hercules sent Bierbauer, who had returned from Curtis Bay, to organize a corporate research facility at Kenvil. Bierbauer's organization, called the Hercules Experimental Station, would have charge of research on commercial as well as military products. The new unit "took tangible form" on November 1, 1917, and was originally housed in several buildings at Kenvil, including Laboratory No. 2. During the next two years, several more structures were added to the Experimental Station, including a multistoried "experimental building" and a dynamic-testing laboratory.[92]

According to the *Mixer,* "the underlying idea in establishing this experimental station was to unify the chemical work and develop the experiments that were formerly carried on at the laboratories of our various plants." The original staff, including at least four "girl chemists," was drawn from the Home Office and the Kenvil laboratory. In 1918, however, personnel from Parlin and Chula Vista were transferred to the station to work on such matters as commercial possibilities for nitrocellulose and derivatives of kelp.[93]

The establishment of the Experimental Station was an important milestone for Hercules. The station's initial work was modest in scope: it dealt with issues of immediate concern such as process improvements and applications of existing chemical knowledge. It was primarily a support organization for the plants. Nonetheless, a handful of researchers focused on longer term problems, including the prospects for adapting military technology to commercial uses. In this respect, the station gave early indications that it would play a more significant role as a source of opportunities for growth in the future.

Conclusion

The signing of the Armistice on November 11, 1918, marked the end not only of the Great War but also of an era of rapid growth and fundamental change at Hercules. The most obvious impact on the company, of course, appeared on the income statement and balance sheet. By 1918, Hercules had become a much wealthier and vastly larger company than it had been four years before. Between 1914 and 1918, the company's sales grew from $7.9 million to $45.5 million, profits from $1.4 million to $2.3 million, total assets from $14.8 million to $42.7 million, and employment from 1,500 to nearly 10,000.[94]

During the war years, Hercules' productive capacity had swollen enormously. The total volume of ordnance production was staggering: the company delivered 46 million pounds of cordite, 71.5 million pounds of TNT, 55.6 million pounds of smokeless powder, and massive amounts of acetone, ammonium nitrate, and other products with a total value of $116.5 million. Hercules had another $40.1 million of military business under contract when the war ended.[95]

Other changes were more subtle and enduring. Hercules was now fully weaned from its dependence on Du Pont's Purchasing, Engineering, and Research departments. At the new Experimental Station in Kenvil, Hercules was cultivating its abilities to improve existing products and develop new ones. At Parlin, the company now enjoyed the capacity to make nitrocellulose and was thereby released from its reliance on Du Pont for this key ingredient of smokeless powders. At Chula Vista, the company possessed an asset of less certain value, since the postwar price of potash and postwar demand for acetone were bound to fall dramatically. Nonetheless, the experience had proved Hercules' ability to pioneer new technologies and to design, engineer, and operate a large-scale chemical facility.

More important, wartime operations at Kenvil, Parlin, Chula Vista, and the Hercules Works had also trained and tested hundreds of young managers and scores of scientists. The company now had a cadre of able and ambitious employees eager to direct their energies to commercial opportunities. And with the wartime expansion of the economy, the commercial business had itself grown rapidly after 1914. The company had enjoyed a particularly strong year in 1917, with sales of dynamite, black powder, and sporting powders surpassing $16 million. Could such growth be sustained? Would it be enough to absorb the people and the capacity from the military business? These were critical questions as Hercules pondered its future at the close of 1918.

CHAPTER 5

DEMOBILIZATION AND DIVERSIFICATION, 1918–1920

The Great War began the transformation of the Hercules Powder Company from an explosives-maker to a chemical company, and from a small company to a large one. Between 1914 and 1918, operation of smokeless powder and solvent plants had required Hercules to develop new abilities in research and analysis, to master new skills in construction and engineering, and to learn new techniques of managing on a scale much greater than ever imagined before the war.

While Hercules rejoiced with the rest of the world on November 11, 1918, peace brought fresh challenges to a company that had grown dramatically as a military supplier. To begin with, the remaining government contracts were canceled almost immediately, leaving manufacturers with millions of pounds of smokeless powder, TNT, and intermediate materials and supplies in inventory. Some of these products could be sold on the commercial market, although the amounts available far outstripped normal industrial demand. The situation was made worse by uncertainties about what the government and competitors would do with their inventories.

A related issue concerned the personnel who had contributed to war production and the assets designed specifically to make military products. Hercules had developed a cadre of outstanding scientists and managers during the war, and it would be difficult to find opportunities for them in the company's commercial explosives plants. The company, moreover, had invested millions in TNT lines at Kenvil and the Hercules Works, in cellulose nitration facilities at Parlin, and in solvent production at Chula Vista. It was not clear that peacetime markets could be found for the chemicals made at Parlin and Chula Vista, or that it would pay to convert the plants to produce commercial products.

Finally, a problem that Hercules had faced at its birth now surfaced again. It had cash and liquid assets in abundance and was concerned about where to invest them. In 1913 and 1914, Hercules had used its money to build the Bacchus Works, acquire Joplin, and expand its smokeless powder business. As a result, the company had become a full-line, nationwide competitor in commercial high explosives. But at the close of 1918, when Hercules had demonstrated (to use Russell Dunham's words) an "overcapacity for making money," it was not clear that plowing all or even most of its cash back into commercial explosives would be prudent.[1] To the contrary, it appeared that other industries offered Hercules at least the promise of higher returns and more stable growth in the long term.

Demobilization

At the end of the war, the total value of Hercules' military contracts canceled by the American, British, and French governments amounted to $37.1 million and covered some 96.6 million pounds of explosives products. These figures do not include the cancellation of the initial contract to make 101 million pounds of pyro cannon powder at Nitro, West Virginia, worth about $3 million.[2]

Hercules' financial exposure, however, was limited for several reasons. First, the contracts to supply ordnance powders had required customers to pay up front for the capacity investments the company had to make. The contracts also included termination clauses that protected Hercules in the event of a sudden cessation of orders. The company would not be liable for any expenditures made to fulfill contracts as of the date of their cancellation, although discussions about how much work had been done and how much money ought to be refunded certainly took place. In the case of Nitro, for example, Hercules had accounted for $9.9 million of a $14.3 million government advance (to cover engineering and operation as well as several production contracts) by the end of the war, and after discussions and adjustments, the company paid back the balance in 1919.[3]

Finally, Hercules had decided at an early date to amortize its investments as rapidly as possible. In January 1917, for example, the directors had authorized the creation of a $2 million fund to cover depreciation of the Chula Vista kelp operations as well as "unforeseen contingencies." The action was taken at about the time the plant produced its initial batch of acetone, and it made the financial consequences of a shutdown later on more bearable.[4] Indeed, Hercules suspended operations at Chula Vista immediately after the Armistice and within a month arranged the sale of its kelp harvesters and barges.[5] The company had not yet determined whether the physical assets at Chula Vista could be adapted to make commercial products—specifically, pharmaceutical chemicals—but it was already clear that the facility could not compete with conventional sources of its primary products, acetone and potash.[6]

If Hercules was prepared to handle the end of its military business, other matters were more serious. "The most important problem presenting itself to us at the present time," asserted George Markell, after compiling the year-end financial results for 1918, "is the rapid and radical reduction of stocks of materials and finished products and the cutting down of our forces to the basis of present and prospective commercial business."[7]

In fact, Hercules had been thinking about peacetime uses of its products and assets for nearly two years. In February 1917—two months before the United States had even entered the conflict—the board asked C.D. Prickett, assistant general manager of the Operating Department, to investigate and monitor uses for surplus stocks and materials after the war.[8] Prickett reported to the board from time to time, but as the war drew to a close, the company made more determined efforts to deal with the problem.

In September 1918, Norman Rood, who had been in charge of commercial explosives sales in Joplin and St. Louis, was called East "for special work in connection with the war." By November, it was clear what his assignment was. Just days before the Armistice, Rood was elected a vice president of the company. Soon thereafter, he took over responsibility from Prickett for disposing of surplus stocks, and he organized the Excess Materials and Supplies (XMS) Department at the Home Office. The department's work was "largely a selling proposition, involving marketing to outside concerns or transferring to our plants to the best advantage."[9]

By February 1919, when XMS was relocated as a division of the Home Office Purchasing Department, more than $1 million in supplies from Parlin, Kenvil, and the Hercules Works were listed for sale. These ranged from small items such as mill supplies, pipe fittings, tools, and miscellaneous items to huge stocks of acid, nitrocellulose, solvents, and other chemicals from the war plants. Superintendents and purchasing agents at the commercial plants were encouraged to scan these lists, which were updated monthly, before buying supplies on the outside. The liquidation of surplus materials and supplies in this manner was neither a small nor a simple chore. The XMS division engaged more than a dozen employees at the Home Office, as well as representatives in each of the plants. The division was not itself disbanded until 1921, nearly three years after the Armistice.[10]

Hercules also began a systematic review of its technical and managerial personnel shortly after the war to decide which employees could be kept on as the company scaled back its operations. Vice President Bacchus requested that the plant superintendents collect information on each employee "relating to name, age, education, previous experience before entering our employ, experience obtained in our employ and estimate of his ability and the desirability of retaining him as a permanent employee from the men who had had him as a subordinate or the superintendent in charge of work being performed." The information was then reviewed by a committee consisting of G.M. Norman, head of the Chemical Department, E.D. Armstrong, superintendent of the Hercules Works, L.N. Bent, manager of the Nitro plant, and W.C. Hunt, assistant superintendent of the Kenvil Works.[11]

The review committee was also charged with studying the organization of the plants and recommending "such alterations as were necessary to bring them to the present business conditions consistent with the highest efficiency." According to Bacchus, the committee's recommendations were not

> as radical possibly as one might expect [because] the men occupying most of the important positions had reached these . . . only after good training and long employment, and are undoubtedly amongst the most valued employees that we have. . . . A great many men who have occupied operating positions to which they had graduated from the Works' laboratories have been returned to the Works laboratory forces at Kenvil, Hercules, and Joplin, replacing the younger chemists who had more recently entered our employ and whose capacities for further advancement had not yet materialized.[12]

Bacchus and the committee found this work to be "a heartbreaking job . . . that has made a profound impression on us all." We do not have exact numbers, but roughly six thousand employees (not counting those at the Nitro facility) were let go during 1919.[13] The assignment was made somewhat easier by the establishment of centralized chemical, research, engineering, and purchasing organizations at Hercules in 1917 and 1918. These services had been provided to the company since 1912 by Du Pont under the final decree of dissolution. At the end of 1919, the Home Office employed 357 people, as compared with 138 before the war. Most of the additional personnel worked in the engineering, chemical, and purchasing (including XMS) departments, although the sales and accounting departments had grown significantly during the war. In addition, the payroll of the Experimental Station included seventy people at the end of 1919.[14]

The new organization of Hercules' business adopted in February 1919 divided the company's operations into distinct groups (see Exhibit 5.1). Bacchus continued to serve as general manager of the Operating Department until a prolonged illness led him first to take an extended leave in the summer of 1919 and then to resign his office the following November. He remained a vice president and director of Hercules, but he was succeeded as general manager by George Markell. The general manager had three assistants: one to work directly with him; the second to be responsible for all lines of commercial explosives, including research and chemical analysis; and the third to be in charge of staff functions, the company's fledgling chemical businesses, and "new plants and businesses" should they be developed.[15]

Charles Higgins, a future president of Hercules, once observed that Hercules came out of World War I with a lot of cash, a fair amount of management talent, and no business.[16] This was a good line, and it was almost literally true. The American economy was about to enter a major recession that would

EXHIBIT 5.1 Hercules' Operating Department Organization, 1919

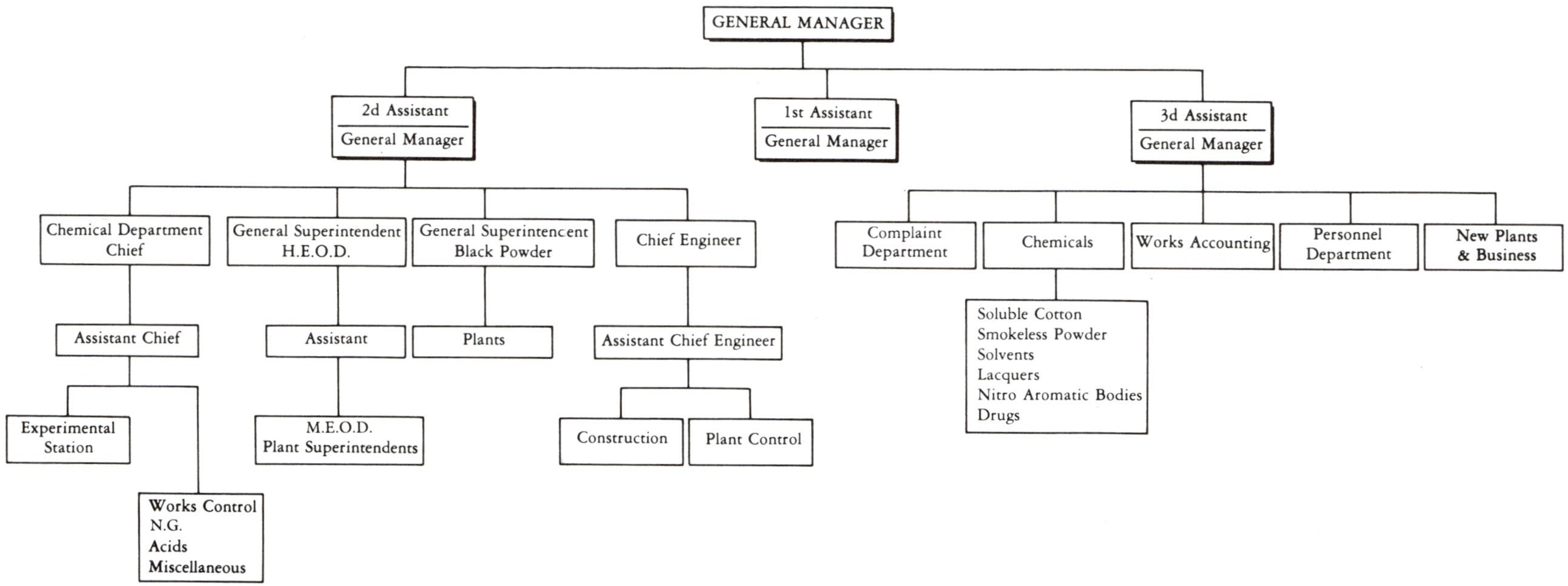

Source: Special Reports to the Directors, 1917–1922, Hall of Records, Hercules Incorporated.

especially trouble the heavy industries that included Hercules' major commercial customers. The worst times for the country would come in 1920 and 1921, but ominous signs were already appearing in the explosives business by Armistice Day.

The questions for Hercules, then, were where and how would it grow? In answering these questions, the company was guided by its own experience and instincts, as well as the examples of other explosives-makers wrestling with the same issues. The questions were also discussed and debated at an extraordinary four-day meeting at the Hotel Traymore in Atlantic City in February 1919. There the current and future leadership of the company—virtually all production superintendents, sales managers, and Home Office staff—gathered to consider the company's current condition and future prospects.[17]

The discussions at Atlantic City and other evidence reveal that Hercules approached the question of its future growth in three ways. First, the company took stock of its position in the explosives industry; that assessment gave it cause for concern. Next, it looked at adapting some of its assets, especially those dedicated to war production, to make new products. In particular, the company sought to develop its capability to make pharmaceutical chemicals, nitroaromatic compounds other than TNT, and commercial forms of nitrocellulose. Finally, Hercules thought through other avenues of growth, such as starting a new business from scratch, or more likely, acquiring a going concern in a new industry. Around the first of February 1919, it created an organizational unit, the Industrial Research Department (IRD), whose mandate was to explore new possibilities outside the company's traditional business.

Crisis in Commercial Explosives

In his opening address to the Atlantic City meetings, President Dunham observed, "We cannot hope for very much increase in our explosives business, except that which comes from the natural increase in consumption."[18] As usual, Dunham's view was understated. According to C.C. Gerow, Vice President Skelly's top assistant, "The industry as a whole is suffering from a lack of business."[19] Statistics bore him out. In 1918, total production of commercial explosives in the United States had dropped 14 percent from the level of the year before, and it would fall another 16 percent in 1919. Total production for 1919—417.6 million pounds—was off 165 million pounds from the 1917 level, and off 33 million pounds from the depressed level of 1914.[20]

At Hercules, production and sales totals fell dramatically as well. In 1919, the company produced its lowest output of commercial high explosives and black blasting powder since 1914, operating its dynamite plants at 54 percent of capacity and its powder mills at 45 percent of capacity. Sales of dynamite and black blasting powder were down by 14 percent and 28 percent, respectively, and prices fell between 15 and 20 percent, depending on the region. Hercules held its market share in black blasting powder (12.28 percent) but dropped a share point in high explosives (to 19.26 percent).[21]

These numbers translated into substantially lower financial returns for the

At the Hotel Traymore in Atlantic City in February 1919, management gathered to ponder Hercules' future: Where would its opportunities lie?

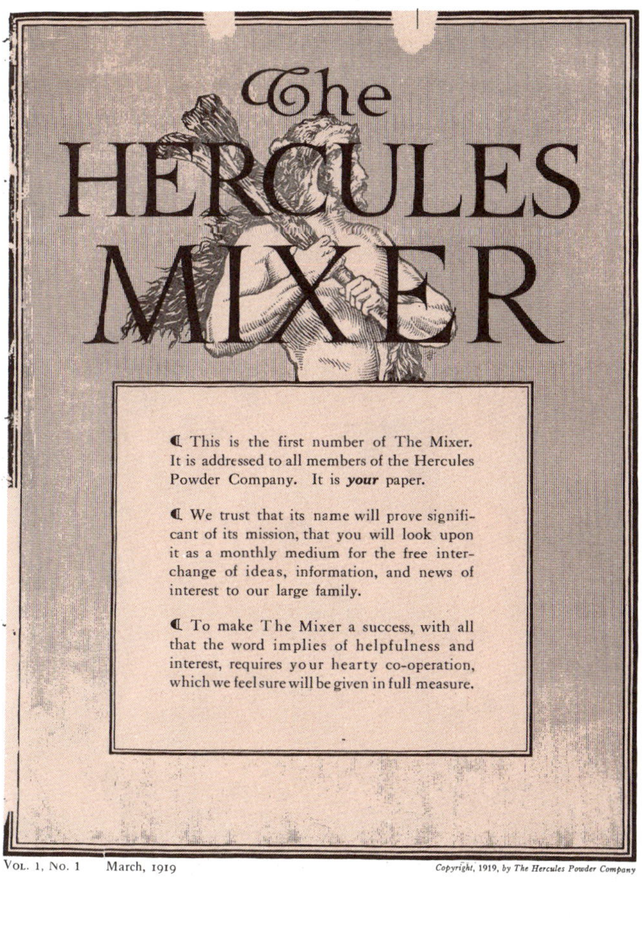

The HERCULES MIXER

℄ This is the first number of The Mixer. It is addressed to all members of the Hercules Powder Company. It is ***your*** paper.

℄ We trust that its name will prove significant of its mission, that you will look upon it as a monthly medium for the free interchange of ideas, information, and news of interest to our large family.

℄ To make The Mixer a success, with all that the word implies of helpfulness and interest, requires your hearty co-operation, which we feel sure will be given in full measure.

Vol. 1, No. 1 March, 1919 *Copyright, 1919, by The Hercules Powder Company*

Started in March 1919, the Hercules Mixer *became a friendly source of company news that continues today.*

In 1920, Hercules took an option to acquire the Yaryan Rosin & Turpentine Company. Yaryan's principal asset was a plant in Brunswick, Georgia (above), for processing stumps into naval stores commodities.

Trade shows were an important vehicle for selling explosives. Right: A display designed to woo farmers.

Gathering stumps: Although stump harvesting failed to stimulate sales of dynamite, stumps eventually proved a profitable source of naval stores chemicals.

Success in naval stores came with the systematic application of chemistry to the industry and with clever marketing. Above: The laboratory at the company's naval stores plant in Hattiesburg, Mississippi.

The antics of "Turp" and "Tine" helped Hercules establish a national brand name for turpentine.

company. Operating income that had averaged well over half a million dollars a month during most of 1918 plummeted at war's end. In the spring of 1919, the company experienced a three-month string of losses, and its average operating income for the year sank to its prewar level.[22]

In these circumstances, the company's sales strategy in explosives involved working to maintain its existing customer base rather than competing vigorously to expand it. In part, this strategy reflected worries about touching off a trade war among competitors by raiding each others' accounts. But it also reflected the belief that it was important to cultivate better relations with current customers that may have felt neglected while Hercules concentrated on the war effort.[23]

The lone exception to an otherwise gloomy forecast for commercial explosives was in the agricultural market. As we noted in Chapter 2, before World War I, the agricultural market for dynamite—for ditch digging, tree planting, stump blasting, drainage, and the like—consumed roughly 5 percent of all U.S. explosives. From its birth, however, Hercules had perceived agriculture as a potentially vast market for dynamite. During the war, the company had organized a systematic campaign to cultivate farmers. The Sales Department mailed out handbooks called *Progressive Cultivation* and *Farm Dynamite* (describing the uses of explosives on the farm), and its salesmen gave farm demonstrations and lectures at agricultural colleges. Skelly was optimistic about the work, reporting that "all these demonstrations were very successful and will undoubtedly result in business for our company."[24]

Customer interest surged early in 1919; hundreds of requests for *Progressive Cultivation* flooded the office. "It is hard to explain the remarkable increase in the number of inquiries on Agricultural Dynamite in any way," said a puzzled Skelly, "except that the whole country has suddenly become much more interested in the subject. We are not doing any more to bring forth the inquiries than we have during the last two years."[25]

Hoping that the agricultural market was at last coming of age, the Sales Department began keeping separate statistics for agricultural sales and assigned several new "agricultural men," including W.R. James, to its major sales offices near farming territory. By the end of 1919, the results were promising. Since April, more than 2 million pounds of dynamite, 1.8 million blasting caps, and 3 million feet of fuse had been sold for agricultural purposes.[26]

Apart from the agricultural market, however, the explosives industry at the close of the war was showing the classic signs of depressed conditions: excess capacity and falling production and price levels. There were opportunities to add business at the margin, but it was not the time to make significant investments to improve efficiency or add capacity. As an outlet for Hercules' "overcapacity at making money," the explosives industry of the immediate postwar era seemed most unpromising.

Swords into Chemicals

During the war, Hercules had launched several investigations into peacetime applications of its wartime capabilities. Its efforts took place at both the plants and the newly finished Experimental Station in Kenvil. The station's primary function was to support Hercules' business in commercial explosives, but it also worked with the plants on "the development of new products and the processes best adapted to manufacturing them." In 1917 and 1918, for example, Experimental Station personnel worked with chemists at Parlin on commercial varieties of nitrocellulose; with their colleagues at Kenvil on the possible development of nitro-aromatic compounds using the TNT lines; and with researchers at Chula Vista on pharmaceutical chemicals derived from seaweed.[27]

The Sales Department also supported those efforts. In October 1917, for example, the Home Office had appointed a salesman to find markets for the chemical by-products produced at the Chula Vista plant such as ethyl acetate and other solvents used to make soluble cotton; acetic anhydride, an ingredient in aspirin tablets and the plastic cellulose acetate; iodine, formulated in tinctures and other medicines; and valeric acid and valerates, elements of various pharmaceuticals.

A few months later, the department took over responsibility for selling a broader line of chemicals, including nitrobenzine and dinitrotoluene (DNT)—products made on the TNT lines and used as ingredients in manufactured dyes—and the by-products and small experimental batches of nitrocellulose products from Parlin. The Parlin products included soluble cotton, various pyroxylin solutions used in the manufacture of artificial leather, lacquers, and other coatings, and Herculoid, a celluloid-like plastic made from cotton linters rather than from wood.[28] The company put two additional salesmen on the job, one in the San Francisco office, the other in New York. Chemical sales, however, were "a distinct side issue" until the end of the war. Total sales of chemical products not related to munitions during 1918 amounted to about $300,000—less than 1 percent of total revenues for the year. The products were also sold at a loss.[29]

In January 1919, Hercules mounted a more determined effort to develop its chemical businesses by creating a full-fledged chemical division in the Sales Department. The division included five employees in Wilmington, three in New York, and salesmen at the branch offices in Pittsburgh, Chicago, and San Francisco.[30] The chemical sales division was expected not only to sell off surplus products and by-products from Chula Vista and Parlin, but also to work with the Experimental Station to explore the market potential for chemicals Hercules might make at its war plants.

Although initial hopes for chemical sales were quite high, by the spring of 1919, problems were beginning to show. The market showed minor interest in nitrobenzine and DNT, while sales of solvents and the pharmaceutical chemicals from Chula Vista began to fall off as imports and more competitive sources came back into play.[31] The Sales Department had better luck with nitrocellulose products, after mounting a major push early in 1919.[32] Sales volumes of soluble cotton and pyroxylin solutions were increasing, but the company was experiencing trouble with Herculoid because the material was available only in low volumes and customers frequently complained about its quality.[33]

The most serious problem was that chemical sales were not returning a profit, a situation that could not be sustained while Hercules' explosives business was also suffering. By May, Markell was alarmed; he argued that it was "absolutely essential that we retrench even more vigorously than we have already done." He urged all departments to make a careful study of their records to ferret out "any unproductive expenses" and called for an especially hard look at the company's R&D efforts:

> While war earnings continued we could afford to carry losing lines and absorb the losses while we hoped for and worked for a change to the profit side of the account, but when profits have reached the small figure that ours have now, we must recognize the fact that we either must show our stockholders that we are operating at a loss or we must eliminate development and experimental expenses other than those small enough to be absorbed even by our reduced earnings or sure enough to justify us in capitalizing the losses by temporarily funding them until they have been turned to profits.

"It seems to me," Markell concluded,

> that with this condition facing us we must recognize the impossibility of continuing the search for new avenues of endeavor when an original development is involved and must concentrate the efforts of our Research Department on the search for an established business which can be merged with ours in such a way that an enlarged income will immediately accompany the use of the enlarged capital.[34]

The board debated Markell's report for three days in early June. The upshot was a decision to begin the dismantling and sale of the Chula Vista plant.[35] On August 12, after a verbal report by Markell (who by this time was acting general manager of the company), the board ordered the Experimental Station to cease all work in connection with pharmaceuticals derived from the Chula Vista operation.[36] Hercules did not immediately suspend work on other chemical compounds. But the nitro-aromatic compounds business based on TNT chemistry was allowed to wither, and the company slowly dismantled its production lines at Kenvil and the Hercules Works in 1919 and 1920.

The company chose to continue to develop the nitrocellulose business, however, although it operated at a loss throughout 1919. There are several reasons why Hercules stayed in this line. First, the Parlin plant supplied a valuable raw material to Kenvil's smokeless powder and dynamite operations, a material that Hercules no longer had to buy from Du Pont or other sources. The Parlin plant, moreover, was designed around a chemistry that Hercules understood already and wished to explore further. The peacetime potential of nitrocellulose was enormous. Annual consumption of commercial nitrocellulose products in 1918 amounted to sixteen million pounds, of which about ten million was captive production by makers of celluloid (plastic) products and photographic film. (These figures did not include consumption of nitrocellulose solvents such as amyl acetate, ethyl propionate, ethyl butyrate, and ethyl acetate, which Hercules had made and worked with during the war.) That left six million pounds up for grabs. The company was extremely interested in entering the commercial nitrocellulose market, which it projected to grow rapidly in the coming decade.[37]

In February 1919, a portion of the Parlin plant had been converted from producing guncotton to making two products that intrigued Hercules: soluble cotton and pyroxylin solutions. The soluble cotton was either sold to celluloid producers or shipped to the Experimental Station at Kenvil where is was mixed with camphor to make Herculoid. Herculoid was then sold as sheets, rods, or tubes to makers of novelty items such as combs, hairpins, eyeglass frames, and ping pong balls. As the Sales Department reported, however, the company would have to solve serious technical and manufacturing problems to succeed with Herculoid.[38]

The pyroxylin solutions business seemed more promising. The products were used by makers of artificial leather, linoleum, wood lacquers, aeroplane dope, and other protective coatings.[38] The key variable was the viscosity of the solutions: thick solutions and pastes were used in artificial materials, and thinner (low-viscosity) solutions were formulated into fine lacquers and coatings. At the close of World War I, the critical technical problem limiting the growth of pyroxylin solutions was the difficulty in manufacturing low-viscosity nitrocellulose in volume. Many companies were attacking the problem, including makers of paints and varnishes and photographic film, as well as other producers of smokeless powder.[39]

In entering the nitrocellulose business, Hercules encountered familiar faces. Among its principal rivals were Du Pont and Atlas. In 1917, Atlas had acquired three companies that made lacquers, enamels, varnishes, artificial leather, and nitrocellulose.[40] Du Pont had become the largest producer of commercial grades of nitrocellulose for sale on the open market. In 1908, it had acquired the Fabrikoid Corporation, a maker of artificial leather and pyroxylin, and during the war it bought several other producers of similar products. In anticipation of the Armistice, Du Pont's Development Department had also studied alternative uses for its smokeless powder plants. In February 1917, Du Pont had decided that its postwar businesses would include dyestuffs and other organic chemicals, vegetable oils, paints and varnishes, water-soluble chemicals, and "industries related to cellulose and cotton purification."[41]

Hercules, Atlas, and Du Pont all came out of World War I with the intention of using their capabilities in nitrocellulose chemistry to enter new businesses. But there were fundamental differences in their strategies. Atlas and Du Pont were vertically integrated, manufacturing not only the basic chemicals but also finished products such as artificial leather, lacquers and varnishes, and fabricated plastics. Du Pont, moreover, controlled its own supply of processed cotton. Hercules took a different tack. As Higgins pointed out, the other companies, by selling both nitrocellulose and products derived from it, were competing with their own customers. That situation Hercules was determined to avoid. Although the company had made special paints and dopes during the war from surplus chemicals at Chula Vista and had explored opportunities in the artificial leather business, it chose to stick to its knitting. Higgins explained:

> Our policy should be . . . as far as possible, to avoid the manufacture of finished lacquers, and to refrain from getting into a position where we would be our customers' competitors. . . . Except in most exceptional circumstances such as I cannot think for the moment would occur, I personally do not think that our interests as a whole would be best served by our making leather cloth. . . . If we specialize and devote our attention to the manufacture of the basic nitro-cellulose, or at best the heavy paste needed as raw material by these trades, I confidently believe the continued operation of our nitro-cellulose plant at Union [Parlin] will be profitable on this basis alone, and that we will have laid a firm and secure foundation for future growth.[42]

Higgins extended the same logic to the development of Herculoid, which he saw not as a finished product but as a medium for fabricators to work with. "Our policy of making only the basic material, which is the big tonnage business, would win us the support of novelty manufacturers and others who, while buying pyroxylin plastics from the manufacturers, find themselves in competition with these same manufacturers who also make certain novelties."[43]

Hercules' initial returns on nitrocellulose products during 1919 showed moderately encouraging signs. While the company lost roughly $84,000 on total chemical sales of $823,000 for the year, it had actually sold 1.5 million pounds of soluble cotton for $437,000, and 1.1 million pounds of pyroxylin solutions for $160,000. Hercules reported a loss on soluble cotton of about $50,000 but made several thousand dollars on some of the pyroxylin solutions. The profits would have to improve, but in terms of sales volume, at least, the new products were in a league with the company's business in sporting powders.[44]

At the end of 1919, Hercules had taken its first tentative steps into the nitrocellulose industry. Its advantages included the plant at Parlin and several years' experience with the basic chemistry. But the company was also a latecomer to the industry, lacked control of its raw materials, and would have to establish distinctive strengths to position itself against larger, more integrated competitors.

A business that would yield greater returns and have a larger impact on the company's fortunes would come from another direction—through the work of Hercules' newly formed Industrial Research Department.

The Industrial Research Department

Despite its name, the purpose of the new, temporary entity was not to conduct scientific research in support of existing or potential product lines, as was the task of the Experimental Station, but rather to investigate potential new lines of business. Hercules hoped to maintain a strong rate of growth—an increasingly difficult task in its traditional business—by utilizing in new ways the substantial resources it had acquired and developed during the war. As Norman Rood, head of the new department, explained at the Atlantic City meetings:

> This department was organized due to the fact that we realize overproduction or at least overcapacity of explosives has placed some limit upon the amount of explosives business we can expect to secure. . . . The purpose of the department briefly stated is to locate, investigate, and report on manufacturing propositions which the Hercules Co. may be interested in acquiring in order to use the facilities it has at hand, to extend its business and increase its profits.[45]

The department's mandate, as articulated by Rood, was first to diversify via acquisition rather than create a new enterprise. This approach would minimize the risk of entering uncharted territory. Second, to avoid threatening its primary business base, the IRD was to spurn involvement in industries that would bring it into direct competition with consumers of explosives. For this reason, Hercules avoided the mining industries outright so as not to "antagonize our best customers." Third, since the underlying motive of the diversification was long-term growth, the IRD would seek a manufacturer of primary products positioned in a healthy or promising industry. Finally, Hercules wanted a good buy. An ideal target would be a company that was "limping badly or badly wounded" because of undercapitalization, poor plant management, inept marketing, or other administrative weaknesses. "It is the opportunity we want more than just the business," Rood explained, adding, "We . . . have a very broad field before us."[46]

The strategy reveals much about how Hercules perceived its strengths at the end of the war. Its operational know-how would improve the efficiency of a floundering and perhaps unprofitable company; its marketing strengths would improve distribution and, if necessary, spur consumer demand; and its financial clout would help overcome the problems associated with lack of capital.

But it is noteworthy that Rood emphasized neither the importance of building on the company's technical strength in cellulose chemistry nor that of diversifying into a business related closely to explosives, because "that field has been rather well covered." Indeed, there is evidence that Hercules was at least as interested in utilizing its mechanical engineering expertise as its chemical engineering expertise. Higgins noted that, "in considering development work," Hercules should remember that its "engineering force is skilled in the design and erection of stills . . . and solvent recovery equipment."[47] In particular, several members of the IRD were veterans of the Chula Vista kelp plant, as well as of Nitro, both complicated exercises in the design and erection of processing equipment and the logistics of handling bulky raw materials.

To achieve its goal, the IRD was allocated an annual budget of $250,000 and staffed with seventeen of the company's most ambitious, skilled, and experienced professionals.[48] All had served under difficult conditions "at high tension and without regard to meal time or bed time." Rood, of course, was an experienced powderman and an established manager at Hercules. The IRD's technical manager and assistant technical manager, Leavitt N. Bent and John S. Shaw, were also men of proven competence. Like Rood, Bent had come to Hercules in 1914 with the Independent Powder Company. During the war, he served as manager of the Chula Vista plant (where, in his honor at the end of the war, fellow employees composed a hymn, "Augured our Kelp in Rages Past"), then transferred to the Nitro project in West Virginia during the "unsettled construction period." He shared the "strenuous times" at Nitro with his assistant, Shaw, who had previously held supervisory positions at the plants in Hercules, California, and Parlin, New Jersey. Shaw had been a powderman since 1906 and had worked at several Du Pont facilities, where he specialized in acid production. Part of Hercules' original management team, Shaw had successively worked at the Hercules, Parlin, and Nitro plants.

Indeed, it could be said that the creation of the department was one of the company's first tangible steps in utilizing excess war resources, since most IRD members had been key administrators at war plants. At least a dozen of the men were alumni of Nitro, where they had headed engineering, traffic, and the production of acid and guncotton. Along with Bent, three others had worked at the Chula Vista plant as manager, assistant superintendent, and chemical director. A.B. Nixon and W.A. Murray had joined Hercules at Kenvil during the cordite days; four others gained experience at the Parlin plant, two prior to 1916 when it was still owned by the Gillespie Company.

The broad, open-ended orientation of the department was reflected in the diversity of its members' professional training, both before and during their tenure at Hercules. The new department included two chemical engineers, two electrical engineers, two mechanical engineers, and one civil engineer, as well as a structural architect and a Ph.D. chemist. While a few members of the group were experienced in functional areas such as maintenance, traffic, and accounting, several possessed manufacturing expertise with acids, oils, sulphur, guncotton, and other constituents of explosives. More important, several had worked in unrelated industries, including lumber, fertilizers, paper, textiles, and railroads.

The department's operating procedures were designed to foster continuous progress and receptivity to unconventional ideas. Attendance at afternoon meetings was mandatory, and copies of daily reports were mailed to those working out of town. Members were directed to present special reports and to cull industry journals, newspapers, and other sources for information on potential ventures. To encourage an atmosphere of "a committee of the whole," each proposal was duly considered and its author guaranteed an opportunity to defend it before final rejection. Mechanisms were also put in place to funnel suggestions into the department from throughout the company.

A Sharply Focused Search

On February 15, 1919, the first and what would prove to be the most important meeting of the Industrial Research Department was held on the eleventh floor of the Finance Building in Philadelphia.[49] As the meeting began, Rood introduced Hercules salesman W.R. James, a specialist in agricultural explosives, who then made a long presentation about the great opportunities he perceived for Hercules in the naval stores industry. (Naval stores were products refined from trees; the industry had retained its name from colonial times, when tar, pitch, turpentine, and the like were used to seal the hulls of wooden ships.)

As those at the meeting knew, Hercules had built and operated a small naval stores pilot plant during the war in southern Mississippi. James and others connected with that experiment had reasoned that sales of such portable plants to farmers and landowners throughout the South would boost dynamite sales in the region. In explaining to the IRD the interdependence of the two types of business, James pointed out that the naval stores industry "operating on a large scale could be made to utilize a vast quantity of our own explosives in the production of the raw material," and he drew an analogy with Du Pont's recent massive investments in General Motors: "General Motors would be immensely profitable to DuPont if they never earned a dividend, because the motor industry uses so much material made by the parent company. I look on the distillation business the same way."

But James had a broader vision. A large naval stores business might not only encourage the sale of Hercules explosives but also generate substantial profits in its own right. In this regard, James focused his remarks on a company he believed was especially promising—the Yaryan Rosin & Turpentine Company of Brunswick, Georgia, and Gulfport, Mississippi—and described its unique market position. For centuries, he explained, naval stores products had been refined from the oleoresins of living trees, the supply of which was rapidly diminishing. The Yaryan company, however, was the leading proprietor of a new process of distilling naval stores products from pine tree stumps, which covered millions of acres throughout the South. Yaryan could be at the forefront of a significant, emerging industry segment.

Moreover, the Yaryan company was apparently mismanaged and financially shaky. Although it had been thriving in recent months, the company's assets had long been in the hands of a receiver considered "very objectionable to the other two receivers as well as to the bond holders, and many of the stockholders." (James had quietly gathered inside information about the company from the attorney handling the bankruptcy.) The Yaryan failure, as James bluntly put it, was due to "rotten financing and very poor management." With tensions running high between the investors and managers, it seemed "an opportune time for a would-be purchaser to acquire advantageously."

Thus, Yaryan's technical promise and market position on the one hand, and its financial and operational shortcomings on the other, made it an ideal acquisition candidate, given the IRD's selection criteria. "Properly managed and financed" under Hercules, James asserted, Yaryan "could be made to control the distillation game in this country."

Naval stores and Yaryan dominated IRD discussions until partway through the third meeting, held at Rood's office on February 24, 1919. At that point, the discussion shifted temporarily as the department considered other possibilities. One member of the department mentioned an "electric manufacturing concern" for sale in Toledo; then another suggested that Hercules might supply the needs of Southern oil refineries for sulfuric acid, while making the blasting of phosphate rock a sideline. A cascade of suggestions followed: virtually every member of the department and several prominent outsiders gave short presentations on a wide variety of industries, enterprises, and processes into which Hercules might diversify (see Exhibit 5.2).[50]

These options ranged from industries closely related to explosives to those not at all related, and as such, they were in keeping with the broadly defined mission of the IRD and the diverse skills of its members. In retrospect, however, it is striking that, according to the IRD's detailed reports, neither these nor any other suggestions not related to naval stores were ever again seriously investigated. The IRD's first ten days marked both the beginning and end of its role as an explorer of possibilities on the American industrial landscape.

The explanation for the apparent dichotomy between the mandate of the department—to explore a "very broad field" of options—and its narrow focus lies with the prehistory of Hercules' large-scale involvement in naval stores. By the time the IRD was first convened, the Hercules Sales Department had for years worked with American farmers to remove stumps with dynamite. In early 1919, those efforts had intensified and, largely through the involvement of the U.S. Department of Agriculture, had led Hercules to enter the naval stores distillation business on a small scale. Through that experimental work, James, Rood, and others at Hercules had acquired a wealth of knowledge about the naval stores business. For one industry, at least, they had essentially begun the work of the IRD before its inception.

It was an appealing head start, given the urgency of Hercules' need to identify a suitable diversification path. Without this particular experience, the department indeed might have seriously considered a broad range of options. Perhaps, too, if further investigation of naval stores had revealed it to be a weak candidate, the IRD might have returned to a wider search.

But that never happened. During the next few months, several members of the department traveled throughout the South, inspecting naval stores operations and submitted reports to Rood and Bent, who themselves made several trips through the pine belt to interview industry leaders and view facilities. The more the IRD delved into all aspects of the naval stores industry—production methods and leading producers, exports and brokers, patents and laws, prices and shipping rates, producers' associations and government regulators—the more promising seemed the opportunities for Hercules.[51]

As the members of the IRD would learn, the naval stores industry had entered a critical period of transition in the late nineteenth and early twentieth centuries. Serious threats to traditional sources of raw materials caused a restructuring of the industry and the rise of new methods of production. The coming of World War I further disrupted the industry's already unstable structure of supply and demand. Because Hercules was weighing its opportunities in the business within this changing context, it is useful to review both the development of the

EXHIBIT 5.2 Industrial Research Department Diversification Proposals, February 24, 1919

- Supply the needs of Southern oil refineries for sulfuric acid, while making the blasting of phosphate rock a sideline
- [Acquire?] electrical manufacturing firm in Toledo
- Shell loading
- Safety fuses and caps
- Oxide of zinc, in the form of French oxide, ordinary oxide, or lithophene
- Floor coverings, such as linoleum
- Cellulose acetate products
- Solidified alcohol
- Shoddy from rags at Union plant
- [Mine?] sulphur in Utah
- Paper and cardboard
- Road construction material
- Potash and ochre
- Tropical oil, vegetable oil, and related foods
- Rubber
- [Acquire?] Heyden Chemical Works
- Pharmaceuticals
- Artificial leather
- Safety fuses
- [Acquire?] Taplex Corporation, New York, makers of foot warmers and heaters for automobiles
- [Manufacture?] bottles for South America
- American green rubber
- Eucalyptus products
- [Mine?] manganese deposits
- Fertilizers
- Water power
- Artificial silk
- "Agricultural scheme" [wood naval stores]
- Nitrocellulose products for boudoir furnishings
- Nitrocellulose for paving bricks
- Concrete blocks, building materials, etc.
- [Acquire?] West Virginia coal property
- Oleum [glycerin]refining
- Manufacture of acids in Texas and Louisiana
- [Mine?] Kentucky sulphur deposits
- Treatment of Oriental rugs
- Oriental rug power looms
- Pot surface paint for boiler and locomotive fronts
- "H" [hydrochloric?] acid
- Oilcloth, linoleum, rubber cloth
- Kerosene carburetors for automobiles
- New process of oil "cracking"

Source: Hercules Powder Company, IRD Report, February 24, 1919.

naval stores industry up through the war and Hercules' tentative involvement in it, before returning to the work of the Industrial Research Department.

THE AMERICAN NAVAL STORES INDUSTRY

Naval stores was among the first industries established by Europeans in North America. So valuable were "Pitch, Tar, Rozin, Planke Knee Timber and other Naval Stores for the use of his Majtys Royal Navy" that during colonial times the British government restricted their export beyond the Empire and encouraged their production with cash bounties. During the Revolutionary era, the regulation and taxation of naval stores in British North America was a hotly disputed issue.[52]

Although production first flourished in New England, by the middle of the nineteenth century virtually all American-made naval stores for both domestic consumption and export were supplied from the rich stands of longleaf yellow pine (*Pinus palustris*) in the Carolinas. For the next several decades, the center of naval stores production continued to drift south through the one-hundred-mile-wide pine belt into Georgia and Florida, then westward through southern Alabama, Mississippi, Louisiana, and Texas.[53]

The drift of the industry reflected not only the relentless march of loggers in search of virgin forests but also the damaging effects of slashing living trees to tap their oleoresins. This "gum dip" was distilled into spirits of turpentine, which was used as a solvent (for wax, fats, resins, sulphur, and phosphorous), an illuminant, and an ingredient in paint, varnish, medicines, and camphene. The distilling process also yielded molten rosin, which was filtered through cotton battening and used primarily for soap-making. Since more rosin was produced than consumed, most of the excess was discarded into pits, lakes, and rivers.[54]

Gum dip gathering began after Christmas; workers chopped scars and deep, crescent-shaped notches ("boxes") into the bases of tree trunks, then returned every few weeks during the spring to gather roughly three pints of oleoresin from each tree and to chip away accumulations of the dried exudate. The first season of boxing would yield "virgin dip," virtually clear oleoresin of the highest purity and desirability. In the remaining few seasons of tapping, the gum dip would grow progressively darker and poorer in quality, while the trees progressively weakened and became vulnerable to the vicissitudes of weather and fire.[55]

Through a different method of recovering naval stores, called "destructive distillation," a limited amount of naval stores products—primarily tar oils and pitch, but not rosin—were recovered from stumps and fallen wood instead of from living trees. The process was appealing to farmers, foresters, and some landowners because it protected living trees and somewhat increased the incentive to clear land.

But destructive distillation had significant drawbacks. It was slow and cumbersome: the wood had to be chopped into small chunks and slow-cooked for days in either kettles or homemade ground "kilns" constructed of wood, pine needles, and earth or clay. Some improvements were made in destructive distillation after the Civil War, but generally the method was plagued with exorbitant operating costs and was considered "wickedly wasteful."[56]

During the late nineteenth century, important changes in markets and in sources of supply began to reshape the naval stores industry. First, demand shifted away from turpentine toward rosin. Rectified spirits of turpentine (alone or combined with alcohol) had been the cheapest fuel for illumination since its development in the 1830s. After the Civil War, new petroleum products captured the turpentine market and made inroads into others. At the same time, rosin came into increasing use, the finest grades for paper size, soap, and varnishes, the medium grades for yellow soap, medicine, and wax, and the crudest forms for ship and brewer's pitch and lubrication. Indeed, rosin became so valuable about the turn of the century that many beds of the discarded substance were excavated. As one expert put it, "The rosin tail began to wag the turpentine dog."[57]

Second, there was a national awakening, thanks largely to government conservation efforts, to the growing problem of deforestation. On the federal level, those efforts had begun in the 1890s and blossomed during the presidency of Theodore Roosevelt (1901–1909), an ardent outdoorsman and conservationist. In 1905, Congress transferred responsibility for forest management from the Department of Interior to the more vigilant Department of Agriculture, creating at the same time the U.S. Forest Service.[58]

Changing federal policy had a direct impact on the naval stores industry. In 1892, for example, a USDA study of the naval stores industry issued a harsh indictment of the "wasteful and careless manner" in which turpentine harvesters conducted their work. Improved practices were called for: better precautions against fire (the leading cause of tree damage), smaller and shallower cuts, and the replacement of boxes with collection cups. The government study also described new methods of steam-distilling naval stores products from waste wood, which, if successful, would ensure that "this industry is capable of the widest extension."[59]

Such government warnings, and the realilty of the nation's rapidly dwindling forests, had two important effects on the naval stores industry, both foreshadowed in the 1892 USDA report. First, some improvements were made in methods of collecting gum dip.[60] Second, important new advances were made in the distillation of pine stumps and waste wood. New techniques that used a combination of steam and solvents had the added benefit of recovering not only turpentine but rosin and pine oil. Pine oil had some limited applications as a disinfectant and as a medium of flotation in the separation of ores for the mining industry.

In the early twentieth century, two companies emerged as the leading purveyors of this new technology: the Yaryan Rosin & Turpentine Company and the Newport Rosin & Turpentine Company. The rise of those companies marked the emergence of wood naval stores as a viable industry segment, one based on a growing source of raw materials. Later, when Hercules sought to get involved in the emerging new industry by acquiring a leading steam-solvent distillery, Yaryan and Newport would attract its closest scrutiny.

The Leading Steam-Solvent Distillers: Yaryan and Newport

The founder of the largest steam-solvent naval stores distillery of the early twentieth century was Homer T. Yaryan, a hard-driving, colorful inventor-entrepreneur. He was born in 1840 in Liberty, Indiana, the son of a prominent lawyer and state senator. Yaryan attended private school, where he excelled in mathematics. He also became enthralled with chemistry and often conducted experiments in a makeshift laboratory at home. At eighteen, he took a job arranged by his father with a Cincinnati chemical manufacturer. After a few years at that position, Yaryan began a career of varied activities that demonstrated not only the advantages of his privileged upbringing but also his special degree of creativity and initiative.[61]

In 1866, Yaryan joined his brother in Nashville to organize an oil-drilling operation, where he developed a method of refining unwanted sulphur compounds from oil, a process later used throughout Canada. After a drastic decline in crude oil prices, Yaryan later recalled, "all we had invested was lost. . . . I was now broke and out of a job." He then took a position with the Internal Revenue Service, rising to chief of secret service by 1876, but two years later he returned to his foremost interest, finding new ways to process raw materials. After designing and constructing a linseed-extracting mill in Richmond, Indiana, he licensed and built mills based on the same process in nine other cities. In 1883, he invented the Yaryan Multiple Effect Evaporator (awarded the Scott Medal of the Franklin Institute in 1886) and sold the foreign rights. In the 1890s, he built several plants based on the new process: "heating with hot water from a central station."[62]

Yaryan's next innovation grew directly out of his extensive experience in extracting and refining. In 1907, he learned of an experimental process to extract tar and turpentine from pine stumps in Michigan. Yaryan tested a sample of the wood and discovered that it contained high concentrations of rosin. Applying the expertise he had acquired from the linseed processing, he developed a steam and gasoline distillation technique (see Chapter 6) that yielded turpentine, rosin, and pine oil from tree stumps; he subsequently took out the first of several patents on the process.[63]

Convinced of the commercial possibilities of the new distillation method, Yaryan invested $25,000 of his own capital and raised an additional $20,000 from outside sources to construct a test plant in Toledo, Ohio. Troubles plagued the experimental operation from the start. When the cost of pulling stumps proved prohibitive, Yaryan ordered a shipment of fat pine from the South. The blades of the stump shredder dulled almost immediately, while fires in the rosin extractors were a common occurrence. But the most serious difficulty was that Yaryan's rosin remained inferior in color, purity, and odor to gum rosin.

"An energetic optimist" who "smoked long, black cigars incessantly and swore like a stevedore," Yaryan remained undaunted. Improving his distillation process to the point where it could be applied commercially on a large scale, he licensed its rights to manufacturers in Cadillac, Michigan. The venture failed in short order, however, owing to "the adverse ratio of operating costs to prevailing product prices."[64]

Still, Yaryan believed that with better access to Southern pine stands and improved stump-pulling methods, the chances for success would be greatly improved. Rejecting one startup offer from a group of Philadelphia investors, Yaryan accepted $120,000 in seed capital from a Toledo stockbroker named J.K. Secor and erected a one-hundred-ton plant on the Mississippi Gulf coast near Gulfport in 1909. Despite construction cost overruns, the plant turned out approximately fourteen thousand barrels of rosin, two thousand of turpentine, and seven hundred of pine oil in that year. Two years later, Yaryan nearly doubled the capacity of the Gulfport plant.[65]

Yaryan completed a second plant near Brunswick, Georgia, at the beginning of 1912. It was designed to handle three hundred tons of wood per day, but capacity was doubled in its first year of operation, making it the largest naval stores distillation facility in the nation. During the 1912–1913 season, the two plants together produced nearly one hundred thousand barrels of rosin, fifteen thousand barrels of turpentine (about 2 and 4 percent of total U.S. output, respectively), and more pine oil than was worth recovering, given its marketability.[66]

In 1912, a second large steam-solvent distiller appeared on the scene. The Bay Minette, Alabama, solvent plant of the Newport Rosin & Turpentine Company was one of many enterprises created by the Schlesingers of Milwaukee, said to be "energetic and competent" businessmen who "stand well in financial circles." The plant was part of a small empire of chemically related businesses built by Ferdinand J. Schlesinger and his sons.[67]

The Schlesingers had entered the naval stores business "in a huff" because they believed that brokers were overcharging them for the rosin required by their paper size factory in Milwaukee. Securing the rights to a steam-solvent distillation process developed by W.B. Harper, the Schlesingers had the plant operating profitably by 1915, when they considered expansion by purchasing Yaryan's business. Instead, for reasons unknown, they erected a second (150-ton) plant at Pensacola, Florida, which also came to be regarded in the naval stores industry as "a model, both in construction and operation." Ironically, the Schlesingers' strategy of vertical integration never worked out; Newport's rosin, like Yaryan's, was too dark for paper size and could only be sold to manufacturers of kraft paper.[68]

During market slumps, such as that before World War I, the Schlesingers' financial clout permitted them to maintain production and stockpile rosin and turpentine until prices rebounded.[69] Yaryan was less secure. In 1913, the American Naval Stores Company, a powerful marketing company that controlled three-quarters of American naval stores exports, was shuttered as a result of a federal antitrust suit. Its closing brought chaos to an already disorganized industry and proved enough to push the Yaryan companies into bankruptcy. But after a court-ordered sale in 1914, Yaryan raised new capital and reorganized the failed company as the Yaryan Rosin & Turpentine Company.[70]

Industry Structure during the First World War

The war years were the final phase in the evolution of the American naval stores industry before Hercules entered the business. The period brought major disruptions in supply and demand, as well as new lessons about the needs and future of the industry. For the IRD, new issues mainly concerned prices. The Hercules analysts believed that concentration of production in naval stores would help stabilize prices.

For centuries, naval stores production had been carried out by hundreds, perhaps thousands, of small producers. Barriers to entry were low, and there was virtually no coordination of production with demand. As a result, radical fluctuations in price were common, as were periods of over- or underproduction.[71]

More than half of all rosin and turpentine was exported before 1914. With the outbreak of World War I, however, the export market was suddenly curtailed, raising the specter of excess capacity and plummeting prices in the United States. In September 1914, the Turpentine Farmers Association was formed in an attempt to boost prices, but the organization was short-lived.[72] Its successor, the Turpentine & Rosin Producers Association, formed in early 1917, was more successful at buttressing prices and establishing product standards.

War conditions forced down production levels by draining Southern manpower and driving thousands of naval stores factors (who borrowed against future harvests) out of business. By Armistice Day, rosin and turpentine were being produced at roughly half their prewar levels—even though they were coming into increasing use in the manufacture of soap, paint and varnish, paper size, and linoleum—as the prices of other commodities skyrocketed.[73]

The result was striking, yet predictable: whereas in August 1914, turpentine had been quoted at about 45¢ per gallon and various grades of rosin sold for between $4 and $6.75 a barrel, by the end of 1918, gum turpentine and steam-distilled wood turpentine were quoted at 71¢ and 65¢ per gallon, respectively, and rosin had quadrupled, commanding from $14.20 to $18 a barrel.[74]

Hercules and Naval Stores before the IRD

Hercules first became directly involved in the naval stores industry several months before the creation of the IRD, albeit for different reasons and on a much smaller scale than it would after the war. The impetus behind that move came from two sources: the USDA's struggle to aid Southern agriculture, and the Hercules Sales Department's campaign to increase sales of explosives for agricultural purposes.

On January 1, 1918, W.R. James was transferred to the Home Office, a move probably intended to give new life and focus to his efforts in agricultural explosives. James wrote to government officials in Washington, asking how Hercules might assist in the war effort and requesting signed endorsements of

Hercules products "to use in our advertisements." He received replies from an impressive roster of government officials, including the director of the Bureau of Mines, the secretary of the interior, the assistant secretary of agriculture, and an official at the Food and Fuel Administration.[75]

James also heard from Bradford Knapp, director of the USDA's Office of Extension Work in the South and chief of the States Relations Service (the department in charge of the county agricultural agents), as well as from Knapp's assistant, J.A. Evans, who wrote:

> There's one thing you might do to cooperate with the Department of Agriculture. I recently attended a Cut-over Land Conference in New Orleans, at which conference the belief was expressed several times that long leaf pine stumps contained valuable resinous matter which can be recovered. If you people can develop a satisfactory process for doing this on a small scale as to offset the cost of land clearing, you will aid materially in solving a problem that had been a serious one to the Department for a good many years.[76]

The challenge inspired James, who soon enlisted the support of Norman Rood in an experimental project. The plan that took shape was for Hercules to design and sell small, portable destructive distillation plants to farmers, who would use the profits from the sale of naval stores products distilled from tree stumps to partially offset the exorbitant cost of clearing land. If successful, not only would sales of Hercules dynamite to farmers and landowners increase, but Southern agriculture would benefit as well. Dynamite seemed ideally suited to the removal of longleaf yellow pine stumps because their deep tap roots decay slowly (thanks to heavy concentrations of oleoresins), making them difficult to remove.[77]

In the spring of 1918, James took an extended sales tour of the Midwest, then headed for Georgia and Louisiana, where he began inspecting naval stores plants. By summer, he was searching for a chemical engineer with the expertise to design and operate a prototype plant so that Hercules could try its hand in the business. It was a difficult task, given the wartime drain on high-level technicians, but the search soon paid off. James secured the services of W.B. Harper, developer of the steam-solvent distillation process the Schlesingers had licensed to enter the business. Harper signed a one-year contract with Hercules on September 16, and the two men began scouting the region for a site on which to construct a pilot plant "as speedily as possible." In October, they chose a spot outside of Gulfport, Mississippi, near the Yaryan plant.[78]

Scaling up

It was at this point that the IRD was created to investigate new directions for Hercules. During the next several months, the department gathered information about the naval stores industry from two important sources beyond its own research: the Hercules Sales Department, especially W.R. James, and the USDA's Southern extension office.

In the first weeks of the IRD's existence, agricultural sales was the center of attention in the Sales Department, which was building its organization to exploit that area. Through James's contacts, Harper and other industry experts spoke to the department and supplied a steady stream of information about naval stores. As the weeks passed, E.I. La Beaume, the company's advertising manager, continued to funnel news from James into the IRD.

The IRD watched the Gulfport experiment with great interest. The plant was operated from the spring until the fall of 1919 (when it was sold to a local wood products company). It produced fuel, charcoal and mixed tar oils; the latter were used in the manufacture of paper, soap, paint, shoe polish, and disinfectants and for mining flotation.[79]

The Sales Department's strategy of selling dynamite with destructive distillation plants ultimately proved to be a dubious success. Only five plants (with a total capacity of thirty-one and a half tons) were completed by the end of 1920, when the postwar depression struck the Southern economy with a vengeance and halted further development. By then, only two-thirds of the year's sales quota of five million pounds of agricultural dynamite had been reached.[80]

Indeed, the experiment was probably more valuable to the IRD than to the Sales Department. The facility had been designed to yield important operating knowledge about the naval stores business through the analysis of inputs and outputs. The experiment allowed several Hercules men, including visiting members of the IRD, to become "thoroughly familiar with the manufacturing process."

The central lesson to emerge from the experiment was that the small daily capacity of the prototype (one ton of stump wood a day) made it "too small to be operated economically," even though it would be transportable. It was estimated, however, that a three-ton plant costing $6,000 would produce a profit of about $50 per acre on average land, and twice as much on very good land. The Sales Department calculated with great enthusiasm that a dozen such plants would generate $150,000 worth of new dynamite sales per year. The IRD, on the other hand, was interested in a nonexplosives business. It no doubt found this practical demonstration of economies of scale (profits per unit would increase with size because operating costs, especially labor, could be held relatively constant) an inducement to think about the benefits of a large-scale naval stores operation.[81]

The USDA's Role

The government's role in Hercules' decision to enter the naval stores industry ultimately extended well beyond J.A. Evans's initial suggestion. To understand the willingness and ability of the USDA's extension service to become involved in this decision-making process, it is necessary to review important changes within the USDA and in Southern agriculture before and during World War I.

The Department of Agriculture grew significantly in size and importance during the first two decades of the twentieth century. Its growth was attributable not only to expanded regulatory and scientific functions but also to the addition of farm management and demonstration work. Prompted by emergency efforts to deal with the boll weevil threat in the South, the first county agents began work in 1904, drawing support from a variety of public and private sources.[82] In 1914, the Smith-Lever Act formalized and stabilized the extension program; county agent responsibilities and support were divided among the USDA, state agricultural colleges, and counties.[83] During the war, the number of extension agents increased dramatically, especially in the South.[84]

The preeminent challenge for the Southern extension office during the late nineteenth and early twentieth centuries was to balance the often conflicting interests of the region's two leading industries, agriculture and forestry. By the early twentieth century, as Southern forests dwindled rapidly, lumber companies with large holdings sought to sell off their lands to farmers and land companies for crops, but with poor results. In Mississippi, writes one historian,

> the widespread movement during the years 1909 to 1915 to convert thin soil pinelands into farms was a total failure. . . . Clearing the land of stumps involved backbreaking toil and failure. The few acres that were cleared rarely produced crops that returned the cost of production. Weary, disillusioned, and bankrupt the [Midwestern farmers who had been enticed to the area by lumber companies] returned to the North.[85]

Such was the "serious problem" Evans referred to in his letter to James. With its extensive network of agents, as well as through contacts with other USDA offices, the Southern extension office assisted the Hercules Sales Department in a variety of ways as it tried to solve the problem. For example, two Ph.D. chemists from the USDA's Bureau of Chemistry helped design, construct, and operate the Gulfport pilot plant. After the plant was sold, James was invited to deliver a paper on the project at the December convention of Southern agricultural extension directors, after which Hercules sales agents at the Chattanooga office received requests for several additional units.[86]

Similarly, the Southern extension office maintained an important presence, directly and indirectly, in the IRD. At the critical first meeting, James explained that the government had long been tracking the naval stores industry, which it believed to be "the most undeveloped industry in the country today." Throughout the IRD's investigations, the USDA remained an influential source of information, both published and unpublished, about the industry.[87]

The USDA's most important input concerned the central question of Southern pine forest depletion. At the end of the war, the Senate, seeking to assess the condition of the nation's timber resources, passed a resolution that directed the secretary of agriculture to prepare an extensive report on the subject. The result, often referred to as the Capper Report (named after Kansas senator Arthur Capper, chairman of a Senate committee on agriculture and a leading farmers' advocate), was released on June 1, 1920.[88]

To say the conclusions of the USDA report were dismal would be to risk understatement: the tone of the Capper Report was one of alarm and imminent crisis. Among its major findings were that three-fifths of America's timber resources were gone and only drastic measures could halt the devastation.[89] The report made some discouraging predictions about the naval stores industry:

> So pronounced is the depletion of the timber upon which our naval-stores industry depends for its supplies that it is commonly regarded as a dying industry in the United States. . . . The production of naval stores in the Southern pine belt will within ten years have been reduced to such an extent that export markets and even our own must look elsewhere for their main supplies. . . . What was once the largest and finest naval-stores forest in existence is about to become a matter of history.[90]

Although the Capper Report was released a few months after the IRD concluded its work, the general thrust of the report's findings was no secret among experts in Southern agriculture, forestry, and naval stores, including the members of the IRD. Indeed, the Capper study "claimed" to be based on "the opinions and estimates of the best-informed men in the industry, men representing every part of the territory and having more than ordinary means of information."[91]

To be sure, through its own investigations, the IRD had heard conflicting views on the topic. At least one leading extension agent disagreed outright with the USDA's dire forecasts, and others expressed the view that although high labor costs, the rising cost of timber leases, and deforestation would continue to inhibit growth, the gum spirits industry was likely to remain important for decades to come.[92]

Much of the debate centered on the prospects for second-growth pine. While motoring through Georgia and northern Florida in the summer of 1919, Rood himself observed that despite widespread damage from fire, wind, and razorback hogs, "turpentine operations were being carried on with at least 50% of the second growth trees." Many industry authorities shared the view of one leading gum producer who believed that "reforestation could not be depended upon to supply the demand for naval stores in the future, though by this means the life of the gum spirit industry at reduced production might be prolonged indefinitely."[93]

Still, the overriding consensus of opinion was that each yellow pine that fell brought the old gum spirits regime closer to extinction and supplied the purveyors of the new distilling technology another inexpensive source of raw material in the form of a tree stump.

The Problem of Entry

Encouraged by its Sales Department and the USDA, Hercules was moving closer to a decision to diversify into the naval stores industry. By the end of

1919, the IRD had compiled and analyzed an impressive corpus of information. Hundreds of detailed reports, varying a great deal in scope and importance, had been filed. Some were brief and informational—abstracts of industry trade journal articles, copies of credit reports and letters, summaries of interviews. Others included in-depth analysis of the history, operations, and costs of potential takeover candidates, with recommendations about how to approach owners, improve processes, and so forth.

Added together, the reports painted a portrait of overwhelming opportunity. The most compelling single factor was the shift from gum to wood distillation at a time when demand for naval stores products was rising through new domestic uses and the resumption of exports. Moreover, by becoming a leading producer, Hercules would realize important economies of scale in production and could help stabilize the market. With the return of normal conditions, abnormally high wartime prices would drop, although never to the levels of before the war, when the industry was in disarray. Leavitt Bent expressed the prevailing view:

> You could see for yourself that literally billions of old stumps were available. Furthermore, it was equally clear that not only the gum yield from the second-growth saplings was small, but also that these slender trees, already weakened by tapping, toppled over like tenpins in the winter windstorms. The set-up—abundant supplies of raw materials and diminishing competition from gum products—appeared to be just about perfection. Nobody could see but that the fancy war prices were going to last forever. Demand for rosin in both soap and paper was growing. Certainly the ridiculously low prices of 1914 would never again be quoted. Everybody agreed on that point and the Government men had bushels of statistics to prove it.[94]

By the end of 1919, Hercules was clearly prepared to enter the naval stores industry. The question was how to go about it. When the IRD was established, Hercules had intended to minimize the risk of diversification by acquiring an ongoing business. However, neither of the two leading steam-solvent distillers, Yaryan and Newport, was for sale, and few attractive alternatives were on the horizon. Therefore, Hercules began to develop plans for the construction of a new plant. In the early months of 1920, the IRD investigated both ways of entering the naval stores business—buying a plant or building its own—to determine which seemed most promising.

Hercules' main hope was to acquire Yaryan, which was known to be struggling.[95] Late in 1919, however, the IRD learned that the Glidden Paint & Varnish Company of Reading, Pennsylvania, had obtained an option on the Yaryan properties. Although IRD analysts knew that Glidden had failed to exercise other options it held on leading destructive distillation companies, the news was a setback.[96]

The unavailability of a suitable acquisition candidate led Hercules to examine an alternative entry strategy early in 1920. Since building a new facility was, for Hercules, riskier than an acquisition, the IRD hoped to engage an experienced industry expert to design and build a complete and operable plant—what today would be called a "turnkey" operation. As Dunham put it on one occasion, "In building a plant we would prefer to employ a good organization now in the business and have the plant turned over to us complete."[97]

Accordingly, in January, Dunham and several members of the IRD met with E.G. deCoriolis of Arthur D. Little, Inc. (ADL), a consulting chemical engineering firm in Cambridge, Massachusetts, with extensive experience in the naval stores industry. In 1911, Little himself had designed a machine for making paper out of Louisiana pine stumps and sawdust for the Great Southern Lumber Company; after the war, ADL selected the site, designed the apparatus, supervised construction, and even operated for four months a $300,000 steam-solvent naval stores plant at Calvert, Alabama, for the National Wood Reduction Company. And ADL was negotiating to build additional naval stores plants for other companies.[98]

At the January meeting and subsequent conferences, deCoriolis displayed a wide-ranging knowledge of the naval stores industry and gained the confidence of the Hercules representatives, who sought assurances that the completed plant would produce according to expectations. DeCoriolis, who would help with site selection, estimated plant construction costs at $400,000, including ADL's fee of $75,000.[99]

Before making a commitment, however, Hercules had to confront the potential obstacle of the Yaryan patents, which, ironically, had long been considered one of the shaky enterprise's key assets. Could the patents now be used to prevent Hercules from operating a steam-solvent naval stores plant of a similar design? DeCoriolis was confident that the Yaryan patents posed no real threat, citing as proof the fact that neither Newport nor National Wood Reduction had been challenged by Yaryan's attorneys. Later meetings with patent attorneys assured IRD leaders that no real danger existed as long as Hercules could establish the existence of either analytical studies of steam distillation antedating Yaryan's patents or processes used in other industries that were analogous to the Yaryan method.[100]

Meanwhile, much effort was expended, but little progress achieved, on the acquisition front. Apart from Yaryan and Newport (which was not for sale), the two remaining potential acquisition candidates were the National Pulp & Turpentine Company of Green Cove Springs, Florida, and the Florida Industrial Corporation (FIC) of Gainesville (see Exhibit 5.3). Initial reports about National Pulp were favorable. Rood reported: "This is a small plant. I believe, however, that the production can be increased and I believe that the quality of the rosin can be improved." He proposed that Hercules operate the plant on a 3–6-month trial basis. But that arrangement never materialized, and Hercules later declined a purchase price of $200,000. After conducting a thorough investigation of the plant, Bent and Shimer concluded that it was not only too small but also "poorly located, poorly arranged, poorly equipped and poorly organized."[101]

Florida Industrial Corporation had been organized by investors predomi-

EXHIBIT 5.3 Leading American Naval Stores Distillers, 1919

	CAPACITY (tons of wood processed per day)
DESTRUCTIVE DISTILLERS	
American Tar & Turpentine Co. (New Orleans, La.)	150–200
Florida Wood Products Co. (Slidell, La.)	100–150
Pensacola Tar & Turpentine Co. (Gull Point, Fla.)	100–150
Chatham Manufacturing Co. (Savannah, Ga.)	—
Georgia Pine Turpentine Co. of New York (Fayetteville, N.C.)	—
Georgia Pine Turpentine Co. of New York (Collins, Ga.)	—
Atlantic Turpentine & Pine Tar Co. (Savannah, Ga.)	—
STEAM- SOLVENT DISTILLERS	
Yaryan Rosin & Turpentine Co. (Gulfport, Miss.)	600
Yaryan Rosin & Turpentine Co. (Brunswick, Ga.)	200
Newport Turpentine & Rosin Co. (Pensacola, Fla.)	175
Newport Rosin & Turpentine Co. (Bay Minette, Ala.)	100
Florida Industrial Corp. (Gainesville, Fla.)	40
National Pulp & Turpentine Co. (Green Cove Springs, Fla.)	30–35
National Wood Reduction Co. (Calvert, Ala.)	—
Wood Reduction Co. (Hattiesburg, Miss.)	—
Pine Products Co. (Lake Butler, Fla.)	—
Pine Products Co. (Ellisville, Miss.)	—
Pine Products Co. (Charon, S.C.)	—
Mackey Pine Co. (Covington, La.)	—

Note: All figures are for February or August 1919, except for Newport's, which are from January 1920. Newport Bay Minette figure reported February 1919 was seventy tons.

Source: IRD Reports, February 15, 26, and 28, 1919; August 1 and 7, 1919; and January 21 and 26, 1920.

nantly from Chicago. Incorporated in 1915, it was on the verge of commencing operations when the United States entered the war and the government commandeered the plant—until the spring of 1919—to extract oil from castor beans. A year later, the little-known steam-solvent distiller attracted the attention of the IRD.

On March 1, 1920, Rood visited Gainesville and met with FIC director B.F. Williamson. Touting his own experience in the industry, Williamson claimed to have constructed plants for a large meat packer that antedated Yaryan's steam-solvent innovations. He explained further that his company had been organized in 1911 for $300,000, and that it operated at a daily capacity of sixty-five to seventy tons of wood. His figures contradicted those in an IRD report, submitted the previous month, that set those amounts at $100,000 and forty tons per day. Whatever the source of the discrepancies, Williamson impressed Rood as being "shrewd" and "evasive," despite his "pleasing personality." Williamson promised to consider a selling price, while Rood parted with the distinct impression that the offer was sure to be high.[102]

As usual when a serious takeover candidate was concerned, IRD analysts followed up initial probes with more thorough investigations. They examined the plant the following month, concluding that its design and operations were sound, and that the company would be a reasonable buy at $300,000. On the same day that Hercules engineers filed their report, however, deCoriolis submitted a strikingly different evaluation, based on his recent, albeit brief, inspection of FIC. He concluded that the plant had poor supplies of both wood and water, was undersized, and was worth no more than between $175,000 and $200,000. Although the IRD continued to compile statistics on the operation of FIC, the company was no longer a serious target for acquisition.[103]

With the demise of this prospect, the IRD's roster of viable acquisition candidates seemed to be exhausted. More than a year had elapsed since the company had begun a concerted quest for a new venture. Eager to begin its new business, the Hercules board reluctantly made its move on March 2, appropriating $500,000 for the construction of a naval stores plant.[104]

At that critical juncture, the IRD received some troubling reports. First, W.R. James, the original inspiration behind Hercules' interest in the Yaryan distillation method, suddenly changed his tune, urging that the company seriously consider destructive distillation, which he now gave "about even chances" with steam-solvent distillation. His switch was apparently prompted by the news that several destructive distillers were thriving, and that one of the largest gum producers, Consolidated Naval Stores of Jacksonville, was planning to build the largest destructive distillation plant in the nation. James was also inspired by a spate of orders for Hercules' small destructive distillation plants. "Surely," he explained, "I have a right to modify my opinions in the light of recent developments."[105]

Moreover, some scathing evaluations of the ADL-designed Calvert plant soon followed. One industry expert claimed the facility "has proven to be a failure in that it has been a disappointment in the matter of yields and quality"; another declared "it had been designed like a Waltham watch to do locomotive duty." Indeed, National Wood Reduction had broken with ADL and planned to construct another plant of a different design.[106]

Although the revelations may have troubled some members of the IRD, they were not enough to halt the department's yearlong momentum. DeCoriolis, who had been scouting Louisiana and Mississippi for a plant location that offered extensive stump lands and ample rail connections, announced on April 6 that he had chosen such a site on the Mississippi Gulf Coast shore near Hattiesburg.[107]

The Hattiesburg Chamber of Commerce was in no small way responsible for his decision. For several months, the organization had conducted an aggressive and largely successful campaign to attract industry to the area. While touting the attributes of their sixteen-thousand-resident community—four railroads, electric streetcars, and a new hotel—several members of the chamber had already met Hercules representatives when they were negotiating for the Yaryan properties.[108]

Hercules' decision also reflected the success of Rood and deCoriolis in securing a contract with the Newman Lumber Company of Hattiesburg for rights to remove longleaf pine stumps from some seventy-five thousand acres of land along the right-of-way of the local railroad it controlled, the Mississippi Central. Hercules agreed to pay the Newman Lumber company 75¢ per acre to remove stumps on its land, with a five-year purchase option. After weeks of bargaining, a contract was signed on May 3, 1920. The next day, Hercules purchased a one-hundred-acre tract in Hattiesburg for the future plant.[109]

At that point, with the die already cast, new developments concerning the Yaryan properties drew attention away from the Hattiesburg plan. Suddenly, Glidden offered to sell its option to Hercules. Soon, several members of the IRD were at the Yaryan plant in Gulfport studying its operations, while Hercules began buying Yaryan stock and installing its own directors and officers. Using the approach it had attempted with National Pulp—operating the plant for a trial period—Hercules signed a contract on May 26 to take over operation of the Yaryan plants on June 1, 1920, including an option to buy all or part of the assets before February 1, 1921.[110]

Suddenly, Hercules had made major commitments to enter the naval stores business by buying *and* building plants. The Yaryan acquisition had remained a strong first choice; that was clear from how eagerly Hercules took up the option. But even though having both projects was arguably more desirable than having only Hattiesburg, it was also arguably less desirable than having only Yaryan. Still, there was substantial latitude for action. A final decision on the Yaryan purchase was months away, and the plant would be operated and scrutinized during the trial period. Hattiesburg, in the meantime, could be scaled down or put on hold.

By the late spring of 1920, the IRD had fulfilled its mandate. The industry selection process had been sparked by the Sales Department's efforts to boost dynamite sales in agricultural markets and the USDA's search for a practical means of converting Southern stump lands into farmlands. The IRD's search for a suitable acquisition had proven more problematic. Its strong commitment to the naval stores industry resulted in the only significant deviation from the original mandate, when it decided that Hercules should build its own plant if it could not buy one. Then, through a quirk of fortune, Hercules entered the business both ways.

Conclusion

The eighteen months after Armistice Day were a time of intense effort and research at Hercules. Much was achieved during that time: The company scaled down its military business, sold off surplus supplies and least-promising assets, cut back its payroll, and positioned itself for a return to peacetime conditions in its base business. It took its first baby step outside explosives by retooling Parlin to make commercial grades of nitrocellulose. In the summer of 1920, however, Hercules was still looking for the right niche in that industry.

Hercules' headlong plunge into the naval stores industry that summer was a far more daring move—more daring, in fact, than the simultaneous diversification efforts of Du Pont and Atlas. Although Hercules had studied the naval stores industry exhaustively, and the industry appeared to satisfy the criteria for diversification defined by Dunham and other senior officers, the company had chosen to take a substantial risk. Whatever sales of dynamite Hercules' quest to clear cutover land might stimulate, naval stores had little in common with explosives: no shared facilities, no shared technologies (beyond construction engineering), no shared customers, and no shared distribution channels. Building the Hattiesburg plant and acquiring the Yaryan option were radical measures, especially when taken in tandem.

Hercules' actions immediately after the war marked another stage in the company's transformation from explosives-maker to chemical company. Its leaders were optimistic that the company was heading in the right direction. At the Atlantic City meetings, Dunham had pointed with pride to the company's record in managing the expansion of wartime business and had suggested it would make a fine precedent for Hercules' growth in the years ahead. The company had come out of the war with significantly expanded and improved facilities and a mountain of cash. Nonetheless, Dunham added that its "biggest asset" was not money or equipment, but "the organization that we have built up. . . . We have proved that this organization is capable of producing goods and transacting all the work involved in handling yearly a volume of $60,000,000 and upwards. We may be able to do the same thing in other fields."[111] Dunham believed that the company's newly acquired expertise in chemistry, engineering, and administration would complement the management abilities it had demonstrated in the explosives business since 1913. He was confident that Hercules would succeed in its new endeavors. The new industries, however, would test the company before they would reward it.

CHAPTER 6

BECOMING A CHEMICAL COMPANY, 1921–1928

In the summer of 1920, Hercules made several decisions that would shape its activities for the next decade. Sensing limits to its growth as an explosives-maker, the company identified two new businesses to pursue: commercial nitrocellulose and naval stores.

Hercules was confronting major challenges in each of its major businesses. In explosives, the market sent discouraging signals. The decade opened with the coal operators and mining industries in recession, and the long-term growth prospects for these industries was a source of increasing concern to Hercules. Would the market allow room for the many existing competitors to prosper? If not, what were the implications for Hercules of overcapacity and heated rivalry? Could the company continue to prosper as a powder company?

The questions concerning commercial nitrocellulose were similarly fundamental. Since the industry was dominated by film producers and plastics, materials, and chemical companies—most of which were larger and more experienced than Hercules—the company had to locate a defensible niche in which to grow. The plant at Parlin, moreover, had been designed to nitrate cotton and make smokeless powder. It could be adapted to make commercial grades of nitrocellulose in volume, but the necessary changes would be both extensive and expensive. Finally, although Hercules' acquisition of the plant in 1915 had represented a form of backward integration to support its smokeless powder operations, the nitrocellulose business was heavily dependent on producers of cotton linters (the residual fuzz attached to the cottonseed after the long fibers have been pulled off), which were generally considered to be the most economical source of raw cellulose. Hercules' managers, then, soon wondered whether they could succeed in the new line without establishing a secure source of cellulose.

In naval stores, the decade greeted Hercules with a series of unpleasant surprises, beginning immediately in 1920, when a national recession devastated the entire industry and forced Hercules to shutter its plants for months on end. Even with economic recovery, however, problems remained. The most disturbing was the realization that the forecasts of pine forest depletion—a view accepted by nearly everyone in 1919 and 1920, and the basis for Hercules' diversification—were completely wrong. Despite all predictions, the gum naval stores industry made a healthy comeback in the 1920s. Even worse for Hercules, the products made by its steam-solvent recovery process had significant drawbacks compared with gum naval stores products: Hercules' rosin came in a single, dark grade with limited uses, and its turpentine carried a pungent odor that painters disliked. The company was forced to reevaluate its operations and reassess its strategy for competing in naval stores.

Finally, Hercules faced the issue of how to manage its three businesses together under a single administrative umbrella. Although it was hardly growing, the explosives business was the principal source of the company's revenues and profits in the 1920s. Its leaders had gained most of their management experience in explosives; after all, Hercules still called itself a powder company. The company's original organization, designed to manage a single product line, remained in place. As the decade wore on, however, the organization was strained by the growing importance of newer lines of business. The new lines, moreover, were fundamentally different from the explosives business, in which success was less dependent on constant improvement in research and engineering than it was on the straightforward ability to lower costs. By 1928, the growth of the company's younger, more complex businesses led Hercules to question its traditional mode of organization and to consider other ways of running its operations.

Consolidation in Explosives

Hercules' strategy of diversification in 1919 and 1920 was a logical result of several factors: a substantial surplus of earnings carried over from the war; the company's sense that it would be difficult to sustain profitable growth in the explosives industry alone; its need to develop commercial applications for its wartime assets and product lines; its desire to find employment and opportunities for people who had served it so well; and its belief that its expertise in R&D, engineering, and administration would be valuable in other businesses.

Two other factors shed light on Hercules' position at the start of the 1920s, however. First, the last years of the war and the years immediately following witnessed a wave of merger and acquisition activity across the chemical industry (see the Chapter 5 discussion of the nitrocellulose industry, in which both Du Pont and Atlas used acquisitions to integrate forward into applications such as film and artificial leather). During this period, Du Pont bought several other chemical companies (including three paint and varnish producers), licensed European technology for manufacturing dyes, and made its extremely significant investments in General Motors.[1] In 1917, the Union Carbide & Carbon Chemicals Corporation was formed as a merger of three companies.

Three years later, Allied Chemical & Dye Corporation was created from the merger of five large chemical companies. Monsanto Chemical Works, which had started out as a producer of saccharin, branched out first into heavy chemicals by acquiring a large acid producer in 1918, and later into coal-tar derivatives by acquiring a half-interest in a leading British producer in 1920. In all, some 500 mergers took place in the American chemical industry in the decade after World War I.[2]

The international moves by Du Pont and Monsanto were part of a worldwide consolidation of the chemical industry at that time. Truly giant chemical combines were being formed in the major European countries. As the decade opened, for example, discussions were under way to merge the eight largest chemical producers in Germany into a single company, I.G. Farben. The deal was completed in 1926. Similar events transpired in Italy, where Montecatini was created by the merger of the leading domestic chemical producers, and in England, where the constituent companies that would form ICI at the end of the 1920s were coming together.[3]

The second factor that provides context for Hercules' actions in the 1920s was its immediate circumstances. The decade dawned with the U.S. economy in one of the worst recessions in its history. Between 1920 and 1921, wholesale prices plunged by 40 percent, and unemployment soared from 4 to nearly 12 percent. These were hard times across the nation, but particularly for the kinds of industries Hercules supplied. The mining industries virtually collapsed; their customers, steel and metal fabrication companies, also suffered heavily, as did the construction industry. Every explosives company had to retrench, but healthier producers such as Hercules were better able to weather the recession than smaller, less-capitalized competitors.

These events were the background not only for Hercules' purchase of the Yaryan properties but also for its next transaction, the purchase of the Aetna Explosives Company in the summer of 1921.

Aetna

The Aetna Explosives Company, the brainchild of Arthur J. Moxham, was incorporated in November 1914. Moxham had been a member of the executive committee at Du Pont for a decade; but he was apparently distressed by the developing feud between Pierre and Alfred du Pont, and he determined to leave the company. He chose to remain in the explosives industry, however, because he believed that the final decree had created a momentary opportunity. The industry was now so unsettled, Moxham reasoned, that a new competitor might prosper if it could establish a national presence quickly. His plan, backed by two members of the du Pont family, Ferdinand L. Belin and Charles A. Belin, was to combine the largest remaining independent makers of black powder and high explosives companies into a single, nationwide organization to compete with Du Pont, Hercules, and Atlas.[4]

The companies that Moxham brought together at the end of 1914 represented a broad range of capabilities and product lines. Aetna Powder Company of Aetna, Indiana (near Chicago), owned two high explosives plants and a black powder plant in the Midwest. Keystone-National Powder Company of Emporium, Pennsylvania, operated two dynamite plants in its home state. The Jefferson Powder Company produced dynamite and black powder at a single facility in Birmingham, Alabama. Miami Powder Company of Goes, Ohio, had been affiliated with Aetna, and its sole asset was a black powder mill in its home town (near Dayton). F.K. Brewster, Inc. of Port Ewen, New York, brought with it a blasting cap plant in Port Ewen and a mercury fulminate (the explosive used in blasting caps) facility in Prescott, Ontario. Finally, Aetna acquired a 60 percent stake in Kingsley Wood Pulp Company of Salisbury, Vermont.[5]

Aetna Explosives Company was capitalized at $7 million, and the consolidated corporation, with a 14 percent market share in high explosives and 5.6 percent of the blasting powder market, was roughly equivalent in size to Atlas Powder Company. Aetna did not succeed in establishing a presence on the West Coast, however. Although Moxham had hoped to include Giant Powder Company of San Francisco in the group, he discovered too late that Atlas had secretly bought a controlling interest in Giant in 1915. In that same year, however, Aetna bought Pluto Powder Company of Ishpeming, Michigan, a high explosives producer that served the ore mining industry of the Great Lakes region.[6]

Although Moxham had been one of the architects of the restructuring of Du Pont after 1902, he did not pursue consolidation with the same intensity at Aetna. Before the separate companies could be integrated, Aetna diverted its attention from commercial explosives to produce munitions for the Allies during the Great War. Like Hercules, Aetna grew rapidly in 1915, booking $30 million in military contracts by the end of the year. The company supplied a broad range of military explosives to the Allies, including smokeless powder, TNT, and picric acid; it also made intermediate products such as guncotton, phenol, benzine, and toluene. Many of these chemicals and explosives were produced at facilities thrown up quickly to satisfy specific contracts. Unlike Hercules, however, which used contract advances to pay for its capacity investments, Aetna chose to finance expansion by borrowing up to 75 percent of the total cost of construction. In addition, Aetna built at least one plant on speculation in 1917. The strategy backfired when the company was unable to obtain sufficient business before the war ended. Moxham retired in 1917, leaving a new management team to cope with a desperate situation. "By this time, however," Aetna's chroniclers noted, "the company had become so badly involved through overexpansion and certain disasters in the munitions business that a receivership was inevitable."[7]

From 1918 to 1921, Aetna operated under the protection of the bankruptcy laws. The court-appointed receivers maintained ordnance production until the war's end and then sold off the munitions plants. The commercial explosives plants, which had languished during the war, were upgraded under the leadership of general manager C.A. Bigelow. But these efforts were not sufficient to put Aetna on a sound footing; moreover, the company, which depended heavily on the coal industry, was hard hit by the postwar recession. The receivers became especially anxious to find a new owner for the properties.

How Hercules learned of Aetna's availability is unknown, but negotiations to acquire the company were under way with J.S. Bache & Company, investment bankers, by the fall of 1920. In November, Hercules agreed to purchase at least 80 percent of the property and assets of Aetna in exchange for 540,000 shares of its own stock.[8]

Hercules apparently had several motives for pursuing Aetna. First, Aetna's plants were located in territories not well served by Hercules. With the acquisition, Hercules would have four more dynamite plants, including one at Birmingham, Alabama, which was especially attractive to Hercules because it had no high explosives capacity in the South.[9] (See Exhibit 6.1.) Second, combination with Aetna would solidify Hercules' position as the second-largest producer in the industry. The company's share of the high explosives market would jump from roughly one-fifth to nearly one-third; its share of the black powder business would improve modestly.[10] These were important considerations for a company pursuing a marketing strategy in explosives that emphasized national distribution and advertising. In addition, the acquisition would increase Hercules' vertical scope. With Aetna's capability of making blasting caps and mercury fulminate, Hercules would no longer have to purchase these products from outside sources. Finally, the acquisition of Aetna would help stabilize competition in the industry by allowing the two companies to combine operations and remove at least some redundant capacity.

While the combination of the two companies made strategic sense to Hercules, the proposed deal reopened questions about the intent of the final decree. In particular, since the court had intended its decision to increase competition in the explosives industry, and since Hercules remained a party "to this cause and subject to the provisions of this [final] decree and bound by the injunctions [therein] granted," it was a serious question as to whether the decree would allow the proposed deal to take place.[11]

Late in 1920, the circuit court of appeals, sitting as the United States Court for the District of Delaware (which Hercules petitioned for permission to acquire Aetna), took the issue into consideration. Five months later, on May 4, 1921, the court rendered its decision in favor of Hercules. The court held that despite the prospective disappearance of an independent competitor, "control of the industry is not so much numerical as it is territorial," and that "actual competition within [Hercules' and Aetna's] respective regions will remain undiminished." Indeed, the court went on to say, "competition will be increased by thus strengthening the Hercules Powder Company in its contest for business against the du Pont Company, its strongest rival."[12]

With this decision, all that remained was to secure final approval of the deal from Aetna's owners and creditors; it was received on June 6. The actual transaction was carried out under the legal umbrella of Hercules Explosives Corporation, a wholly owned subsidiary of Hercules Powder Company created specifically to acquire Aetna. The new subsidiary was capitalized at $7.6 million (the final purchase price of Aetna), including $5.44 million in cash and the remainder in stock of Hercules Powder Company. The new subsidiary also assumed Aetna's liabilities, for which it issued bonds amounting to $3.9 million and placed an additional $1.35 million in an escrow account.[13]

EXHIBIT 6.1 **Combined Explosives Operations of Hercules and Aetna, 1921**

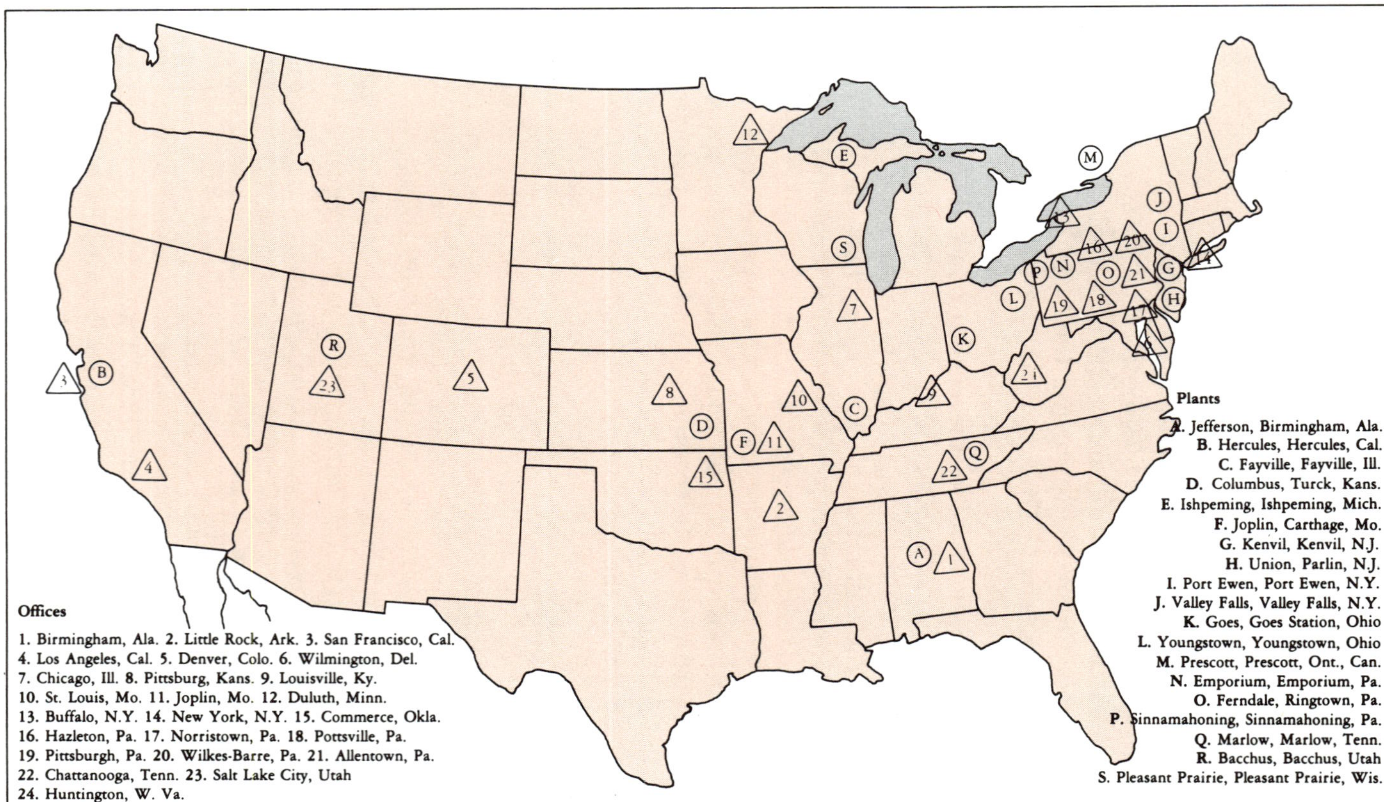

Source: Hercules Mixer 3 (January–December 1921), 10–11.

The operational issues involved in integrating Aetna's business would absorb Hercules' attention for the next several years. The increased size of the company's explosives business led to a reorganization of the Operating Department in the fall of 1921. George Markell remained general manager, and C.D. Prickett was promoted to serve as assistant general manager. Markell gained two other top assistants: Leavitt Bent, in charge of naval stores production, and John Shaw as "industrial assistant." Arthur Pine Van Gelder was designated general superintendent for high explosives, with J.J. Geer as his counterpart for black powder. The Aetna plants continued to report through a separate organization headed by Bigelow until January 1922, when Van Gelder retired and Bigelow moved up to become assistant general manager for high explosives manufacturing.[14]

Hercules also sought to rationalize operations and reduce personnel in the wake of the acquisition. Sales offices of the two companies were combined, and the number of brands was reduced sharply. In general, the old Aetna brands were phased out in favor of Hercules' products, a policy that achieved considerable savings. In February 1922, for example, Hercules and Aetna offered thirty-seven brands of permissible powders (for use in underground coal mines) alone; a year later, the total was down to twelve.[15] At the same time, Markell appointed a committee to investigate the number of salaried employees and their compensation levels at each plant and at the Home Office. Implementation of the committee's recommendations, endorsed by the board in the spring of 1922, resulted in annual savings of about $150,000.[16]

The integration of the Aetna plants proved a daunting assignment. Because it had been periodically short of funds, Aetna had not maintained most of its plants, which featured old-fashioned equipment. The Fayville plant, for example, operated water-driven mixing machines that the *Mixer* described kindly as "an old type of installation."[17]

Hercules began to take stock of its new assets by transferring managers from its own operations to the Aetna facilities. "Aetna was really a mess," recalls Leon Babcock, who moved from a position as acid supervisor at Joplin to the Emporium and Fayville dynamite plants in 1921 and 1922. On his arrival at Emporium, he discovered that the plant had no accurate records of its inventory, a problem that occupied all his time for several weeks. At Fayville, the situation was worse. An explosion had virtually destroyed the plant shortly after the acquisition, and it took nearly a year to get it restarted. Babcock does not remember Fayville fondly: "It was a horrible place to live, right on the Mississippi River. The humidity, dirt, and dirty people . . . hobos walking the track every day knocking on the back door and scaring our wives to death begging for food."[18] Albert "Slick" Ely, another young Hercules chemist, was sent to the Birmingham plant, where he witnessed firsthand the effects of chronic underinvestment: "If you ever saw a bunch of junk, that was this plant . . . [as well as] every other [Aetna plant except Port Ewen] that I heard anything about."[19]

Hercules' efforts to rationalize its new operations were made against a background of stagnant levels of sales and production throughout the explosives industry. The recession was particularly acute in the West and the Midwest:

Hercules shuttered its own Bacchus Works between May 1921 and August 1922, and the Hercules Works operated at less than 50 percent capacity. In the Midwest, the Fayville plant operated at 20 percent of capacity or less between 1922 (when it was restarted) and 1924. Hercules decided to close Fayville in 1924 and to supply its accounts from either Joplin (also running at less than half its rated capacity) or Emporium.[20]

The situation was less bleak in the South. Few Southern coal operators were unionized, and their facilities remained open when mines in the East and Midwest were closed by strikes. The iron ore and limestone mines of the Birmingham region also offered a growing market for dynamite. The Hercules naval stores camps promised to consume substantial amounts of dynamite as that business grew after 1922. The Aetna facility at Birmingham, however, was the smallest dynamite works in the corporation. At a rated capacity of nine million pounds per year, it was roughly half the size of Bacchus or Joplin. The Birmingham plant, moreover, lacked equipment for making gelatin dynamites (which were becoming increasingly popular in the 1920s), and it had no room to expand. At the start of 1924, Hercules purchased 1,280 acres in the nearby town of McAdory, Alabama, as the site of a new dynamite plant to serve the Birmingham region.[21]

Hercules' Engineering Department began building the new plant, called "Bessemer," in September 1924 on a budget of nearly $800,000. Bessemer was started up the following June, at which time the old Aetna plant at Birmingham was closed down. The new facility had capacity to make 10.4 million pounds of high explosives annually and was designed to be doubled in size, should conditions warrant an expansion. Bessemer also boasted "the most modern machinery and equipment," including plants for making nitric acid and ammonium nitrate, and it was "able to manufacture and pack every size and grade of commercial high explosives that [Hercules] sells."[22]

A Maturing Market

The nation's economic recovery in the mid-1920s led to substantial improvement in explosives production—but not in the fortunes of explosives producers. In 1923, Hercules' sales of dynamite and other high explosives reached their highest level since 1917—nearly ninety million pounds. For the next five years, however, sales remained essentially flat.[23] Hercules' problem—and that of every other explosives-maker—was a maturing market. As George Markell said succinctly in 1923, "There has been no increase at all in the country's total consumption of explosives [since 1913]."[24]

Markell overstated his case—total sales of high explosives had grown about 15 percent in the decade after Hercules' founding—but his point that the nature of competition in the industry had changed was essentially correct. Once the company had assimilated the Aetna properties, its opportunities to grow in explosives were severely limited. Hercules' principal customers were the coal operators and the copper industry, and neither prospered during the 1920s. The coal industry steadily lost sales from year to year as its customers switched to oil as their favored energy source. The copper industry fared only a little better—it did not suffer decline, but neither did it grow after 1924. Of the other major segments—quarrying, construction, and agriculture—only the construction business grew at a healthy rate in this period.[25]

A second problem for Hercules was substantial overcapacity in the industry, a situation that held prices down and squeezed margins, even as production picked up with the recovery. Writing in Hercules' annual report for 1923, Russell Dunham lamented, "There is in the country today a capacity for the production of explosives nearly double the present requirements of consumers, so that it is likely that the margin of profit will continue to decline rather than increase."[26] This was a prophetic comment: in 1926, Hercules achieved a record year for high explosives sales ($16.5 million), but its rate of return (12 percent) was the lowest in recent history (see Exhibit 6.2).

The situation with black blasting powder was even worse. Sales fell by 50 percent between 1923 and 1928 as permissibles, pellet powders, and other chemical explosives replaced black blasting powder for most applications. The decline was speeded by the U.S. Bureau of Mines' campaign to outlaw black powder in underground mines after 1924. As a result of such pressure and other factors, Hercules closed many of its powder mills in the decade after World War I: Schaghticoke in 1920, Ferndale in 1921, Goes in 1924, and the powder lines at Bacchus and Birmingham in 1924 and 1925, respectively.[27]

The lone bright spot among Hercules' explosives businesses was smokeless powder. Although production virtually ceased immediately after the war as the loading companies used up surplus stocks, the market picked up rapidly in the mid-1920s. Between 1922 and 1928, sales of Hercules' smokeless brands more than doubled as the company increased its penetration of major customers such as Winchester and Remington. Smokeless sporting powders, however, remained a small business, accounting for less than 5 percent of Hercules' overall sales and profits in 1928.[28]

In sum, Hercules ran up against the limits of its growth as a powder company during the 1920s. "Our industry has changed from one rapidly expanding and full of opportunity for improvement to one slowly developing where opportunity for improvement must be diligently sought," as Markell put it. The implications of lower growth were clear to him: "This means an entirely different outlook for our Company and all its employees. . . . Our job is now . . . the intense cultivation of every opportunity, no matter how small, whereby we can improve efficiency, lower cost, and give better service."[29]

To compete successfully in such a market, Hercules developed several creative approaches to selling explosives. In 1923, for example, it launched a clever nationwide advertising campaign that featured milestones in the construction of public works and emphasized the contributions of the explosives industry to modern civilization. The company even reprinted the advertisements in the form of a book, *Conquering the Earth,* for its customers. Hercules also launched an attractive monthly magazine, *The Explosives Engineer,* which was aimed not at promoting Hercules products but at stimulating demand. It offered advice and information about new blasting techniques and products and discussed issues of

EXHIBIT 6.2 Hercules' Sales and Profits by Department, 1922–1930

PRODUCT LINE	1922	1923	1924	1925	1926	1927	1928	1929	1930
EXPLOSIVES									
Net sales	$14,671,317	$17,388,850	$15,984,986	$15.150,399	$16,549,712	$15,815,103	$14,785,240	$16,160,525	$13,520,677
Profits	2,646,465	2,900,947	2,444,716	1.876,924	1,978,486	1,892,261	2,167,737	3,048,198	1,807,731
Return on sales	18.0%	16.7%	15.3%	12.4%	12.0%	12.0%	14.7%	18.9%	13.4%
Percentage of total sales	87.5	86.6	85.3	71.0	65.5	66.7	59.6	53.9	58.2
Percentage of total profits	103.7	108.0	103.5	56.6	45.9	58.7	53.7	57.3	77.1
CELLULOSE PRODUCTS									
Net sales	$ 818,539	$ 1,349,446	$ 1,116,312	$ 2.729,779	$ 3,796,784	$ 3,932,683	$ 6,472,444	$ 7,438,186	$ 5,996,081
Profits	16,694	46,353	168,777	705,792	1,210,362	1,097,964	1,915,124	1,958,453	1,242,895
Return on sales	2.0%	3.4%	15.1%	25.9%	31.9%	27.9%	29.6%	26.3%	20.7%
Percentage of total sales	4.9	6.7	6.0	12.8	15.0	16.6	26.1	24.8	25.8
Percentage of total profits	0.7	1.7	7.1	21.3	28.1	34.0	47.5	36.8	53.0
NAVAL STORES									
Net sales	$ 1,269,226	$ 1,332,322	$ 1,639,742	$ 3.450,564	$ 4,925,968	$ 3,985,200	$ 3,565,661	$ 4,049,766	$ 2,948,617
Profits	− 110,117	− 260,857	− 250,544	735,121	1,118,780	235,899	− 46,880	90,124	− 915,510
Return on sales	−8.7%	−19.6%	−15.3%	21.3%	22.7%	5.9%	−1.3%	2.2%	−31.0%
Percentage of total sales	7.6	6.6	8.7	16.2	19.5	16.8	14.4	13.5	12.7
Percentage of total profits	−4.3	−9.7	−10.6	22.2	26.0	7.3	−1.2	1.7	−39.1
VIRGINIA CELLULOSE									
Net sales	—	—	—	—	—	—	—	$ 3,180,831	$ 2,671,871
Profits	—	—	—	—	—	—	—	$ 225,722	$ 208,967
Return on sales	—	—	—	—	—	—	—	7.1%	7.8%
Percentage of total sales	—	—	—	—	—	—	—	10.6	11.5
Percentage of total profits	—	—	—	—	—	—	—	4.2	8.9
TOTAL SALES	$16,759,082	$20,070,619	$18,741,042	$21.330,742	$25,272,465	$23,723,987	$24,823,347	$30,008,457	$23,222,960
TOTAL PROFITS	2,553,042	2,686,443	2,362,949	3.317,837	4,307,628	3,226,124	4,035,981	5,322,497	2,344,083
RETURN ON SALES	15.2%	13.4%	12.6%	15.6%	17.0%	13.6%	16.3%	17.7%	10.1%

Source: Special Reports, 1931, Hall of Records, Hercules Incorporated.

general concern to the explosives and explosives-consuming industries, such as safety and trends in public policy. In 1925, *The Explosives Engineer* cooperated with the Bureau of Mines to offer an annual trophy to promote safety in the use of explosives. (Along with their prizes, the first winners of the trophy received a commendation from Secretary of Commerce Herbert Hoover.)[30]

To increase sales to quarries, Hercules tried branding a method of blasting. As developed by Ernest Symmes, "Hercoblasting" was a process of filling deep columns drilled into stone with black powder, which was then touched off by a fast-burning fuse. Ignited in this manner, the black powder delivered a force comparable to high explosives at 30 percent lower cost.[31]

Such tactics had little impact on Hercules' overall position in the explosives industry, which actually declined from a peak of 31.1 percent the year of the Aetna acquisition to 22.3 percent in 1928.[32] These numbers tell a story in two parts. The first is that of a mature market and tough competition from nationwide rivals such as Du Pont and Atlas, as well as from producers supplying particular geographic markets, such as Apache Powder Company in the Rocky Mountains.

The second part of the story hints at Hercules' long-term thinking about the powder business. After building the Bessemer plant, Hercules was reluctant to make major investments to grow in the explosives industry.[33] It would—and did—invest to maintain its best facilities, improve its products, and lower its costs. For example, the company increased its capacity to manufacture ammonium nitrate dynamites and gelatin dynamites when evidence mounted that customers preferred them. Hercules also committed to a new process for oxidizing nitric acid from ammonia at Kenvil in 1928.[34] But after Aetna, Hercules no longer gave serious consideration to further acquisitions in explosives—as Du Pont did, for example, when it purchased the explosives business of Grasselli Chemical Company in 1928.

At the end of the 1920s, Hercules remained the country's second-largest producer of explosives, but it was locked in a difficult, wearying struggle for market share. Prices and margins were falling in its major product lines, and the long-term outlook offered little hope of change. Fortunately, however, the company was less dependent on the powder business than it had been immediately after the war, and it could focus on better opportunities elsewhere.

NICHES IN NITROCELLULOSE

Hercules' reluctance to invest in explosives after the mid-1920s reflected the conservative financial principles its management had held since 1913. The company simply would not pour money into long-term projects that would not pay "a fair return."[35] The explosives business had generated a lot of cash, however, and Hercules used those funds to develop newer lines of business.

Hercules' approach with nitrocellulose, which it had been making since

1915, was cautious. In the immediate postwar years, Hercules was unsure of the value of its plant in Parlin. On the one hand, the plant could supply Hercules' nitrocellulose needs for smokeless sporting powders and (as the decade wore on) its requirements for nitrocellulose plasticizers used in gelatin dynamites. On the other hand, these uses would not nearly absorb the plant's potential output. The company's challenge, then, was to find and develop new markets for commercial nitrocellulose.

At the start of the 1920s, Hercules' future as a producer of nitrocellulose did not appear to be particularly bright. Despite the high growth rates that Charles Higgins had projected for the business at the Atlantic City meetings (see Chapter 5), big companies such as Eastman Kodak, Celluloid Corporation, and Du Pont already dominated the industry. To complicate matters further, a mountain of pyrocellulose smokeless powder was left over from the war. If ways could be found to transform the surplus material into commercial products—a problem that attracted many chemists—the market would be saturated, perhaps for years.[36]

At the start of the 1920s, Hercules was groping for the right path into the industry. Parlin produced at least fifteen distinct nitrocellulose products and solutions in those years.[37] The key differences among them involved the nitrogen content of the nitrocellulose and the type and viscosity of the solution in which it was dissolved. Nitrocellulose was manufactured in many grades, but there were four general categories of it, defined by the amount of nitrogen in the compound. Starting at the lowest level, grades containing 10.7 to 11.2 percent nitrogen were generally dissolved in mixtures of alcohol and camphor and found their way into celluloid plastics. At the next level (nitrogen content of 11.2 to 11.7 percent), grades could be dissolved in ether-alcohol mixtures and were used to make photographic film and fine materials such as artificial silk. Nitrocellulose with a still higher level of nitrogen (11.8 to 12.3 percent) was normally dissolved in common ester solvents such as ethyl, amyl, or butyl acetate for use in making pyroxylin lacquers and artificial leather. Finally, nitrocellulose with the highest nitrogen content (12.4 to 13.5 percent) was pyrocellulose, or guncotton. When dissolved in ether alcohol or acetone, guncotton formed the base material for smokeless powders.[38]

Parlin was capable of making the entire range of nitrocellulose grades and could also deliver them in a wide variety of viscosities. Viscosity was calculated by measuring the time it took a 5⁄16-inch ball bearing to drop through a standard column of nitrocellulose solution under standard conditions. The quicker the fall, the lower the viscosity, and the lower the viscosity, the thinner the application. For example, four-second solutions were used for lacquers, and fifteen-second solutions for films. High-viscosity solutions (from twenty to two thousand seconds) were called "dope cottons" and were used to make tough materials such as artificial leather and linoleum.[39]

Of the many types of nitrocellulose made at Parlin in the early 1920s, only the grades used in explosives were in steady demand. Even these products—which accounted for about one-third of the plant's average output of roughly two hundred thousand pounds per month during 1922—required little of Parlin's capacity, and the plant operated only during a portion of the year. Most nonexplosives products were made in small batches or experimental runs as Hercules continued its search for commercial opportunities.[40]

The plant itself was a ramshackle affair. In early 1924, Leavitt Bent lamented "the general state of decrepitude of the temporary type of construction which is characteristic of a good many of the buildings on the plant." Since Parlin had been built "as a purely war proposition," he pointed out, "all the buildings are of a very temporary nature and were designed to last, at the most, only a few years." Hercules had upgraded the plant during the war but had concentrated on improving only the nitrating house and the acid-mixing equipment. As a result, Bent noted, "this portion of the plant . . . is really the only [area] which has not necessitated very extensive and elaborate repairs on account of the character of the original type of construction."[41]

In the first years after the war, Hercules' efforts to sell nitrocellulose gave little hint of the significant business it would become. As we saw in Chapter 5, Hercules lumped nitrocellulose together with other chemical products, which were marketed through the chemical sales division of the Sales Department. In the early postwar years, the company's chemical products included not only nitrocellulose but also acids, surplus smokeless powder, TNT oils, wood pulp, and other miscellaneous products made at Kenvil and other explosives plants.[42] In 1921, total sales of all chemical products amounted to less than half a million dollars, on which Hercules lost slightly more than $75,000.[43]

That was the last year the company lost money on its chemical products, however. In 1922, sales climbed past $800,000, and the company earned a profit of nearly $17,000. Much of its success resulted from increasing sales of dope cotton to producers of artificial leather. The next year, sales reached $1.3 million, with profits of $46,000. An agreement to supply half of Ford's requirements for dope cotton, which went into upholstered seats on the Model T, was responsible for the increase.[44]

At that point, Hercules stood on the threshold of an astonishing burst of growth. Between 1924 and 1928, sales of cellulose products skyrocketed to $6.5 million; the company earned nearly $2 million in profits on the business—a return on sales of nearly 30 percent (see Exhibit 6.2). The growth of celluloid and dope cotton applications (including a particularly viscous form of dope cotton that was sold to U.S. Industrial Chemical Company as the major ingredient of Sterno canned heat) accounted for a small fraction of new sales. The sudden popularity of quick-drying lacquers and paints for mass-produced automobiles and furniture drove the major new market. The new coatings, in turn, were made possible by the development of low-viscosity nitrocellulose that was inexpensive and of high quality. And Hercules chemists and engineers played a key role in this breakthrough.

The automobile revolution that swept across America in the decade after World War I was the major stimulus to demand for low-viscosity lacquers. Between 1918 and 1929, production of automobiles in the United States soared from less than one million per year to nearly five million. A principal reason for the rapid growth was the strategy pioneered at General Motors of segmenting customers and offering "a car for every purse and purpose." To wean the public away from Ford's popular, low-priced Model Ts in the early 1920s, GM sought to offer cars in new styles and colors each year.[45] (Before the mid-1920s, mass-produced cars came only in black because black baking enamel, which could be oven-dried quickly, was by far the most efficient coating available. This fact explains Henry Ford's famous comment that his customers could have a Model T in any color they wanted—as long as it was black.)[46]

While GM was searching for ways to differentiate its low-priced cars from the Model T, the automaker's part-owner, Du Pont, was simultaneously wrestling with the problem of converting surplus smokeless powder to commercial grades of nitrocellulose. In 1922, Du Pont researchers found that treating scrap smokeless powder with sodium acetate as a hydrolyzing agent resulted in a low-viscosity nitrocellulose that could be used to make quick-drying lacquers and paints (subsequently sold under the Duco trademark). GM and Du Pont collaborated to test the new lacquer on cars built for the 1923 model year. Favorable response was nearly overwhelming: not only did customers flock to showrooms to buy the new colored models, but the fast-drying finishes, which were sprayed on car bodies as they rumbled down the assembly line, also led to significant productivity improvements and savings in labor costs in GM's factories. Within three years, GM overtook Ford as the largest automaker in the world.[47]

This milestone in automotive marketing coincided with Hercules' efforts to develop low-viscosity nitrocellulose. Under the prodding of Charles Higgins, the company pursued a different approach from Du Pont's. Rather than process scrap smokeless powder, Hercules sought to manufacture the product directly from raw materials.[48] The company's challenge, then, was not so much to create a new product as to find new ways of making it economically at high levels of quality.

The conventional approach to lowering the viscosity of nitrocellulose came as the final step in the manufacturing process: after the raw cellulose had been nitrated, purified, and dehydrated, and while it existed in solution, it was treated with certain chemicals under pressure to reduce its viscosity. (This process was essentially how Du Pont transformed scrap smokeless powder into low-viscosity nitrocellulose.) The problem with this approach—which was to bedevil Du Pont for several years—was that it tended to reintroduce impurities that could discolor or cause spots in fine lacquers and paints.

At Hercules, chemists and engineers saw a way around the problem: they focused on reducing viscosity during the manufacturing process itself. The fundamental breakthrough was to combine viscosity reduction with the purification of freshly nitrated cotton. At the same time that the Hercules process washed away acids and other impurities, it "digested" the nitrocotton into a low-viscosity form by boiling the nitrocotton in water under very high pressures at high temperatures in steel tubs lined with acid-resistant brick. By varying the boiling time and the number of repetitions of this step, Hercules was able to produce low-viscosity nitrocellulose of extremely high purity.[49]

The Hercules breakthrough came in 1924, just as demand for quick-drying lacquers was taking off. The new digesting process, developed at Parlin under the leadership of A.B. Nixon, was "the outstanding event of the year," according to Leavitt Bent, who reported the story to the board of directors. Bent went on to summarize the significance of the achievement: "This method of viscosity reduction has been of great value to us . . . and has been successful in enabling us to produce a product of such quality and at such a small increase in cost as to permit us to secure the bulk of the business for material of this character." Indeed, Hercules quickly applied for patents on the process.[50]

Hercules suddenly found itself in an enviable position as a key supplier of a high-quality product in high demand. In keeping with the company's strategy of avoiding competition with customers, Hercules chose not to take the next step and manufacture coatings itself. Rather, it sold its low-viscosity nitrocellulose to the largest paint and lacquer companies in the country: Glidden, Sherwin-Williams, Egyptian Lacquer, and even Du Pont. These companies, in turn, supplied coatings to furniture-makers, construction companies, and the automakers. Hercules also shipped substantial volumes of low-viscosity nitrocellulose to Ford, which belatedly came to the realization that it would have to copy GM's strategy of offering its models in different colors.[51]

By 1925, Parlin was operating round-the-clock, six days a week, to keep pace with demand. Hercules stepped up its investments in the facility, spending more than $1 million to replace temporary wooden structures dating from the war with modern buildings of brick and steel.[52] Capacity was increased to 1.3 million pounds per month, and the company even began to consider building a second nitrocellulose plant in the Midwest to serve the automobile industry. Hercules engineers continued to press the company's technological advantage, finding ways to increase the scale and throughput and improve the safety of operations at Parlin. In 1928, for example, M.G. Milliken developed a continuous system of nitration and digestion, a step that automated materials handling and resulted in much greater efficiency and lower levels of personal injuries.[53]

Hercules' rapid growth in cellulose products raised one nagging problem, however: its reliance on a handful of producers for cotton linters, which accounted for more than half of its raw materials cost.[54] Hercules was also growing concerned about the quality of cotton linters delivered to Parlin: the cleaner the linters before nitration, the simpler the process for digesting it into low-viscosity forms. In the mid-1920s, Charles Higgins stated the essential problem:

> Admitting the desirability and necessity of our controlling the preparation of cotton linters for our own nitrating operations, we were confronted with two alternatives—shall we build our own purifying plant and develop the necessary skill and experience in this operation requisite to the producing of satisfactory products, or shall we acquire some existing organization with facilities and experience in the preparation of this material already developed?[55]

A Secure Source of Cellulose

In February 1926, as Hercules leaders were pondering Higgins' question, the company received a letter from Philip B. Stull, president of the Virginia Cellulose Company (VCC) of Hopewell, Virginia. He proposed "some sort of deal" between the two companies, whereby Hercules would guarantee to buy a minimum volume of cotton linters at a set price, "thereby insuring a high rate of production" at VCC's plant in Hopewell.[56] At the time, Stull's company was the principal source of cotton linters for Parlin, and the letter prompted a series of discussions between the two companies that resulted in Hercules' acquisition of the VCC in the summer of 1926.

When it approached Hercules, VCC was barely two years old, and its principal asset, the plant for processing cotton linters in Hopewell, was barely ten. The plant had been built during the war by Du Pont to supply chemical cotton to the company's giant smokeless powder works nearby. After the war, however, demand for processed linters evaporated, and the purification plant was closed. Nevertheless, in 1920, Du Pont found a buyer: Stamsocott Company. Stamsocott was a joint venture involving three cotton oil processors: East St. Louis Cotton Oil Company (a subsidiary of Armour & Company, the meat-packing giant), American Cotton Oil Company, and Southern Cotton Oil Company. (The name "Stamsocott" derived from *St.* Louis, *Am*erican, *So*uthern, and *cott*on.)[57] The driving force behind the joint venture was John Walter Stull, president of East St. Louis Cotton Oil; his plan was to use linters as a substitute for rags in making fine paper.

No less an authority than Arthur D. Little had encouraged Stull at the start of this venture. Nonetheless, Stamsocott came to nought. Linters tended to clog papermaking machinery, and a poor cotton harvest in 1922 drove up cotton linter prices. At the same time, other sources of paper pulp were becoming cheaper. Late in 1923, on the verge of bankruptcy, the Stamsocott partners dissolved their agreement and sold the Hopewell plant "at a salvage price" to several young men, including Philip Stull (son of John Walter), who organized Virginia Cellulose Company.[58]

"Starting on the proverbial shoestring," as a chronicler of the venture put it, the new company commenced operations in January 1924. The younger Stull abandoned plans to supply paper companies and turned instead to producers of commercial nitrocellulose. Sales in the first year amounted to $541,000, and the company achieved earnings of about $80,000. The tremendous and growing demand for low-viscosity nitrocellulose across the country benefited VCC handsomely: in 1925, the company's sales climbed to $1.3 million, with a net profit after taxes of $180,000. VCC's major accounts included several producers of rayon, such as American Cellulose and British Celanese, but its largest customer (by a factor of more than two) was Hercules.[59]

Stull's letter of February 1926 inquiring about "some sort of deal" was inspired by rumors that Hercules was preparing to build its own cotton linter plant. VCC originally hoped to arrange a merger or partnership, but Hercules refused. As recounted by Higgins, "We told them that in our opinion, this merger proposal would not interest the Hercules Powder Company at all, but if they were interested in making another proposal which would give us control of the preparation of our raw material, we would be glad to consider it before proceeding further with our own plans for the erection of such a plant." Stull had little choice but to agree to sell VCC outright. On July 1, Hercules exchanged stock with a value of just over $718,000 for the entire stock and business of Virginia Cellulose Company.[60]

In purchasing VCC, Hercules achieved its goal of providing a secure source of cotton linters for Parlin. It also acquired a promising business in its own right. By 1928, sales of the subsidiary reached $3 million, with shipments to Parlin accounting for about 40 percent of volume. Of the remaining business, rayon producers were the largest customers. After the acquisition, VCC drew on Hercules capital to rebuild and modernize its operations, Stull asserted, "We believe we can now safely say that we have the most modern plant of its kind in the country."[61]

Chemical revolution in naval stores

Despite the Industrial Research Department's long months of exhaustive study and analysis of the naval stores industry, and despite the careful attention Hercules gave to the timing and mode of its entry, the new business proved full of unexpected—and mostly unpleasant—surprises. They became evident almost immediately upon Hercules' assumption of control of the Yaryan plants, when it took stock of what it had acquired; the surprises continued throughout the 1920s. Over time, however, their character changed as Hercules chemists, engineers, and salesmen turned misfortune into opportunity. By the end of the decade, Hercules had established a promising business in naval stores—though the business was rather different from what the company had hoped for initially.

Learning the Ropes

After taking over the Yaryan plants in the fall of 1920, Hercules retained existing plant staffs and some managers, filling other top positions from its own ranks, including the disbanded IRD. At Brunswick, for example, the former plant manager, G.C. Smith, was retained at his post because of his "remarkable showing" in 1919, while Joseph E. Lockwood, who had supervised the construction of the plant for Yaryan, was made sales manager. Former IRD members C.M. Sherwood, A.A. Shimer, and J.P. McLean were put in charge of research and development, process engineering, and wood procurement, respectively. C.A. Lambert of Hercules, meanwhile, served as resident superintendent at Hattiesburg during its construction.[62]

Several Hercules veterans transferred from other assignments would also play key roles in the struggle to make the new business a success: Albrecht H. Reu, plant chemist at Brunswick; Arthur Langmeier, head of product development and sales; and, especially, A.S. "Schubert" Kloss, who managed Brunswick (and later Georgia) operations for nearly two decades, gaining the respect of his peers as "a man of outstanding character . . . and true leadership."[63]

Leavitt Bent, now forty-three and assistant general manager of Hercules, occupied the difficult post of general manager at Brunswick until 1928, when he became general manager of all naval stores operations. Because the 1920s would prove to be the most turbulent and challenging for the company's naval stores business, Bent's "faith and vision," along with the continuing commitment of other Hercules managers and workers in naval stores, proved critical.[64]

Both of the Yaryan factories were near vast cutover lands, major ports, and railroad junctions. Brunswick, the largest naval stores plant in the world, was situated near the Sea Island coastal town of the same name, a U.S. port of entry with major shipping facilities and four railroads. Its neighboring stump lands, some seventy-eight thousand acres, were worked by 350 men in four company-operated and nine independent wood camps and were traversed by the rail lines of the Brunswick Terminal & Warehouse Company. The Gulfport plant, two miles from town (population 6,000) in the middle of the Mississippi Gulf coast, had access to the L&N Railroad and Gulf & Ship Island lines, and its wood camps regularly employed a hundred men.[65]

The manufacturing process that transformed cutover lands into wood rosin, wood turpentine, and pine oil began at the tree stump. A hole was drilled diagonally into the base of the stump and loaded with dynamite, which was detonated (from a distance). The fragments and larger pieces of wood (the largest were cut into four-foot lengths) were collected and hauled by mule wagon to railroads or tram lines for shipment to the plants.[66]

There, the wood chunks were loaded onto a conveyor that fed them into "hogs"—mechanical chippers that reduced the wood to chips approximately 2 inches long and ½–1 inch thick. Then a hammer mill, or shredder, pulverized the chips into match-sized chips, increasing their ratio of surface area to mass. Another set of conveyors fed the milled wood into "extractors"—vertically standing, cylindrical steel tanks 7–14 tons in capacity.

Now began the first part of the steam-solvent process: steaming the milled wood to remove the turpentine and most of the pine oil. The resulting mixture, known as crude turpentine, was "driven off" with live steam for several hours, then isolated from the water contained in the mixture in a separator. Finally, it was distilled repeatedly in ordinary evaporators, or pot stills, to produce commercial grades of turpentine and pine oil. Pure wood turpentine, or "P.W. Turps," was obtained by the first of several distillation cycles.[67]

The steamed chips remaining in the extractors entered the second major phase of the process: solvent treatment to recover rosin and the remaining pine oil. (In the early 1920s, gasoline was the solvent of choice because, unlike other solvents, it tended to dissolve the lighter resinous materials selectively, resulting in a more desirable product.) The steamed chips were boiled in steam-heated solvent for four or five hours before the solution (containing approximately 15 percent rosin) was drained out of the bottom of the extractors. The chips were steamed for several more hours to recover as much of the solvent as possible for recycling, although losses of 8–10 gallons of gasoline per ton of wood processed were common. Finally, after removal of the solvent and terpene components by distillation the rosin was packed in barrels for shipment, while the spent chips were "pulled" out of the bottom of the extractors with hand-held tools (a grueling task) and burned as fuel.[68]

As we will see, at virtually every stage, Hercules engineers improved the process and equipment it had acquired from Yaryan. At some stages of the operation—solvent recovery, wood retrieval and handling, boiler efficiency, extractor pressures and solvents, fractional distillation, and even barreling operations—the improvements were dramatic.

But before many of these improvements could be made, and before the true condition of either the former Yaryan plants or the new facility ADL was constructing at Hattiesburg could be tested under full operating conditions, the market had to demand Hercules naval stores products. The timing could hardly have been worse.

Crisis

Hercules prepared to make its fortune in the naval stores industry at the exact moment the agricultural sector of the economy suffered one of the worst depressions in American history. By early 1921, Hercules had idled both of its plants. Although Brunswick was restarted in September, drastic austerity measures were instituted: the plant staff was slashed to twenty men, wages were reduced, Lockwood's downtown sales office was closed, and Bent was recalled to the Home Office. For a time the plant continued to build its wood reserves, but camp operations were curtailed by the summer. (Many of the laid-off plant workers cultivated vegetables and fished local waters to ease them through the slump.)[69]

The depression began to lift in 1922, accompanied by general industrial recovery. In the naval stores industry, turpentine prices rose and naval stores producers eagerly awaited the return of prewar prices. However, rosin continued to sell at depressed rates, and overall returns to producers were discouraging. Hercules operated Brunswick, its larger plant, at reduced capacity, and Gulfport remained idle at a cost of nearly $10,000 a year for security and insurance.[70]

No longer could the postwar depression be blamed for the languishing naval stores market. At the end of 1923, Hercules directors were compelled to report to stockholders that the performance of their company's new naval stores "has again been disappointing."

> During the last several years the gum production has exceeded the consumption. Under these circumstances, although your Company has broadened the market for its Naval Stores products, and has improved manufacturing efficiency in this line, it has been unable to obtain a profit from the sale of such products at prevailing prices. This cannot go on indefinitely, but it is impossible to predict when the relation between production and consumption will adjust itself to a point that will result in fair selling prices.[71]

What caused the imbalance in the naval stores market? And why was demand for Hercules products in particular so weak? Several factors were at work: the poor condition of the company's naval stores plants; unanticipated customer resistance to the use of wood rosin and wood turpentine; product substitution by synthetic products; and the unexpected good health of the gum naval stores industry. The first two challenges were fairly apparent before Hercules entered the naval stores business and were glaringly obvious by the middle of the decade. The second two were much more difficult to anticipate and gradual in making their appearance.

The Brunswick plant was plagued with operational difficulties from the outset of Hercules' involvement. In the first year under the company's control, plant operators complained of "numerous small fires," "constantly bursting" extractor coils, decrepit boilers, and repairs "of considerable magnitude." Well into 1923, shutdowns to correct faulty equipment were common. Moreover, conditions were worse at Gulfport. Although only the smaller plant was operated for a few months during 1923 to use up wood stocks, its operating costs during that stint ran one-third higher than at Brunswick.[72]

Hercules must have expected few such problems from the new plant that ADL had designed at Hattiesburg. But its hopes were quickly shattered. A few months after the plant began operating in mid-1923, engineers gave discouraging reports: Hattiesburg was "very poorly constructed" and in need of widespread repairs.[73] Upon hearing the news, perhaps some former IRD members ruefully recalled the reports, filed just before Hercules decided to build Hattiesburg, that cited difficulties with other naval stores plants designed and constructed by ADL. Hercules not only had overpaid for the Yaryan properties (see Chapter 5) but had also paid for a new plant that was apparently second-class.

Meanwhile, customers were proving resistant to the attractions of wood rosin and turpentine. To be sure, such resistance had been prevalent since Yaryan and the Schlesingers began extracting pine stumps before the war. Hercules was aware of the issue but initially was not greatly concerned about it. In his major report to the board recommending the Yaryan purchase, George Markell had acknowledged that "wood rosin will always sell at a discount as compared with gum rosin, unless some method, mechanical or chemical, can be developed to remove the color of the wood rosin and make it in every respect comparable with the corresponding gum spirit grade"; but he thought this effort would be unnecessary. Naval stores experts at the time believed that the price differential at the lower grades was vanishing, and that new customers were thus being attracted to the wood products.[74]

The problem with this reasoning was twofold. First, experts had never had much luck predicting the capricious naval stores market, and their luck was not about to change. As prices continued to fall, gum naval stores became relatively more attractive. Second, and more important, Hercules chemists knew that wood rosin and turpentine were effectively equivalent to like grades of gum products. Customers did not. Relying on their senses, they spurned wood rosin because of its red hue and wood turpentine for its foul odor. As we will see, not until Hercules conducted extensive market research did it fully realize the strength of customer resistance to wood naval stores products.

Like most large corporations, Hercules would use modern marketing techniques to promote its wares. But if the decade of the 1920s was an age of advertising, it was also a revolutionary period for industrial chemistry. Many of the significant advances made in the area of lacquers and synthetic resins posed an acute challenge to the naval stores industry. Beginning with Du Pont's Duco in 1923, pyroxylin solutions began to supplant turpentine as a paint solvent. New synthetic organic solvents (amyl alcohol, amyl acetate, ethyl acetate, butanol, butyl acetate, and several aliphatic compounds) made inroads as lacquer solvents, while synthetic resins (oil-soluble phenolics and the alkyds) challenged rosin as an additive to varnish.[75]

Some of the new base materials, plasticizers, and solvents took the marketplace by storm (sales of butyl acetate, for example, increased one hundredfold between 1923 and 1929). Hercules, of course, benefited in its nitrocellulose business from the coatings revolution. But the company also stood to lose a great deal of revenue in naval stores. It is impossible to calculate the extent to which new lacquers and synthetic resins were substituted for traditional products such as rosin and turpentine, but it was certainly enough to "[stir] the old paint and varnish industry to the core."[76]

Still another unanticipated factor hurting Hercules' naval stores business in the 1920s was the surprising resilience of the Southern pine forests. Reforestation in the South not only continued to support the needs of the gum naval stores and lumber industries throughout the 1920s but also contributed later to a major new growth industry in the region—paper pulp. The Southern forests continued to flourish, and new growth has outraced demand ever since. Because anticipation of crisis was the single most compelling reason that Hercules entered the naval stores business, the issue of reforestation deserves special attention.

Early in the decade, as unexpected signs of healthy reforestation became evident, some industry experts began to soften their alarmist predictions. Evidence of second-growth longleaf pine forests led some to decry "the inaccuracy of the too commonly accepted idea that the longleaf pine will not reproduce itself."[77]

Then attention shifted to a new potential savior: the slash pine (*Pinus caribaea*). Known as the "heaviest, hardest, and strongest" of American pines, the slash had long thrived in marginal wetlands, where fires were rare, barely attracting the notice of turpentine farmers. Thanks to its far-floating seeds, the species began springing up with unexpected rapidity in longleaf stump lands. By mid-decade, careful observations and tests of slash pine trees and forests revealed that the slash was ideally suited as a source for both gum naval stores and lumber. According to USDA studies, slash pines grew at a much faster rate than longleaf pines and could be tapped without serious harm after only fifteen to twenty-five years. In addition, they produced more gum and had a longer running season than longleafs, and their oleoresins yielded higher quality rosin.[78] As one naval stores expert put it, the slash pine's "habit of taking over lands from which the long-leaf has been cut gives it the role of rescuer in the business of naval stores, almost murdered by the neglect and wasteful methods of former years."[79]

The Capper Report, less than a decade old, now had a hollow ring, even among Forest Service officials. In 1927, U.S. Chief Forester W.B. Greeley declared that prospects for the naval stores industry were "very encouraging. . . .

We have all quit regarding the naval stores industry and the pine lumber industry in the South as dying institutions." The following year, the American Forestry Association called the future of naval stores "assured."[80]

How had it happened? Certainly, government-business cooperation in the adoption of conservation-oriented silviculture methods played an important role. By the early 1920s, government warnings, the spread of the cup-and-gutter method, and the actions of forestry schools and associations began to "arouse broad public concern" and show results. While the postwar depression slowed the pace of lumbering in the South, both the federal and state governments set up forest stations, commissions, and boards in the South. The passage of the McSweeney–McNary Act in 1928 funded an extensive, ten-year silviculture research program and established several regional forest experiment stations. Perhaps the greatest gains were made in the area of forest fire prevention, which became a "virtual crusade" under Greeley.[81]

In addition, large industrial users of wood contributed to forest revitalization by reducing consumption. Fearful of diminishing lumber supplies and plagued by high labor costs, they invested in research in wood preservation and the use of wood substitutes such as steel, concrete, and plastics.[82] Railroads, the largest consumers of wood at the turn of the century, successfully treated ties, cars, and bridges to protect them against fire, insects, rotting, and shrinkage. On the production side, formerly inaccessible timber stands were brought within reach, and their bounty hauled more efficiently, by new construction roads and internal-combustion trucks, while new accounting and management methods also improved the efficiency of wood-handling operations.[83]

Finally, strong evidence suggests "considerable error" on the part of government officials in estimating the physical production of lumber and the extent of usable timber. Because of confusion in the late nineteenth and early twentieth centuries about the size and species of trees and the difficulties of measuring their locations, the Forest Service's predictions, each "built upon the assumptions of the previous reports," collapsed like a "house of cards." The Capper Report was perhaps the most glaring example of such error. For example, it misinterpreted wood hoarding by railroads after the turn of the century as a reflection of their fear of depletion, whereas the railroads had learned to build inventories to better season wood.[84]

A novice in the business, Hercules had followed the government's advice and was as surprised as the nation's high-ranking forestry officials when depletion proved to be a myth. The situation developing, however, was exactly opposite to what Hercules had planned for. As gum naval stores regained its footing, moreover, another question loomed over the naval stores industry. As one industry leader queried: "What are we going to do with this prospective increase in production?"[85]

Faced with these difficulties in the early 1920s, Hercules fashioned a long-term strategy to transform its naval stores business into a success. According to Leavitt Bent, the company did not attempt to boost naval stores prices by restricting output because it did not wield enough influence in the marketplace. Instead, Hercules concluded that the current dilemma was a result not only of poor market conditions but also of production and marketing difficulties. "Production capacity far exceeded the market, costs were high, and a very decided prejudice was held by the consumer against wood rosin and wood turpentine. To meet this situation," Bent recalled,

> a very definite policy was decided upon. First, to cut production commensurate with sales, and then to embark upon an extensive research program having as its objective improvement of process and quality of products and the development of new markets and new products which could be sold.[86]

The mission was notable not only because it would prove effective, but also because it represented a considerable, long-term commitment of the company's resources. The strategy had three major components: improving plant operations and wood procurement through innovations in mechanical engineering; broadening markets through sophisticated marketing; and creating and improving products through chemical research.

In each area, the greatest gains of the decade were achieved between 1925 and 1927, in part because the market was strong during that period, bringing in large revenues for new repairs and construction, marketing, and research.[87] It also simply took time for Hercules to sow and cultivate its naval stores rescue programs, and for them to bear fruit.

Engineering for Efficiency

The history of technology tends to focus on the revolutionary impact of "great inventions," such as the electric light, the telephone, the airplane, and the computer. In many key industries, however, incremental advances in technology have played an important role. "A large portion of total growth in productivity takes the form of a slow and often invisible accretion of individually small improvements in innovations," notes one scholar.[88] Indeed, the cumulative effect of such improvements to an invention often overshadows the gains in productivity brought about by the original invention itself.[89]

The engineering history of the Hercules naval stores plants in the 1920s well illustrates this concept. The basic Yaryan process was intact; the challenge was to increase the efficiency of the plants, and with it, the company's competitiveness.

At Brunswick, the challenge was placed in the hands of G.E. Ramer, a former Babcock & Wilcox mechanical engineer who joined Hercules in 1915 and was made plant manager in 1922. That year alone, to improve boiler performance, repair crews were retrained, boiler leaks were tracked relentlessly and repaired, crown sheets were replaced and tubes relined with noncorrosive metals, fires were better regulated, and new systems for water purification and mechanical feeding were installed. Elsewhere in the plant, experiments were conducted to increase the load factor in the mill room by running a single turbogenerator, a hog, and a shredder almost continuously, and small fires in the superheating stages, once a common occurrence, were prevented by cooling spent chips before pulling them.[90]

In 1925, Kloss reported that "for the first time since we started the Naval Stores industry, our plants were unable to take care of the demand for rosin and turpentine."[91] The upsurge permitted plant operators to evaluate their improvements under full-capacity conditions. With continuous production, economies of scale and engineering improvements caused unprecedented operating efficiencies. Brunswick, for example, reduced its solvent losses by one-third during 1925 and produced eleven-thousand round barrels of rosin while using eight thousand fewer tons of wood than in the previous year. Moreover, Hercules used this prosperous phase as an opportunity to improve and expand its naval stores production facilities. In 1925 and 1926, the company spent $735,633.69 on capital improvements at the two plants (more than 90 percent of it for Brunswick), and $174,271.76 on new stump leases, equipment, and roads at the naval stores wood camps (70 percent for Brunswick).[92]

Seen from the perspective of 1928, the efficiency gains at the two plants were impressive (see Exhibit 6.3). Yields of all products at both plants increased, and solvent losses and unit costs decreased substantially. The greatest gains at Brunswick were achieved during its first two years; when compared with Hattiesburg for the period from 1924 to 1928, rates of improvement at the two plants were remarkably similar.

Beyond the walls of the plants, Hercules engineers also transformed the process of wood procurement, one of the most difficult and labor-intensive stages in the manufacturing process. Advances in wood handling evolved differently at the two sites because of the indigenous trees and terrain conditions at the Georgia and Mississippi wood camps. The central goal throughout the decade, especially as labor costs climbed in the mid-1920s, was "to replace men with machines."[93] Doing so mainly involved the replacement of stump blasting with a variety of mechanical stump-pulling devices—an ironic turn of events, given Hercules' original interest in the naval stores industry. Apart from being cheaper and faster, pulling had the added advantage of retrieving the entire taproot, which was rich in oleoresins.[94]

As early as 1922, tractors were used to pry blasted stumps out of the ground. By the middle of the decade, Brunswick engineers were testing "almost every kind of stump pulling device" and were experimenting with several of their own design. Some of the devices were ingenious, if impractical. For a time, stumps were pulled by a cable and motor anchored to a larger tree, but this method often severed the top of the target stump. Later, an A-frame cable apparatus was used in conjunction with crawler-type tractors, but the cable tended to slip off the stump and was cumbersome to move through the cutover lands.[95]

By 1927, Hercules had settled on a giant stump excavator designed by Phil Powelson: a forty-five-ton crawler-type bulldozer outfitted with a fifty-foot boom that suspended a "nutcracker" dragline. This apparatus grasped the stump, wrenched it from the ground, and swung it to one side so that the lateral roots could be trimmed. At an average cost of $17,000 apiece, the new dragline

EXHIBIT 6.3 Trends in Output and Efficiency at the Hercules Naval Stores Plants, 1922–1928

	BRUNSWICK				HATTIESBURG		
	1922	*1928*	*Change*	*Change since 1924*	*1924*	*1928*	*Change*
Rosin production	81,829	156,145	91%	152%	29,154	76,885	264%
Turp. production	581,267	1,012,303	74	120	352,889	691,245	196
Pine oil production	386,758	1,158,074	299	178	140,997	430,105	305
Rosin yield	0.593	0.859	145	113	0.771	0.888	115
Turp. yield	4.53	5.64	125	108	8.91	7.99	112
Pine oil yield	3.39	6.40	189	136	6.56	4.97	132
Solvent loss	5.11	3.70	138	138	5.86	4.13	142
Cost per unit	3.43	2.71	127	103	3.26	2.56	127

Note: Rosin production is measured in round barrels, turpentine and pine oil production in gallons; yields, solvent losses, and costs per unit are per ton of wood processed.

Source: Operating Department, Annual Reports for 1922 to 1927, and Naval Stores Department, Annual Report for 1928.

excavators were hardly an inconsequential expense, but their long-term cost-saving potential was obvious. Between late 1927 and late 1928, Hercules bought six machines for the Brunswick camps.[96]

Hattiesburg's stump lands—hilly and with older, closely cut stumps—thwarted the use of the excavators. There, too, engineers experimented with some interesting pulling devices. In 1926, it looked as if "an entirely new method of harvesting wood" had been developed: a pile wedge was used in a hammerlike fashion to split stumps into pieces below the surface. Throughout the decade, however, Hattiesburg's wood camps continued to rely on a combination of dynamite and small tractors.[97]

In the 1920s, unit production costs at the wood camps rose and fell largely in response to the changing labor market. In 1928, the excavators were just beginning to affect this dependence on the labor market and would clearly demonstrate their value in subsequent years.

Marketing

Before Hercules fashioned a comprehensive marketing plan, it faced sensitive issues regarding its reputation. For one thing, although it was uneconomical to run the plants in the postwar slump, the company hoped to compete in naval stores by emphasizing the dependability and reliability of its products, especially compared with those of small-scale gum producers. But to ensure a dependable supply of products and to maintain its reputation, the company resumed production as soon as supplies to existing customers were jeopardized. Lockwood was not speaking rhetorically when he explained in the *Naval Stores Review,* "We are opening our plant solely to provide a continuous dependable supply to customers regardless of profit or loss at any particular time."[98] In the same vein, the company did not switch the brand name of its naval stores products from Yaryan to Hercules for several months until it could ensure their quality.[99]

Hercules developed a somewhat different approach for marketing each of its three major commodities. To promote the use of rosin, the company drew on its considerable experience in the marketing of explosives and quickly established a strong technical service program. Just as explosives specialists consulted with industrial explosives users on product selection, use, and safety, Hercules naval stores representatives worked closely with manufacturers of paper, linoleum, polishes, printing inks, and the like and invited customers "to present their problems for the solution of highly skilled experts."[100]

Turpentine, meanwhile, presented a puzzling marketing predicament.[101] Chemical tests showed that Hercules wood turpentine was equal or superior to gum turpentine when used in the manufacture of paints and varnishes, its biggest market. Yet the product moved slowly, and feedback from distributors, jobbers, manufacturers, and salesmen alike was unfavorable. Although the plants were running below capacity, 11 percent of the product weight and one-quarter of the product value of every ton of wood the plants processed was composed of turpentine. Stocks were piling up, threatening to force the curtailment of rosin and pine oil production. As a result, Hercules decided in 1924 that a marketing strategy "entirely new in the national distribution of turpentine" was needed.

The first step was to investigate the distribution channels, users, and applications of the product. Hercules worked with the National Industrial Advertisers Association, "several government bureaus," and a leading manufacturer of mineral spirits. The findings were enlightening, even surprising. The majority of bulk turpentine sold in the United States went not to manufacturers of paint and varnish, as previously thought; rather, large distributors passed these products on directly to retailers and property owners. Among retailers, moreover, turpentine was more likely to be sold in hardware than paint stores.

A direct-mail survey to thousands of end-users produced another revelation: many painters shunned wood turpentine because its odor was "different" from that of traditional gum varieties; indeed, it was often confused with inferior mineral substitutes such as commercial spirits and terebenthene, which had long plagued the market. "Consumer ignorance and prejudice" were "blocking the ultimate outlet" for Hercules wood turpentine.

The marketing strategy that evolved from the findings had three major components: new packaging, new advertising, and new channels of distribution. So comprehensive and successful was the marketing plan that, as documented by N.S. Greensfelder, the company's advertising manager, and J.G. Pollard, Jr., a market research expert at the Harvard Business School, it became the first in a series of case studies published by the National Industrial Advertisers Association.

The unifying theme of the program was that "an intelligent marketing plan must begin with the customer's point of view." Consider the experience, as Hercules marketers did, of a typical turpentine customer before 1924. After purchasing several cans of paint at a local hardware store, he orders turpentine as a thinner. The clerk rummages around for an empty container, finds one, perhaps without a lid, and fills it from a vat. By the time the transaction has been made, the hands and clothes of both proprietor and consumer have been soiled.

The customer has also purchased a product of dubious quality. According to the Bureau of Chemistry, in the early 1920s, roughly one-quarter of all U.S. turpentine was adulterated with mineral spirits (and more than half of all rosin was misgraded). "Against these practices," one official remarked, "neither the producer nor the smaller ultimate consumer has the least redress." To safeguard the integrity of their products and to drive out low-cost competition from substitutes, in 1922 naval stores industry interests submitted a bill that would "prevent fraud and deception" and that called for the establishment of federal product standards. The Naval Stores Act, passed the following year, established product grades and fines and prohibited the use of misleading terms such as "turpentine substitute" and "paint spirits."[102]

However, several weaknesses in the law made for passive and ineffective enforcement. No provision was made for the regular inspection of rosin; there were only two agents authorized for the task nationwide, and customers had to submit samples at their own expense. Turpentine producers, meanwhile, were not required to label their products according to government standards. Active enforcement of the law did not begin until 1926 (the bureau was occupied in developing tests for adulteration), and it was not until the end of the decade that compliance improved significantly.[103]

Hercules started offering its turpentine in attractively lithographed five-gallon, one-gallon, and one-quart cans. Shipping crates for the five-gallon size were convertible into tilting dispensers. Emblazoned on the front of each can was the Hercules name (a familiar sign of quality to millions of powder customers), as well as a guarantee that the container's contents complied with the standards established by the U.S. Naval Stores Act of 1923.

The key messages—convenience and reliability—were reinforced by an extensive national advertising campaign. Again, paint and varnish jobbers and hardware and paint retailers, rather than manufacturers, were targeted. In addition to advertising in the trade magazines of these constituencies, Hercules launched a massive direct-mail campaign. Tens of thousands of retailers and jobber salesmen received copies of the "Hercules Guarantee," which explained how Hercules turpentine was manufactured and pointed out its selling advantages, including better quality and uniformity resulting from advanced distillation facilities. To enliven its advertising, Hercules hired a cartoonist. He created two whimsical, folksy-talking painters named "Turp and Tine," who became a regular feature in Hercules ads. When the company expanded its advertising into film, *The Doings of Turp and Tine* became a sales orientation vehicle for distributors throughout the United States and as far away as Japan.[104]

Finally, Hercules revolutionized its turpentine marketing through improved distribution. As noted, the company had shifted its emphasis to end-users in an effort to "pull" its turpentine through the distribution chain. However, because there were fifty thousand hardware and paint and varnish retailers, who typically placed small orders, Hercules concentrated on performing "considerable missionary work" rather than attempting direct sales. The company's extensive network of explosives and sporting powder salesmen performed that work.

Relations with distributors were also strengthened. Jobber margins, once slim and unpredictable, were increased from between 2¢ and 4¢ to a standard 10¢ for drums and 15¢ for cans, and weekly selling prices were set, instead of being allowed to fluctuate as often as daily. With these policies and the addition of several new naval stores intermediaries, Hercules was able to build its distributor network to more than six hundred by the end of the decade.

The turpentine campaign was costly—direct advertising costs for the product jumped from $4,000 in 1923 to $35,000 in 1927—but its results were dramatic. Ten months after the campaign began, turpentine stocks at the plants had dwindled to almost nothing, prompting new production by 1926. To be sure, much of the demand was a result of economic recovery. However, the positive effects of the company's direct efforts to build brand loyalty among consumers could be seen in the marked increase in domestic noncarload orders, which soon surpassed previous sales worldwide. Consumer awakening to the consistency and reliability of Hercules turpentine was also reflected in a shrinking price differential between wood and gum turpentine. With a comprehensive marketing strategy, Hercules had succeeded in transforming turpentine from a commodity into a national brand-name product.

The challenge of marketing pine oil was overcoming its general unfamiliarity to potential consumers. Unlike rosin and turpentine, it was not a gum naval stores product with a long history in the marketplace. Hercules made progress on two fronts. First, consumers already using the products were introduced to new applications. For many years, mining operators had relied on pine oil as a frothing reagent in the separation of ores by flotation. By the end of the decade, product applications had grown more sophisticated, "from the original acid circuit oil flotation to the present highly developed flotation in which the various minerals are separated into proper groups for smelting." Similarly, by 1928, makers of disinfectants had learned that nontoxic, nonirritating, nonstaining, sweet-smelling Hercules pine oil was effective against seventeen disease-causing bacteria, including typhoid and scarlet fever. The product also continued to sell well as an adjunct to turpentine, being used to mask odors and dissolve mineral oils.[105]

Second, Hercules pioneered entirely new catagories of consumption for pine oil in the 1920s. To identify new potential uses, Hercules elicited feedback from customers and supported research fellowships at a number of colleges.[106] From that program, several new and valuable applications emerged. As an excellent solvent, penetrant, and emulsifier, pine oil gained widespread use in the textile industry. By removing gums, waxes, grease, and dirt from cotton, wool, and natural and artificial silks and lowering the surface tension of these materials, it began replacing less effective mineral and vegetable oils in preparing fabrics for dyeing. These properties, along with its disinfectant and deodorant qualities, made pine oil a valuable ingredient in laundry, cleaning, and scouring soaps and in factory deodorants. It could also help reclaim rubber (to its prevulcanized state) from old tires and other sources.[107]

The middle of the decade was a heyday for marketing. Advertising surged when the budget was increased from an average of $15,700 for 1922–1924 to $33,000, $66,000, then leveling off to $60,000 in subsequent years. If there was a unifying theme to Hercules' naval stores promotions during the 1920s, it was the company's advantages as a highly technical, large-scale producer. Advertisements spoke of a "great manufacturing concern" that produced "Guaranteed Turpentine" and did so "Under Chemical Control." Greensfelder put it succinctly in his 1925 year-end report: "Our advertising in the various trade publications reaching consumers of our turpentine, rosin, and pine oil [has] the underlying idea of emphasizing the size and reliability of our organization and the quality and uniformity of our products."[108]

Distribution channels proliferated as well. Hercules opened a European service department, the Holland Limited Company, in 1926. By 1928, Hercules had naval stores sales offices in Chicago, St. Louis, New York, Salt Lake City, San Francisco, Birmingham, Denver, and Los Angeles, five Canadian distributors, and forty-seven product representatives distributed evenly throughout the continental United States. Never was it more convenient to purchase Hercules naval stores products.[109]

As marketers mastered the art of selling wood rosin, turpentine, and pine oil, a few began to test their skills at a new task: selling the growing array of new wood chemical products emerging from the company's research laboratories.

Dark to Pale Rosins, Turpentine to Terpenes

Research in naval stores began at the Experimental Station in 1920. The first task was to overcome the handicaps of Hercules wood rosin: its dark color and low melting point. The solution, of course, lay in the chemical composition of the substance, but no chemist fully understood it, certainly not those at Hercules. As one company researcher recalled, "We were handicapped in having no background in experimental data such as we possessed in the field of explosives."[110]

Research was also carried out at Brunswick after Hercules took over the Yaryan properties. The company was intent on improving its products through science; when the plant was shut down during the postwar depression, the research staff stayed on, erecting a "Golden rosin" plant to test fuller's earth as a purifier. A laboratory was completed at Brunswick in 1923. There, Albrecht Reu designed equipment and supervised tests, while C.M. Sherwood, then Arthur Langmeier (after Sherwood's death in the spring of 1923), headed naval stores research at Kenvil.[111]

By that time, turpentine and pine oil were receiving a good share of attention. Turpentine research had three objectives: decreasing the odor, increasing yields, and finding new uses for the product. To reach these goals, chemists again attempted to probe the liquid's chemical structure. (Turpentine as then produced was composed of hydrocarbons, known as terpenes, in the following proportions: 70 percent pinene, 25 percent dipentene, and 2 percent *para*-menthane.) In 1922, using a converted still, researchers obtained some fractions that acted as good solvents for certain grades of nitrocellulose. But in 1924, after investing in a $12,000 rectifying column to fractionate turpentine, and using alkali as a purifier, they finally succeeded in removing "the most objectionable odors originally present in our P.W. Turpentine."[112]

Pine oil also began to reveal its secrets. Knowing it to be a mixture of terpene alcohols (it was found at that time to contain 55 percent *alpha*-terpineol and 12 percent borneol), researchers sought to prepare its esters, which could have a variety of uses in coatings and other industries. In 1923, several derivatives were developed, most notably "Solvenol" solvent, a clear, water-free additive for turpentine substitutes produced "by the catalytic dehydration of pine oil and Herco."[113]

In 1924, Langmeier transferred to a sales function at Wilmington but continued to act as liaison between the research staff and the Home Office. His post at Kenvil was filled by V.R. Croswell, whose staff was increasing rapidly thanks to important technical advances and an expanding budget. By 1926, eight naval stores researchers were occupying a good portion of Kenvil's laboratories.[114]

The work was now directed toward three major goals: the purification of rosin, the fractionation of pine oil, and the development of a roller crusher. The first two goals were seen as "fundamental" developments "of such nature as to radically change the character of our Naval Stores industry." Indeed, a successful purification process promised to increase the market for rosin so greatly that it would be "the sole justification for materially increasing the capacity of our Naval Stores plants." The simultaneous fractionation of pine oil would permit markets for it to grow in proportion to the invigorated rosin demand. Finally, a successful roller crusher might not only increase plant yields but also produce spent chips that could be used by wood pulp manufacturers, thereby converting a waste by-product into a profitable raw material.[115]

If there was a growing sense of urgency and excitement about the work, it was justified. Hercules chemists had already made significant improvements to the company's basic naval stores products and had isolated several constituents with market potential. On March 6, 1926, however, came the greatest technical breakthrough of the decade: pale wood rosins.

The same day, Harry E. Kaiser and Roy S. Hancock of Hercules applied for a U.S. patent for their process of manufacturing pale wood rosin using a solution of low-grade rosin, gasoline, and—the key ingredient—furfural. Other related patent applications soon followed, two by Irvin W. Humphrey (for processes using aniline and chlorohydrin in the solvent), another by Kaiser and Hancock (again with furfural), and one by George Norman (using liquid sulfur dioxide). Altogether, Hercules was granted six sequential patents for rosin refining on May 28, 1929.[116]

The first Kaiser and Hancock patent (No. 1,715,083) was the critical one. As the inventors described it:

> Our process comprises the formation of a solution of gasoline, rosin, and furfural, its cooling or refrigeration with resultant separation of furfural and coloring bodies, and the recovery of high grade rosin from the remaining solution. The process may be carried out in a single operation, or in two or more operations.[117]

In the cooling stage, the furfural would carry most of the coloring bodies to the bottom of the vessel by means of gravity separation, leaving a solution of high-grade rosin at the top. Furfural, in the lively words of Williams Haynes, was "that war-baby chemical prepared first from corncobs and now made commercially by the Quaker Oats Company of the waste oat hulls."[118]

Early in 1927, the Research Department recommended that the "Pexite"® process, as it was called, be scaled up to plant-level production. On April 15, the directors appropriated $211,267 to erect a purification facility for Pexite® at Brunswick. The new plant was designed to refine two hundred barrels of Hercules FF-grade rosin a day. (Rosin is graded according to its color; according to a system established by the USDA, pale grades are X, WW, WG, and N; medium grades are M, K, and I; and dark or common grades are H, G, F, E, D, and B. In general, pale grades were used for paper and fine varnishes; medium grades found use in soaps and waxes; and common grades were used in pitch for ship building.) In addition to the refined rosin, the process yielded another product, Belro® resin, which was dark in color and had a low acid number, but good heat stability. For many years, Belro® resin was used in abattoirs as a depilatory for hogs. The principal product of the process, Pexite® pale rosin, was equivalent in color to I-grade gum rosin but was cleaner and more uniform and reacted more steadily under heat. By September, the Pexite® rosin plant was turning out product graded as high as N, with an impressive yield of 98 percent.[119]

Hercules announced these products in the spring of 1928. That year, the board of directors proudly reported to stockholders that the company had "developed and patented a process for the manufacture of pale types of rosin, which should considerably extend its field."[120]

This was no overstatement. During the previous five years, the middle grades of rosin—M, K, and I—were the only ones to hold a market share in the double digits, and together they accounted for an average of 60 percent of the total rosin market. Meanwhile, F gum rosin (equivalent to FF wood rosin, the lightest grade hitherto produced by Hercules) had averaged just above 7 percent of the market in the same period, and there had been no significant shifts in the relative importance of these categories.[121]

With its dark and new pale rosins, Hercules could satisfy an estimated two-thirds of all rosin uses. As the inventors themselves explained:

> Wood rosin embodying our invention is adapted for use in sizing the higher grades of paper and in the manufacture of the better grades of varnish, and indeed is so thoroughly purified that it is adapted for use in the manufacture of high grade rosin soaps and limed varnishes of satisfactory color, as well as for most, if not all, other purposes for which gum rosin is adapted.[122]

Even Joseph Lockwood, known for his "naturally . . . sanguine temperament," called the commercialization of I-grade wood rosin "the most important development in rosin, since the first commercial production of wood rosin in 1910."[123]

Along with the Pexite® rosin process, Hercules researchers achieved remarkable results in the fractionation of pine oil and turpentine in 1928. J.L. Jones, a Georgia Tech graduate and specialist in fractional distillation hired by Hercules in 1927, was central to this development. As Reu recounts, Jones "devised a concept of fractionation as carried on in a plant column which was quite contrary to many of the precepts held dear for many years."[124] From Brunswick and Hattiesburg now came not only rosin, turpentine, and pine oil but also many of their building blocks: *alpha*-terpineol, abietic acid, ethyl abietate, borneol, dipentene, dipolymer, and pinene.

In sum, by the end of the 1920s, Hercules operated two naval stores businesses, distinct in their products, markets, scale, and prospects. The original naval stores business, which involved production of commodities, was characterized by both continuity and change in the 1920s. Total volume of rosin, turpentine, and pine oil output had grown by 200–300 percent, and Hercules was still the world's largest producer of these products. Their uses had not changed radically, but the company's rosin sales had become increasingly concentrated among a few large customers. Half of the rosin sold in 1922 was used in the manufacture of linoleum and paper and was distributed among a score of buyers. By the end of the decade, aside from the Pexite® rosin Hercules sold, half of its rosin went to only two customers: Paper Makers Chemical Corporation (a rosin size manufacturer) and Armstrong Cork Company (a maker of fibrous insulation). As we shall see, this interdependency would draw Hercules into a new industry in the 1930s.[125]

The outlook for wood naval stores producers in 1928 was as perplexing as ever. They had improved their position at the expense of the gum industry, increasing their market share from 10 to 17 percent since the mid-1920s. Their gains, however, helped spark unprecedented competition. The largest rival to challenge Yaryan (later Hercules) and Newport since the early days of the industry was the Acme Products Company of De Quincey, Louisiana. Acme manufactured an exceptional, noncrystallizing brand of FF rosin ("Solros") and was backed by Gillican-Chipley, the most powerful gum naval stores producer and distributor in the United States. Still, Hercules maintained a comfortable lead in the field, and many smaller wood producers dropped out during market slumps.[126]

In 1928, however, Newport (which had spent $400,000 to double the size of its Pensacola plants the previous year) bought out Acme for nearly $2 million. The purchase brought its total market share to within a slim margin of Hercules' share (see Exhibit 6.4). In addition, Dixie Pine Products opened a steam-solvent plant at Hattiesburg.[127]

It was not a propitious time to face heightened competition. Once again, the wrong combination of weather conditions, output, and demand was driving down naval stores prices. Gum producers were flooding the market with their largest output since 1909. Compelled by economic necessity, Hercules scaled back production. Once again, it was beginning to lose money in naval stores.[128]

But there was also Hercules' second naval stores business: the manufacture of the chemical components and derivatives of the pine stump. Seven such products had reached the market by 1928. Hercules sold 15,000–20,000 pounds each of *alpha*-terpineol, abietic acid, and ethyl abietate in 1928, and lesser quantities of borneol, and dipolymer. Pinene was the clear leader at 100,000 gallons, all of it to synthetic camphor producers.[129] These specialty chemicals also faced growing competition. But they seemed to lead Hercules in a promising new direction as it continued its transformation into a chemical company.

THE CHEMICAL COMPANY

By the mid-1920s, Hercules was clearly a company in transition. Signs of change were visible everywhere: in the growth of its new businesses, in the

EXHIBIT 6.4 Shipments of Wood Naval Stores Products by Hercules' Leading Competitors, 1928

	ROSIN		*TURPENTINE*		*PINE OIL*	
	Barrels	*Percentage*	*Gallons*	*Percentage*	*Gallons*	*Percentage*
Hercules	214,233	55%	1,935,510	48%	1,297,559	54%
Newport	95,008	24	1,019,310	25	590,010	25
Acme	49,380	13	681,520	17	319,607	13
Others	32,232	8	418,778	10	176,734	7

Source: Naval Stores Department, Annual Report for 1928.

increasing scale and complexity of its operations, and in the primacy each business gave to engineering, research, and chemical control. Such trends were most noticeable in the newer lines of business, but the explosives plants also required more sophisticated technology simply to keep up. The ammonia oxidation plants for making nitric acid, for instance, cost $1 million apiece; moreover, it took considerable technical skill to construct and operate them safely and efficiently.

The increasingly technological basis of competition in each line of business was reflected in the company's spending for R&D. Between 1924 and 1928, the Experimental Station's annual budget grew from $145,321 to $363,577, from 0.7 percent of sales to 1.2 percent of sales; the number of personnel at the station grew from 76 to 130.[130] The character of Hercules' research activities was also transformed during the 1920s. At the start of the decade, the explosives plants, which used the station to help improve product quality or lower manufacturing costs, had accounted for most of the research budget. By 1928, however, naval stores research accounted for the largest part of the budget, and its projects involved basic, rather than applied, research in organic chemistry.

In 1926, Hercules broadened its research activities still further by organizing the Development Department at the Home Office, with Charles Higgins in charge. The new unit was given a general mandate to study and report on issues of long-term significance to Hercules, including additional opportunities for diversification. Among the questions taken up during its first year, for example, were potential markets for spent chips from the naval stores operations, the pros and cons of building a second nitrocellulose plant in the Midwest, and a potential investment in a venture to manufacture rayon.[131]

These years saw another change at Hercules. The company started its first international operations in the 1920s. In a sense, foreign expansion began with the Aetna acquisition, which included a small plant in Canada for producing fulminate of mercury.[132] In 1925, Hercules made a more aggressive commitment to foreign production by forming a fifty-fifty joint venture with Du Pont's Compania Mexicana de Explosivos (CME) to purchase and operate a dynamite plant in the Mexican state of Torreon. The plant, about halfway between Mexico City and El Paso, Texas, produced gelatin dynamites and distributed other explosives for local consumption. Although CME sold more than six million pounds of dynamite in 1926—on a par with output at Hercules' Ishpeming plant—it actually imported more than half of that amount from the United States. By 1928, however, with help from Hercules and Du Pont engineers, local production climbed to nearly ten million pounds.[133]

A still more ambitious step overseas followed: the organization of N.V. Hercules Powder Company in Rotterdam on September 1, 1925. The new company grew out of a relationship the company had established with Petrus W. Meyeringh, president of a local chemical company, two years earlier. Hercules was exporting significant amounts of naval stores products, and it had arranged with Meyeringh to distribute them across Europe. By 1925, Meyeringh "indicated the need for a European organization with warehouse facilities to expedite deliveries, with enough salesmen to cover the territory more frequently, and with servicemen to aid in obtaining best results with Hercules rosins, turpentine, and other chemicals."[134]

In its first year, N.V. Hercules achieved sales of $700,000. In the next two years, however, it broadened its offerings to include nitrocellulose and chemical cotton and its once tiny staff spilled over from a two-room suite to occupy an entire floor of a commercial building in Rotterdam.[135] By 1928, more than $3 million in sales (roughly 10 percent of Hercules' overall sales) originated overseas, with N.V. Hercules responsible for the vast majority of that business.[136]

Among other changes at Hercules in the mid-1920s, the saddest was the untimely death of George Markell on March 14, 1926. Just forty years old, he had been hospitalized for several months for treatment of rheumatic fever, when he contracted fatal pneumonia in early March. The board paid tribute to its youngest member: "He wrought large for the prosperity and welfare of the company; he never shirked his task nor hesitated to assist others. Blessed with great and unusual mental and moral capacity and gifted with an indefatigable industry, he always rendered service of the highest order." Markell was not replaced on the board, and at the company's annual meeting on March 24, President Dunham assumed the additional title and responsibilities of general manager.[137]

Reorganization

The increased volume and complexity of Hercules' business in the late 1920s put an increasing burden on its board of directors. Throughout its growth, Hercules had made no concessions in its management reporting systems. Indeed, senior management's appetite for detail was as voracious as ever. By 1928, though, preparation for monthly board meetings required the assimilation of a dozen departmental reports, several of which ran to nearly a hundred pages in length, and all of which were crammed with significant data. Indeed, it often required one or more extra meetings per month to discuss the material submitted.[138]

The thickest reports came from the Sales and Operating departments, whose assistant general managers collected information on all three major lines of business. Yet much of their information was repetitive: the operating reports discussed trends in sales and demand, while the sales reports went over issues in production volume and quality control. The demands of each major business, moreover, were different—a circumstance the company's reporting systems tended to disguise. The explosives business required close attention to costs and monitoring of competitors' actions. In nitrocellulose, the problem was to fund development of new products and establish new accounts with a broad array of industrial consumers. Establishing new accounts was especially a problem as long as nitrocellulose was regarded as "the very interesting pet of the chemical sales division."[139] The situation in naval stores was similar in that the business was becoming an aggressive consumer of R&D funds. In sales, however, naval stores were unlike nitrocellulose in that they required much greater spending on advertising and more attention to cultivating channels of distribution.

EXHIBIT 6.5 Hercules' Multidivisional Organizational Structure, 1928

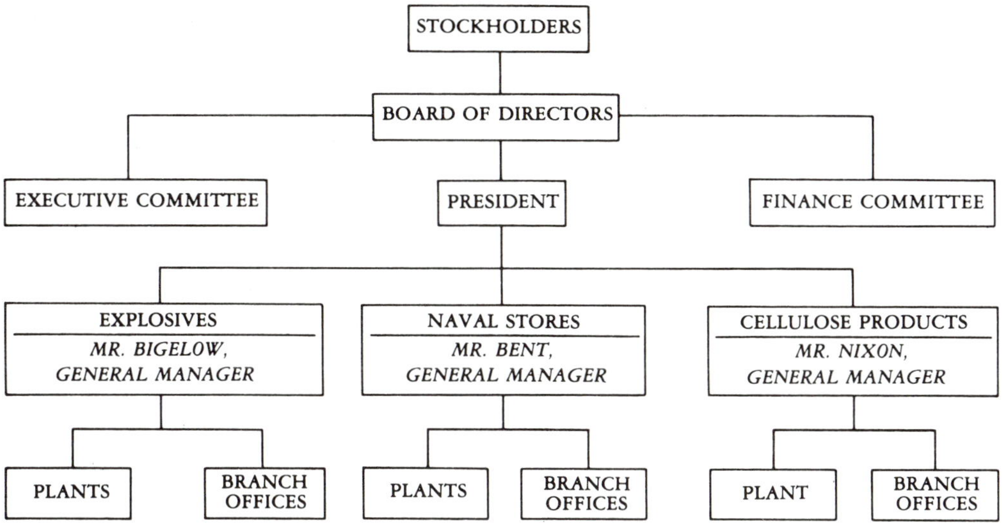

In the fall of 1928, the accumulated burden of overseeing its varied activities finally drove the board not only to expand itself, but also to reorganize the company. On September 4, the board elected five new members—Bent, Bigelow, Higgins, Hoopes, and Norman—and divided itself into two groups: the executive committee, charged with reviewing the company's operations; and the finance committee, responsible for managing Hercules' relations with shareowners and banks, as well as supervising its general financial condition.[140]

On September 1, the company also abandoned the functional organizational structure that had supported it from birth and adopted a new, multidivisional structure that highlighted its distinct lines of business (see Exhibit 6.5). Each new department combined responsibility for operations and sales under a single general manager. Bigelow was placed in charge of the company's new Explosives Department, Bent at the head of the Naval Stores Department, and Nixon at the top of the Cellulose Products Department.[141]

In making organizational changes, Hercules was following the example of other chemical companies, including Du Pont.[142] Yet no particular crisis, the typical motivation for change in other companies, triggered the reorganization. Rather, the new structure at Hercules appears to have been the result of an evolutionary process as the company slowly and cautiously involved itself in new activities. The reorganization of 1928 thus institutionalized a transformation that had been long under way at Hercules. No longer simply a powder company, it was now competing in a wide range of industries, and increasingly on the basis of its skills in chemical research and engineering. In short, Hercules had become a chemical company.

EXHIBIT 6.6 Hercules' Financials, 1921–1928

	1921	*1922*	*1923*	*1924*	*1925*	*1926*	*1927*	*1928*
Total sales	$16,091,390	$18,728,886	$22,260,796	$20,862,603	$23,669,009	$28,453,496	$27,961,494	$30,559,877
Net income	820,964	2,264,895	2,508,669	2,156,902	2,999,369	3,433,419	3,203,896	4,038,980
Total assets	37,255,601	39,115,752	40,281,961	40,898,494	42,824,991	45,922,536	44,914,019	48,006,175
Permanent	22,568,399	22,911,966	23,756,932	24,373,020	25,818,540	26,814,425	27,936,603	30,487,721
Current	14,657,201	16,203,785	16,525,029	16,525,474	17,006,451	19,108,111	16,977,416	17,518,454
Total equity	16,340,500	24,386,900	24,475,000	24,695,000	24,872,200	25,839,200	26,124,100	26,124,100

Source: Hercules Powder Company, Annual Reports for 1921–1928.

EXHIBIT 6.7 Diversification at Hercules, 1922–1930

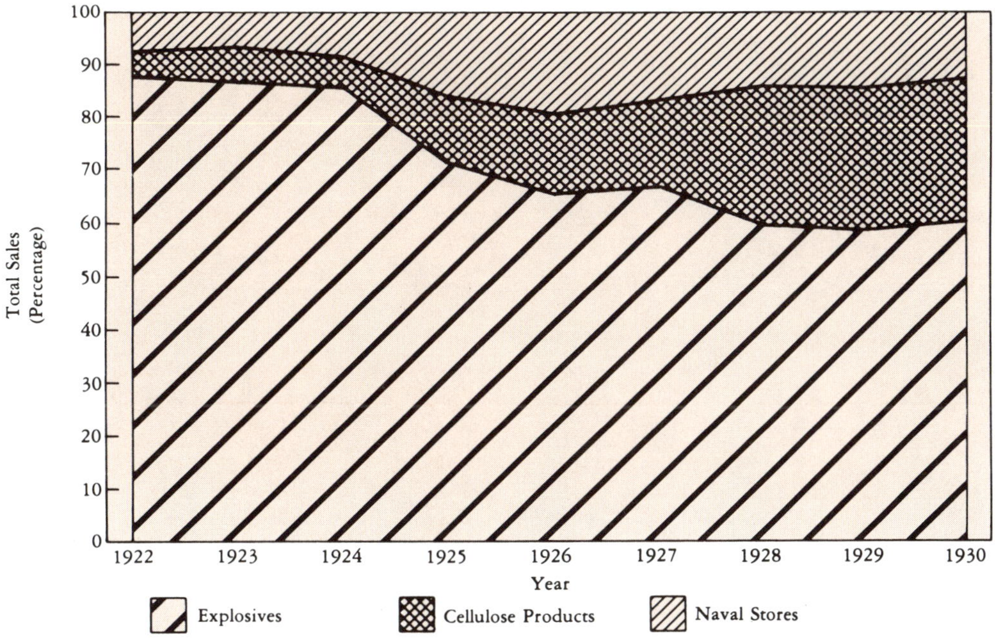

Source: Special Reports, 1931, Hall of Records.

Conclusion

The reorganization of 1928 brought a decade of fundamental change at Hercules to a close. At the start of the 1920s, the company was still wrestling with problems of demobilization from the war effort and wondering where its future lay. Although the company had placed a large bet on wood naval stores, and a smaller one on commercial nitrocellulose, these new businesses contributed less than 10 percent of the company's total sales of $16 million in 1921.

By late in the decade, however, Hercules had become a different sort of company. Its growth to $30 million in sales was but one aspect of the change (see Exhibits 6.6 and 6.7). Russell Dunham, remarking on the company's transformation in the annual report for 1927, noted that naval stores and nitrocellulose accounted for one-third of Hercules' permanent investment. Dunham also commented on another aspect of the changes at Hercules. Referring to "developments taking place in the chemical industry," he described a new role for research in the company. R&D was now designed "not only to ensure the maintenance of [Hercules'] position, but also to enable it further to take advantage of the opportunities for further expansion of its business."[143] Hercules, in other words, had changed its competitive strategy. The company would build its future around its new lines of business, where opportunities in the chemical industry held out promise of healthy growth.

CHAPTER 7

SPECIALIZING IN CHEMISTRY, 1929–1939

The reorganization of 1928 was a fitting symbol of the transformation of Hercules Powder Company from an explosives-maker to a chemical company. Yet the transformation was hardly complete once the new structure was in place. In the following decade, the pace of change at the company accelerated as it committed more funds to research and spun new applications out of the basic chemistry of its product lines.

The 1930s were also a time of growing diversity in the company, a result, in part, of the increasing complexity and specialization of Hercules' businesses: by the end of the decade, the company was supplying more than two hundred products to dozens of industries. Hercules' new character also reflected the correspondingly high technical demands placed on it by its major customers of the period: the makers of protective coatings, artificial fibers, and paper.

Finally, the diversity at Hercules stemmed from a technical revolution in chemistry and its ripple effects across modern industrial economies. As a writer in *Wilmington* magazine described a new era of research at Hercules in 1930,

> We are now entering an era in history most significant for the fact that man will make what he wants instead of looking for it. For many years the science of chemistry devoted itself chiefly toward tearing things apart and analyzing them. Now chemical ingenuity has turned toward synthesis, the building together of elementary materials into products as good or better than those found in nature and the creation of products entirely new in substance and usefulness.[1]

In the 1920s and 1930s, "chemicalization" spread throughout the economy as manufacturers of many goods saw the advantages of working with refined rather than natural materials: uniformity, standardization, reliability, controlled availability, and often, lower cost. In addition, manufacturers benefited from increased technical knowledge about their businesses: a better understanding of basic chemical reactions allowed them to make dramatic improvements in products and processes and to open new areas for growth and diversification. Chemically controlled processes replaced traditional empirical methods in industry after industry: textiles, materials, paints and varnishes, paper, printing, adhesives, perfumes, rubber, glass, cement, asphalt, and many others.[2]

The new role of industrial chemistry was a powerful stimulus to the growth and change of chemical companies such as Hercules. Its mastery of chemical knowledge and technology became a formidable competitive weapon, enabling the company to supply an expanding range of industries, increase the value of its products, and raise customers' dependence on it for support and service. In short, the revolution in industrial chemistry and the chemicalization of industry supported and validated Hercules' evolution into a specialty chemicals company.

Each of Hercules' major product lines was affected by—and prospered in—the new environment. In high explosives, for example, the new process of synthesizing ammonia by "fixing" nitrogen from the air, then oxidizing the synthetic ammonia to make nitric acid, changed the economics of the business. Not only was the explosives-maker's age-old dependence on Chilean nitrates finally broken, but ammonium nitrate suddenly became a much cheaper source of high energy than nitroglycerin. The chemistry of the process was straightforward, but oxidation of ammonia required heavy investment and sophisticated technology. Opportunities to grow in new areas such as fertilizers and other ammonia-related products were created, but competition from oil and gas producers and others invested in the new process also arose.

In cellulose products, Hercules not only consolidated its leading position as a supplier of nitrocellulose but also became an important producer of other chemicals such as cellulose acetate and chlorinated rubber. The initial markets for these products were in protective coatings and synthetic fibers. Over time, however, the company's research on the film-forming characteristics of cellulose ethers and esters and its developing ties with European chemical producers led to new opportunities in the rapidly growing plastics industry.

In naval stores, the efforts of Hercules scientists and salesmen to develop new products and find new markets bore fruit in the late 1930s. The derivatives of rosin, turpentine, and pine oil grew more valuable than the commodities themselves and satisfied a wide range of commercial needs in industries such as paper, printing, adhesives, rubber, and asphalt. In 1928, naval stores was Hercules' most troubled product line; a decade later, it was one of the strongest and steadiest contributors to corporate earnings, and parent to a new operating unit, the Synthetics Department.

Such changes in Hercules' businesses reflected the company's heightened emphasis on research. The Experiment Station, relocated from Kenvil to Wilmington in 1931, worked with the departments to develop new products and areas of business. Hercules was also beginning to understand, at a deeper level than ever before, the outlines of a new relationship with its customers. A key source of the new understanding was the Paper Makers Chemical Corporation (PMC), which Hercules acquired in 1931. At that time, PMC was already a sizable but unfocused company; it would prove significant to Hercules, however, in ways beyond its scale and scope. Since its founding in 1900, the company had grown from making a single product (rosin size, a chemical used to improve paper's resistance to liquids so that, for example, ink will not blot) to supplying, through acquisitions, a broad range of chemicals to the paper industry. The company had positioned itself as an important and valued supplier to its customers, and its strongest products became a source of enduring profits.

In the 1930s, Hercules' efforts to grow and change were, of course, greatly hampered by the extraordinarily hostile business climate of the Great Depression. Nonetheless, the company's growing diversity helped moderate the impact of terrible times. Indeed, advances in research and marketing—not adverse economic conditions—were the dominant themes of the depression decade, as Hercules continued its transformation into a specialty chemicals producer.

Managing the Chemical Company

The strategic and organizational changes at Hercules at the close of the 1920s occurred against the backdrop of the evolution of large companies in general, and large chemical companies in particular. In moving to a multidivisional structure, with department ("operating unit") general managers responsible for production and sales of families of products, Hercules was following the great administrative revolutions at Du Pont, General Motors, and other companies of the period. It was no accident that this structure appeared first in the chemical industry, which was fast becoming the most diversified industry in the American economy.[3] At Hercules, manufacturing products as different as explosives, cellulose products, and naval stores, and serving markets as disparate as mining, protective coatings, and paper, simply put too much stress on the company's traditional functional organization. The old Operating Department was poorly organized to track expenses and plan output in such varied businesses; so, too, the old Sales Department experienced difficulties in responding to the needs of customers in a dozen different industries.

Yet Hercules' definition of the new strategy—to grow in distinct lines of business—and the design of its new structure left open many questions. How independent and decentralized should the businesses be? Under the old structure, a single corporate staff had provided for common needs such as advertising, legal work, purchasing, accounting, traffic, research, and engineering. Should the new operating units create their own staffs for these functions? Or should they continue to draw on the corporate pool? If they chose the latter, how would the cost of corporate services be allocated? And how would conflicts between and among the operating units and the staff be resolved?

A debate over the role of research and development put these questions into sharp relief. Under the old structure, research had been decentralized, with the Experimental Station at Kenvil concentrating on explosives and cellulose research while cooperating on naval stores research with the plant laboratories at Brunswick and Hattiesburg. At issue after 1928 was whether the company ought to replace these disparate operations with one centralized research station, perhaps near Wilmington, to support more directly the work of the whole corporation.

These questions, so familiar to executives in modern decentralized and diversified companies, were fresh in the late 1920s, and they proved somewhat problematic to Hercules after its reorganization. By the mid-1930s, however, the company had developed an approach to managing its business that would endure for more than four decades.

Working out the Details

The center of power in the reorganized Hercules was the board's executive committee, which consisted of President Dunham, C.D. Prickett, and Frederick Stark, and vice presidents James Skelly, T.W. Bacchus, and Norman Rood. None of the vice presidents had operating responsibilities; instead, they reviewed the overall affairs of the company and occasionally focused on specific assignments. The executive committee met each Tuesday morning at ten o'clock to review reports and decide on appropriations requests below a limit of $100,000.[4]

Responsibility for managing day-to-day business fell to the general managers of the operating departments: C.A. Bigelow (Explosives), Leavitt Bent (Naval Stores), and A.B. Nixon (Cellulose Products). In addition, Philip Stull served as head of Virginia Cellulose Company, which operated as a separate subsidiary until 1931. The general managers were supported by staff units: the Legal, Technical, Development, Advertising, Traffic, and Safety and Service departments.

In the first months after the reorganization, Hercules worked out the details of how the levels of governance would interact and interrelate. Although other companies (including Du Pont) had passed through similar reorganizations, Hercules appears to have identified and solved problems without reference to other companies' experiences. In deciding on how to set prices for "interdepartmental transfers of important materials," for example, Hercules' treasurer, C.C. Hoopes, studied the company's records as the basis of his recommendation to charge "fair sales prices," based on costs, plus an average profit for the unit volume actually transferred.[5]

Similarly, Hercules' managers worked out their own solutions to the problems of allocating indirect expenses (corporate overhead) to the operating units, and the degree of centralization of the corporate staffs. To address the first problem, the board appointed a committee consisting of Hoopes, Bent, Bigelow, and Nixon to study the matter. The group met occasionally during the second half of 1929 and finally agreed on a formula: overhead costs would be allocated on the basis of actual usage. Each staff head met with the committee and was asked to "make a distribution of the time of each [employee] in his

Between the wars, Hercules spun a wide variety of valuable chemicals from an unlikely source: cotton linters, the short fuzzy fibers attached to cottonseeds after ginning.

Process breakthroughs made Hercules a leading supplier of low-viscosity nitrocellulose to the protective coatings industry. In the late 1920s, M.G. Milliken developed a process for continuous digestion that was safer and much more cost-effective than batch digestion and resulted in a product of higher quality and greater uniformity. The Milliken digester at Parlin included a 200-foot tower to help regulate pressure in the system.

Members of the "class of '29," a group of researchers and managers who led Hercules into many new areas pose at a reunion in 1954 at Hercules Country Club. From left: E.G. (Pete) Peterson, A.L. (Col)· Rummelsburg, E.G. (Ed) Crum, A.M. (Alf) Boll, D.H. (Don) Sheffield, W.W. (Bill) Delaney, C.E. (Clell) Tyler, Emile (Bud) Progoff, F.B. (Forrest) Evans, L.C. Pritchett, R.W. (Bob) Lawrence, C.T. (Pat) Butler, C.M. (Carl) Linder, A.L. (Whitey) Larson, Fred K. Shankweiler, R.T. (Dick) Yates, E.M. (Ernest) Waxbom, M.R. (Monty) Budd, W.T. (Bill) Bishop, L.W. (the Boss) Babcock, W.E. (Walt) Gloor, W.H. (Bill) Stevenson, and J.S. (John) Tinsley. Babcock (fourth from right) had been assistant director of the Technical Department at Kenvil when most of the group was hired. In 1954, he was head of the Personnel Department and a director of Hercules.

Between Hercules' founding and 1930, the board of directors doubled in size, its growth a reflection of the company's increasingly complex business. Posing at an anniversary dinner honoring C.D. Prickett's fiftieth year of service in the explosives industry, the directors included: (seated from left) Messrs. Dunham, Prickett, and Bacchus; Standing from left: Messrs. Rood, Hoopes, Rheuby, Skelly, Norman, Bigelow, Bent, and Higgins.

The opening of the Experiment Station near Wilmington in 1931 signaled the company's growing emphasis on improving and developing products through systematic research.

Acquisition of Paper Makers Chemical Company in 1931 helped ensure a steady market for rosin size and opened new lines of business. PMC's plants (including one at Savannah, Georgia, above) required upgrading and investment before returning a steady profit.

department, and to make a report to the committee showing in considerable detail the various activities of his department, subdividing each one to the four general departments." The statistics were then translated into dollars "by applying salary rates together with rent of space occupied, telephones, depreciation, etc." The new system was started in January 1930, with Hoopes claiming that "it is more accurate" than the method it replaced.[6]

Hoopes also worked with the corporate secretary, Edward B. Morrow, on improving the company's reporting systems. The board had raised the question of whether the staff heads should continue to report to the executive committee, or whether each general manager should report on "the activities of all auxiliary departments, so far as they affect his particular department." After studying the question, Hoopes and Morrow recommended that the staff heads continue to report to the executive committee, thereby affording the top officers "general supervision over their activities from a company policy standpoint." Hoopes and Morrow also observed, however, that there was "considerable room for improving the type of report" that usually came in. They pointed out that the reports "consist, mostly, of a stereotyped mass of data, in probably a great deal too much detail for Executive Committee consideration." They therefore urged the staff units to report quarterly rather than monthly and suggested that each report "have the personal thought and attention of the department head in its preparation." To avoid having a crush of reports at the end of each three-month period, Hoopes and Morrow recommended that three departments file their reports in February, May, August, and November, and that the others report in March, June, September, and December.

Hoopes and Morrow also set guidelines for the content of each department report. The new formats were designed to highlight significant trends in each area, including the activities of competitors, suppliers, and other stakeholders.[7]

The Place of Research

The reorganization of 1928 raised one other highly significant issue: the role of research and development. Where should responsibility be lodged? Centralized at corporate headquarters, where senior management could more easily maintain control and assess the value of research in creating new business opportunities? Decentralized in the departments, which were, after all, technically disparate and concerned primarily with improving existing products and processes? Or in some combination of centralization and decentralization?[8] In 1928, the answer to this question was not straightforward, and it took years and several key personnel changes to resolve it.

In the chemical industry at that time, research was often obtained from sources outside of companies, including universities, specialized consulting firms such as Arthur D. Little, Inc., and public and private research institutes.[9] In 1928, few chemical companies operated centralized R&D laboratories, the most notable exception being Du Pont; indeed, among industrial companies generally, only a handful—for example, AT&T, General Electric, Alcoa, and General

Motors—maintained such facilities. In seeking an answer to the question of how best to organize and administer R&D, then, Hercules had few models to go by, although Du Pont was a powerful model for at least some centralization. In the late 1920s, under the dynamic leadership of Charles M.A. Stine, Du Pont's Chemical Department was evolving beyond "the gutted remnant of decentralization" that followed the company's postwar restructuring into "a vital force" for growth and change at the center of the company.[10] A far larger and more diversified company than Hercules, Du Pont cast a long shadow in Wilmington.

In October 1928, with Du Pont's example in mind, President Dunham appointed a group including George M. Norman (head of the Technical Department), Harry E. Kaiser (head of the Experimental Station at Kenvil), and R.E. Zink (head of the Engineering Department) to consider "the proposition of relocating the entire Experimental Station in some locality which will be easily accessible to the Home Office." The three men responded quickly with a memo weighing the pros and cons of building a new laboratory near Wilmington versus upgrading the existing facility at Kenvil. According to Zink's calculations, Kenvil could be modified to serve the whole company with an investment of $325,000; building a new station in Wilmington would cost about $800,000. Assuming that most of the existing properties in Kenvil would have to be written off if the laboratory were closed, Zink estimated that the net cost of moving to Wilmington would be roughly $400,000. That seemed like a lot of money to Kaiser, who pointed out that the interest on the sum would amount to "about one-seventh of the cost of running the Station for one year."[11]

Despite Kaiser's inclination to keep the laboratory at Kenvil, Dunham appointed a committee consisting of Norman and the three general managers of the operating departments to study the issue further. The group met twice in early November and recommended unanimously that the Engineering Department start preparing detailed estimates for building a new research station near Wilmington. The committee came around to this decision for several reasons. First—and uncharacteristically—it consulted an outside expert, Du Pont's Charles Stine. Stine described the organization of research at his company and declared himself "altogether in favor of a Central Experimental Station located conveniently to the Home Office." In support, Stine cited several reasons: the increased likelihood of pioneering major new developments; the ability "to employ a higher grade of special research man because there are more problems to occupy his attention"; the greater value of a large, centralized library over many, dispersed smaller ones; and the "great advantage" of being able to send people between facilities "on short notice and at frequent intervals."[12]

The committee agreed with Stine's reasoning, although it noted—a bit wistfully—that Kenvil has "a very much better climate" than Wilmington, "especially in summer time." ("Moving the Station from Kenvil," remarked Norman drily, "is certainly going to reduce attendance at New York musical shows considerably.") What clinched the move, however, was the belief that "frequent contact between heads of divisions in the Home Office and the Experimental Station is extremely important . . . and will become more important as time passes . . . because of the growth of the company."[13]

In January 1929, Hercules began full-scale engineering studies for a new research facility. Throughout the late winter and spring, several employees scouted for appropriate sites near Wilmington. Finally, in June, the decision was made: Hercules purchased from Alfred D. Peoples a farm consisting of two hundred acres of wooded hills about five miles west of the city. The site (which Norman had surveyed on horseback) was accessible by car along the Lancaster Pike and served by a branch of the Baltimore & Ohio Railroad. Total cost of the property, including "landscaping, roads, trestle, and railroad," came to $121,825. In August, Hercules purchased another one hundred acres of adjacent land for $32,580.[14]

Preparation of the site—planning, grading, building roads, drilling wells, and installing conduits for water and power—took nearly a year. Finally, on July 30, 1930, President Dunham laid the cornerstone and spoke of "the importance of well directed chemical and scientific research in a business such as ours." He offered the hope "that in these surroundings Hercules men and women in the years to come may find a common rallying place where useful things can be done for the benefit of our company and the world at large."[15]

The new Experiment Station, completed at a total cost of over $1 million, was ready to be occupied early in 1931.[16] The complex consisted of eleven buildings, including a three-story main laboratory, a powerhouse, a machine shop, storage areas, and special structures for testing explosives. The main laboratory boasted many of the latest features in industrial construction, including fireproof brick walls, stoneware pipes, chromium steel fixtures, and state-of-the-art systems for controlling air movement and humidity. Individual laboratories were designed to provide researchers "with a quiet, safe, and completely equipped place to work." Each lab had at least two exits, with doors that had no knobs or latches.[17]

The new Experiment Station, like the reorganization of 1928, was a sign of new times at Hercules. As Norman put it in support of the move from Kenvil to Wilmington:

> During the past seven years, due to the expansion of our Company into lines of business requiring intensive development and to a better appreciation of the importance of technical studies as related to our business, the Station organization has quadrupled in size.
>
> It is our constant endeavor to find and employ only the best trained and keenest minds to be found in the highest grade technical schools in this country. The competition for such men is very intense, especially since such large organizations as the Bell Telephone Laboratories, the General Electric Co., and the Standard Oil Companies, etc., are seeking the outstanding young technical men for their research organizations. In order to attract and hold such men it is essential to provide the best possible facilities for work.
>
> Our objective, as we understand it for the long pull, is to help build a Company that will endure and make profits, but, in addition, we should contribute to the material welfare of all mankind. We believe a

> sound, well-planned research policy, fostering the creative spirit of outstanding minds, is one of the soundest and most certain methods of achieving this objective.[18]

For several years, Hercules frequently had helped acclimate new professional hires by sending them for several months to the Kenvil Experimental Station.[19] After 1929, that practice became standard. For example, the "class of '29," which included Donald Sheffield, Ed Crum, Pat Butler, and Arthur Larson, was to leave its mark on Hercules' history.[20] Norman could also remark on the recent hiring of "a number of men having outstanding research ability" for permanent assignment at the Experiment Station. Several young chemists—O.A. Pickett, David Wiggam, Wyly "Josh" Billing, W.A. Kirklin, Joseph Borglin, Eugene J. Lorand, Harold Spurlin, and Walter E. Gloor—were to play roles in many of the company's significant technical achievements over the next three decades.

The decision to relocate the Experimental Station in 1929 coincided with a larger debate on the role of research in the company. At issue were the relationship of the station to each of the major businesses—especially explosives—and the nature of research at the station itself. Before 1931, research had tended to concentrate on improving processes and product quality and uniformity. These activities had been somewhat decentralized; work on high explosives was done at Kenvil and Port Ewen (blasting caps), on nitrocellulose at Kenvil and Parlin, on cellulose purification at Hopewell, and on naval stores at Brunswick and Hattiesburg. The central lab at Kenvil had functioned primarily as a service organization in support of the plants. The bulk of its work involved Requests for Investigation (RIs) from the plants. RIs commonly involved brief, focused studies of chemical samples, procedures for handling materials, or other specific inquiries.[21]

Hercules' decision to build the new station in Wilmington raised the immediate question of the new facility's scope. At least some work on process improvements would still have to be done at the plants. Moreover, a business like explosives, in which products and their uses changed slowly, required less intensive research than more technical businesses like nitrocellulose and (increasingly) naval stores. Because Hercules already had significant investments in research facilities at Kenvil, and because the work was dangerous, the company decided against conducting research on explosives in Wilmington.

Research policy continued to stir debate, however. In the spring of 1930, the executive committee asked George Norman to prepare a report "covering experimental work and cost of same for the four major branches of the company's business." The study was not intended to be a critical review but rather to serve as a basis for future decision making. Although Norman did not draw any conclusions, his report showed that the research needs of the major businesses were distinctly different. He also showed that the company was not generally engaged in exploring the basic chemistry of its major products, nor was it devoting much effort to analyzing competitors' products or potential substitutes for its own.[22]

In 1930, the executive committee created a technical development committee to deal with general questions of research policy and with research investigations that either did not fall into clear categories or involved greater expense than any one department might be able to bear. The committee was headed by Charles Higgins (who had been elected vice president of the company two years earlier) and included Norman, Kaiser, J.N. McVey (Higgins's replacement as head of the Development Department), and the general managers of the operating departments.

Disturbing signs were appearing that the company needed greater coordination of and focus in its research efforts. Although Hercules could point to some important recent advances, such as improved yields in the production of nitric acid by the ammonia oxidation process, it was also experiencing difficulties in other areas.[23] In particular, progress on new cellulose products was frustratingly slow, and many smaller projects, such as the search for a new blasting cap initiator, were plagued by delays, cost overruns, and inadequate results. By early 1933, such problems so disturbed O.A. Pickett, head of Physical Chemistry, that he prepared a long, probing, and well-written critique of research and development at Hercules.[24]

Pickett lodged four main charges against the status quo: the "lack of a clear-cut basic research policy"; "failure to get sufficient basic data prior to semi-plant scale operations"; "too much division of authority in the execution of research"; and "failure to centralize research—too much research, both laboratory and semi-plant, is done on the plants."

On the first point, complaining that "a lot of money is spent on research in attempting to develop hoped-for quick money makers; on improvements, often sketchy in character; on isolated bits of sales-service work; and on quick solutions (?) of customer troubles," Pickett argued that Hercules would be far better off if it focused on accumulating "more basic knowledge" of its major products. In explosives, for example, he urged that a systematic review be made of recent work on the mechanics of detonation, a step that "none of us has ever taken the time to do," and one that would achieve far better results than the standard practice of "rushing into the library and skimming through the literature in between test tube experiments." (He sketched out similar basic studies to be done in the other product lines; see Exhibit 7.1.) Pickett believed that Hercules' research would be far more efficient if it was better grounded in basic chemistry. He went on:

> We need the knowledge these investigations would give in order to make progress intelligently, logically, and economically. We kid ourselves so long as we do not build up such foundations of basic knowledge along with other work we may think stands better immediate chances of proving profitable. The truth is, these latter types of work would stand a far better chance of success if built upon a properly prepared foundation of basic knowledge.

Pickett's other charges centered on the management of research in the company, which he believed was far too decentralized. As he put it, "Our present

EXHIBIT 7.1 O.A. Pickett's Proposed Program for Hercules' Basic Research

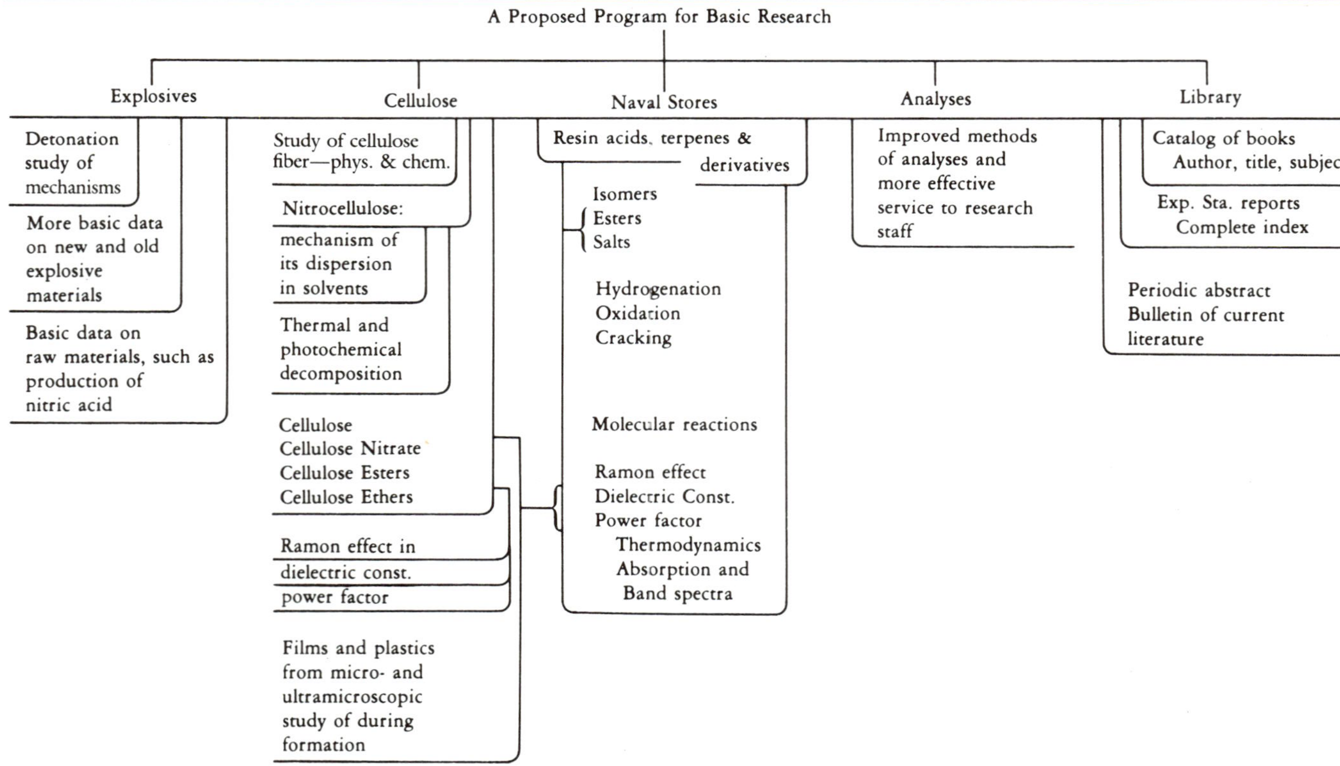

Source: George M. Norman papers, Hagley Museum and Library, accession 1150.

manner of administering research makes it everybody's business and no one's responsibility." He pointed out that researchers at the Experiment Station too often had to drop their work to provide quick answers to the Home Office, the operating departments, the sales force, and the plant laboratories. As a result, he claimed,

> We encourage our research men to be more concerned about the bookkeeping of research accounts than about getting results. Instead of surrounding them with an atmosphere inducive to creative thinking and acting, we dangle before their eyes worries as to whether they will be allowed a reasonable chance to finish a decent job. Instead of looking to the head of the research organization for direction, instruction, and inspiration, they nervously anticipate bolts of authority hurled from the hands of any assistant to the assistant keeper of Thor's thunder room.

The solution to these problems, Pickett believed, was a thorough reorganization of research at the company. Citing the examples of AT&T, General Electric, General Motors, and Du Pont, he advocated the complete centralization of research under a single manager who would control an annual budget and be accountable to the board of directors. He recommended that at least 10 percent of the annual budget for research be devoted to basic research, adding that "until we build up greater reserves of basic knowledge than we now possess relative to our raw materials a somewhat larger proportion is justified."[25]

He also proposed that the internal organization of the Experiment Station no longer correspond so tightly with the operating departments, to help remove the "Cellulose, Explosives, and Naval Stores gang isolation" and overcome "an accelerating retrogression into a set of 'water-tight compartments' with less and less free circulation between compartments." Pickett advocated, instead, setting up departments based on types of research problems: basic research, laboratory development, and semiplant development, with support from testing (a combination of the old Physical Chemistry and Analytical Chemistry departments), library, maintenance, and office (see Exhibits 7.2 and 7.3).

Pickett's report amounted to a devastating indictment of existing research policy and management. His argument was powerful: the technical content of Hercules' business was increasing every day; the company was having trouble in developing new products; many large companies had already centralized R&D; wouldn't Hercules suffer if it failed to follow suit? The readers of the document were sympathetic. Delivered first to Norman, the report probably made its way to Higgins and Bent as well. Within months, a major reorientation of research and development was under way. In June 1933, the technical development committee heard reports on the need for basic research on the detonation of explosives.[26] Soon thereafter, Harry Kaiser resigned as director of the Experiment Station and Pickett was named acting director. The following spring, the assignment was made permanent.

The Experiment Station was reorganized along the lines Pickett had sug-

EXHIBIT 7.2 O.A. Pickett's Proposed Plan for the Administration of Research at Hercules

Director of Research
Directly accountable to Board of Directors for administration and success of research

A Basic Research	B Development	C Plant Aids C1	 C2	D Sales Aids
See Chart 1 [Exhibit 7.1] Funds to be spent based on annual budget Problems selected and worked on to be basic to company's health and continued growth None of funds to be spent on actual development of plant processes	Work done on annual budget basis. Total sum spent not to exceed a specified sum for given year. Program to be followed in all development work: 1. conception of idea 2. preliminary laboratory testing of idea 3. investigate sales possibilities 4. literature study & laboratory study for basic data 5. design and build semiplant unit 6. operate unit for development and operating data 7. check and OK product for trade samples 8. supply data for designing plant unit 9. test and turn over to plant successfully operating unit	Work to be done on annual budget basis Plant laboratory inspections Check samples Operating control specifications Methods of analysis	Costs under control of company general manager save in cases where job is transferred to (A) or (B) because of basic or broad interests Solving of new or old operating problems for the plant	Sales managers directly responsible for expenditures save in cases of transfer of job to (A) or (B) because of basic or broad interests Customer problems Complaints, etc.

Source: George M. Norman papers, Hagley Museum and Library, accession 1150.

EXHIBIT 7.3 O.A. Pickett's Proposed Scheme of Administration for Hercules' Experiment Station

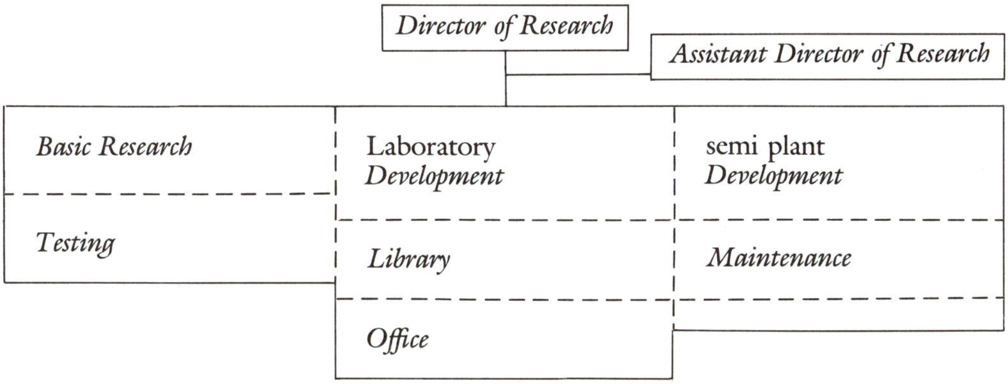

Source: George M. Norman papers, Hagley Museum and Library, accession 1150.

gested. The Explosives, Cellulose Products, Naval Stores, and Physical Chemistry departments of the station were abolished; in their place came separate new departments for basic research, development, and semiplant operations. Researchers from the old Physical Chemistry Department were reassigned to either the new departments or the supporting Analytical Department. Thereafter, personnel would be "placed in keeping with the type of work to be done."[27] Finally, the consolidation of research at the center was completed with the opening of a new explosives laboratory at the Experiment Station in 1934.

By the mid-1930s, Hercules had its own "vital force" for growth and change in place at the center of the company. Henceforth, the Experiment Station assumed a significant role in developing all of the company's lines of business.

HERCULES AND THE GREAT DEPRESSION

In the fall of 1928, Hercules' finance committee recommended that the company follow its reorganization with a plan to increase its capitalization and, in particular, that its common stock, which was given a par value of $100 in 1912, be replaced by a new issue of no-par common stock to be exchanged on a four-to-one basis. The recommendation, endorsed by the board and approved at the annual meeting in 1929, resulted in the creation of 1.6 million new shares, of which 80,000 were set aside for employee subscriptions. The finance committee also recommended that Hercules apply to have the new stock listed on the New York Stock Exchange. Taking these actions, explained Dunham, was intended "to improve the investment value" of the stock "by increasing its marketability."[28]

Hercules' stock first appeared on the New York Stock Exchange on July

11, 1929, closing at a price of $130. At that very moment, economic warning signs were in the air. Factory production was already declining across the United States, and unemployment was on the rise. By late summer, the signs were becoming more evident still: the Federal Reserve Bank raised interest rates by a full point, a move soon made by the Bank of England as well. On October 14, stock prices began to fall; ten days later, on "Black Thursday," nearly thirteen million shares—an unprecedented volume—were traded. By the following Tuesday, October 29, the *New York Times* industrial average plummeted forty-three points, wiping out the entire gains of the previous year. By mid-November, the index of industrial output stood at half the level of the previous August.

So began the worst depression in the nation's history (and the worst in modern world history), a dismal period that would endure for a decade and, at its low point, result in a 30 percent drop in the U.S. gross national product, a 32 percent plunge in wholesale prices, roughly nine thousand bank failures, and a national unemployment level of 25 percent.[29]

At Hercules, the effects of the crisis were prolonged and severe. The company saw its total volume of business cut in half between 1929 and 1932; employment dropped from 4,261 to 2,690. Hercules tried to hold on to as many employees as it could, but the environment could not be ignored. Even those who survived the payroll cuts—including the directors and top officers—saw their salaries and wages slashed as much as 30 percent, and many employees found themselves working a three-day week to boot. The company's new benefit programs had to be modified. The employee stock subscription plan was amended to reflect lower stock prices, and a merit payroll plan was suspended for the duration of the crisis. Hercules' brand-new pension plan had to be changed a year after it took effect because there was not enough money available to fund it.[30]

The Great Depression affected virtually every product line, although at different rates (see Exhibits 7.4 and 7.5). In 1930, for instance, explosives "showed less curtailment than the average of manufacturing industries of the country" and even "benefited from the activity in certain construction lines." During the next two years, however, production in the mining industries dropped an average of 40 percent, and by 1932, sales of explosives were the lowest in the company's history.[31] Hercules negotiated a mutual supply agreement with Atlas for black blasting powder: Hercules furnished Atlas's powder needs west of the Mississippi River, and Atlas reciprocated in the East. The agreement allowed Hercules to close permanently its powder mills in Youngstown, Ohio, and Pleasant Prairie, Wisconsin. Soon thereafter, the executive committee voted to close the Emporium dynamite plant and to serve customers in the region from Kenvil.[32]

The Great Depression wrought havoc almost immediately in the Cellulose Products Department. Demand for nitrocellulose lacquers—by far Hercules' best-selling product in this line, and one still heavily dependent on the auto industry—fell dramatically in the second half of 1929. By 1932, lacquer sales had plunged 70 percent from peak levels before the crisis, and wages and em-

EXHIBIT 7.4 Hercules' Financials, 1929–1939 (Dollars in Thousands, Except per Share)

	1929	*1930*	*1931*	*1932*	*1933*	*1934*	*1935*	*1936*	*1937*	*1938*	*1939*
Net sales	$29,314	$23,499	$18,412	$15,946	$20,043	$25,795	$29,670	$36,741	$44,559	$32,830	$41,010
Earnings before interest and taxes	4,748	2,571	1,474	1,048	2,745	3,176	3,615	5,191	5,682	3,621	6,665
Total assets	45,399	46,065	45,859	44,898	46,869	47,990	318,621	49,623	56,356	57,209	61,915
Fixed[a]	26,907	27,997	30,228	30,085	30,295	30,238	301,992	31,359	34,906	35,333	39,700
Current	17,913	17,411	14,833	14,086	15,931	17,298	16,209	17,834	21,137	21,221	21,691
Other	579	657	798	727	643	454	420	430	313	655	524
Total equity	26,374	26,501	26,580	26,580	26,580	26,580	24,775	24,775	26,565	26,565	26,565
Stockholders' investment	34,329	33,964	32,184	29,322	29,502	29,404	27,398	27,893	33,625	34,214	35,261
R&D Expense	541	609	478	452	461	639	727	792	1,178	897	932
Dividends/share	$4.00	$3.00	$3.00	$2.00	$2.25	$3.50	$3.50	$5.25	$3.00	$1.50	$2.85
Earnings per share	$5.95	$2.61	$1.04	$0.24	$2.79	$3.94	$4.23	$6.33	$3.23	$1.95	$3.65
Employees	4,261	3,789	3,344	2,690	3,072	4,484	4,572	5,013	5,954	5,163	5,786
Return on sales	16.2%	10.9%	8.0%	6.6%	13.7%	12.3%	12.2%	14.1%	12.8%	11.0%	16.3%
Return on assets	10.5	5.6	3.2	2.3	5.9	6.6	1.1	10.5	10.1	6.3	10.8
Return on equity	18.0	9.7	5.5	3.9	10.3	11.9	14.6	21.0	21.4	13.6	25.1
Return on stockholders' investment	13.8	7.6	4.5	3.6	9.3	10.8	13.2	18.6	16.9	10.6	18.9
R&D/Sales	1.8	2.6	2.5	2.8	2.3	2.5	2.5	2.2	2.6	2.7	2.3

Source: Hercules Powder Company, Annual Report for 1952. Total equity, dividends/share, and EPS figures are from annual reports by year.

EXHIBIT 7.5 Hercules' Sales and Profits by Department, 1929–1939 (Dollars in Thousands)

PRODUCT LINE/YEAR	*1929*	*1930*	*1931*	*1932*	*1933*	*1934*	*1935*	*1936*	*1937*	*1938*	*1939*
EXPLOSIVES											
Net sales	$16,161	$13,521	$10,019	$6,891	$7,506	$9,209	$8,993	$10,477	$12,368	$11,067	$13,096
Net profit	3,048	1,808	704	32	767	1,013	637	1,273	1,721	1,774	2,596
Return on sales	18.9%	13.4%	7.0%	0.5%	10.2%	11.0%	7.1%	12.2%	13.9%	16.0%	19.8%
Percent of total sales	53.9	58.2	53.0	41.0	35.0	33.9	28.9	28.0	27.7	33.9	31.9
Percent of total profits	57.3	77.1	80.6	5.1	26.7	30.0	16.6	20.4	25.8	48.0	33.8
CELLULOSE PRODUCTS											
Net sales	$7,438	$5,996	$4,494	$2,812	$3,727	$4,686	$5,265	$6,536	$6,636	$5,590	$8,335
Net profit	1,958	1,243	986	548	1,063	1,336	1,355	1,931	1,605	1,411	2,669
Return on sales	26.3%	20.7%	21.9%	19.5%	28.5%	28.5%	25.7%	29.5%	24.2%	25.2%	32.0%
Percent of total sales	24.8	25.8	23.8	16.8	17.4	17.2	16.9	17.5	14.9	17.1	20.3
Percent of total profits	36.8	53.0	112.9	87.0	36.9	39.6	35.4	31.0	24.1	38.2	34.7
NAVAL STORES											
Net sales	$4,050	$2,949	$1,978	$1,455	$2,458	$3,118	$3,962	$5,354	$6,397	$4,254	$5,406
Net profit	90	(916)	(1,150)	(534)	14	119	264	849	1,537	(196)	507
Return on sales	2.2%	−31.1%	−58.1%	−36.7%	0.6%	3.8%	6.7%	15.9%	24.0%	−4.6%	9.4%
Percent of total sales	13.5	12.7	10.5	8.7	11.5	11.5	12.7	14.3	14.3	13.0	13.2
Percent of total profits	1.7	−39.1	−131.7	−84.8	0.5	3.5	6.9	13.6	23.1	−5.3	6.6

VIRGINIA CELLULOSE											
Net sales	$3,181	$2,672	$2,401	$1,970	$2,588	$4,052	$5,264	$5,580	$6,380	$3,585	$4,386
Net profit	226	209	333	411	539	450	779	1,006	850	854	1,152
Return on sales	7.1%	7.8%	13.9%	20.9%	20.8%	11.1%	14.8%	18.0%	13.3%	23.8%	26.3%
Percent of total sales	10.6	11.5	12.7	11.7	12.1	14.9	16.9	14.9	14.3	11.0	10.7
Percent of total profits	4.2	8.9	38.1	65.2	18.7	13.3	20.4	16.1	12.8	23.1	15.0
PAPER MAKERS CHEMICAL											
Net sales				$3,659	$5,163	$6,118	$7,604	$10,727	$12,863	$7,995	$9,399
Net profit				173	494	454	792	1,216	599	(66)	763
Return on sales				4.7%	9.6%	7.4%	10.4%	11.3%	4.7%	−0.8%	8.1%
Percent of total sales				21.8	24.1	22.5	24.5	19.5	28.8	24.5	22.9
Percent of total profits				27.5	17.2	13.5	20.7	19.5	9.0	−1.8	9.9
SYNTHETICS											
Net sales								$69	$82	$131	$388
Net profit								(39)	(92)	(83)	1
Return on sales								−56.5%	−112.2%	−63.4%	0.3%
Percent of total sales								0.2	0.2	0.4	0.9
Percent of total profits								−0.6	−1.4	−2.2	0.0
TOTAL NET SALES	$30,008	$23,223	$18,892	$16,787	$21,442	$27,183	$31,088	$37,355	$44,617	$32,622	$41,010
TOTAL NET PROFIT	5,322	2,344	873	630	2,877	3,372	3,827	6,236	6,660	3,694	7,688
RETURN ON SALES	17.7%	10.1%	4.6%	3.8%	13.4%	12.4%	12.3%	16.7%	14.9%	11.3%	18.7%

Note: Total net sales figure does not include interdepartmental sales after 1935.

Source: Departmental Reports to Executive Committee 1929–1935; Treasurer's Reports, 1937–1943, Hall of Records, Hercules Incorporated.

ployment levels at Parlin had been slashed repeatedly. This misfortune was principally attributable to the nationwide decline in auto sales, but it also reflected the appearance of substitutes for nitrocellulose in lacquers and other products: paint and varnish makers switched to coatings using synthetic resins or acrylics, and film and plastics companies adopted cellulose acetate as their preferred raw material.[33] (See below.)

Ironically, this sudden reversal helped save Hercules from what might have been an expensive mistake: in 1930, before the full dimensions of the crisis were apparent, the Cellulose Products Department received approval to build a nitrocellulose plant in the Midwest. The company actually purchased a site for the plant in Port Huron, Michigan, although it abandoned the project before construction work was begun.[34] At the same time, although the company had been "loath to enter the cellulose acetate market," the relatively strong performance of that product (compared with nitrocellulose) spurred Hercules to begin serious inquiries into manufacturing it.[35]

The impact of the Great Depression was also immediate in naval stores: losses began late in 1929 and reached nearly $1 million the next year. By 1932, rosin sales volume had fallen by one-third, and net prices by two-thirds.[36] To make matters worse, Hercules was coming out of a shaky decade in which it had earned a profit in naval stores in two years only. Since 1920, Hercules had invested $7.75 million in the plants ($4 million since their initial acquisition or construction), had spent $1.32 million on naval stores research, and had lost a total of $2,148,243.[37] The executive committee was soon asking probing questions: Could the business be saved? If so, when should the probationary period end? Or should the company stanch its wounds and exit?

In the early 1930s, Hercules considered the future of its naval stores business from several angles. Two proposals received the most serious attention. The first grew out of a scheme that the Development Department had worked on for several years. As championed by Charles Higgins, it called for Hercules to produce gum naval stores on a large scale. By operating a large "central still" (which would be the first of its kind in the industry), the company probably hoped to realize economies of scale in production and marketing, as it had done in wood naval stores. The plan might also allow Hercules to be a "stabilizing influence on naval stores prices," which had declined steadily during the 1920s because of overproduction.[38]

In 1928, under Higgins's prodding, Hercules had formed a joint venture, called the Pine Products Refining Company, with Gillican-Chipley, the leading French gum distiller. The venture imported an experimental still from France and took a purchase option on a gum refinery in Savannah to serve as the nucleus of a new plant. When the depression hit, however, the project lost momentum. Although Bent and Lockwood backed Higgins in urging Hercules to become "an important factor" in gum naval stores, the executive committee concluded that "it would be difficult if not impossible, even controlling a large part of the output, to stabilize prices" in the current environment. Although the committee authorized gum naval stores work on a "small and experimental" scale, that option was never again pursued seriously.[39]

To cure its ailing naval stores business, Hercules was forced to weigh other options. Late in 1931, for example, the company considered licensing a process from the British Xylonite Company for making synthetic camphor from turpentine. That option was ultimately rejected, however, because the executive committee feared the synthetic camphor market might suffer from the same agriculturally related problems that plagued the market for natural camphor: overproduction and unstable prices.[40] Still another option called for Hercules to merge its naval stores business with that of another steam-solvent producer. The executive committee briefly discussed the possibility of joining forces with Newport and other rivals, but the idea apparently was soon dropped, perhaps out of nervousness over potential antitrust issues.[41]

Although the company rededicated its energies to the development of the wood end of the naval stores business in 1931, the economic situation continued to deteriorate.[42] According to Albrecht Reu, superintendent of the Brunswick plant, "The Hercules Board took a hard look at the naval stores business . . . and [was] about determined to withdraw from that type of enterprise." The plant managers of Brunswick and Hattiesburg were summoned to Wilmington to formulate plans to pull out of the business.

> These two men, faced with the possible permanent closing of the two plants on which they had spent so many hard years, did some real thinking and figured far into the night. The next morning they offered to the directors a plan for keeping the plants alive, although just barely so, which would not cost Hercules any more than the cost of insurance, fire protection, and watchman service on the plants if they were to be idle.[43]

The rescue plan called for drastic measures—deep cuts in personnel, wages, and operations—and the directors agreed to its terms. Working hours were reduced by half, wages by as much as four-fifths. At Brunswick, only one chemist and one engineer were retained. Personnel who were let go—including long-term managers such as V.R. Croswell—were offered severance pay and assistance in locating new jobs.[44]

These actions, which apparently helped mollify the executive committee, coincided with another event: the possibility of a new acquisition that would help boost sales of Hercules naval stores products.

Paper Makers Chemical Corporation

On September 15, 1931, the executive committee mulled over an important offer. C.K. Williams and C.H. Knight of Easton, Pennsylvania, and R.L. Snell of Holyoke, Massachusetts, proposed to sell three companies they owned: Paper Makers Chemical Corporation (PMC) (incorporated in Delaware), Vera Chemical Company of Canada, and Georgia-Louisiana Corporation. The three enterprises produced and distributed a variety of industrial chemicals, especially those used in the manufacture of paper.[45]

The offer called for Hercules to exchange $2,432,975 worth of "cash and/ or Liberty Bonds, and 10,000 shares of no-par value common stock" for the total assets of the three companies; these figures reflected the established value of the companies as of May 31, 1931, but they were subject to revision pending a reassessment. The executive committee approved the plan, and six weeks later, on Halloween, Hercules consummated the deal.[46]

PMC originated in 1900 as a partnership between John and C.H. Knight (father and son), C.K. Williams, and George Noble, a paper chemist. Its first facilities—a rosin size plant, office, and laboratory—were built in a converted grist mill at the foot of Mount Jefferson in Easton, Pennsylvania, a mid-sized community on the eastern edge of the state along the banks of the Delaware River. In 1910, the company added a second plant and product line in Easton by building a plant to make "satin white" (a reaction product of pure white lime and iron-free aluminum sulfate used to color and coat paper). William J. Lawrence, a Canadian-born chemical consultant, was hired to run the new operation.[47]

In the next few years, PMC embarked on an aggressive expansion program. First, the company added more paper chemicals to its line by opening new facilities near leading paper manufacturers to reduce shipping costs and improve service. The new facilities were run under the auspices of a separate entity set up by the same investors, Western Paper Makers Chemical Company, which built a plant in Kalamazoo, Michigan, the nation's leading papermaking center.[48] Soon thereafter, PMC added plants in Holyoke, Massachusetts; Pensacola and Jacksonville, Florida; and Savannah, Georgia. These facilities operated on rosin dross and other waste materials from each region and supplied each region's paper trade.[49]

The PMC family of companies also grew by acquisition. In 1921, it bought out the chemical plants of Vera Chemical Company of Milwaukee.[50] By 1928, it had purchased several paper size plants in New York and New England, Superior Pine Products Company of Georgia (which supplied naval stores from a 200,000-acre pine forest), and the American, Canadian, and British plants of Vera Chemical and its affiliates. It also built a new plant in Portland, Oregon, to serve its active paper region. In 1928, the disparate properties were gathered into a single operating company, Paper Makers Chemical Corporation, under Lawrence as president and general manager. With total capitalization of $8 million, the new PMC was the largest paper chemicals maker in the world.[51] It manufactured 125 paper chemicals and other products and acted as a distributor for another fifty related products.[52]

Before the consolidation could be completed, however, the Great Depression rocked the paper industry and sent fortunes tumbling. Total production of paper and paperboard in the United States plummeted by 30 percent. PMC's sales of rosin size totaled $5.1 million on a volume of 163 million pounds in 1929. Two years later, it sold three-quarters of that amount and earned less than half the 1929 total in profits. Sales of linseed oil dropped 23.5 percent, and turpentine sales 11.3 percent.[53] The owners, who were carrying a heavy burden of debt to finance the consolidation, started looking for a way out.

The attractions of PMC for Hercules were obvious: not only would it broaden and diversify the company's capabilities and bring it a dominant position in an important market for chemicals, but it would also stimulate rosin sales. PMC, in fact, was the biggest single consumer of Hercules naval stores.[54] Over time, wrote Dunham in Hercules' annual report, "a close community of interest has developed between the two companies," making the acquisition "a logical integration of your company's business."[55] PMC's chemists, for example, had much to share with Hercules' naval stores chemists about rosin size. Distributors of the two organizations' products sold to many of the same kinds of customers—makers of paper and cardboard, soaps and disinfectants, coatings and varnishes, solvents and insecticides, textiles, rubber, and synthetics.

And despite the depression, the paper industry was one of the largest industries in the economy, ranking seventh in terms of invested capital in 1919. Between 1899 and 1929, production of paper products had increased more than 500 percent, while their total value had risen from $127 million to $1.2 billion. Growth was particularly rapid in the South, where a new pulping process allowed yellow pine to become the major source of kraft paper (a heavy brown wrapping paper used in packaging). Southern paper production, much of it located near Brunswick and Savannah, actually increased during the depression.[56]

Moves such as the acquisition of PMC helped Hercules diversify its risk by entering businesses less exposed to the terrible impact of the depression. Hercules' leaders, individually and collectively, also explored other options. Skelly, for example, supported President Roosevelt and served on a subcommittee of the Chemical Alliance (a national association of chemical producers) seeking a way out of the crisis under the National Recovery Administration (NRA). Other directors were less sanguine about government intervention and opposed efforts to cooperate in the drafting of codes for naval stores products. In this respect, Hercules was a microcosm of the chemical industry, which was bitterly divided over the terms as well as the legality of government-sanctioned cooperation. Indeed, efforts to draft a comprehensive industry code came to nought even before the Supreme Court ruled the NRA unconstitutional in May 1935.[57]

A likelier source of relief, in the board's view, was the company's own actions. In 1933, to help understand the causes of the crisis, the directors and senior managers began meeting monthly with Dr. Lionel Edie, a New York economist and consultant, for wide-ranging discussions about economic and political conditions.[58] Although the meetings produced no specific changes in how Hercules managed its operations, they nonetheless sensitized Hercules to long-term economic trends and emphasized the importance of careful analysis and interpretation of economic data. Escaping the depression would obviously require changes in circumstances beyond the company's control. But in some respects, at least, Hercules could speed its own recovery by pursuing a strategy that built on its newfound strength in research as well as its traditional expertise in marketing.

THE SPECIALTY CHEMICALS COMPANY

In the spring of 1933, Hercules' fortunes began to revive along with the slow recovery of the world economy. The company posted sales increases ranging from 10 to 40 percent by department, and, for the first time since 1929, net earnings exceeded dividend payments.[59] Yet Hercules emerged from the Great Depression a much-changed company. The outward signs were obvious: the new Experiment Station, the new PMC subsidiary, and the happy reopening of factories, rehiring of employees, and restoration of salaries, wages, benefits, and the full working week. By the middle of the decade, the company could even begin to help repay its Wilmington employees for their sacrifices by building a golf course and country club for them on land adjacent to the Experiment Station.

As the crisis had a different effect on different businesses, so did the revival: the new environment offered different problems and opportunities to each of Hercules' lines. All of them, however, shared four important and interrelated characteristics: the increasing scale of operations to ensure economic viability; the greater capital intensity of most facilities; the mounting importance of research and engineering to ensure success; and the growing significance of technical sales and support to customers. Together, these characteristics helped define Hercules as a producer of specialty chemicals.

The Age of Cellulose

The businesses that emerged from the Great Depression in the strongest position were cellulose and cellulose products. "The star in the theatre of synthesis at the present time is 'Cellulose,' " exclaimed a Wilmington writer, who added, "We can truthfully say that we are entering upon the 'Age of Cellulose,' for the epochs of man have been conveniently named for the outstanding raw material that he has learned to use."[60]

Hercules was well positioned to play a leading role in the "theatre of synthesis," beginning with production of purified cellulose. Chemical cotton had been the company's lone bright spot during the depression, largely because of the phenomenal popularity of rayon. In 1925, U.S. producers made 58.2 million pounds. Output more than doubled by 1930, doubled again by 1935, and nearly doubled again by 1939 (when the total stood at 458.8 million pounds). By then, rayon accounted for more than 10 percent of all textile fiber sales in the country. The starting point for manufacture of the artificial fiber was purified cellulose.[61]

The largest rayon manufacturers at the start of the period, American Viscose (a subsidiary of the British firm Courtauld's) and Du Pont, used the "viscose process" for making rayon. (Other types of rayon could be made from nitrocellulose, cuprammonium solutions, or, as will become clear, cellulose acetate). The viscose process involved reacting purified cellulose with caustic soda. The resulting substance, alkali cellulose, was then treated with carbon bisulfide to produce sodium cellulose xanthate. This chemical in a solution of water or weak caustic soda dissolved to form a viscous, molasses-like liquid (hence the name "viscose"). The viscose solution was then spun into an acid solution, which decomposed the viscose and allowed the regeneration of the cellulose in fibers or other forms.[62]

The first viscose rayon started with wood pulp as its base material, although both American Viscose and Du Pont began mixing chemical cotton with the pulp in the mid-1920s to improve the flexibility and appearance of the resulting fiber. At the same time, Celanese Corporation of America (a subsidiary of the Celanese Company of Switzerland) was commercializing a different process for making rayon using cellulose acetate. "Acetate rayon," as it came to be known, was produced by reacting a mixture of wood pulp and chemical cotton with acetic acid and acetic anhydride in the presence of a catalyst. The solution was then washed repeatedly under different conditions to ripen, precipitate, and purify the cellulose acetate. After the material was dried and flaked, it was combined with acetone and oils to create a "spinning dope" mixture from which fibers could be drawn.[63]

Although acetate rayon was at first more difficult to dye and weave than other rayons or natural fibers, it had a crucial advantage: it ignited much less easily. Within a few years, American Viscose, Du Pont, and other producers followed Celanese into acetate production. Between 1925 and 1940, the new fiber grew from 3 percent to one-third of all synthetic fibers (see Exhibit 7.6). The growth of the rayon industry—both viscose and acetate—represented a bonanza to Hercules' Virginia Cellulose Department, which became, under the leadership of Philip Stull, the largest producer of chemical cotton in the world. By 1936, rayon manufacturers accounted for more than two-thirds of the department's total sales of $5.6 million. Of that amount, acetate rayon accounted for by far the largest part.[64] The equally spectacular development of artificial fibers in Europe led Hercules in the summer of 1937 to form a joint venture with Bleachers' Association to build a chemical cotton plant at Holden Vale in Great Britain.[65]

EXHIBIT 7.6 Production and Price of Viscose and Acetate Rayon, 1920–1950

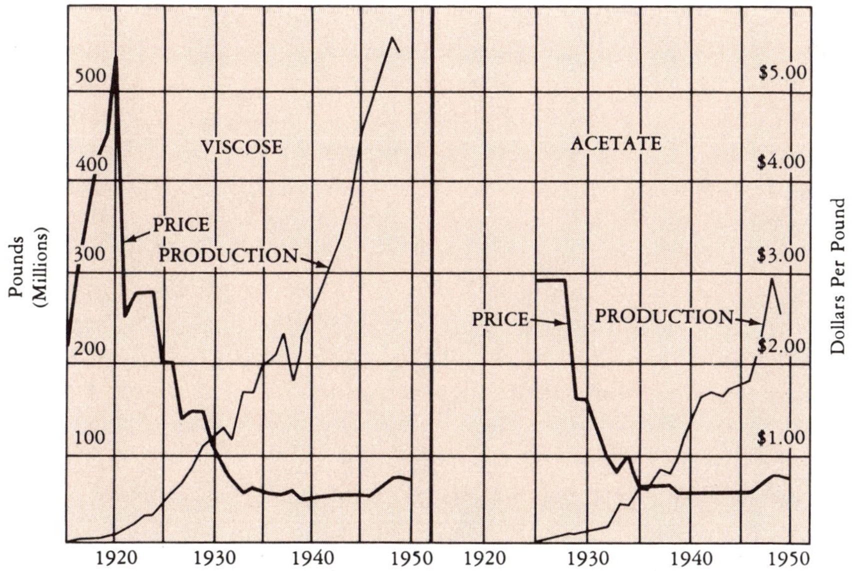

Source: Williams Haynes, *Cellulose: The Chemical That Grows* (Garden City, N.Y., 1953), 145.

Cellulose acetate proved increasingly popular in other forms as well. When delivered as a film or powder, the product substituted directly for nitrocellulose. Eastman Kodak, for example, converted its photographic films from nitrocellulose to cellulose acetate in the mid-1920s to avoid problems of flammability. A decade later, as he watched cellulose acetate erode the market for nitrocellulose in many applications, President Dunham asked Leavitt Bent to investigate the development of the newer product. Early in 1936, Bent reported back with disturbing news: the historical constraint on the growth of cellulose acetate, he believed, was the high cost of acetic anhydride (conventionally made from sodium acetate produced from the destructive distillation of hardwood). The high demand for acetate fibers, however, had led not only to significant improvements in the production of acetic anhydride by traditional means but to wholly new processes for making it from coal and petroleum. As a result, Bent found, the price of acetic anhydride had fallen dramatically, and "cellulose acetate can be produced at a cost not much greater, if any, than cellulose nitrate."[66]

Bent's study reinforced the gnawing sense in the Cellulose Products Department that many nitrocellulose applications it promoted were at grave risk (see Exhibit 7.7). Although Hercules had worked with cellulose acetate since World War I (when it made small quantities as a nonflammable dope for coating airplanes) and had experimented with a lackluster hybrid product, cellulose nitroacetate, as the basis of photographic film, its relationships with key customers such as Celluloid Corporation, Agfa Ansco Company, and Bakelite Corporation were clearly in jeopardy.

Accordingly, Hercules' Development Department launched a search for the best available technology for making cellulose acetate. The department, which had close ties to Europe through its members Max Riemersma and Dr. Emil Ott, reported back in the spring of 1936 with a recommendation to license a process in use by I.G. Farben, the German chemical giant. In June, the board appropriated $1.4 million to build a six-story cellulose acetate plant with an annual capacity of 2.4 million pounds at Parlin.[67]

Over the next three years, Hercules spent an additional $2.1 million to bring total capacity up to five million pounds per year. Engineers at Parlin also improved the I.G. Farben process in significant ways. In 1939, for example, Hercules cut four hours off the time required to react cellulose with acetic acid and acetic anhydride. In addition, Hercules engineers succeeded in increasing the purity and color of the finished product. According to M.G. Milliken, general manager of Cellulose Products, such efforts not only lowered manufacturing costs and impressed existing customers but also led to entirely new applications. He predicted that Hercules could sell the entire rated capacity of the plant to plastics fabricators in 1940—if it chose to do so.[68] (See Exhibits 7.8 and 7.9.)

Hercules' strategy for entering the cellulose acetate market was replicated in the way it handled other cellulose products, such as ethylcellulose, cellulose acetobutyrate (Hercose C), and cellulose acetopropionate (Hercose AP). In each case, the company stepped up development of the product—often through the licensing of German technology—when it perceived either threats to its existing nitrocellulose business or opportunities to use an increasingly familiar chemistry

EXHIBIT 7.7 Hercules' Nitrocellulose Sales to Consuming Industries, 1928–1939

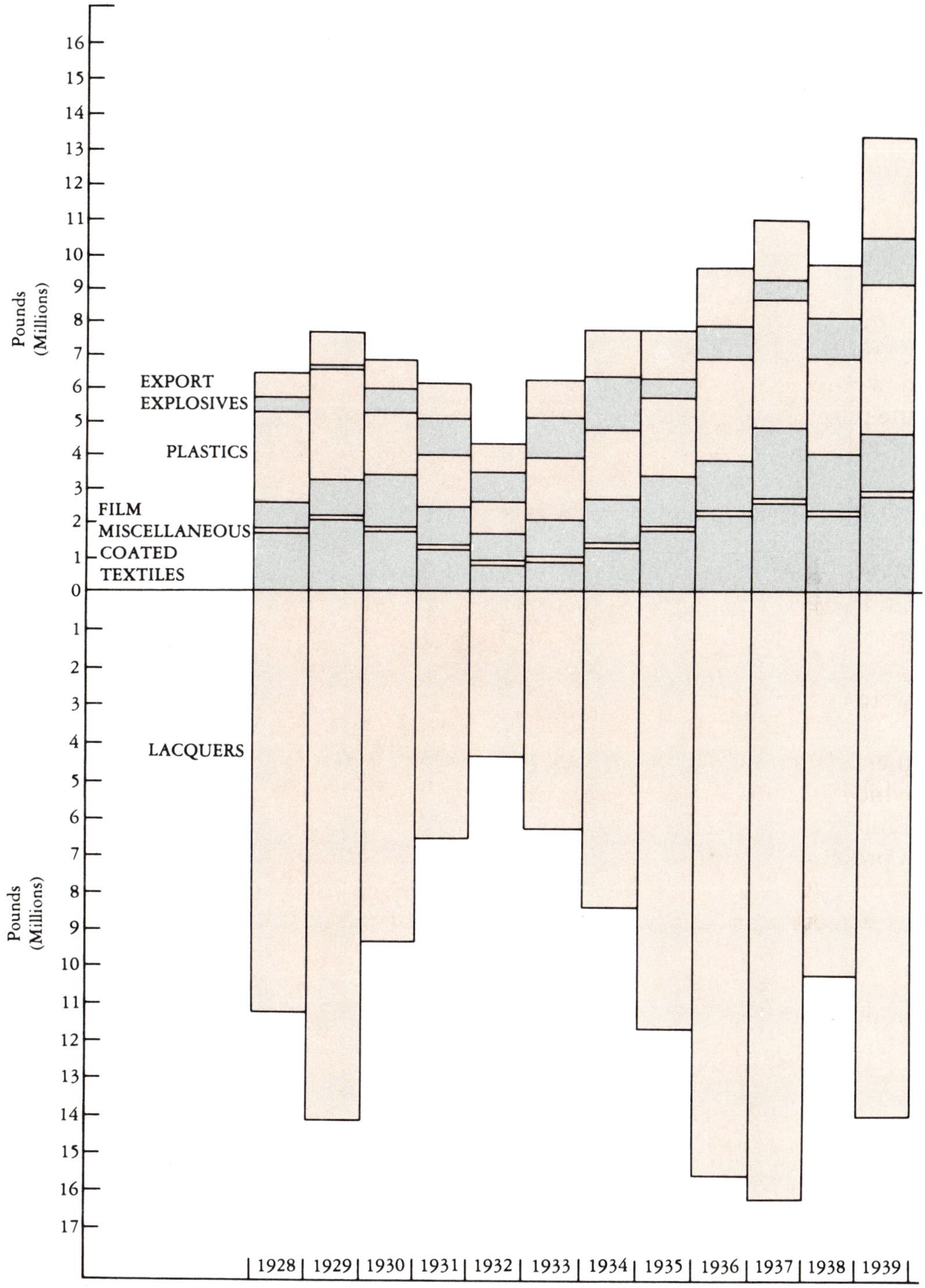

Source: Cellulose Products Department, Annual Report for 1939.

EXHIBIT 7.8 Hercules' Sales of Cellulose Acetate, by Consuming Industry, 1936–1939

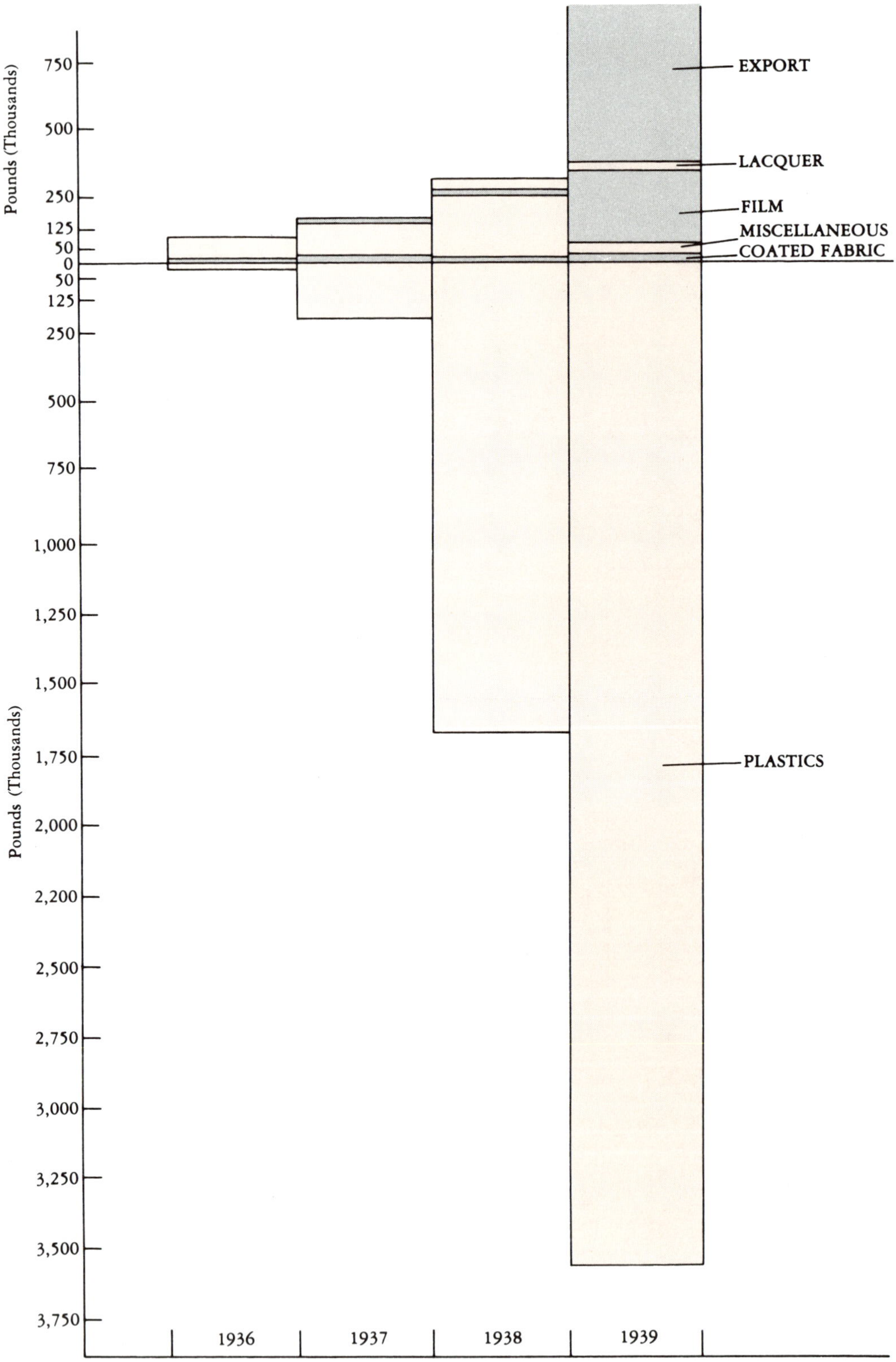

Source: Cellulose Products Department, Annual Report for 1939.

EXHIBIT 7.9 New Products of Hercules' Cellulose Products Department, 1936–1939

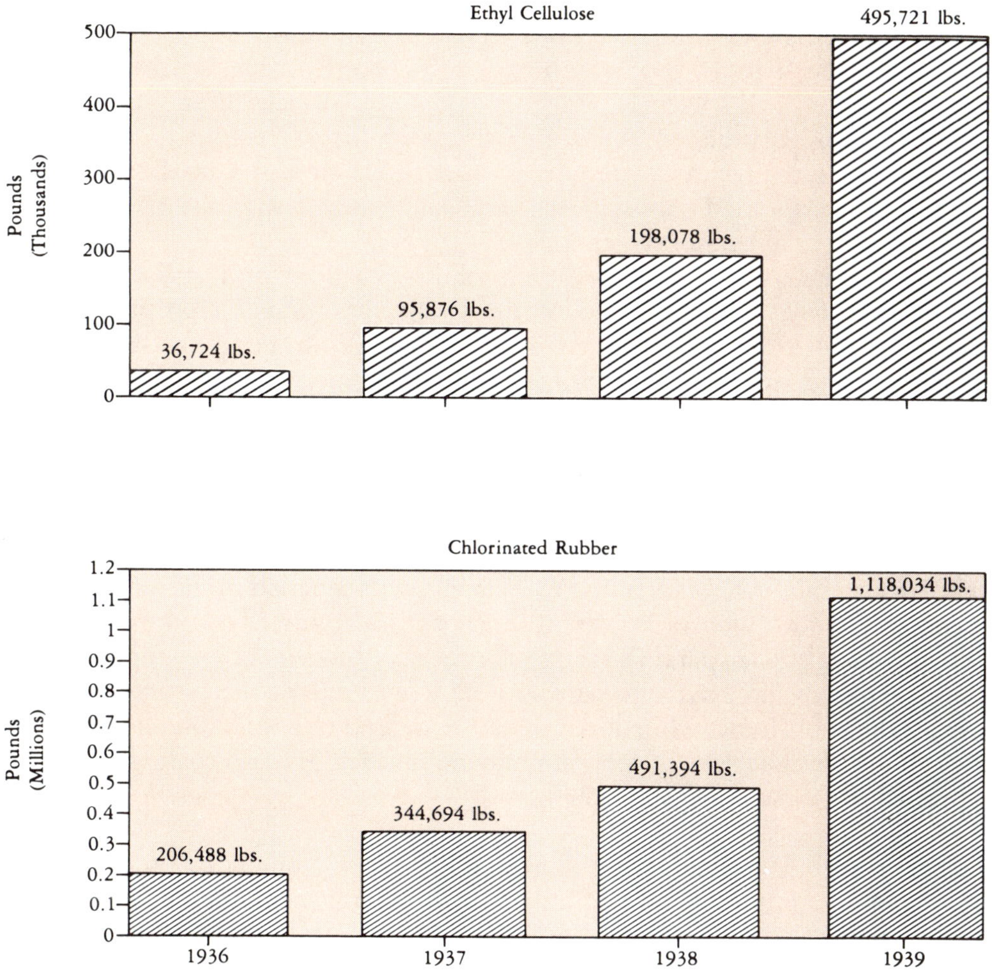

Source: Cellulose Products Department, Annual Report for 1939.

to sell ancillary products. In each instance, both the peril and the promise involved Hercules' sales to makers of protective coatings. To understand the company's reactions to threats to its nitrocellulose business, as well as its eventual decision to promote a broad range of cellulose products, it is helpful to consider not only its position in the protective coatings industry but also changes in that industry.

"The Paint, Varnish, and Lacquer Industry, or the Protective Coatings Industry," wrote an anonymous Herculite, "can be said to be two industries, one old and one new." The distinction in the writer's mind was between the "old industry," in which paint consisted of white lead pigment in a linseed oil base, and the "new industry" geared to the modern world of mass production, with products that hastened drying, improved durability, hardened finishes, or provided color to almost any surface under almost any conditions.[69]

The "new" protective coatings industry was the direct result of the revolution in industrial chemistry after World War I. Paints, varnishes, and lacquers were no longer simple products mixed and stirred on-site by the user, but complex mixtures of solvents, pigments, resins, plasticizers, drying agents, and other additives formulated by manufacturers to match customers' specific requirements. As the technical complexity of coatings increased, so did the overall market, which doubled between 1919 and 1939 and grew at a much faster rate than the industrial economy generally.

By far the fastest growing segment of this expanding industry, and the one in which Hercules was entrenched in the 1930s, was industrial coatings: quick-drying, specialized lacquers and enamels for mass-produced goods such as cars, appliances, and furniture. Uniformity and consistency of coating was of paramount importance for such products; these criteria were much more likely to be satisfied with synthetic than with natural ingredients. Manufacturers of coatings came increasingly to rely on the control provided by chemical companies such as Hercules.

That nitrocellulose might be replaced by other chemicals in lacquers, and that lacquers might be supplanted by enamels (as happened in the auto industry during the 1930s), were ever-present concerns to Hercules. To address them, the company sought to expand the range of products and services it supplied to the coatings makers and gradually came around to a broader conception of its cellulose products business: while it continued to defend nitrocellulose, it also began to push the development of other ingredients and additives to coatings. Milliken expressed the new strategy in 1937:

> It is a fixed policy of the Hercules Powder Company to produce a wide range of cellulose products, thus making available to users all modifications of cellulosic esters and ethers which show possibilities of market development. It is not our intention to concentrate on any specific cellulosic material: rather, the object is to make available a wide selection of cellulose materials so that each manufacturer can determine which products best meet his requirements. Technical service is available for all cellulose products. In this way, the individual manufacturer has at his command a fund of technical data and background to help in evaluating each cellulosic material according to specific needs. Active research work is continually being done on a large scale in an effort to search out and produce new cellulosic materials with practical properties, as well as new applications of the present products.[70]

As Milliken suggested, Hercules was anxious to strengthen further its ties to the coatings industry by providing technical service and support to its customers. In 1936, the company started a new magazine, *Hercules Chemist,* which was aimed at keeping customers abreast of new developments. At the same time,

Hercules launched a six-month "industrial finishes school" for coatings manufacturers because "it was felt that more widespread basic information on lacquer technology was essential if the lacquer industry was to have a healthy development and growth."[71]

The company also looked outside cellulose chemistry for products that would appeal to the coatings industry. Between 1933 and 1937, for example, the Cellulose Products Department and the Development Department worked together on a project to commercialize zinc sulfide pigments, which produced bright colors in paints.[72] Hercules ultimately abandoned the project because zinc sulfide could not compete with less expensive titanium dioxide pigments. But another foray into a "foreign" chemistry proved extremely successful: chlorinated rubber was a substance that improved paint's resistance to acids, alkalies, and salts and adhered well to metals and concrete surfaces.

Hercules first learned of the product, which was sold in Europe under the trademark Tornesit (derived from the German town of Tornesch, where chlorinated rubber was developed), early in 1932 through the efforts of the Development Department. The following year, Hercules licensed the process to manufacture the additive from its European developers and their American representatives: Deutsche Tornesit GmbH, New York–Hamburger Gummi–Waaren Compagnie, and Chadeloid Chemical Company of New York. A plant was built at Parlin in 1934, and production was started almost immediately. Within three years, Hercules' investment was repaid; ready applications were found for Tornesit in chemical-resistant finishes, in waterproof adhesives for paper, cloth, and linoleum, and in certain printing inks.[73]

From Naval Stores to Synthetics

The years of the Great Depression were most decisive in the history of Hercules naval stores. When the decade opened, hard times and the weight of history nearly led to the demise of the Naval Stores Department. When it closed, however, the business was firmly on its feet and the world leader in the industry. In 1939, Hercules achieved an all-time volume record and earned a profit of $650,000.[74] Moreover, although the smaller wood naval stores producers increased their overall market share approximately 10 percent (to 20 percent), Hercules maintained its leading position relative to its chief rival, Newport Industries (successor to Newport Rosin & Turpentine Co.), and surged ahead in turpentine sales.[75]

The turnaround had many sources: the expansion of international sales, increased coordination of domestic markets, gains in productive efficiency, and reaping the fruits of the nation's most sophisticated naval stores research program. To be sure, there had been key breakthroughs during the decade, but what mattered more was the cumulative effect of gains in these diverse areas and the company's ability to coordinate them.

To begin with, Hercules improved its naval stores product mix. The biggest challenge was to substitute FF rosin for lighter-colored higher priced Pexite® rosin. That was especially critical early in the decade, when the smaller wood extractors, who lacked the ability to produce pale rosins, flooded the market (already loaded with F-grade gum rosin) with the darker grade and cut prices. (Between 1926 and 1932, the price of F-grade rosin fell from $14.70 to $2.45 a barrel.) As early as 1934, Hercules was converting two-thirds of the FF rosin produced at Hattiesburg and Brunswick into Pexite® rosin. Moreover, Hercules succeeded in increasing its sales of FF rosin overseas during the worst years of the glut.[76]

Indeed, Hercules, like other wood naval stores producers, strengthened its naval stores exports during the 1930s. By 1939, half of the company's FF and Pexite® rosins, one-third of its pine oil, and one-quarter of its turpentine and Belro® resin were sold outside the United States. Most of these products were sold in Europe, which remained the leading consumer of Hercules naval stores exports, although shipments to South America amounted to one-fourth of all Hercules rosin sold abroad in 1939.[77]

On the whole, the wood naval stores producers made dramatic gains relative to their gum competitors during the 1930s, especially in the export market. Between 1929 and 1939, exports of gum rosin declined from 1.24 million barrels to 594,381, while exports of wood rosin rose from 196,888 to 341,285; as gum turpentine exports shrank from 326,091 barrels to 196,831, exports of wood turpentine climbed from 17,419 to 40,457. Despite prices that were attractive to customers, the gum naval stores sector suffered severely at the hands of still lower priced foreign gum commodities, more standardized foreign and domestic wood products, and natural and synthetic substitutes. In addition, gum rosin enjoyed little or no price advantage over wood rosin (the margin decreased in the higher grades), and wood rosin was widely regarded as more uniform and clean.[78]

At the end of the 1930s, the noted industry authority, Thomas Gamble, commented on this important transformation. The gum sector, he argued, would find salvation only when its small, individualized stills were replaced by large, capital-intensive distillation plants, and when extensive research and careful chemical control of processes were used to improve the quality and diversity of products. These attributes, of course, had long been Hercules hallmarks.[79]

The most serious challenge to Hercules' preeminence in wood naval stores came from its old rival, Newport. In the summer of 1929—barely a year after Hercules had introduced Pexite® rosin—Newport announced an alternative process for making pale wood rosin from activated fuller's earth.[80] During the next several years, Newport patented processes for liming and alkali-treating rosin and featured popular pale rosin brands such as Munn (grade M) and Tenex (grade X).[81] Hercules challenged some of Newport's innovations as patent violations. The most important case came in 1938, when Hercules charged that Newport had infringed one of its critical Kaiser and Hancock rosin-refining patents.[82]

Hercules' relationship with its rivals was often tempered by its desire for stability in the marketplace. Some cooperation was achieved through trade associations. Lockwood was elected president of the Pine Institute of America (which acted as an industrywide information clearinghouse) in 1929, the year Hercules attempted to establish both domestic and foreign trade associations for steam-solvent producers. It succeeded at the latter in 1935 by forming the Wood Naval Stores Export Association (exempt from antitrust prosecution under the new Webb-Pomerene Export Law) to act as the sole foreign distributor of FF rosin and pine oil. With Hercules in the commanding role, the organization increased export market coordination and provided the company with additional revenues from commissions.[83]

For the most part, however, Hercules, Newport, and their smaller rivals remained locked in a close contest to develop and market products first, better, and cheaper. The early 1930s were a heyday of naval stores innovation at the new Research Center. (Ten of the fourteen new products that emerged from Hercules labs in 1932 and 1933 were in the naval stores area.)[84] Still, technical breakthroughs are not always commercially successful, and new products are not always research-driven. Some emerged through market demand, and still others were by-products of the manufacturing process. The strength of the Naval Stores Department was an ability to take advantage of opportunities in whatever guise they appeared.

Consider the very different paths of development of two of the decade's most successful naval stores products: Staybelite® hydrogenated rosin and Vinsol® resin. Staybelite® was envisioned in the marketplace and born in the laboratory. From the beginning, pale rosins had been plagued by their propensity to discolor and grow brittle with age. In 1938, after a dozen years of research, Hercules overcame these problems and launched the world's first highly saturated, hydrogenated rosin, dubbed Staybelite® in recognition of its ability to resist the oxidation that caused deterioration.[85] That and other characteristics made Staybelite® resin highly valued by makers of adhesives, laminants, paper coatings and size, rubber, and certain synthetics. Before going into commercial production at Hattiesburg, Hercules worked out production problems by running a pilot plant and anticipated marketing problems through numerous consumer tests. In short, Staybelite® began as a consumer need and followed a circuitous path from marketing to research, then back to production, marketing, and the customer.[86]

In contrast, Vinsol® resin emerged from the manufacturing process through serendipity. Reginald Rockwell, first at Hattiesburg (as assistant superintendent) and then at Brunswick (as assistant manager), spearheaded the development of the solvite process of rosin recovery, which substituted benzene for gasoline as the solvent. Whereas gasoline insoluble materials had been left behind in the spent chips, the new solvent permitted the recovery of a dark, hard, high-melting natural resin dubbed Vinsol® (from "V. Insol," the chemists' abbreviation for very insoluble). Further investigation revealed other special characteristics, including an unusual resistance to breakdown under high voltages.[87]

Unfortunately, few customers knew of Vinsol® resin attributes. At Hattiesburg, surplus resin began to pile up and was burned as a "waste by-product and an irritating nuisance." The problem spread to Brunswick, after its conversion to the solvite process in 1938; there, workers jokingly measured the product in acres, five feet deep, and lit up the night sky around the plant by burning Vinsol® resin.[88]

But marketers, working with researchers, targeted uses for the product as an extender for phenolic molding resins, a foundry core binder, an ingredient of inks, and its sodium salt as an emulsifier for asphalt. Because the special insulating, bonding, electrical, and petroleum oil–resisting properties of Vinsol® made it ideally suited as a component in underground electric cable joints, Hercules secured contracts with Detroit Edison and other electric utilities. What was once the company's "No. 1 Naval Stores problem" was slowly transformed into a substantial source of income.[89]

Although these and other new products were sold in bulk by the end of the decade, the company's growing number of "special products" were still meeting with dubious commercial success. Of the vast array of terpene compounds and rosin derivatives offered during the early 1930s, only a handful (including *alpha*-terpineol, abietic acid, ethyl abietate, Daintex,® Hercosol® rubber reclaimer, and borneol) contributed more than a few thousand dollars to annual sales. By 1936, twenty-nine special products were adding $200,000 to annual revenues, but this still represented less than 4 percent of total naval stores sales.[90] Except for this special group, the company's naval stores products no longer could be easily segregated into basic commodities and specialty chemicals. By 1939, the bulk of Hercules naval stores business was composed not simply of rosin, turpentine, and pine oil but of more refined products for specialized uses: Pexite® rosin; Belro,® Vinsol,® and Staybelite® resins; and pinene, Dipentene,® Solvenol terpene hydrocarbon solvents, and Daintex.®[91]

In February 1936, Hercules transferred responsibility for its more specialized naval stores derivatives to a new unit, the Synthetics Department, under the leadership of Wyly "Josh" Billing. Hired by Hercules in 1917, Billing took leave to earn a Ph.D. from Columbia, then returned to serve as a research chemist at Kenvil, as chief chemist at Hopewell, as a member of the Development Department, and, after 1934, as assistant director of the Experiment Station.[92]

At its inception, the new department was a marketing organization formed to sell Abalyn® methyl ester of rosin, Hercolyn® hydrogenated ester of rosin, Petrex® synthetic resins, and other naval stores derivatives used as plasticizers, solvents, and other constituents of protective coatings, inks, and adhesives.[93] These products were among the most chemically refined and highest in value-added that Hercules had to offer. More important, they possessed very specialized characteristics and their successful application required a high level of sophistication on the part of potential consumers.

In short, the synthetics were the most "special" of the company's specialty chemicals. The premier challenge for the new department would be to build sales in narrowly defined markets where product differentiation was both difficult to establish and essential for success. Hercules had great hopes for its cache of expensive new synthetics. Through them, Billing, his team, and Hercules would learn important lessons about the problems associated with being a supplier of specialty chemicals in a competitive marketplace.

The department's most important product at the outset (accounting for nearly one-third of its revenues) was Abalyn® methyl ester of rosin. Abalyn® first appeared in 1932, after Hercules researchers succeeded in reducing the acid in rosin (to get ester gum of higher value) by reacting it with methyl alcohol rather than the usual glycerin. The process imparted special properties to the product, including its pale color and faint odor, high boiling point, water and alkali resistance, and solvency and solubility—qualities that made it an excellent plasticizer for cements, adhesives, and nitrocellulose lacquers and a useful solvent for most natural and synthetic resins, rubbers, and drying oils. Its sister product, Hercolyn,® shared many of these properties but differed primarily in its resistance to oxidation. (Hercolyn® also contributed far less to sales and profits.)[94]

Petrex® resins, developed by E.G. Peterson (a member of the class of '29), seemed the most promising of the early synthetics. A complex combination of acidic materials derived from Hercules terpenes, Petrex® resins could be used as part of the building blocks of alkyd synthetic resins. Hard, tough, and flexible, Petrex® resins (thanks to their terpene roots) also possessed the high gloss, good pigment wetting, and even-flowing and leveling qualities sought in a good coating. Paler, more adhesive, and more resistant to discoloration, water, and alkali than other acids, they could be combined with other polybasic resins to produce a wide range of synthetic resins.[95] The department also marketed several specialized resins named Petrex® 1, 2, 3, 4, 5, 6, 13, and 22, which were used primarily in furniture lacquers, shellac plasticizers, and paper lacquers and hot melt coatings, and which at first accounted for a small fraction of total sales and no profits.[96]

The early marketing of these products, especially Petrex,® presented interesting challenges. One was to properly identify markets—not an easy task in an economy in the throes of a synthetic chemicals revolution. At first, Petrex® was seen as a strong rival of phthalic anhydride for interior coatings.[97] An oversupply of phthalic anhydride, combined with the "breakdown of the patent situation on alkyd resins," prompted many paint and varnish makers to manufacture their own synthetic additives. That situation, in turn, shaped Hercules' thinking about where in the vertical chain from producer to consumer it should direct its efforts to market Petrex.® The decision was to market synthetics ingredients by educating paint and varnish makers about how to fabricate synthetic resins (using Petrex®). "If it is proved that Petrex is technically and economically sound," Billing explained, "it lets us enter the extensive varnish and protective coating field with a raw material instead of a semi-finished product like a resin solution." Hercules would be able to reduce sales and production expenses by concentrating on one rather than dozens of products, while customers would realize several advantages from making their own synthetic resins, including better control of properties and materials costs.[98]

By 1938, the phthalic anhydride substitution strategy was seen as "erroneous," and new directions were charted. Abalyn® and Hercolyn,® too, displayed "no striking utility" in filling "a definite need of industry." The perennial marketing challenge—described by one marketing expert as a search for the "unique selling position"—was particularly difficult because buyers were being inundated with hundreds of new synthetics, many almost indistinguishable from the others, pouring out of the world's chemical laboratories.

At the end of the decade, Hercules' synthetics business remained small and highly specialized; the company was still groping for new markets. An interesting possibility, targeted at a special three-day conference in Wilmington to discuss technical service and sales issues relating to synthetics, was the textile industry.[99] Whether efforts to develop that market would pay off was a question for the future. Nonetheless, the need for the synthetics group to splinter off from naval stores epitomized the ongoing transformation of the company from a processor of commodities from raw materials to a fashioner of man-made chemical building blocks for industry.

Paper Chemicals

The twin needs to develop a deeper understanding of basic chemistry in existing products and to build strong technical ties to targeted industries emerged in Hercules paper chemicals business as well. In many respects, the paper industry was an ideal customer for a chemical company to serve. Papermaking consists of two stages: pulping, in which plant matter is broken down and its cellulose fibers are isolated and treated; and papermaking itself, whereby pulp is formed and bonded into uniform sheets. Both processes involve heavy consumption of specialty chemicals. Indeed, it is fair to say that paper is a tissue of chemicals; Hercules positioned itself in the 1930s to supply a significant proportion of them.

The acquisition of PMC was a pivotal event in the history of Hercules. As John M. "Jack" Martin, former chairman of the company, recalls, "This had to be the first major turning point in Hercules in my view because we broadened our raw material base. We opened up new areas of chemical potential activity, things for our Research Department to work on and we were able to penetrate markets that we had not been able to touch before."[100]

Before Hercules could realize its new opportunities in paper chemistry, however, it had first to deal with the sprawling and chaotically organized PMC. When Hercules bought it, the company operated sixteen plants distributed throughout the major papermaking regions in the United States and Canada and a single facility in England. The company also maintained a laboratory at Kalamazoo with nine technicians; its plants at Easton, Pennsylvania; Milwaukee, Wisconsin; Marrero, Louisiana; Portland, Oregon; and Atlanta, Georgia; also had chemists.[101] PMC's direct sales force, headed by the colorful salesman Jimmy Foxgrover, included fourteen salesmen in paper chemicals, eighteen in general industrial chemicals, and two technical representatives stationed at various plants. In addition, PMC sold some chemicals through jobbers and distributors that did not serve as Hercules representatives.[102]

PMC produced at least 180 chemicals in 15 distinct groups: acids (10), alcohols (5), alkalies (32), alums (5), ammonia (5), calcium (4), cleaners and cleansers (25), disinfectants and insecticides (8), foundry compounds (3), oils

(7), pigments (4), salts (5), soaps (20), soap powders (4), and miscellaneous (43).[103] The leading product by far was rosin size, the vast bulk of which was sold to papermakers. Other important products included satin white, sulfonated castor oil, sulfonated tallow, turkey red oil, aluminum sulfate, silicate of soda, casein, and foam killers. Most of these chemicals had specific applications in the paper industry, although some were used in coatings for objects as varied as textile fibers and window shades, or for other purposes such as treating water.

For several years after the acquisition, Hercules gave PMC's strong-minded general manager, William J. Lawrence, a free hand in running his business. When business picked up after the depression, Hercules' board supported Lawrence's requests to add new product lines by purchasing the assets of three small companies: Universal By-Products, a West Coast processor of casein (April 1934), Dairy Products, a casein producer in Chicago (September 1935), and Providence Drysalters, a maker of sizing chemicals and oils for the paper and textile industries, based in Rhode Island (January 1936).[104]

In 1936, PMC enjoyed its best year ever: gross profits were nearly $1.7 million, up more than one-third from the previous year.[105] Prosperity was short-lived, however; the paper industry suffered during the decade's second major economic downturn, which began in the summer of 1937 and lasted slightly more than a year. Revenue in the paper industry plummeted by 40 percent in the second half of 1937 alone.[106] To complicate matters, when the Hercules executive committee showed signs of abandoning its hands-off policy toward the department, Lawrence resigned abruptly in May 1937. ("He was an entrepreneur who had his own ideas about how to run business and it didn't fit with Hercules' philosophy," recalls one who knew him.)[107] Philip Stull replaced Lawrence as general manager of PMC. Soon thereafter, Hercules relocated PMC's executive offices from Kalamazoo to Wilmington, a move that for a time engendered "a certain amount of unsatisfactory discussion in the paper trade, particularly in the Middle West."[108]

Stull's immediate response to the economic downturn was to diversify output at several plants. This strategy, difficult in most industries, was relatively easy for PMC because of the small size of most of its facilities, the variety of products made at many of them, and the simplicity of its processes. At least two casein plants were converted to the manufacture of skim milk, for example, because demand for that foodstuff was holding up. In addition, when it was discovered that the casein plants in Carthage, Watertown, and Elroy suffered from a "lack of ability to diversify operations," they were modified to make other products, such as whey, dried skim milk, and acid, or rennet casein.[109]

In 1937, under Stull's prodding, the PMC Department undertook a searching evaluation of its businesses. Several distinct weaknesses were observed: the Industrial Chemicals Department, for example, was found to have an "urgent need" for more technical men, both in operations and in sales and service, to bring a "greater degree of technical control." Two important, more general recommendations came out of the 1937 study, however, and Hercules began to institute them the following year: the consolidation of plants, and the standardization of manufacturing operations.

By shutting down "small unprofitable locations," Hercules hoped to realize "major economies" in the PMC Department. The plan was not executed timidly. In 1938 alone, the plants at Lockport and Easton, Pennsylvania, were permanently shuttered, and the manufacture of size was discontinued at Carthage, Jacksonville, and Stoneham. Plans were also made for the "discontinuance of such locations as Providence, Rhode Island, and discontinuance of size operations at Albany, New York, and Pensacola, Florida."

The standardization program was also sweeping in its impact. Arthur C. Dreshfield, PMC's "practical and technical" manager and head of the Kalamazoo laboratory, described the change:

> In prior years, our manufacturing operations were conducted with a minimum of technical and engineering supervision. As a result, local plant practices developed spontaneously without uniformity or correlation with practices existent elsewhere in the Division. Heterogeneity best characterizes our prior manufacturing technique. In 1937 it was concluded that operating practices should be correlated, coordinated, unified and improved. Nineteen thirty-eight has been the first attempt to put such plans to practice. . . . [It marks] the beginning of a transition period, the duration of which will extend several years into the future.[110]

At the end of the decade, such measures took effect: PMC sales, which had fallen by 40 percent between 1937 and 1938, slowly recovered. The department returned to profitability as well. A key aspect of PMC's business that did not change under Stull, however, was its sales strategy. The flamboyant Foxgrover (whose outlandish dress—green-and-red-striped pants, blue-and-white hat, purple coat, orange socks, alligator shoes, and "over 500 garish ties"—became a trademark) maintained a high esprit de corps among PMC salesmen, who were renowned for maintaining close relationships with their customers.[111]

Explosives

The mounting importance of R&D across product lines and increasingly tight relationships with key customers were the hallmarks of the decade and manifested themselves even in Hercules' first business, commercial explosives. On the demand side, the period was marked by the growing significance of ammonium nitrate as a leading energetic constitutent of high explosives: in 1928, ammonium nitrate accounted for about 28 percent of the total ingredients; a decade later, its share climbed to nearly 40 percent.[112] Ammonium nitrate had several key advantages over nitroglycerin, including better performance in cold and wet conditions. The chief benefit, however, was cost: ammonium nitrate was far cheaper to make than nitroglycerin.

Customers' preference for ammonium nitrate had important implications for Hercules. For one thing, it brought the company into competition with other manufacturers of the product—chiefly fertilizer companies—that had the advantage of large-scale operations. For another, Hercules was obliged to reconsider its growing dependence on manufacturers of synthetic ammonia, the key ingredient for producing nitric acid via the ammonia oxidation process in operation at Parlin, Bessemer, and Hercules Works. East of the Rockies, Hercules purchased its ammonia from Du Pont and Allied Chemical; in the West, from Shell Chemical Company. As a large-scale purchaser of a key raw material, Hercules was eager to find ways to supply its own needs. The opportunity to enter the growing market for synthetic ammonia in agricultural and industrial markets was an additional motivation.[113]

In 1932, Hercules began a multiyear investigation of existing ammonia-production technology with an eye toward developing a low-cost process. Although the company had limited experience with production of the key ingredients of ammonia, hydrogen and nitrogen, at its ammonia oxidation plants the company was rapidly accumulating expertise with natural gas, the feedstock used by Shell Chemical. Shell's process involved producing hydrogen from natural gas and obtaining nitrogen from the air. At Hercules, engineers sought to produce both hydrogen and nitrogen from the same natural gas source.[114]

Under chemical engineer James H. Shapleigh, Hercules developed a process to produce hydrogen from natural gas by catalytic cracking. The key breakthrough was a new furnacing process that enabled continuous reaction of natural gas with steam under high pressure in the presence of a catalyst and resulted in hydrogen of high purity. Soon thereafter, Hercules developed a proprietary technique for producing nitrogen from flue gas produced in the cracking process. The final step was to combine the hydrogen and nitrogen to produce ammonia. After evaluating existing technologies, Hercules chose to license the high-pressure process for synthesizing ammonia from the French company, L'Air Liquide.[115]

By the end of the decade, Hercules was sufficiently confident of its ability to process natural gas into hydrogen and nitrogen to invest in a commercial-scale ammonia synthesis operation at Hercules Works, a site chosen for its proximity to cheap natural gas sources and to markets for ammonia fertilizers. In February 1939, the company appropriated more than $1 million for a plant with capacity to manufacture eight thousand tons of anhydrous ammonia a year.[116]

Hercules' venture into ammonia synthesis would prove highly significant in later years. At Hercules Works the company refined and extended its engineering skills and developed new expertise in chemical processes involving extremely high pressures and wide temperature ranges. The venture also opened new markets such as fertilizers and other uses of ammonia, its constituents, and their by-products.

In commercial explosives, more efficient processes and chemical controls allowed Hercules to lower manufacturing costs and tailor products for specific uses and conditions. These were important advantages in a market that remained troubled despite signs of economic recovery in the late 1930s. Not only did many explosives-consuming industries—especially coal—continue to decline or convert to other methods of mining, but they were also learning to become more efficient with the help of technical sales and service representatives. As George Norman put it, "What would be a natural growth is largely offset by the industry teaching the users how to do their work with less and less explosives." The industry's principal growth market in the 1930s was the construction segment; public works projects such as the Los Angeles water system and the Pennsylvania Turnpike were consuming massive amounts of explosives.[117]

Conclusion

By the end of the depression decade, Hercules was again transformed, this time from a maker of commodity chemicals to a producer of specialty chemicals—a supplier of specific, technical, and high-value products to industrial customers. Signs of the change were evident across the company: the increasing scale and expense of operations; the higher technology of most product lines; the mounting importance of technical agreements and licenses; the growing emphasis on chemical control and standardization; and the greater significance of technical sales and support. An important symbol of Hercules' transformation was the creation of a separate Patent Department on March 31, 1939, to monitor and protect the company's technical achievements.[118]

Yet another sign of a new era of Hercules was Russell Dunham's resignation in 1939 as president of the Hercules Powder Company after more than twenty-six years on the job. Although he remained chairman of the board of directors and chairman of the finance committee, his decision to step aside symbolized the changing character of the company he had led: sixty-eight-year-old Dunham, the accountant and financial manager, was replaced by fifty-year-old Charles Higgins, the cellulose chemist.

The change in leadership, moreover, was part of a general changing of the guard at the company. Of the company's first leaders, only Dunham, Bacchus, and Prickett remained active in 1939. Although a writer from *Fortune* observed (a bit unfairly) that "none of these gentlemen lend themselves very vividly to portraiture," the first generation left important legacies to the company: a strategy of careful growth, a prudent (if not conservative) approach to financial management, administrative systems that exposed significant details for managing the business, an appreciation for the value of hard work, and a genuine concern for the safety and welfare of the employees.[119]

The transition at the top was arranged long in advance and proceeded in an orderly fashion. Higgins—who happened to be Dunham's son-in-law—was groomed carefully to become president. Elected a vice president in 1928, he joined the executive committee a year later. He became the clear heir apparent in March 1932, when he was elected vice chairman of the executive committee and was designated ahead of Skelly and Bacchus to act for the chairman and president in his absence. His final step in preparation for the succession was to serve as chairman of the finance committee, starting in November 1936.[120]

Henceforth, Hercules was in the hands of its second generation: Higgins, Bent, Nixon, Morrow, and others hired since the company's founding. As they looked ahead, there was ample reason for optimism about the business. At the same time, though, the news from Europe was deeply troubling.

CHAPTER 8

Defense Contractor, 1940–1945

When Charles Higgins succeeded Russell Dunham as president of the Hercules Powder Company, the world's major powers were already preparing for the second full-scale war in less than a quarter-century. The signs of troubled times appeared early at Hercules, which in the late 1930s found itself cast once again in the role of a major munitions supplier.

During World War I, Hercules had started its major transformation from an explosives-maker to a chemical company. Sales and employment expanded nearly tenfold, and Hercules quickly demonstrated the ability to develop and manage new chemical technologies on a large scale. The direct impact of World War II on the company was much less dramatic. The rapid buildup of the 1940s certainly swelled sales and employment and again changed the mix of business. This time, however, change did not betoken transformation: Hercules simply did what it already knew how to do, albeit on a much expanded scale.

Hercules' role as a military supplier during the two world wars also differed in other significant respects. No longer were the major European powers dependent on foreign sources of munitions. As a result, although Hercules supplied some propellants directly to the British early in World War II, its principal customer was the U.S. government. By the early 1940s, moreover, the United States (and the allies it supplied through the Lend-Lease program) had agreed on specific standards for ordnance production. For Hercules, standardization meant that the pace of technological change in military supplies was slower, and that the company had less call to innovate than it had during World War I. Finally, the specialization of weaponry had reduced opportunities to apply military technology to commercial markets. This specialization, coupled with widespread dissatisfaction among munitions suppliers and the government with aspects of their relationship during World War I, led to changes in the management of defense production. The U.S. Army Ordnance Department preferred not to advance money to private businesses for expanding or upgrading their commercial plants; instead, it chose to pay for new facilities that the government owned and that private companies designed, built, and operated. The model for the relationship between Hercules and the government, in other words, was not what had happened during World War I at Kenvil, Parlin, or the Hercules Works; rather, it was the experience at Nitro, West Virginia. Between 1940 and 1945, Hercules supervised the construction and provided the management of six government-owned, company-operated (GOCO) plants.

Although World War II did not transform Hercules, it nonetheless accelerated trends already developing in the company's business. In particular, the war stimulated tremendous demand for coatings and synthetic fibers, as well as for new, "strategic" materials such as plastics. To meet the demand, Hercules redoubled its efforts to improve its technological capabilities and increase production capacity in its commercial operations. The company also continued to develop new, specialized chemicals from cellulose and rosin. Despite frustrating wartime shortages and market restrictions, Hercules strengthened its position in specialty chemicals and also made forays into promising new markets for agricultural chemicals and synthetic rubber. By the end of the war, the company had deliberately positioned itself to prosper in a peacetime economy.

ORDNANCE SUPPLIER

When Hitler's armies stormed into Poland on September 1, 1939, the governments of England and France responded with declarations of war. The invasion of Poland was less the start of a new conflict, however, than an escalation of isolated but increasingly aggressive military actions around the world. The 1930s were troubled years: prior to the declarations of war, the Japanese had invaded Manchuria and China, the Italians had marched into Ethiopia and Albania, the Germans had annexed the Rhineland, Austria, and part of Czechoslovakia, and the Fascists, with support from Hitler and Mussolini, had triumphed in Spain. The events of September 1939 changed the terms of these conflicts and enlarged them.

In the United States, the government watched the developments in Europe with growing apprehension but took only limited actions to prepare for war. After World War I, Hercules virtually abandoned the manufacture of military powders, although it maintained open lines of communication to the Ordnance Department. In 1929, for example, the Ordnance Department developed a plan of preparedness in the event of national emergency that listed Hercules as a potential supplier of cannon powder, ammonium nitrate, dynamite, and certain explosives ingredients. The plan was revised and updated several times in the following decade, although periodic budget cuts limited the Ordnance Department's ability to coordinate and fund new developments. By 1939, when the Ordnance Department opened an office in Wilmington, Hercules was designated an authorized supplier of cannon powder for 37mm, 75mm, and 81mm guns and propellant and explosive charges for trench mortars, hand grenades, and pyrotechnic signals. The company made it clear, however, that it would require additional equipment and construction to scale up to meet an emergency of two years' duration.[1]

During the interwar years, Hercules' attitude toward military supply was ambivalent. Military buildups around the world had occurred while currents of pacifism and isolationism swirled across the United States. Since the Armistice, a substantial body of public opinion held that international financiers, large munitions suppliers, and other "Merchants of Death" not only had reaped enormous profits from the Great War but also, in the pursuit of private gain, had actually helped push the United States into the conflict.[2] Such views gained currency between 1934 and 1936 during sensational hearings in Congress presided over by Republican Sen. Gerald P. Nye from North Dakota. The Nye Committee probed the wartime activities of Du Pont, J.P. Morgan, Bethlehem Steel, and several other large companies. Although Hercules was not directly implicated (nor were Du Pont and other companies sanctioned in any way), it nonetheless felt the reverberations: the company was concerned to avoid exposing itself to the possibility of such scrutiny in the future.[3] Moreover, Congress would not make it easy for U.S. companies to supply foreign governments again. The Neutrality Acts of 1935 and 1937, passed in the wake of the Nye hearings, forbade private companies to export smokeless powder and other munitions to nations at war.[4]

Hercules was also ambivalent because changes in military powders made commercial spinoffs less likely. After World War I, the U.S. military continued to seek higher velocities and more powerful payloads. Not only were these objectives inappropriate to commercial markets, but the process of meeting them had grown more specialized. For Hercules, engaging wholeheartedly in ordnance supply would have entailed developing new and specialized skills.

After World War I, most innovations in smokeless powder were developed by the U.S. Army Ordnance Department at the Picatinny Arsenal in New Jersey or under contract to Du Pont. The Ordnance Department worked with Du Pont chemists to improve the hygroscopic (moisture-resisting) properties of artillery powder and to also reduce the flash that typically accompanied firing. The result was new formulas for artillery powders, designated NH and FNH types. The government also investigated other, more economical methods to produce the powders: for instance, rather than being extruded, they could be rolled into paper-thin sheets, which could then be slit and cut into grains.

In 1937, the Ordnance Department designated Hercules as a potential supplier of rolled artillery powders in the event of national emergency, but it lacked the funds to help the company acquire the new technology. Although Hercules asked repeatedly for permission to experiment with the new process, it did not actually negotiate a contract to do so until 1938, when it installed rolling equipment at Kenvil and produced a small batch of powder for trench mortars.[5]

The contract was the start of an increasingly tight relationship between the Ordnance Department and the company. In 1938, for example, Hercules collaborated with government chemists to rework surplus smokeless powder left over from the Great War. The company also transformed 1.3 million pounds of surplus cannon powder into the FNH type and agreed to rework another 2.5 million pounds in 1939 for the Picatinny Arsenal. In the summer of 1939 the company appropriated more than $100,000 to refurbish its smokeless powder line at Kenvil. At the same time, Hercules gave the military access to a proprietary process developed by Parlin chemist Raphael L. Stern. In the mid-1930s, Stern had developed and patented a mechanical process that allowed wood pulp to be substituted for cotton linters in the manufacture of many grades of nitrocellulose, including smokeless powder. In January 1939, Hercules agreed to license the process for use at the Picatinny Arsenal at cost plus 10 percent, with a royalty of half a cent for each pound of powder produced. The company agreed to suspend the royalty, however, in the event of a national emergency.[6]

Edging into War

On September 5, 1939—just days after Hitler's invasion of Poland—the implications of a European war became apparent to Hercules. The executive committee met with representatives of the Explosives and Legal departments to discuss plans to increase production of military powders. In the weeks that followed, Hercules licensed Du Pont's processes for making FNH flashless powders and poured money into new capacity for Kenvil and Parlin. The company arranged a $2 million contract to produce eight hundred thousand pounds of No. 155 howitzer powder, two million pounds of No. 155 gun powder, and one million pounds of 75mm cannon powder. By the end of the year, Hercules was delivering four hundred thousand pounds of smokeless powder per month to the government.[7]

That fall, Congress amended the second Neutrality Act to permit export of munitions to nations approved by the president. Almost immediately, Hercules arranged to supply munitions to friendly governments. The first contract was for the production of nine hundred thousand pounds of Hi-Vel No. 2 double-based powder at 65¢ per pound for use in the Far East. A much more significant deal was arranged soon thereafter. After being contacted by the British Purchasing Commission, Hercules agreed in the spring of 1940 to make up to sixty million pounds of single-based powder at a price of 56¢ per pound. The powder was to be made at Kenvil, which, along with Parlin, required still more expansion. Production of smokeless powder at Kenvil was organized in two lines. The A-Line filled orders for the U.S. military, and a new B-Line was built to supply the British. To help pay for the expansion the British Purchasing Commission advanced $1.6 million on the initial contract.[8]

Hercules also held discussions with the French Purchasing Commission, first about selling surplus smokeless powder from Parlin, and then about expanding Parlin's capacity. Before the negotiations could be completed, however, France fell to the Germans in June 1940, and the discussions became moot.[9]

Tragedy at Kenvil

That summer, as England battened down for the Battle for Britain, Hercules prepared to make still more powder for the British by acquiring an abandoned

The worst accident in Hercules' history occurred on September 12, 1940. An explosion in the solvent recovery area at Kenvil resulted in 55 deaths and more than 100 casualties. Shown here: The solvent recovery area, before and after the blast.

Above: A government-owned, company-operated smokeless powder facility, Radford Ordnance Works started production seven months after ground was broken in the fall of 1940. Below: During World War II, Radford turned out more than 600,000,000 pounds of powder for guns and cannons ranging from .30 caliber to 90mm.

Hercules supplied powder for small rockets such as the bazooka. Behind the mask: President Higgins readies to fire.

Transporting nitroglycerin was — and is — a dangerous job. Here, a Radford employee walks an "angel buggy" (the portentous nickname for an "NG" [nitroglycerin] cart), along a wooden track.

Above: A worker at Radford cuts strands of .50 caliber powder extruded from a press to proper length. Below: Women performed many critical operations at the ordnance plants. At Sunflower Ordnance Works, inspectors examine sticks of rocket powder prepared in the shape of a cross, a design allowing increased surface area for burning.

Two employees at Badger Ordnance Works use X-ray equipment to check for flaws in rocket powder.

"There's plenty of future in explosives, cellulose, and rosin," said Charles Higgins after World War II. At a management retreat in Virginia Beach in 1946, the company endorsed Higgins's strategy. Seated in the front are the board of directors. In the middle, Higgins is flanked by A.B. Nixon and Leavitt Bent.

textile factory on 270 acres in Belvidere, New Jersey. The company's plan was to reconfigure the plant to make up to 24,000 pounds of smokeless powder daily for small arms and to include the capability to supply its own needs for nitrocellulose. Hercules announced the Belvidere project on September 12, 1940.[10] On that day, however, the company's attention was diverted by the worst accident in its history.

About 1:00 P.M., a few hours after the Belvidere announcement, safety inspectors at Kenvil made an unannounced visit to the solvent recovery area for the plant's two powder lines.[11] The area consisted of three buildings that were designed to process large batches of powder. Two of them, the B-1 and B-2 buildings, designed for the new B-line, had been completed only a week before. The inspectors had no reason to be alarmed; they were making a routine search for matches. To their relief, they found none among the seventy-five men working in the buildings.

Solvent recovery was one of the final operations in the production of smokeless powder. The powder had already been mixed and cut into grains, although traces of ether-alcohol solvent still remained. To remove and recover the solvent, the powder was poured into large tanks. (The capacity of each B-Line tank was 18,400 pounds; each B-Line building contained five rooms with two tanks in each. On September 12, seven of the ten tanks in the B-1 building were fully or partly loaded; the tanks in the B-2 building were empty.)

Solvent recovery involved two steps. In the first, a batch of powder was placed in a tank for three days or more, while a warm air-vapor mixture was blown through the powder from the top. At the bottom of the tank, the mixture passed through a condenser to liquefy the solvents, which were then drained off. The powder was ready for the next step: "water-drying" in the same tank. Warm running water was piped in from the bottom of the tank and allowed to overflow at the top, carrying off the remaining solvent. The treatment lasted four to seven days, after which time the tank would be drained and the powder moved to other locations for finishing and packing. Solvent recovery was not thought to be an especially dangerous operation, although extreme caution was always mandated when employees worked near so much powder.

Shortly before 1:30 P.M. on September 12, witnesses reported a minor "puff" or "shot" in the corner bay of the B-1 building's second floor. Within seconds, a fire broke out, its flames shooting through the roof to a height of about fifty feet. The building's automatic sprinklers came on, but the fire could not be contained. Burning increased very rapidly, and people in the area reported hearing "a loud roaring noise." Fifteen to thirty seconds after the first puff was spotted, according to one witness, "the whole structure seemed to fall apart. Part of it rose into the air. There was a terrific explosion." Three seconds later, a loaded tank from the B-1 building was blown more than one hundred feet into the B-2 building and exploded on contact. After another second or two, the third major explosion, caused by fires spread by the first two, rocked the nearby A-Line solvent recovery building.

The destruction of the area was nearly total. Capt. Clarence Robey, the pilot of a TWA airliner that was flying over Kenvil at the time, saw "great flames" shoot up into the air: "Seconds later, clouds of smoke billowed into the sky and rose to a layer of broken clouds at about 6,000 feet. . . .The whole area looked as though it was being subjected to a terrific bombing from the air."[12]

The impact of the three explosions and the fires, which raged into the night, was devastating: fifty people were killed instantly, and more than a hundred were injured, including five whose wounds ultimately proved fatal. The violence of the blasts was so great that victims could be identified only by their badge numbers. Every structure save one in a radius of six hundred feet from the B-1 building was completely destroyed. (The lone survivor, the Steam and Air Dry House, was protected by barricades.) Buildings as far away as eighteen hundred feet were severely damaged, and bricks, pipes, and other fragments from the explosion were hurled more than two thousand feet away.

The damage carried well beyond the plant. A *New York Times* reporter, who arrived on the scene within hours, noted that

> windows in stores and houses in Dover, 6 miles away, were smashed. Throughout the small community of Kenvil, windows were broken, furniture was knocked down and broken, telephone wires were torn loose, power service was disrupted, and the whole countryside was rocked.

The blast was felt as far away as Poughkeepsie, New York, and it was recorded on the seismograph at Fordham University in New York City, about seventy miles to the east.

Given the time and circumstances of the tragedy, speculation about its cause centered on sabotage. (If it was sabotage, it was also suicide; a person starting such a fire would surely perish in the explosion.) Reporters noted that a German-American Bund (a pro-Nazi organization) camp was located only a dozen miles from Kenvil, and that several Hercules employees lived there. The FBI was called in to investigate but turned up no hard evidence of conspiracy. The company's own analysis, conducted by smokeless powder experts Col. Henry Marsh and Bernhard Troxler, focused on the origins of the fire and the circumstances that turned a fire into an explosion.

The fire was found to have started in the No. 2 tank in the first bay of the B-1 building. The tank was fully loaded with 20mm smokeless powder (a type manufactured for the British) that had been undergoing the process of solvent recovery for twenty-three hours. "At that stage," according to Troxler, "the vapors [in the tank] were thought to be so rich in solvent that they were outside the explosive limit." He had difficulty imagining how they might have ignited. Indeed, it was impossible to determine which caught fire first—the vapors or the powder. Although Troxler did not rule out sabotage as a possible explanation, he thought that "a cool flame explosion in the system," perhaps started by the vapors coming in contact with a hot bearing in the motor of the circulating fan, "could lift the lid from the tank and allow strong combustion to develop, leading rapidly to ignition of the powder."

Hercules was also mystified as to why the powder had detonated after it caught fire. It had been commonly believed that smokeless powder, being a propellant, would burn rather than explode. The solvent recovery tanks were designed to accommodate the pressures of burning in case of accident. But as Troxler put it (with considerable understatement), "this belief was shown to be untenable" by the Kenvil explosions. Subsequent tests proved the point that burning rates increased dramatically in large concentrations of smokeless powder, and "even with very light confinement, violent explosions may occur when sufficiently large quantities of high potential smokeless powder are burned."[13]

These findings had major implications for the design of new powder lines in the reconstructed Kenvil plant and at government ordnance plants during the war. "The entire concept of the treatment of smokeless powder was thus changed," wrote one expert. The powder was henceforth "recognized as a high explosive under certain conditions," and plants were redesigned to disperse buildings in compliance with standards for distances involving high explosives. Hercules engineers paid particular attention to the redesign of the solvent recovery process. The practice of combining solvent recovery with water-drying was discontinued; the two operations were thereafter conducted in separate locations. In addition, new solvent recovery buildings were made smaller, tanks were shrunk and redesigned to alleviate pressure at the top, and many other safety precautions were taken.[14]

The final death toll at Kenvil was 55 people, with 104 sustaining serious injuries. Total damages from the accident, including Hercules' monetary settlements with survivors and heirs, cleanup and rebuilding of the smokeless powder operations, and repairs to property beyond the plant, totaled more than $1.1 million. In addition, Hercules may have lost as much as four million pounds of powder production in 1940 because of the accident. Although the company was partly reimbursed by the British Purchasing Commission, its losses were substantial. The plant was not reopened to make smokeless powder until April 1941.[15]

Radford

The explosion at Kenvil occurred amid a vast increase in smokeless powder production in the United States. In the late 1930s, total output of smokeless powder and TNT was about one hundred thousand pounds each per day—"roughly the quantity to maintain an army of only 100,000 combat troops in the field of active combat for a single day."[16] As it became apparent that the federal arsenals at Picatinny, New Jersey, and Frankford, Pennsylvania, could not possibly supply the military's need for ammunition in a war, the Ordnance Department planned to build several new smokeless powder plants. On September 1, 1939, the day that Hitler's blitzkrieg struck Poland, representatives of the Ordnance Department met with railway and local officials in Radford, a small city along the New River in the western part of Virginia, to approve the location of one such plant. (The site was near the location of a powder mill that had supplied George Washington's troops during the Revolutionary War.)

The following spring, as the British Expeditionary Force evacuated Dunkirk and Paris fell to the Germans, the need for new powder capacity became more urgent. By the summer of 1940, President Roosevelt had asked for—and received—defense appropriations totaling nearly $9 billion, with another $8 billion to follow by autumn. (To put these numbers in perspective, the budget for the entire War Department in 1939 was $462.3 million.)[17] On July 2, 1940, Congress passed legislation authorizing the secretary of war

> to construct military posts, plants, buildings, and facilities for the development, manufacture, maintenance and storage of military equipment, munitions and supplies and to operate and maintain any such plants, buildings, and facilities "either by means of Government personnel or through the agency of selected qualified commercial manufacturers under contracts entered into with them."

This was the beginning of an unprecedented buildup: between the summer of 1940 and the end of 1942, the War Department authorized the construction of more than sixty GOCO plants, representing a total capital outlay of roughly $3 billion, with an annual operating cost of $1 billion.[18]

The third GOCO plant to come onstream was at Radford, Virginia. On August 16, 1940, the Ordnance Department selected the country's second-largest producer of smokeless powder as the contractor for the job—Hercules.[19] In return for its efforts, the company received payment on a cost-plus-fixed-fee basis. The arrangement—standard at all GOCO plants—included what the official historian of the Ordnance Department called ". . . rather liberal provisions. Each company was reimbursed at regular intervals for approved expenses in operating the plant, and in addition was paid a fee based on the number of rounds of ammunition or pounds of explosives produced." (Fees were graduated depending on the nature of the assignment and ranged from 3 to 6.5 percent. During the war, the fees tended to decline because the government paid lower percentages to operate facilities than to build them.)[20]

The design of the Radford facility was based on a composite of Ordnance Department specifications, Hercules' operations at Kenvil and Parlin, and information known about Du Pont's smokeless powder plants. Radford was vertically integrated and made its own acids and nitrocellulose, as well as finished powders. The government initially ordered fifty million pounds of powder, to be produced on three lines operating around-the-clock and turning out two hundred thousand pounds per day. Total cost of the operation was projected to be $25 million.

Two weeks after construction began, however, the contract was amended to take account of the Kenvil disaster. Because Kenvil could no longer make smokeless powder, the government poured another $10 million into Radford to increase capacity to three hundred thousand pounds of powder per day. Subsequent changes and modifications, including the redesign of the solvent recovery process and the construction of an area for making rolled powders, added still more costs. By the time the plant's third line began producing powder in the fall of 1941, the government's total investment amounted to nearly $51 million.[21]

Construction proceeded rapidly. The Norfolk & Western Railway began to lay out track on the property on September 7, 1940. Construction of the first powder line started two months later. The plant was dedicated on March 14, 1941, and produced its first finished powder three weeks later. It is hard to overestimate the magnitude of this achievement. Under the supervision of Hercules and the New York construction firm Mason & Hanger Company, about 23,000 people were involved in building Radford at the peak of activity. By any measure, it was a mammoth operation, sprawling over 2,400 acres and featuring more than 400 manufacturing buildings (most of permanent construction), 26 storage areas, and 156 other structures, including laboratories, office buildings, bunkhouses, firehouses, cafeterias, guard stations, and a hospital. Forty-three miles of roads, 17 miles of fence, 12 miles of railroad track, and 800 miles of telephone wire crisscrossed the plant.[22]

About 9,000 people—more than the prewar employment of the entire Hercules Powder Company—worked at Radford as it came up to full capacity late in 1941. The labor force was drawn from the local region, in which three chemical companies (among other employers) operated: Tennessee Eastman, American Viscose, and Celanese. Hercules supplied managers and supervisors, most of whom—including the first plant manager, H.V. Chase—had learned their trade at Kenvil or Parlin. In addition, the U.S. Army assigned a small group of officers to represent its interests in the "approval of designs, subcontracts, purchases, expenditure of funds, steps taken to protect life and property, and in their inspection activities to insure that contractual obligations were being fulfilled."[23]

In 1941, 38.2 million pounds of powder were produced at Radford—a prelude to the huge expansion of government business to follow.

GOCO Contractor

Radford Ordnance Works was the first of six GOCO plants that Hercules engineered and operated as the conflicts in Europe and Asia escalated into global war. In March 1941, the Lend-Lease Act allowed the president to furnish "surplus" military supplies to nations whose fortunes were deemed vital to the interests of the United States. In practical terms, Lend-Lease entailed massive support for Great Britain and placed still heavier demands on U.S. munitions manufacturers. Well before the U.S. declaration of war in December, Hercules started work on the New River bag-loading plant (to pack ammunition for certain caliber artillery) near Radford and started plans for a TNT plant near Chattanooga, Tennessee (Volunteer Ordnance Works), and an anhydrous ammonia plant on the Mississippi River at Louisiana, Missouri (Missouri Ordnance Works). After Pearl Harbor, the company contracted to run additional smokeless powder plants in Baraboo, Wisconsin (Badger Ordnance Works), and eastern Kansas (Sunflower Ordnance Works).[24] In addition, Hercules produced military powders at Kenvil throughout the war, and at Belvidere (for the British) until early 1944. The company's chemical cotton plant at Hopewell and nitrocellulose plant at Parlin were also largely given over to supplying raw materials and intermediate products for smokeless powder.

In all, between 1941 and 1945, Hercules produced 1.2 billion pounds of smokeless powder, 828.6 million pounds of TNT, 27.7 million pounds of Pentolite (a special high explosive—see below), and 159.2 tons of anhydrous ammonia (see Exhibits 8.1 and 8.2). This output represented more than eleven times the company's production of ammunition and explosives during World War I. In the Second World War, Hercules produced more than one-third of all smokeless powder made in the United States and about one-fifth of all TNT. A measure of the responsibility placed on the company was that it managed three of the five largest GOCO plants in the United States—Sunflower, Badger, and Radford.[25]

The massive output of ammunition was accompanied by an equally impressive growth in manpower levels. In June 1940, shortly before starting work on Radford, Hercules employed 7,300 people in all of its plants. By April 1945, the peak of wartime activity, employment at the GOCO plants reached 38,000, and more than 45,000 people worked for the company as a whole. About one-third were women.[26] (See Exhibit 8.3.)

Hercules' ability to staff and manage the GOCO operation was sorely strained by rapid mobilization. The task was made still more daunting by the Selective Service programs, which summoned most able-bodied males into the armed forces, and by intense competition for skilled employees among the many companies whose business expanded greatly during the war. Hercules was further handicapped by the remote locations of many of its GOCO plants. The company had to take aggressive measures to attract employees. For example, for Sunflower Hercules opened personnel offices in Kansas City and Topeka; made extensive use of newspaper and radio advertising; sent recruiting teams into towns and cities in Kansas, Missouri, and Arkansas; broadcast appeals through sound trucks in towns within a hundred miles of the plant; appealed to existing employees to interest friends and family members; hosted speeches by local veterans to stimulate patriotism; and created citizens' emergency committees in eighteen nearby towns to help obtain new employees and discourage terminations. The company also appointed extra supervisors in personnel and employee relations and hired special counselors for women and black workers. Finally, with financial support from the government, the company built housing and barracks, provided canteen service for meals, opened pharmacies and grocery stores, and helped subsidize transportation to and from the plant.[27] Such measures had the desired effect: although recruitment and employee turnover were constant preoccupations of the GOCO plant managers, Hercules met or exceeded its production schedules throughout the war.

This performance reflected the care that Hercules put into management and personnel, starting at the top. William R. Ellis, general manager of the Explosives Department, moved his most capable and promising executives into key positions at the GOCO plants. At Radford, the first plant manager, H.V. Chase, had been a smokeless powder supervisor during World War I and later manager of the Kenvil plant. His assistant (and later, successor), Andrew Van Beek, had "considerable experience in acid and dynamite manufacture." The operating manager, Pat Butler, had been an assistant superintendent at Kenvil.

EXHIBIT 8.1 Hercules' Production of Military Explosives in World War II, 1941–1945

	Badger	*Radford*	*Sunflower*	*Volunteer*	*Kenvil*	*Belvidere*	*Total (lbs)*
Small arms, single-based	102,592,615	73,307,772	11,682,847		40,546,165	20,515,712	248,645,111
Cannon, single-based	146,933,885	430,162,510	71,503,113		16,653,730	94,509,491	759,762,729
Cannon, double-based		52,999,820	1,950,097		21,071,454		76,021,371
Cannon mortar, double-based		11,602,990			943,230		12,546,220
Rocket propellant, solventless	14,432,405	2,912,876	81,842,748				99,188,029
Rocket propellant, solvent		11,150,494	15,521,490		528,454		27,200,438
E.C. blankfire	1,644,500				2,384,546		4,029,046
Experimental		585,682					585,682
Pentolite		27,661,415					27,661,415
TNT				828,640,744			828,640,744
Totals	265,603,405	610,383,559	182,500,295	828,640,744	82,127,579	115,025,203	2,084,280,785
Anhydrous ammonia manufactured by Missouri Ordnance Works (tons)							159,171

Source: Historical Report of Sunflower Ordnance Works, 47.

EXHIBIT 8.2 Bag Loading by New River Ordnance Plant, 1941–1945

	Number	*Pounds of Powder*
Howitzer and gun charges	21,836,033	122,220,161
Rolled powder increments packaged	239,709,092	3,373,089
Flash reducers	84,993	
Total	261,630,118	125,593,250

Source: Historical Report of Sunflower Ordnance Works, 47.

Similarly, employees responsible for technical matters, engineering, safety, and other operations at the plant were seasoned veterans of the dynamite or smokeless powder businesses. At the supervisory level, the company brought in many experienced hands from Kenvil or Parlin. Newcomers were put through training programs that ran from 6 to 12 weeks and consisted of "class-room work covering theories, policies, etc.," as well as observation of the work of regular employees.[28] As other GOCO plants were commissioned, they were staffed by personnel experienced or trained at Radford or (especially in the case of Missouri Ordnance Works) the commercial explosives plants.

Throughout the war, Hercules focused training of its employees on the line around the issue of safety and accident prevention. Safety rules and regulations (including prohibitions against carrying matches) were published in employee handbooks, displayed prominently on bulletin boards, featured in plant newsletters, and discussed with every employee hired. Although the immense volume of activity and the constant flow of new personnel through the GOCO plants were sources of concern, the company suffered few fatalities or production stoppages because of accidents, despite massive everyday involvement with explosive materials.[29]

Employees of the GOCO operations generally have warm recollections of life at the plants. In part, their memories reflect pride and patriotism, as well as a sense of having risen to meet—and overcome—challenges. (Such feelings are well founded: during the war, more than 95 percent of employees purchased war savings bonds via payroll deduction. The company also received eighteen Army-Navy "E" awards for maintaining excellent production records.)[30] Morale was generally high, and people worked hard. The plants ran twenty-four hours a day, seven days a week, and most employees worked at least six days a week. The pace took its toll, but there were also sources of relief: at Sunflower, recalls Duard H. "Tex" Little, one shift finished at midnight and came on again the following morning at eight o'clock; in between, many workers stayed at the plant to relax and party. According to Paul Steele, a supervisor who worked at Radford and Badger, employees banded together to train each other and make suggestions to help meet production goals. At Radford, for example, female operators who

EXHIBIT 8.3 Explosives Department Employees, 1940–1945

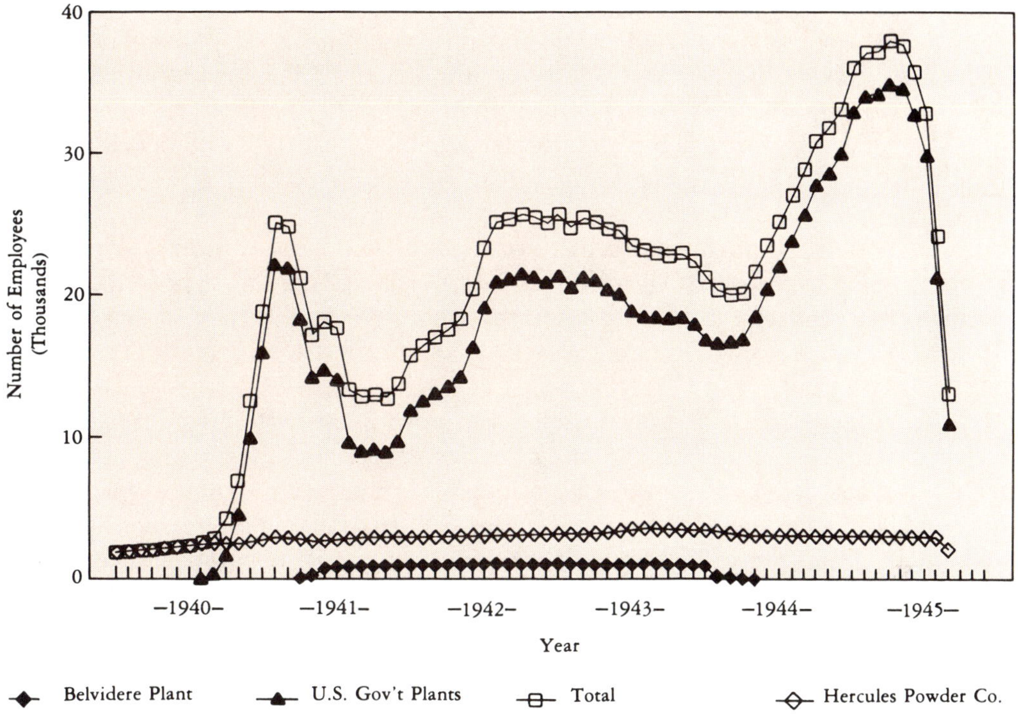

Source: Historical Report of Missouri Ordnance Works, 29.

packed grains learned to pack them in bunches, a technique that resulted in a fiftyfold increase in productivity. "It was just a matter of the people getting used to the work. Once they did, they really figured out how to work productively."[31]

The overall experience, says Jack Martin (who was at Radford and Sunflower), "really brought out the best in a lot of people. And it was another advantage for Hercules in that a lot of young people learned how to manage in large-scale operations."[32] Indeed, as had occurred during World War I, placing major responsibilities in the hands of young employees was to pay dividends to Hercules long into the future. Young managers such as Martin, Werner C. Brown, Alexander F. Giacco, John R. Ryan, John E. Greer, Tex Little, and Richard Winer moved rapidly into important positions at the GOCO plants and, after the war, into important managerial and technical positions with the company.

New Weapons of War

For the most part, Hercules understood well the products manufactured at the GOCO plants. The company had been making smokeless powder since 1913 and was the largest producer of TNT in the United States during World War I. Not only did Hercules master existing techniques of ordnance production, but it also participated in the development of new munitions processes and products throughout the war. Since the armed forces wanted greater firepower and the delivery of more destructive payloads, Hercules had to work closely with the services, national defense researchers, and other munitions suppliers. The company was responsible for several important breakthroughs in manufacturing processes: a new method of washing nitrocellulose to improve the chemical's stability; new procedures for quality control and testing; and development at Radford and Sunflower of X-ray and ultrasonic inspection of large-grained powders.[33]

The most important product innovations in ammunition during the war were the development of solid propellants for rockets, more powerful high explosives to fill shells and bombs, and more sophisticated fuzes and detonating devices. In each instance, Hercules played a significant role as a prime contractor or subcontractor.

Although rockets had been used as weapons in ancient China, in the West they were used principally as signaling devices or carriers of flares until the early 1940s. For the delivery of projectiles or explosive charges, the military preferred rifled artillery, which was far more accurate and controllable than contemporary rockets. In the Battle of Britain in 1940, however, the British demonstrated the effectiveness of rockets as antiaircraft weapons: they could achieve greater altitudes and deliver more power than artillery and, if fired in sufficient numbers, compensated for their lack of accuracy. At the same time, the U.S. Army was looking for a more effective antitank device than the grenade-launcher, which lobbed its payload toward the target and suffered heavy recoil, making it difficult to aim and use. The Army's solution to these problems was the development of the hand-held "bazooka" rocket in 1941 and 1942. Thereafter, small rockets were increasingly prominent in land, sea, and air battles in every theater of the war.[34]

Producing propellant for such rockets required considerable manufacturing ingenuity. The powders had to be double-based because single-based powders lacked sufficient energy. The prewar technology for making double-based powders, however, was not suitable for rocket powders, which required large grains or sticks (ranging from several inches to a foot or more in length) of exceptional purity, uniformity, and quality. Conventional techniques for solvent recovery, for example, could not remove all traces of solvent, which could affect the length and steadiness of burning. Manufacture of large grains, moreover, presented special problems. Even a tiny crack or imperfection in a grain could have disastrous consequences, such as an erratic flight trajectory or a premature explosion.[35]

For small rockets, such as bazookas, the conventional extrusion process could be used—provided that great care was taken to ensure quality. For larger rockets, new powdermaking techniques were required. In 1939, the Smokeless Powder Development Group at Kenvil had found a way to make powder without solvents. The new process involved plasticizing the ingredients under heat and pressure into a paste (as opposed to the traditional emulsification using solvents), which was then rolled into sheets. The Hercules process was used initially to make a thin-grained powder for mortars, but it soon proved adaptable for making rocket powders.

Early in 1942, Hercules began cooperating with the National Defense Research Council on a pilot program to make solventless rocket powders at the California Institute of Technology in Pasadena. That fall, Dr. Robert Cairns, director of the Experiment Station, and David Bruce from Belvidere went to England to study methods of extruding dry solventless powder to make large grains. The result of their study and the work at Cal Tech was a development program for rocket powders at Radford. For the remainder of the war, the facility made both solventless powder for large rockets and solvent powder for bazookas. In the summer of 1942, Sunflower Ordnance Works was directed to produce a mammoth order of solventless rocket powder for the Russians defending Stalingrad. Massive orders for rocket powder for the U.S. Navy followed. As a result of these contracts, Sunflower became the largest producer of rocket powder (both solvent and solventless) in the United States, with a total wartime output of more than ninety-seven million pounds.[36]

Hercules also helped develop new, extremely powerful high explosives such as pentaerythritol tetranitrate (PETN). Although too unstable to be used in pure form in shells, a mixture of PETN and TNT—commonly known as Pentolite—carried 40 percent greater destructive power than TNT alone. This was an important advantage for military tacticians looking for new ways to penetrate tanks and other armored vehicles and heavily defended positions. In 1942, Hercules engineers at Radford pioneered a process for making Pentolite in a slurry form that greatly facilitated its loading into shells and bombs. Thereafter, the company became a leading producer of the new explosive, which found wide application in detonators and in payloads of bazookas, grenades, and antitank shells.[37]

In developing new propellants and explosives such as rocket powders and Pentolite, Hercules acted as a prime contractor to the Ordnance Department. The company also served as a subcontractor in the development of other critical innovations in ammunition. The most significant of these was the radio proximity fuze (also known as the VT fuze), developed between 1941 and 1943 by researchers at three universities in cooperation with the Army, the Navy, and the British.[38]

The problems of how and when to detonate a shell or bomb to achieve maximum destructive power were complex. Both types of fuze had drawbacks. Conventional fuzes were set to go off either on impact or after a specified time delay. Shells exploding on impact with the ground diminished the destructive effects of fragmentation; as had happened in World War I, defenders could withstand an artillery barrage merely by digging in. The impact fuze was even worse as an antiaircraft weapon because it required a direct hit, which was extremely difficult to accomplish, given the fire control technology of the period. Time-delay fuzes helped overcome these problems but were subject to others, such as difficulties in estimating and setting the proper delay (especially under battle conditions) and limitations on the length of delay—a severe disadvantage for antiaircraft purposes.

The solution to these problems was the radio proximity fuze, which detonated at a specified distance from the target. The device used reflection of short-wave radio pulses to sense nearness to its objective; when the shell or bomb came close enough, the fuze fired to set off the main charge. The VT fuze was ideal for use against aircraft, submarines, and ground troops and targets. As the *Hercules Mixer* put it, the innovation "elminates fuze setting, excludes errors inherent in time mechanisms, and makes possible maximum damage at split-second speed."[39] The VT fuze was first used against aircraft and flying bombs—among other achievements, it is credited with minimizing the damage to ships from Japanese kamikaze missions. Indeed, at the end of the war, the U.S. Navy rated the radio proximity fuze second only to the atomic bomb as a critical Allied technological advantage. The fuze also functioned efficiently to deter the German counteroffensive in December 1944, and helped pave the way for Allied troops to enter the Rhineland. No less an authority than General Patton attributed victory in the Battle of the Bulge to what he called "the funny fuze."[40]

As a subcontractor to several prime production contractors, Hercules made several key contributions to the development of the VT fuze. First, at Port Ewen, the company designed and manufactured the miniature electrical detonators used in the device. Second, at Hopewell, the company produced the ethyl cellulose that was injection-molded to form the plastic nose cone of the fuze. In addition, Hercules chemists provided technical assistance in the development of a hot-melt "potting material" to protect the sensitive mechanisms inside the fuze from the stresses of firing and flight.[41]

Although Hercules played a significant role in the development of many of the new military technologies of World War II, these had a lesser long-term impact on the company than might be imagined. In the first place, the military innovations of World War II were qualitatively different from those of World War I. For the most part, the new technologies—radar, the radio proximity fuze, rocketry, artificial rubber, and, at the very end of the war, the jet engine and the atomic bomb—drew heavily on sciences unfamiliar to Hercules. (The Second World War has been called "the Physicists' War—in contrast to the First World War, "the Chemists' War.")[42] As a result, Hercules' place as a technological leader between 1940 and 1945 was relatively less prominent than it had been between 1914 and 1918, when it pioneered new methods of making double-based powders on a large scale and developed new processes for making solvents. During World War II, Hercules found itself increasingly in the position of either a subcontractor or a participant on large development teams.

Second, the new technologies that Hercules helped produce during World War II were highly specialized to meet military requirements and found few commerical applications once the fighting was over. Military powders were far too powerful for sporting uses, and peacetime uses of rocket powders or expensively produced miniature detonators were hard to imagine. Hercules' experiences at the GOCO plants during World War II positioned the company to be a postwar defense contractor—for ammunition and rocket propellants, for example—but did not materially add to its knowledge base as a commercial enterprise.

This is not to say that Hercules learned little of value during World War II. Rather, its direct participation in the war effort had less enduring consequence than its indirect involvement as a producer of commercial chemicals and strategic materials while the war was pulling the U.S. economy out of a dozen years of depression.

What was true in technology was also true in economics: although Hercules was amply rewarded for its efforts in the war, earnings from the GOCO plants did not distort its overall financial performance, as had happened during World War I. Although revenues swelled from $52.4 million in 1940 to $120.9 million in 1943, the peak year of activity, the company's military business netted about $5.3 million, or less than a quarter of total earnings. The largest fees were earned at the biggest plants, the ammunition facilities at Radford, Sunflower, and Badger. (See Exhibits 8.4 and 8.5.) In the last two years of the war, both sales and earnings declined, although both remained at levels far higher than anything seen before the 1940s.

THE COMMERCIAL BUSINESSES DURING THE WAR

After 1940, the U.S. economy began to thrive in a manner not seen since the mid-1920s, with high growth rates of industrial output and GNP, full employment, and—because of government controls—stable prices. In such an environment, Hercules, like most industrial companies, flourished.[43]

Wartime expansion affected every product line. Although its impact was greatest on the parts of the company that directly contributed to military production (explosives, cellulose products, and chemical cotton), the expansion boosted sales of many nonmilitary cellulose products, as well as the naval stores, paper chemicals, and synthetic businesses (see Exhibit 8.6). The fastest growth came in Hercules chemicals supplied to the protective coatings, plastics, and artificial fibers industries, with sales to the construction industry, agriculture, and the new artificial rubber industry close behind.

A Trip through Hercules Land

In 1942, Hercules launched a new and imaginative advertising campaign to illustrate and promote its diversified chemical businesses. The campaign featured a map of "Hercules Land," whose circular shape was sliced into six segments reflecting the company's major product lines; in each segment, Hercules listed its principal products. Surrounding Hercules Land was a wide world labeled "Unexplored" (see Exhibit 8.7). The advertisements were quite unlike any Hercules had used before the war. The difference could be seen in the overt patriotism of the Hercules Land ads, but it was also present in Hercules' new self-image as a science-based chemical company. The Hercules of the early 1940s was an

EXHIBIT 8.4 Fees Received by Hercules from U.S. Government, 1941–1945

Plant/Fee	*1941*	*1942*	*1943*	*1944*	*1945*
Radner					
Construction	$1,170,706	$209,562	$78,180	—	$31,543
Operation	634,955	3,087,306	2,478,734	$1,868,841	1,505,156
Termination	—	—	—	—	51,747
New River					
Construction	105,047	—	1,717	586	—
Operation	30,300	201,000	108,750	—	20,000
Administration	60,000	30,000	—	—	—
Missouri					
Construction	4,947	77,786	19,267	—	—
Operation	—	2,992	125,967	118,546	77,789
Administration	51,200	122,880	71,680	—	—
Termination	—	—	—	—	7,400
Volunteer					
Construction	7,795	127,316	10,946	19,422	—
Operation	—	233,475	1,067,898	1,074,250	736,014
Administration	60,000	144,000	12,000	—	—
Badger					
Construction	—	254,086	47,666	48,095	42,598
Operation	—	48,000	739,888	1,307,600	1,149,861
Termination	—	—	—	—	73,330
Sunflower					
Construction	—	167,917	130,725	65,432	31,809
Operation	—	—	385,779	1,515,945	1,913,834
Total	$2,124,950	$4,706,320	$5,279,197	$6,018,717	$5,641,081
Direct charges to fees	$169,982	$254,183	$279,907	$169,091	$206,816
Net fee before allocation of indirect charges	2,294,968	4,452,137	4,999,290	5,849,626	5,434,265
Amount credited to Explosives Department	297,000	574,500	765,600	732,000	726,000

Source: Explosives Department, Annual Report for 1945.

EXHIBIT 8.5 Hercules' Financials, 1940–1945 (Dollars in Thousands, Except per Share)

	1940	*1941*	*1942*	*1943*	*1944*	*1945*
Net sales	$52,429	$85,612	$111,378	$120,948	$105,678	$100,556
Earnings before interest and taxes	10,082	20,262	23,630	22,501	16,842	14,395
Total assets	76,354	88,135	103,343	111,102	104,703	106,697
Fixed[a]	41,230	44,691	46,691	50,035	52,077	56,026
Current	34,451	42,676	54,289	57,103	47,906	50,303
Other	673	768	2,363	3,964	4,720	368
Total equity	26,565	26.565	26,565	26,565	26,565	26,565
Stockholders' investment	36,791	38,415	40,611	42,299	43,306	44,624
R&D Expense	1,173	1,403	2,071	2,887	2,953	2,746
Dividends/share	$2.85	$3.00	$2.50	$2.50	$2.50	$2.50
Earnings per share (common)	$4.01	$4.23	$3.81	$3.93	$3.26	$3.36
Employees	7,011	9,401	11,642	12,085	10,401	9,831
Return on sales	19.2%	23.7%	21.2%	18.6%	15.9%	14.3%
Return on assets	13.2	23.0	22.9	20.3	16.1	13.5
Return on equity	38.0	76.3	89.0	84.7	63.4	54.2
Return on stockholders' investment	27.4	52.7	58.2	53.2	38.9	32.3
R&D/Sales	2.2	1.6	1.9	2.4	2.8	2.7

a. Gross fixed assets, including depreciation.

Source: Hercules Powder Company, Annual Reports for 1940–1945.

EXHIBIT 8.6 Hercules' Sales and Profits by Department, 1940–1945 (Dollars in Thousands)

Product Line	*1940*	*1941*	*1942*	*1943*	*1944*	*1945*
EXPLOSIVES						
Net sales	$16,201	$31,579	$47,031	$48,656	$30,226	$24,801
Net profit	3,799	10,954	14,153	12,344	6,308	4,043
Return on sales	23.4%	34.7%	30.1%	25.4%	20.9%	16.3%
Percentage of total sales	31.7	38.3	42.9	41.5	30.3	26.1
Percentage of total profits	36.3	51.3	54.7	53.9	45.0	29.5
CELLULOSE PRODUCTS						
Net sales	$12,499	$18,108	$20,955	$21,019	$20,901	$22,920
Net profit	4,015	4,938	5,536	4,271	3,308	3,490
Return on sales	32.1%	27.3%	26.4%	20.3%	15.8%	15.2%
Percentage of total sales	24.5	21.9	19.1	17.9	21.0	24.1
Percentage of total profits	38.3	23.1	21.4	18.6	23.6	25.5
NAVAL STORES						
Net sales	$5,292	$7,463	$8,814	$12,083	$11,369	$11,807
Net profit	335	1,658	2,278	2,944	1,346	2,251
Return on sales	6.7%	22.2%	25.8%	24.4%	11.8%	19.1%
Percentage of total sales	10.4	9.0	8.0	10.3	11.4	12.4
Percentage of total profits	3.4	7.8	8.8	12.8	9.6	16.4

VIRGINIA CELLULOSE						
Net sales	$6,933	$11,100	$15,861	$16,466	$15,864	$15,002
Net profit	1,794	2,270	2,279	1,868	1,443	1,294
Return on sales	25.9%	20.5%	14.4%	11.3%	9.1%	8.6%
Percentage of total sales	13.6	13.4	14.5	14.0	15.9	15.8
Percentage of total profits	17.1	10.6	8.8	8.1	10.3	9.5
PAPER MAKERS CHEMICAL						
Net sales	$9,456	$13,044	$14,348	$15,094	$15,797	$14,163
Net profit	563	1,481	1,392	1,104	872	1,435
Return on sales	5.9%	11.4%	9.7%	7.3%	5.5%	10.1%
Percentage of total sales	18.7	15.8	13.1	12.9	15.9	14.9
Percentage of total profits	5.4	6.9	5.4	4.8	6.2	10.5
SYNTHETICS						
Net sales	$642	$1,263	$2,651	$3,921	$5,502	$6,222
Net profit	−47	63	245	390	739	1,179
Return on sales	−7.3%	5.0%	9.2%	9.9%	13.4%	18.9%
Percentage of total sales	1.3	1.5	2.4	3.3	5.5	6.6
Percentage of total profits	−0.4	0.3	0.9	1.7	5.3	8.6
TOTAL NET SALES	$51,113	$82,557	$109,660	$117,239	$99,659	$94,915
TOTAL NET PROFIT	10,479	21,364	25,883	22,921	14,016	13,692
RETURN ON SALES	20.5%	25.9%	23.6%	19.6%	14.1%	14.4%

Note: Total net sales figure does not include sales to other Hercules departments.

Source: Special Reports to Directors, 1940–1943; Reports to the executive committee, 1944–1945, Hall of Records, Hercules Incorporated.

EXHIBIT 8.7 Hercules Advertisement, 1942

Here may lie more keys to Victory...

This is a map of Hercules Land. Let us explore it together thoughtfully . . . in terms of *your* production and ours. For somewhere within the known or unknown boundaries of this amazing world of chemistry may lie more aids to surer, swifter victory.

This is a land where anything can happen. Where the wood of pine trees is made to yield chemicals for reclaiming rubber . . . and resins that speed the construction of air bases. Where the tiny fibers around a cotton seed become plastics of countless purposes . . . or swift-drying coatings that slash days from production schedules. Where the properties of unobtainable materials are matched by the magic of synthetic chemistry. Where each new re-arrangement of molecules leads to endless new uses for textiles, rubber, protective coatings, paper . . . uses for war, and uses later for peace.

This is a land where, night and day, thousands of Hercules men and women are beating schedules on vast quantities of military explosives . . . and of industrial explosives to rush vital construction and to speed the mining of metal, stone, and coal. In this sector, too, we are eager to learn of new materials, new processes or equipment which may help us to contribute further to the winning of the war.

Let us explore Hercules Land together, for here your products and ours . . . for both war and peace . . . may find an unimagined future. And who can say how much sooner V-Day will come if you, with your special knowledge of what you make . . . and we, with our special knowledge of the chemistry of cellulose, of terpenes, rosins, synthetics, explosives . . . now contribute to each other's knowledge, with imagination and resourcefulness?

The entrance to Hercules Land is through the doors of our Experiment Station . . . open to all who, like ourselves, have victory as their goal . . . to all those whose laboratories and plants, like ours, are places where well enough is something never to be let alone.

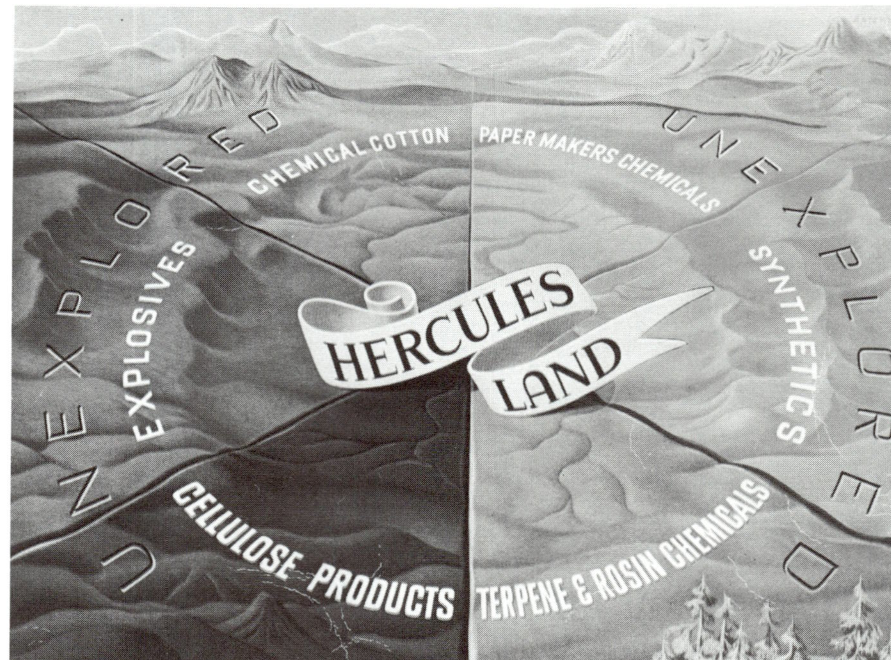

CAN YOU HELP US? If you have evolved any new material, process, or equipment which can help to speed *our* war production, write to us, so that we can explore the possibilities with you. The knowledge we both gain *now* can not only speed V-Day, but can also equip us both to speed *after* V-Day employment and recovery.

CAN WE HELP YOU? If you need a chemical material with specific properties, to help in *your* war production, tell us what your product is, and the properties you need. We will send literature and information—and, if possible, samples—of Hercules materials which may aid you. Hercules Powder Company, Wilmington, Delaware.

HERCULES

CHEMICAL MATERIALS FOR INDUSTRY

HERCULES POWDER COMPANY · WILMINGTON, DELAWARE

Source: Hercules Chemist, July 1942.

optimistic institution, confident that its Experiment Station would open doors to an endlessly bright future.

In 1942, Hercules had ample cause for optimism: the changes wrought in the company in the previous decade were playing out remarkably well. The major new products of the 1930s—cellulose acetate, ethylcellulose, chlorinated rubber, Staybelite® and Vinsol® resins, and a host of synthetic resins—were establishing themselves solidly in the marketplace. Hercules had gained greater control over once troublesome products such as naval stores commodities and many chemicals in the PMC Department. Finally, the company began to open new lines of inquiry and investigate new markets for products based on cellulose and rosin, as well as other sources of raw materials. A renewed emphasis on R&D under Dr. Emil Ott, director of research after 1939, supported Hercules' pursuit of these new opportunities.

Cellulose and Cellulose Derivatives

The surge in defense construction between 1940 and 1943 not only benefited Hercules directly as a GOCO contractor, but also indirectly as a supplier of chemicals to supporting industries such as protective coatings and construction. With the advent of mobilization, for example, demand for chlorinated rubber (marketed under the trademark Parlon® after May 1940), which served as a basic constituent of weather-resistant paint on military bases and battleships, skyrocketed, rising 37 percent in 1940 and 47 percent a year later. By then, Parlon® was also established as an essential ingredient of flameproofing mixtures for treating canvas and other textiles used by the military. In July 1942, when natural rubber went on allocation, sales of Parlon® suddenly plunged. Some months earlier, however, researchers at the Experiment Station, anticipating the problem, had developed chlorinated paraffin as a substitute. Hercules subsequently adopted the trademark Clorafin® for a family of plasticizers and resins based on chlorinated paraffin and thus retained its leading position in flameproofing and weatherproofing applications.[44]

Other Hercules products, including ethylcellulose and cellulose acetate, were classified as strategic materials during the war. When the company first developed ethylcellulose in the late 1930s, its primary use was as a grinding base for paint pigments. By the early 1940s, however, it was showing versatility as a plastic: for example, it supplemented (or substituted for) phenolformaldehyde plastics in phonograph records. Because it was flexible at low temperatures and had good electrical properties, it also found wide application as an insulator for wiring in buildings and airplanes. By 1941, according to M.G. Milliken, general manager of the Cellulose Products Department, direct military purchases accounted for half of the product's sales, and "we could probably sell twice the amount we make." Two years later, with the advent of techniques for molding ethylcellulose powder into sheet plastics, the product went entirely on allocation for defense purposes. The largest single use was for the nose cone of the VT fuze; however, the toughness and impact strength of ethylcellulose plastic through a wide range of temperatures had made it valuable for many other applications. In the summer of 1944, at the government's request, Hercules agreed to increase by 60 percent its capacity to make the product at Hopewell, at a total cost of $1.4 million.[45]

Demand for cellulose acetate was even more impressive. In 1940, Hercules sold nearly eight million pounds; four years later, the company sold more than eighteen million pounds, of which the vast majority went for wartime and "essential civilian applications." In 1944, Milliken estimated that sales could have been 75–100 percent higher if only the company had the capacity. Traditional markets for the product—photographic film, rayon, and lacquers—grew steadily during the war, but cellulose acetate proved especially popular as a plastic material. In 1942, plastics applications accounted for about two-thirds of the company's sales; by 1945, more than 90 percent of sales went into plastics. The widespread adoption of injection-molding techniques and the rapid emergence of a plastics fabrication industry spurred the change: by the mid-1940s, cellulose acetate plastics were being used in everything from personal items like combs and toothbrushes to molded parts for applicances and motor vehicles. In aircraft, where weight was at a premium, cellulose acetate was increasingly used in panels, switches, knobs, and levers. The material also found military applications such as in gun stocks, pistol grips, and knife handles.[46]

The mushrooming popularity of cellulose acetate posed a challenge to Hercules. The company reached the rated capacity at Parlin (fifteen million pounds) at the end of 1942, although engineers were resourceful in increasing throughput by lowering acetylation times and improving other steps in the process. Nonetheless, Hercules debated at length the question of adding more capacity. The question involved two key issues: expense, which could range in the millions and could represent the largest single capital investment in the company's history; and concern about the price and availability of the critical reagent, acetic anhydride. By 1943, the first issue was settled: Hercules had ample cash, and demand for cellulose acetate seemed likely to continue soaring in the future.

The sourcing of acetic anhydride, which accounted for more than one-third of the final manufacturing cost of cellulose acetate, proved a more complex problem. Hercules bought acetic anhydride from Carbide & Carbon Chemicals Corporation (forerunners of today's Union Carbide), which in the late 1930s had developed a proprietary process for reacting acetic acid with ketene (a derivative of the petrochemical propylene) to make the reactant. The Carbide & Carbon process was the most efficient available; nonetheless, if Hercules were to increase substantially its production of cellulose actetate, it would thereby also increase its reliance on a powerful supplier. To avoid this situation, the company mounted a research investigation in the spring of 1942 to find an acceptable alternative for making acetic anhydride. (Subsequently, Hercules exchanged technical know-how on the problem with Tennessee Eastman.)[47] By late 1943, with "the cost of acetic anhydride. . .increasing markedly," Hercules was ready to make its move: it committed funds for a pilot plant to produce acetic anhydride from acetic acid using Tennessee Eastman's cracking process. A year later, Hercules authorized $1.1 million to construct a commercial-scale acetic anhydride plant at Parlin with an annual capacity of thirty-two million pounds. This was the prelude to Hercules' major bet for the postwar world: a $7 million appropriation in the spring of 1945 to double production of cellulose acetate.[48]

The impact of World War II on Hercules' chemical cotton business was mixed. The early years of mobilization brought rapid growth: sales of the Virginia Cellulose Department more than doubled between 1940 and 1943. The Hopewell plant ran at full capacity to help feed the tremendous demand for nitrocellulose, as well as healthy growth markets for cellulose derivatives and rayon. These demands were met, wrote the department's general manager, Lloyd Kitchel, "by ironing out production bottlenecks throughout the plant without any large outlay of capital."[49]

After 1943, however, the department's sales declined, and profitability suffered throughout the war. Costs proved difficult to control: federal agricultural policies kept the price of cotton linters at high levels, and Hercules was forced to raise wages substantially to keep skilled workers, owing to competition from defecting to the shipyards at Newport News or to other chemical plants nearby. The loss of its profitable export business to Europe was another problem.[50] Third, and most serious, were improvements in the purification of wood pulp by Hercules' competitors and customers. The improvements, combined with the federal War Production Board's decision to restrict use of cotton linters in products other than smokeless powder, inspired more and more makers of nonexplosive cellulose products to convert to wood pulp. (The government itself followed suit in 1942, when the GOCO plants began nitrating wood pulp to make smokeless powder using the process developed by Hercules' Stern in the 1930s.) Early in 1943, Kitchel predicted glumly that

> the recapture of the commercial business will be difficult. The quality advantages which chemical cotton pulp enjoyed over wood pulp in the past have been largely wiped out so that we must consider ourselves on a strictly price competitive basis with wood pulp in the cellulose acetate industry. Some slight premium will undoubtedly be paid for cotton over wood pulp but it will not be large.[51]

But Kitchel also saw opportunity in the skewed workings of the wartime economy. In the 1930s, Du Pont and several rubber companies had found that viscose rayon (made from chemical cotton) had superior attributes as a tire cord: it saved rubber and increased the longevity of tires.[52] With the wartime rubber shortage, the rayon tire cord came into its own and pulled chemical cotton along with it. In 1943, the War Production Board sponsored a crash program to produce 240 million pounds of high-tenacity rayon cord per year—a level eight times that of the year before. In response, Hercules quickly added new sheeting capacity at Hopewell. By the end of the war, more than 40 percent of the department's sales went into viscose rayon—up from just 13 percent in 1940.[53]

In sum, World War II had a profoundly disorienting effect on Hercules chemical cotton business. Through a combination of market distortions and government policies, wood pulp became a competitive alternative to chemical cotton in most of the company's prewar markets. To meet this threat, and to cement ties with its remaining commercial customers, Hercules strengthened its research staff at Hopewell. After 1942, plant chemists and department salesmen worked with customers in plastics and artificial fibers to improve the purity and quality of chemical cotton.

Rosin, Terpenes, and Synthetics

During the war, the same economic forces that helped revive and transform Hercules' business in cellulose and its derivatives swept across the company's Naval Stores, Paper Makers Chemical, and Synthetics departments. Although the company's products in these lines were not, for the most part, vital to defense production, their fortunes surged with the reviving national economy. For example, sales of resins and plasticizers, as with sales of Parlon® and Clorafin,® climbed with the rising demand for protective coatings for military bases and products. The disruption of foreign trade and shortages of key materials also created both problems and opportunities for many Hercules products. To cite but one instance: because varnish manufacturers could no longer obtain tung oil from China, they turned to pentaerythritol-abietate resins made by Hercules and other suppliers.

The contrary tugs of the booming but distorted economy were felt most dramatically in the Naval Stores Department. Sales and profits climbed sharply from 1940 to 1943 but declined thereafter, principally because the department was unable to sustain the momentum of mobilization. Although exports were "wiped out" (to use General Manager Al Forster's phrase) with the German occupation of most of Western Europe after the spring of 1940, defense mobilization in the United States more than offset that loss. By 1943, however, demand for naval stores products in construction was falling, and sales of basic commodities were limited by federal allocations or price ceilings.

In the early years of the war, Vinsol® resin became the department's star performer. In 1940, Hercules sold 16.3 million pounds of Vinsol® resin; three years later, sales reached 43.1 million pounds.[54] Not only did it serve well as an electrical insulator, but it also substituted for shellac in coatings and plastics. Combined with other chemicals in a product called Truline Binder, Vinsol® resin proved valuable in coating metal parts and pieces in foundries. In pulverized form, Vinsol® resin was widely used as a stabilizer for roadbeds and as an emulsifier for asphalt and cement. Indeed, one of its most celebrated applications was as a soil stabilizer in aircraft runways.[55]

Mobilization stimulated demand for other naval stores and synthetics products, too. Sales of naval stores commodities—rosin, turpentine, and pine oil—soared in the early years of the war. Sales of refined products like Staybelite® resin fared even better. Staybelite® resin, for example, had grown rapidly in its prewar markets as a constituent of synthetic resins for adhesives and inks. After 1942, through the efforts of Hercules technical salesman C.H. Boys, sales to the synthetic rubber industry surpassed these traditional markets.[56] A similar pattern emerged in Hercules synthetics. Pentalyn® synthetic resins—used in tung oil substitutes—experienced phenomenal growth and propelled the tiny Synthetics Department to profitability for the first time. Demand for Pentalyn® resins was so strong by the summer of 1941 that Hercules spent $275,000 to acquire its principal source of phenolic (pentaerythritol) resins: the assets of John D. Lewis, Inc. of Providence, Rhode Island. The transaction also included valuable trademarks and an expanded product line in ester gums.[57]

Such booming growth proved difficult to sustain. As early as 1943, Forster was complaining about "the restrictive impact" of the war: "increased difficulties in procuring operating supplies of all sorts and construction materials, increasing scarcity of labor, wood shippers going out of business, greatly expanded Government control, shift in demand for our products in industry," and other problems. The building of airfields, army camps, and war plants was virtually completed, and Forster saw little hope of gaining business in commercial construction until the war's end. In 1944, the Brunswick plant, which had run flat out since 1940, was forced to curtail production, and Forster predicted that rosin would shortly go on allocation.[58]

Amid these difficulties, the Naval Stores and Research departments collaborated on developing new markets for terpenes and synthetics. One of the most promising opportunities was the growing need for insecticides to help increase production of food and essential materials like cotton, and to stem the effects of insect-borne diseases in combat areas. Hercules brought long but limited experience to insecticides: since its entry into the naval stores industry in 1920, it had sold pine oil for use in formulations designed to kill insects. In the 1920s and 1930s, Hercules worked on insecticides in cooperation with the USDA and several Southern universities. In the late 1930s, Naval Stores researcher Donald H. Sheffield developed a powerful terpene-based chemical—the eponymous DHS Activator—for use with pyrethrum and rotenone as an insecticide. Sales of the product built up slowly, then fell dramatically in 1942 when both pyrethrum and rotenone went on allocation. Fortunately, Hercules had developed Thanite, a terpene-based insecticide that did not require scarce chemicals, and that was primarily effective as a knock-down agent against flies.[59] Thanite proved popular immediately: sales shot up from 397 gallons in that year to 365,643 gallons in 1944. By that time, however, the product was revealing its limitations: it was irritating to human skin and had an obnoxious odor. To maintain its position in a very attractive market, Hercules worked on modifying Thanite and developing other new insecticides. In 1943, the company hired several entomologists to work with its chemists on insecticides, and screening of new compounds was conducted at the University of Delaware.

A breakthrough came within a year. George Buntin, a research chemist in the Naval Stores Department, discovered that camphene (a derivative of pine oil) when chlorinated to a specified degree proved extremely effective in killing flies. Soon thereafter, Hercules submitted samples to the U.S. Department of Agriculture for screening against a wide range of agricultural pests. The results were promising enough to justify development work to determine the optimum composition of the product and practical methods for its production. This product, eventually marketed under the trademark Toxaphene, would turn out to be one of the company's most successful in the decade after the war.[60]

Hercules' experience in insecticides paid off in other respects, too. In 1944, the U.S. government asked Hercules and several other chemical companies to install capacity to make dichlorodiphenyltrichloroethane (DDT), "an extremely interesting and urgently needed insecticide." The government provided funds to build a DDT plant at Parlin, where Hercules produced more than six hundred thousand pounds of the chemical in aerosol form by the end of the war. DDT proved far more popular than Thanite, for which it substituted in most applications. In 1945, Hercules arranged to buy the government's DDT plant at Parlin, anticipating a bright future as a supplier of insecticides.[61]

During the war, Hercules also developed several other significant chemicals derived from rosin and pine oil: Poly-pale® resin for protective coatings, *para*-cymene for essential oils, and Resin 731 (a resin soap) for synthetic rubber. Of the new products, Resin 731 grew the fastest, not only because of its intrinsic value, but also because it served as the basis of an even more attractive product, Dresinate® 731 rosin soap (the name derives from "disproportionated resins"—the product family included many different sodium and potassium salts of modified rosins), made by the PMC Department.[62] In 1942, PMC launched its "Dresinate® program" to sell "water-soluble or dispersible resinates to all fields outside of the paper trade." According to Philip Stull, general manager of the department,

> the program embraces two distinct objectives. First it is intended to procure a reasonable share of business from concerns who have been buying rosin and saponifying it themselves. It is the special objective of the program to move Dresinate into fields which have previously used gum rosin. The second intention is to introduce Dresinate into fields where saponified rosin is not now being used but where other materials have heretofore acheived purposes for which water-soluble resinates are suitable.[63]

The program proceeded "in an exploratory way" to investigate potential markets. By far the most promising was synthetic rubber: under a federal crash program to find alternatives to natural rubber, production of synthetic rubber soared one hundredfold between 1941 and 1945 (from 8,383 long tons to 830,780 long tons). By the end of the war, more than 85 percent of all rubber consumed in the United States was synthetic, and Hercules' Dresinate® emulsifiers played a major role in this remarkable success story.[64]

Of the several types of synthetic rubber produced during the war, the most popular general-purpose variety was made from a copolymer of butadiene (a hydrocarbon available from a number of sources that is closely akin to isoprene, the monomer in natural rubber) and styrene, a chemical normally made from benzene and ethylene. Called GR-S (government rubber—styrene), the synthetic was used primarily to make tires. The first step in the manufacture of GR-S involved mixing the butadiene and styrene monomers in glass-lined tanks and emulsifying the mixture in a solution of purified soap suds. From there, a peroxide catalyst and modifying agent were added, and the mixture was heated and agitated for about 15 hours, during which time about two-thirds of the mixture was polymerized. Thereafter, agents were added to destroy the catalysts and retard oxidation, unreacted monomers were recovered and recycled, and the synthetic latex was transferred to coagulation tanks and eventually readied for shipment.[65]

In the war, many improvements were made to the process, including the widespread adoption of Dresinate® 731 as an emulsifying agent in the first stage. The rubber manufacturers had initially tried fatty-acid soaps and even commercial detergents as emulsifiers. The problem with such soaps, however, was that the resulting rubber lacked tack—it would not stick to itself—and tires made from it tended to shred or fall apart after minimal wear. Conventional tackifiers such as rosin failed to solve the problem, because they interfered with the peroxide catalysts used in polymerizing the butadiene-styrene mixture. Researchers at Hercules' Experiment Station, including Joseph Borglin, however, saw that a rosin derivative might be tailored to impart the desired tack to rubber without reacting with the catalysts. Accordingly, they developed a process for disproportionating rosin to produce Resin 731, which could be made into a soap by treatment with caustic soda. As the properties of the soap, Dresinate® 731, became established, it proved an extremely successful product.[66]

In 1942, PMC managers estimated the total market for Dresinates® at about 90 million pounds based upon successful substitution for fatty-acid soaps as emulsifying agents. The department also predicted that Dresinates® would find valuable applications in alkaline cleaners, the petroleum industry and associated chemical compounders, flotation, soaps, disinfectants, linoleum, textiles, and many other markets. These projections bore out almost immediately. In 1942, PMC sold 2.8 million pounds of Dresinate,® most for use in detergents and floor coverings. A year later, as the markets for synthetic rubber took off, sales reached 7.9 million pounds, with substantial increases in sales in the "emulsifying field." By 1945, the company had sold 21.5 million pounds of solid and liquid Dresinate,® of which nearly 40 percent was accounted for by sales of Dresinate® 731 to the synthetic rubber industry.[67]

Upgrading Research

Although many of Hercules' most successful commercial products flowed from department-based R&D, in the early 1940s, the company also developed a reputation for conducting basic research of the highest order. The energetic director of research, Emil Ott, was primarily responsible for enhancing Hercules' scientific and technical capabilites.

Born and educated in Switzerland, Ott had emigrated to the United States in 1927. He worked briefly for Stauffer Chemical Company in San Francisco before moving to Baltimore to become a postdoctoral research fellow, and eventually assistant professor of chemistry at Johns Hopkins University. In 1933, Ott, an expert in the emerging field of "applying X-rays and X-ray diffraction analysis to chemical systems," became a consultant to Hercules, and soon thereafter, the company recruited him as a full-time research chemist. Assigned first to the Development Department, where he worked closely with George Norman on sulphonated chemicals for the rubber industry, Ott was named director of the Research Division at the Experiment Station in 1937. In this position, he gained exposure to senior management by reporting regularly to the technical development committee. Two years later, at age 37, Ott was promoted to the new position of director of research for the company.[68]

As head of research, recall his colleagues, Ott guided the transformation of the Experiment Station from an "applications and technical service organization" into a "research organization."[69] He interviewed all technical personnel hired at the station and increased recruiting of Ph.D. chemists, among them, his former student at Johns Hopkins and eventual successor at Hercules, Bob Cairns, as well as John Long, who later became a corporate vice president. To stimulate creative research and the flow of information, he instituted regular meetings of the technical staff to report on research under way or to hear academic scientists discuss their latest work. To enhance the department's reputation, he encouraged technical personnel to become active in professional associations and to publish. Setting an example, in 1943, Ott, with the help of his staff, especially Dr. Harold Spurlin and Mildred Grafflin, edited an encyclopedic summary of available knowledge about cellulose and cellulose derivatives.[70]

Although research at Hercules continued to reflect the priorities of the operating departments, in the early 1940s, Ott also urged the company to consider "developments along lines which are less closely connected with our existing business." He pointed out that the company's wartime operations "scarcely permit any additional peacetime developments to accrue therefrom," and speculated that Hercules might be "in great danger of losing its relative position in the chemical field" unless it opened new areas of research. In particular, Ott believed that it was critically important for the company to move beyond its historical dependence on agricultural raw materials into such fast-developing areas as chemicals derived from coal, natural gas, and petroleum. To achieve this goal, he argued, it would be necessary to spend greater sums on research, a policy followed by the fastest-growing companies in the chemical industry such as Du Pont, Monsanto, Dow, and American Cyanamid.[71]

In 1940, Ott established a General Research Group under Spurlin to conduct basic research and explore scientific and technical issues of potential long-term interest to the company. In developing such an approach, Ott was probably influenced by the example of other chemical companies such as Du Pont. But he was also realizing the goals of his predecessors, especially Norman and O.A. Pickett, and, in some respects, carrying on the work of the old Development Department, which had been disbanded in 1937. The General Research Group, recalls long-time Hercules researcher, Dr. Reginald Ivett, "aimed at getting into new fields, developing new knowledge, which it would eventually turn over to a department as soon as the department became interested."[72]

After 1940, Hercules increased the portion of its research budget allocated to explore "new products for new outlets" from 15 percent to 25 percent of the total research budget.[73] Although researchers under Ott and Spurlin sometimes ranged far afield, looking at new developments such as synthetic rubber and petrochemical polymers, most work continued to concentrate on rosin and cellulose, with an eye toward gaining a better understanding of fundamental chemistry and improving processes for making derivatives. Indeed, Ott probably saw this research as a necessary first step; he may also have sensed a reluctance on the part of senior management to investigate areas foreign to the company.[74] In any event, Hercules' investigation of the composition of rosin, for example, led eventually to new products for the PMC Department such as fortified rosin size, as well as many improved synthetic resins and esters for the Naval Stores and Synthetics departments.[75] The General Research Group also helped develop and improve such new products of the 1940s as cellulose plastics, Thanite, Dresinates,® and Pentalyn® and Poly-pale® resins.

Where possible, Ott also backed efforts to investigate new technologies and sources of raw materials. For example, the extensive testing of Resin 731 drew Hercules' researchers into close collaboration with the rubber companies and into the world of large-scale polymerization. Similarly, Ott encouraged the expansion of a promising line of research in the Naval Stores Department that eventually opened up vast new opportunities after the war. The department's research on the air-oxidation of terpenes to make synthetic pine oil and other products was broadened in scope to include a general study of air-oxidation of chemicals similar in composition to terpenes, including the hydrocarbons cumene and cymene. In the late 1940s and early 1950s, Hercules' research on the air-oxidation of cumene led not only to catalysts for the synthetic rubber industry, but also to several promising postwar ventures into the petrochemicals.[76] Finally, Ott, Spurlin, and others were keenly interested in upgrading the company's capabilities in fractionation and distillation processes, as well as in exploring new technologies such as high-pressure synthesis of ammonia, methanol, and petrochemical intermediates.[77]

Ott's insistence on high standards and his willingness to entertain new possibilities not only redirected research at the company, but also energized a generation of promising young scientists. Their skills and enthusiasm, in turn, would help push Hercules in new directions after the war.

Conclusion

World War II had a profound and diverse impact on Hercules. As had happened during the First World War, the company became a major producer of munitions for the United States and the Allies. Its sales, profits, and employment levels swelled to record peaks, and its young managers, scientists, and employees received invaluable training in their work on large, complex projects. Unlike in World War I, however, Hercules was not transformed by its experience as a munitions supplier during World War II. By the early 1940s, the company was no longer heavily dependent on explosives and closely related technologies. The war years accentuated and accelerated trends already under way at Hercules: an increasing ability to tailor specialized chemicals for particular markets, a continuing resourcefulness in developing new products from cellulose and naval stores, and a growing reliance on R&D. As President Higgins remarked in 1944, the company's "newer lines"—by which he meant nonexplosives—"based upon cellulose, rosin, and terpenes . . . now constitute, eliminating wartime distortions, over 80 percent of our total business."[78]

Although Hercules acquired familiarity with one new military technology of note—rocket powders—and also gained valuable experience in working with the federal government, the company never became wholly dedicated to defense production. Rather, Hercules leaders believed that the company's future would depend on sustaining its ability to work wonders with cotton linters and pine stumps. The record of the commercial businesses—plastics, synthetics, and insecticides—during World War II afforded reasons to feel confident. Nonetheless, questions remained: how long—and how far—could the company depend on its base in naval stores and cellulose products?

CHAPTER 9

JOINING THE PETROCHEMICALS REVOLUTION, 1946–1954

In April 1946, Hercules management gathered for a companywide meeting "to project . . . a greater future" and "promote a better understanding of the company, its products, and departments." The three-day meeting, held at Virginia Beach, Virginia, was similar in character to the one held at Atlantic City in 1919. This time, however, the company's circumstances were far different. By 1945, Hercules was already a diversified chemical company. Although it had played a prominent role as a military contractor, the company had also developed its commercial business apace, planning at an earlier stage of the war for the return to peacetime conditions. As a result, when ordnance production halted in the late summer of 1945, the company was well prepared to prosper in a peacetime economy. Indeed, despite a brief but sharp postwar recession, Hercules' net sales actually increased during the first year after the war, and its net earnings dipped only slightly (see Exhibit 9.1).

At Virginia Beach, then, Hercules leaders did not concern themselves with unanswered questions about the future, or with potential avenues of diversification. Rather, they took as a given that the company would move forward by continuing to execute its prewar strategy. Indeed, management forecast "a tremendous increase in business in the plastics, protective coating, synthetic fibre, paper, textile, insecticide, and mineral industries" and predicted that, based on its existing product lines, "the outlook for the company is an exceedingly bright one." In his opening address, President Higgins pointed out that "of our one hundred million dollar sales a year, 70 to 80 million dollars is new business which did not exist 27 years ago [in 1919]. That sounds like healthy growth, doesn't it?" The conclusion seemed inescapable: "Since the rate of growth is currently increasing rapidly, there is reason for optimism in the future."[1]

EXHIBIT 9.1 Hercules' Financials, 1946–1954 (Dollars in Thousands, Except per Share)

	1946	1947	1948	1949	1950	1951	1952	1953	1954
Net sales	$100,728	$131,270	$129,267	$120,977	$160,231	$216,849	$181,517	$190,202	$187,548
Earnings before interest and taxes	14,327	22,044	17,752	16,690	29,210	43,926	32,116	31,338	31,217
Total assets	70,051	80,704	84,039	89,270	106,960	131,622	126,006	131,245	135,242
Fixed[a]	26,788	34,815	35,944	36,862	38,313	46,298	50,341	52,876	62,411
Current	42,467	45,239	47,013	51,901	68,179	83,794	73,944	76,819	70,833
Other	796	650	1,083	508	468	1,531	1,721	1,550	1,998
Stockholders' equity	48,646	55,877	60,852	63,862	69,702	75,350	78,439	81,995	89,227
R&D Expense	N/A	N/A	N/A	4,237	4,570	5,433	6,394	7,905	7,578
Dividends/share	$1.50	$2.00	$2.25	$2.60	$3.30	$3.00	$3.00	$3.00	$3.00
Earnings per share (common)	$3.03	$4.75	$3.99	$3.60	$5.31	$4.96	$4.03	$4.20	$5.10
Number of employees	10,240	10,299	10,263	9,473	9,781	11,177	10,679	10,689	10,943
Return on sales	14.2%	16.8%	13.7%	13.8%	18.2%	20.3%	17.7%	16.5%	16.6%
Return on assets	20.5	27.3	21.1	18.7	27.3	33.4	25.5	23.9	23.1
Return on equity	29.5	39.5	29.2	26.1	41.9	58.3	40.9	38.2	35.0
R&D/Sales				3.5	2.9	2.5	3.5	4.2	4.0

a. Gross fixed assets, including depreciation.

Source: Hercules Powder Company, Annual Reports for 1946–1954.

Hercules' postwar strategy thus called for expansion within known boundaries. It was a vision animated by confidence in the company's capabilities. After the war, such confidence seemed well founded. After all, Hercules' strategy as a supplier of specialty chemicals had proved successful. The company was vertically integrated in explosives, cellulose products, and naval stores and controlled the production chain from the gathering and synthesis of raw materials through the distribution of chemical products to its customers. Since the late 1930s, moreover, the Experiment Station and the operating departments had developed a stream of successful products—plastics, synthetics, additives, and agricultural chemicals—that promised to carry the company forward. In 1946, it seemed reasonable to expect that such successes would be repeated again and again in new products derived from chemistry the company had already mastered.

In the decade following the war, this expectation was put to the test as Hercules pushed the frontiers of its major product lines. In short order, however, the company began to encounter limits to growth in each of its existing businesses. By the early 1950s, it was no longer clear that Hercules could sustain indefinitely its historical levels of growth and profitability. In dealing with the problem, the company had to work through two issues. First, key leaders on the executive committee who continued to have faith in the traditional businesses had to be convinced of the need to diversify. Second, Hercules had to identify and take advantage of an appropriate diversification opportunity.

In the early 1950s, the opportunity was not difficult to spot: petrochemicals was the fastest growing and most promising segment of the chemical industry. In many respects, petrochemicals represented a logical extension of Hercules' business in plastics and synthetic fibers. The new area, however, was not a small, underdeveloped, and neglected industry such as naval stores, but rather a complex, expensive, and increasingly competitive field that attracted investment from the largest chemical, rubber, and oil companies in the world. As a latecomer to petrochemicals, and a relatively small participant, Hercules had to choose its point of entry carefully. By the end of 1954, under President Higgins and his successor, Albert E. Forster, the company had at last found its way in, following three separate paths that branched off from its existing businesses.

A STRATEGY FOR THE POSTWAR WORLD

Coming out of the war, Hercules had many assets: experienced leadership; an enthusiastic cadre of younger employees seasoned by the wartime challenges; a diversified production base; excellent research and technical capabilities; promising new products among its cellulose and rosin derivatives; a strong balance sheet; and a proven formula for success as a specialty chemicals company. That it also had liabilities—the loss of significant revenues from ordnance production; "a case of personnel indigestion" (as one executive put it) resulting from the huge increase in wartime employment; an old and mature product line in commercial explosives; and several departments (Naval Stores, PMC, and Virginia

Cellulose) still straining to achieve consistent performance—the company's leaders were well aware.[2] But in 1946, the opportunities appeared to clearly outweigh the concerns.

The concerns, moreover, seemed remediable. At the dawn of the postwar era, Hercules leaders were confident that lost revenues could be easily replaced, that places could be found for the best of the younger personnel, that the commercial explosives business would continue to generate cash even as the company's dependence on it diminished, and that the struggling departments would soon find ways to record steady results.

If in some respects Hercules' postwar strategy appears conservative, it was also prudent. The legacy of the company's first generation of managers, including Russell Dunham, who remained active on the board, was very much alive. The executives now at the top—Higgins, Bent, Nixon, Stull, Meyeringh, Morrow, and others—were also conditioned by the company's earlier successes, especially in rosin and cellulose chemistry. Hercules' acceptance of significant risk when it entered the naval stores business after World War I was a rare exception to the company's general pattern of growth. Hercules leaders, moreover, were painfully aware of the long, hard struggle to succeed in a business not fully understood at the outset. Hercules eventually prospered in naval stores and cellulose products by pursuing a common strategy: patient and methodical refinement of processes and products, investment in R&D and technical support, and careful cultivation of customers. In general, the company was conservative in its approach to risk, opting to expand only when an increased volume of business virtually prepaid the investment.

Hercules' strategy also reflected its assessment of the external environment. In 1946, the company—like many others in the United States—feared the return of economic depression. It is easy to underemphasize this widespread concern, especially in view of the longest and greatest economic expansion in modern history that followed. After the war, however, many Americans vividly remembered the long years of economic turmoil that had preceded it. Some also recalled the severe recession that had followed World War I. It was not illogical to expect that the peace would again be accompanied by depression. Fears of a return to hard times were deeply felt and pervasive, a phenomenon John Kenneth Galbraith has termed "depression psychosis." At Hercules, decision makers who shared this concern tempered their optimism and were reluctant to contemplate rapid expansion.[3]

So the company moved forward cautiously, with an eye on its past. As President Higgins insisted, "There's plenty of future in explosives, cellulose, and rosin."[4]

Defense Contracting in the Cold War

Production at all six Hercules GOCO plants ceased by the end of 1945, leaving the company with the unenviable task of releasing thousands of employees from work that had all but consumed them for years. Although the company hoped to keep as many technical employees as possible, its commercial explosives operations were not large enough to create meaningful employment opportunities for more than a handful of people. The company sought to retain its ablest production supervisors from the ordnance plants, as well as people involved in advanced technologies such as ammonia synthesis and rocket fuel development. In choosing who could stay on, Hercules was guided not only by the wish to hold on to its best and brightest but also by input from the newly established Personnel Department. Guided by Leon W. Babcock, the department developed systematic employment policies and career development plans for managerial and technical employees.

The early onset of the Cold War helped ease the process of reducing manpower. During 1946, for example, three GOCO plants—Radford Arsenal, Sunflower Ordnance Works, and Missouri Ordnance Works—were reopened to make fertilizer grades of ammonium nitrate. Output from these facilities was shipped overseas as part of U.S. efforts to help rebuild the economies of Western Europe. Although Hercules' part in this work ended in 1948, the company continued to operate Radford on the basis of a contract with the U.S. Navy to make propellant for antiaircraft rockets.[5]

In the early postwar years, another government-owned installation, the Navy's Allegany Ballistics Laboratory (ABL), appeared to offer more promising and enduring opportunities. Located in eastern West Virginia, near Cumberland, Maryland, ABL was designed as an R&D facility for solid rocket propellants; the navy was seeking an experienced propellant manufacturer to manage it. Hercules was engaged in late 1945 to operate the facility. Its research for the Navy continued after the war, and several hundred Hercules employees, including many from Radford and Sunflower, moved to ABL in 1946.[6]

Initially, Hercules concentrated at ABL on developing propellant for the Bumblebee, a large, short-burning rocket for antiaircraft protection. The key technological problem was to find ways to design and manufacture propellant for rockets significantly larger and faster than those used in the war. For example, the extrusion process, used to make propellant grains for bazookas, was limited by the press size. Although the Navy experimented with clustered rockets powered by extruded propellant, a wholly new process based on wartime research by Dr. John Kincaid and Dr. Henry Shuey at the government's Explosives Research Laboratory at Bruceton, Pennsylvania, provided a better solution. The process consisted of treating extruded double-based grains with a solvent (nitroglycerin), a plasticizer, and a stabilizer, then heating the resulting mixture to form a homogeneous mass that could be cast into molds to make large-diameter grains. Whereas the largest extruded grain was six inches in diameter, the first cast double-based propellant was sixteen inches in diameter. Better still was "the lack of any restriction on the size of configuration casting."[7]

ABL's work on cast double-based propellant, led by Drs. Lyman Bonner and Richard Winer, culminated in a successful test-flight of a booster rocket in August 1947. Thereafter, ABL participated in the development of most of the tactical rockets used by all three branches of the service in the early postwar era: the Nike, Honest John, Sparrow, Matador, Terrier and Talos guided missiles, as well as jet-assisted take-off (JATO) booster rockets for bombers and other aircraft. During these years, Hercules researchers Winer, Ralph Preckle, and Rudolph Steinberger were responsible for significant advances in propellant design and for achieving characteristics such as uniformity, low-temperature coefficients, and high-impulse power.[8]

Hercules researchers at ABL also worked closely with other government contractors on the design and development of rocket motor cases. Early rocket motors were housed in cases made of steel or aluminum. In the late 1940s, however, Richard Young, a consultant to M.W. Kellogg Company and a frequent visitor to ABL, began advocating the use of lightweight cases made of wound fiberglass. In 1951, a Nike missile using a Young-designed fiberglass case and Hercules propellant was test-fired successfully. Although many technical problems remained in bonding propellant to filament-wound cases, the pairing of the two technologies would have momentous consequences for Hercules later in the decade (see Chapter 10).

During the Korean War, Hercules' involvement in defense contracting escalated rapidly when Radford and Sunflower were reactivated and rehabilitated.[9] The facilities, which featured many wooden structures and temporary buildings, had deteriorated badly during six years of inactivity, and their renovation required round-the-clock efforts. Radford came back on line in the spring of 1951, followed several months later by Sunflower. In the emergency, the plants produced a wide range of powders, from small arms to mortar, artillery, and rockets. The new business provided a brief but significant boost to corporate earnings: in 1952, the peak year of the war, Hercules earned more than $6.6 million in profits from its GOCO operations.[10] After the signing of the Armistice in the summer of 1953, production fell off dramatically at Radford and Sunflower; the latter was eventually placed on standby status. At both plants, Hercules remained to supervise continuing operations and maintain facilities and equipment in working order.

The Indian Summer of Commercial Explosives

No longer the foundation of Hercules' business, the Explosives Department nevertheless made solid gains during the decade after World War II, quite apart from the sudden expansion of business during the Korean War. Net sales and profits roughly doubled (to $44.3 million and $5.9 million, respectively), and during peacetime the department consistently accounted for about 15–20 percent of total company profits (see Exhibit 9.2). By the early 1950s, however, a series of disquieting signals foretold the rapid decline of the business.

The department's creditable postwar performance in large measure reflected the nation's voracious appetite for thousands of miles of new roads and highways, millions of barrels of oil for automobiles, millions of tons of coal for power plants, and massive amounts of minerals such as iron, copper, and zinc. Accordingly, Hercules targeted its R&D and marketing efforts on new techniques for open-pit mining, quarrying, and seismography. The department's greatest problem was chronic shortages of raw materials and, occasionally, labor. Glycerin, anhydrous ammonia, and ethylene glycol were in especially short

EXHIBIT 9.2 Hercules' Sales and Profits by Department, 1946–1955 (Dollars in Thousands)

Product Line	*1946*	*1947*	*1948*	*1949*	*1950*	*1951*	*1952*	*1953*	*1954*	*1955*
EXPLOSIVES										
Net sales	$20,237	$27,332	$31,090	$29,196	$32,847	$39,408	$43,516	$44,484	$44,346	$50,835
Net profit	2,841	4,308	4,929	4,772	6,843	9,260	8,912	7,338	5,910	7,679
Return on sales	14.0%	15.8%	15.9%	16.3%	20.8%	23.5%	20.5%	16.5%	13.3%	15.1%
Percentage of total sales	20.3	20.9	24.1	24.2	20.6	18.3	24.5	23.9	24.1	22.8
Percentage of total profits	16.0	16.8	22.6	23.6	20.6	19.5	31.9	27.3	21.7	19.8
CELLULOSE PRODUCTS										
Net sales	$28,072	$32,972	$33,525	$30,184	$44,213	$55,854	$46,712	$49,800	$48,357	$45,843
Net profit	3,241	2,476	2,979	3,323	7,937	8,861	6,434	6,646	6,122	7,431
Return on Sales	11.5%	7.5%	8.9%	11.0%	18.0%	15.9%	13.8%	13.3%	12.7%	16.2%
Percentage of total sales	28.2	25.3	26.0	25.1	27.7	26.0	26.3	26.8	26.3	20.5
Percentage of total profits	18.3	9.7	13.6	16.4	23.9	18.7	23.0	24.8	22.4	19.1
NAVAL STORES										
Net sales	$15,279	$20,116	$16,779	$17,726	$25,806	$32,804	$21,358	$24,245	$28,448	$36,561
Net profit	8,032	14,600	4,848	3,620	6,795	11,318	190	1,574	2,251	4,446
Return on sales	52.6%	72.6%	28.9%	20.4%	26.3%	34.5%	0.9%	6.5%	7.9%	12.2%
Percentage of total sales	15.3	15.4	13.0	14.7	16.2	15.2	12.0	13.0	15.5	16.4
Percentage of total profits	45.2	56.9	22.2	17.9	20.5	23.9	0.7	5.9	8.2	11.4
VIRGINIA CELLULOSE										
Net sales	$7,782	$15,047	$14,438	$14,650	$17,385	$34,148	$21,803	$21,294	$17,332	$33,454
Net profit	793	2,000	1,964	3,159	3,517	4,945	3,644	3,042	3,196	6,816
Return on sales	10.2%	13.3%	13.6%	21.6%	20.2%	14.5%	16.7%	14.3%	18.4%	20.4%
Percentage of total sales	7.8	11.5	11.2	12.2	10.9	15.9	12.3	11.5	9.4	15.0
Percentage of total profits	4.5	7.8	9.0	15.6	10.6	10.4	13.1	11.3	11.7	17.5
PAPER MAKERS CHEMICAL										
Net sales	$18,816	$23,763	$22,443	$19,934	$24,532	$34,230	$26,784	$27,242	$25,160	$32,759
Net profit	1,504	2,005	5,403	4,475	4,884	8,837	6,965	6,302	6,318	8,993
Return on sales	8.0%	8.4%	24.1%	22.4%	19.9%	25.8%	26.0%	23.1%	25.1%	27.5%
Percentage of total sales	18.9	18.2	17.4	16.5	15.4	15.9	15.1	14.7	13.7	14.7
Percentage of total profits	8.5	7.8	24.7	22.1	14.7	18.6	25.0	23.5	23.2	23.1
SYNTHETICS										
Net sales	$9,436	$11,352	$10,521	$8,792	$14,920	$18,754	$17,234	$18,771	$20,156	$23,852
Net profit	1,345	258	1,712	900	3,181	4,187	1,769	1,937	3,493	3,487
Return on sales	14.3%	2.3%	16.3%	10.2%	21.3%	22.3%	10.3%	10.3%	17.3%	14.6%
Percentage of total sales	9.5	8.7	8.2	7.3	9.3	8.7	9.7	10.1	11.0	10.7
Percentage of total profits	7.6	1.0	7.8	4.4	9.6	8.8	6.3	7.2	12.8	9.0
Total net sales	$99,622	$130,582	$128,796	$120,482	$159,703	$215,198	$177,407	$185,836	$183,799	$223,304
Total net profit	17,756	25,647	21,835	20,249	33,157	47,408	27,914	26,839	27,290	38,852
Return on sales	17.8%	19.6%	17.0%	16.8%	20.8%	22.0%	15.7%	14.4%	14.8%	17.4%

Source: Hercules Powder Company, Department Annual Reports, 1946–1955.

supply during the late 1940s. As a result, Hercules sought to meet its own needs by integrating backward. In the early 1950s, for example, the company erected a new anhydrous ammonia facility at the Hercules Works in California. Soon thereafter, it built the first synthetic ammonia plant in the Southeast in Birmingham, Alabama, through a joint venture with the Alabama By-Products Company.[11]

Sales of dynamite for civilian uses accounted for almost all of Hercules' growth in commercial explosives. Sales of smokeless powder remained steady, but sales of black powder trickled down to a few thousand kegs, and the product was discontinued after 1950. The department's intermediate chemical products fared well. By 1954, sales of anhydrous ammonia, ammonium sulfate, and ammonia-based fertilizers yielded more than $4 million in profit.[12]

In retrospect, the postwar decade was an Indian summer for commercial explosives, the final successful phase of a business that would soon be dealt a crushing blow. The problem, increasingly manifest, was a revolutionary development in the manufacture of high explosives: growing numbers of customers found that ammonium nitrate explosives with no nitroglycerin component were highly effective. The realization came haltingly to industry experts, including Hercules' Jack Martin. Although the nitroglycerin content of explosives had been declining for years, the notion that it could be dispensed with altogether seemed absurd, even in the face of evidence to the contrary. "One day up at Kenvil," the incredulous Martin recalled,

> we were shipping so much stuff [ammonia dynamites], just pouring it through the packing machines. The next day, when we looked at the reports, however, we found that we had failed to put any nitroglycerin in the stuff. Oh God, I scrambled around. I called up to [the sales office at] Hazelton, and said, "What the hell happened to that batch?" They said, "What batch?" We told them the order number and so on. "Oh," he said, "we shot it this morning." I said, "How did it go?" He said, "It worked fine."[13]

The displacement of nitroglycerin in commercial explosives gathered momentum in the late 1940s and early 1950s in the wake of several spectacular accidents involving fertilizer-grade ammonium nitrate (FGAN). On April 16, 1947, the most disastrous explosion in the history of the U.S. chemical industry occurred at Texas City, Texas, when a ship laden with twenty-five hundred tons of ammonium nitrate fertilizer exploded offshore, detonating a nearby Monsanto styrene plant, damaging or destroying thirty-four hundred nearby residences, and killing more than five hundred people. Investigations of this and other incidents involving the fertilizer, as well as news of the success of iron ore producers blasting with FGAN on the Mesabi Range in Minnesota, helped spread the word of "nature's little deadly secret": when mixed with fuel oil or carbon black and "shot" with a large enough charge in a confined vessel, ammonium nitrate–fuel oil (ANFO) mixtures, selling for roughly $5 a ton, could do much of the work of nitroglycerin explosives, which cost fifteen to twenty times more.

Throughout the 1950s, an increasing number of quarry operators and others practicing large-diameter fixed blasting switched to ANFO explosives. The trend was accelerated by the invention in the 1950s of new "prilling" techniques, which produced small, porous particles of ammonium nitrate that were easily poured into boreholes and absorbed oil more readily than large crystals. The result was a "ridiculously simple material" that, according to the *Explosive Engineer*, "has to be regarded as the greatest advance to date in modern explosives technology."[14]

Before the war, the sudden decline of a major business segment would have devastated Hercules. By the mid-1950s, however, the success of its diversification strategy had reduced the company's vulnerability. Most of Hercules' attention—and its hopes for future growth—were focused on its newer businesses: cellulose products and naval stores.

From the Cottonseed

Hercules cellulose businesses fared well after the war, but not as well as predicted at Virginia Beach. The problem was twofold: rising and uncontrollable raw materials costs at one end of the production chain, and uneven demand at the other. The squeeze between rising costs and uncertain markets affected virtually all cellulose-related businesses, but especially the Virginia Cellulose Department (see Exhibit 9.2).

In the early postwar years, shortages of chlorine and other chemicals severely hampered Virginia Cellulose's performance. Although such problems were overcome in time, a more serious constraint was fluctuating and generally rising prices for cotton linters. Poor cotton crops, including a disastrous harvest in 1950, forced the department to scramble for alternative sources of cellulose. At one point, during the Korean War, the unit even recycled old mattresses. In these circumstances, Virginia Cellulose stepped up its work on purifying wood pulp to serve internal and external markets.

The boom in ordnance production during the Korean War was a major stimulus to the department's growth. This was fortunate, because the department's major external market, for rayon, showed troubling signs after the war. In 1946, 41 percent of Hercules' output of chemical cotton went into viscose rayon, but the proportion dropped below 29 percent in 1947 as rayon manufacturers switched from chemical cotton to less expensive wood pulp. At the same time, after three decades of rapid growth, demand for rayon (both viscose and acetate) turned flat after 1950 as other synthetic fibers—nylon, acrylic, and polyester—came on the market.

Similar dynamics affected performance at Virginia Cellulose's sister unit, the Cellulose Products Department, whose sales and profits nearly doubled between 1945 and 1954. Its rate of return on sales consistently ran just above 25 percent (see Exhibit 9.2). The postwar resurgence of the protective coatings industry helped nitrocellulose remain the department's leading contract in sales volume and profits. (The robust health of the coatings industry generally benefited the department's line of chlorinated products as well. Parlon® chlorinated rubber experienced rapid growth, especially as more natural rubber became available.) About two-thirds of Hercules nitrocellulose was consumed in coatings and lacquers, a performance boosted by heavy advertising and aggressive promotions. After 1947, however, sales to large users such as Celanese, Monsanto, and Du Pont declined because of new entrants and the availability of alternative sources. Demand for explosives-grade nitrocellulose, which more than doubled after 1950, helped offset the decline and sustain operations at Parlin.[15]

Among the department's newer offerings, ethylcellulose proved slow to penetrate the market for plastics. Despite increased sales to manufacturers of phonograph records, the product's overall performance was lackluster. Ethylcellulose had attractive properties, but for many applications it was not cost-competitive with new petrochemical plastics such as polyvinyl chloride, polyethylene, and polystyrene. Unable to generate sufficient demand for ethylcellulose, Hercules lost money on the product in five out of the nine years (and on balance) between 1946 and 1954.

Cellulose acetate came closer to fulfilling its early promise. With demand outstripping supply, Hercules constructed at Parlin in late 1946 a new cellulose acetate plant that doubled previous capacity. More than one-quarter of the company's output was used in the manufacture of rayon; in 1948 alone, sales to makers of the synthetic fiber jumped from about 1.6 million to 13.5 million pounds. In 1949, Hercules' investments in cellulose acetate started paying off: the company reported a profit of more than $316,000.

Even so, cellulose acetate, like ethylcellulose, encountered stiff competition from petrochemical plastics, especially polystyrene: in 1947, sales of this plastic totaled 147 million pounds, up more than half from the previous year. In response, Hercules developed new flame- and heat-resistant acetates—polystyrene is highly flammable—and launched a special plastics advertising program promoting the relative advantages of cellulose acetate. The program not only helped drive sales of the product above projections but also educated Hercules: the company discovered "many more interesting applications for plastics" than previously appreciated, and realized it would be unable to exploit many of them as a supplier of cellulose derivatives. "In plastics," complained J.J.B. "Jack" Fulenwider, head of Cellulose Products in 1949, "we continue to lose position."[16]

This frustration moved the company to rethink its strategy for plastics. Since the late 1930s, Hercules had sold cellulose acetate and ethylcellulose in the form of molding powder for plastics manufacturers. However, the company found that many of these fabricators were producing new products without undertaking significant design and development work. Sensing an opportunity, Hercules retained the Detroit industrial design firm of Sundberg–Ferar to create designs for its molding powder customers. It also worked closely with Monsanto to develop a new cellulose acetate flake, which the competitor promised to purchase at a rate of about two hundred thousand pounds a year. These efforts, a kind of quasi-forward integration, soon paid off. The new molding powders (Hercocel A and E) sold briskly, finding their way into plastic toys, tool and utensil handles, and other durable plastics applications.[17]

In sum, during the late 1940s and early 1950s, Higgins's faith in the commercial potential of cellulose chemistry proved justified, but also subject to limits. Among the department's older products, nitrocellulose remained in demand for both commercial and military applications, although its uses were specialized. Cellulose acetate, however, already displayed weaknesses in the face of competition from alternative plastics, and ethylcellulose seemed destined to become a slow-growth specialty product. Fortunately, Hercules' goals for growth and profitability in Cellulose Products were met by a relative newcomer among cellulose derivatives with the cumbersome name sodium carboxymethylcellulose, known more simply as CMC.

The Rise of CMC

CMC was first formulated in Germany during World War I. The product subsequently found commercial application as a thickener and as an ingredient in wallpaper paste throughout Europe, where vegetable gums and starches were less abundant and more expensive than in the United States. In the 1930s, German chemical companies, seeking independence from natural gums imported from Asia and the Far East, entered into large-scale production of CMC as part of Hitler's dual plan of military expansion and economic self-sufficiency.[18]

The availability of refined, controlled, and uniform substitutes for natural gums also proved attractive to customers in the United States. Many companies imported CMC as a replacement for more expensive, natural water-soluble materials such as Irish moss, gelatin, tragacanth gum, and sodium alginate. Before the war, for example, Hercules purchased German CMC for specialized uses at its plant in Hopewell, Virginia.[19]

The coming of war stimulated demand for CMC and emphasized the need for U.S.-based production. In the early 1940s, Hercules briefly studied the potential market for CMC but was spurred to action only after Dow Chemical began production of a sister water-soluble cellulose ester, methylcellulose. Accordingly, cellulose chemist Eugene D. Klug and other researchers at the Experiment Station stepped up efforts to develop a CMC manufacturing process.[20]

CMC was made by reacting purified alkali cellulose with sodium chloroacetate, then washing (with aqueous methanol of ethyl alcohol), drying, and grinding the material. The result was a "white, odorless, tasteless, non-toxic solid" that was also very hydrophilic: that is, it would suspend and actually thicken water much more efficiently than natural gums and starches, and it would remain water-soluble at extreme temperatures better than other known water-soluble polymers, including methylcellulose. CMC's high viscosity also made it an excellent film-forming and stabilizing agent.[21]

The product's greatest virtue was its versatility. By lengthening the polymer's molecular backbone, chemists could increase its viscosity; by purifying the physiologically stable substance, they could produce grades suitable for foods, pharmaceuticals, and cosmetics. At the outset, a wide variety of end uses were clear: textile printing pastes, resin emulsion paints, synthetic rubber, package coatings, drilling "muds" (oil companies used CMC to keep dirt and debris suspended away from drill bits), "soapless soaps," ceramics, adhesives, and anti-sticking compounds. In addition, Hercules found after beginning commercial production that "each month . . . several new applications appeared."[22]

In 1945, Hercules scaled up from pilot plant to commercial production, manufacturing more than 208,000 pounds that year. Although output (mostly at Parlin) nearly doubled the following year, competition emerged quickly: Dow, Du Pont, Standard Chemical Company of Canada, Sylvania Division of American Viscose, and Wyandotte Chemical Corporation all announced plans to build CMC capacity. Hercules was not yet operating its CMC business in the black, but it acted to consolidate its lead by announcing a large-capacity expansion at Hopewell. As the new plant went on line early in 1947, there were hopeful signs: Hercules seemed to be offering the highest quality CMC on the market; the company significantly reduced its rate of loss; and Sylvania and Dow soon bowed out of the competition.[23]

By 1948, demand was soaring, and the business turned profitable (see Exhibit 9.3). But success had its drawbacks. Awaiting further plant expansions, the company sold out of CMC and began allocating sales "to the outlets where we thought we would do the most good over the long pull." By the time Hercules was able to meet demand in mid-1949, it had curtailed supplies to textile printers (hoping that market would remain an "ace in the hole") and severely cut sales to detergent-makers. The move backfired, however, when Procter & Gamble began making its own CMC. At the same time, Du Pont captured a greater share of the petroleum market.[24]

Even so, Hercules' expertise in cellulose chemistry—its ability to "broaden the market [through] modified CMC characteristics"—allowed it continuously to augment and diversify its list of customers. New uses included additives to food (CMC was an excellent thickening agent in ice cream), cosmetics, laxatives, toothpaste, jellies, antacids, and antiobesity tablets. CMC also found applications in X-ray contrast media, penicillin suspensions, and emulsions of many kinds.[25]

The largest customers for CMC were petroleum companies and food processors, which accounted for roughly one-third and one-fifth of sales, respectively. Other major customers—manufacturers of paper, detergents, textiles, cosmetics, chemicals, and pharmaceuticals—accounted for the rest. Exports climbed to 12 percent of total sales.

In the early 1950s, CMC became the Cellulose Products Department's second-most profitable product (after nitrocellulose). Its development helped bolster a business that increasingly displayed signs of maturity. On the demand side, the rising popularity of new petrochemical plastics and the decline of rayon posed serious threats. As for raw materials, supplies and prices of cotton linters remained subject to the vagaries of weather and federal price supports, circum-

EXHIBIT 9.3 Hercules' CMC Sales Volume and Profits, 1946–1954

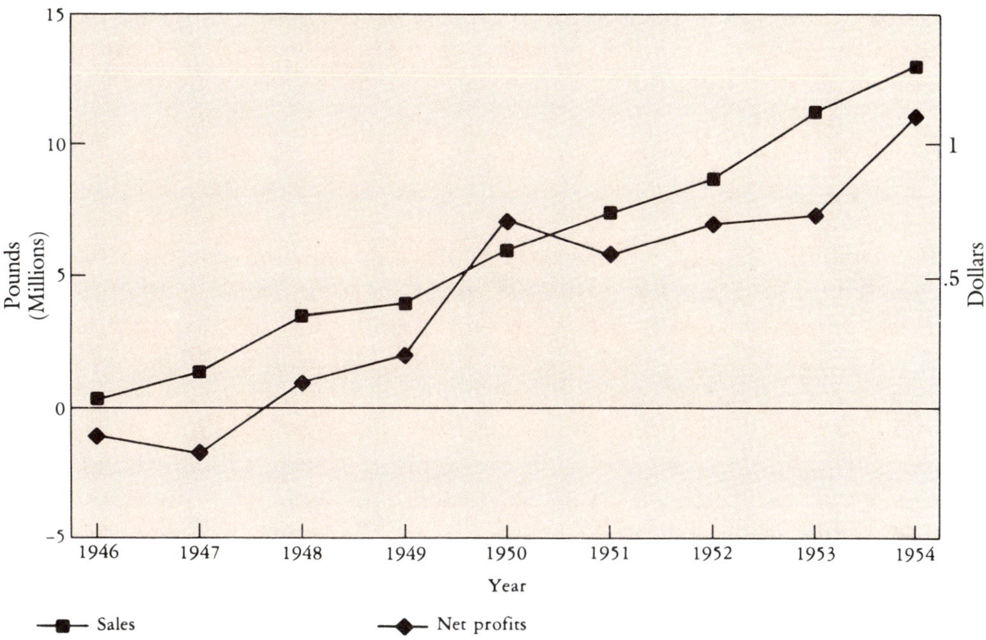

Source: Cellulose Products Department, Annual Reports for 1946–1954.

stances that led many customers to prefer wood pulp rather than chemical cotton. Although the department generally held its position in sales and profits, by the early 1950s, opportunities for long-term growth in cellulose derivatives seemed much less promising than they had immediately after the war.

From the Pine Stump

For the Naval Stores Department, the postwar decade was a similar experience, opening brightly but closing on a note of concern. Although sales more than doubled, profits declined sharply after 1951 because of general economic conditions as well as reasons specific to the business. Return on sales fluctuated from a peak of nearly 73 percent in 1947 to a low of less than 1 percent in 1952.[26] (See Exhibit 9.2.)

The pattern of growth and the key challenges in the company's naval stores business were strikingly parallel to those in Cellulose Products. Like the Cellulose Products Department, the Naval Stores Department made respectable yet undramatic gains in its traditional lines; but it also introduced one tremendously successful new product: Toxaphene, an insecticide. And like Cellulose Products, the Naval Stores Department became increasingly subject to, and concerned with, its dependence upon agricultural raw materials.

Two distinct—and at times opposing—goals occupied the department after the war. To secure market leadership, it sought to develop increasingly refined and specialized products. At the same time, however, the department's role as a supplier of commodity raw materials, especially pale rosins, to PMC and Synthetics (which, in turn, converted them into specialty chemicals) expanded. Generally, the internal market took priority over external sales, making Naval Stores as dependent on the success of other Hercules departments as on its own efforts.

To expand sales on the outside, Hercules continued to develop new naval stores products and uses. In the late 1940s, for example, the department scaled up its production of rosin amine (used for flotation and secondary petroleum recovery, textiles, algaecides, and other uses), but the company did not complete a full-scale plant at Hattiesburg until 1953.[27] Meanwhile, sales of Vinsol® resin and Truline binders to the foundry, steel, cement, and plastics industries brought in more than $1 million immediately after the war but slumped in the late 1940s. The downturn was partially offset by demand for Vinsol® resin from large makers of phonograph records such as RCA and Columbia Records, although these customers soon switched to less expensive substitutes.

Another specialty chemical—anethole—also experienced spotty growth. One of the most expensive naval stores derivatives, this sweet-smelling oil competed mainly against imports of oil of anise from China. Anethole sales, like those of many other naval stores products, plummeted in 1952 because of high inventories, declining sales of essential oils, and competition from substitutes such as Dow's *para*-hydroxybenzaldehyde. Imports and product substitution also put downward pressure on prices; as Hercules sold more and more anethole, it earned fewer and fewer dollars.

The department also strove for greater specialization in its traditional products. In 1953, production of the scarcely refined FF rosin (Hercules' original naval stores product) was phased out in the face of competition from products such as Newport's new, low-priced, partially polymerized rosin called Newtrex. "With Newport and Crosby [another competitor] now offering polymerized rosins," department general manager Al Forster explained in 1951, "we realize that it is necessary to get even further away from straight rosin, if possible." The company redoubled its efforts to sell products such as limed Poly-pale® rosin, about one-third of which was sold to manufacturers of floor coverings and the remainder to distributors and makers of inks, paints and varnishes, and synthetic resins. Poly-pales,® too, experienced a severe profit squeeze during the 1950s.

The department's greatest successes in specialty chemicals came indirectly through the other departments it supplied. Beginning in 1946, for example, half of all Hercules pale rosins were consumed internally (about two-thirds for paper chemicals and one-third for synthetics). Nearly all its Resin 731, its third-largest product at that time, was used by other departments to make Dresinate® 731 resin for synthetic rubber. In 1952, Forster complained that his department could have sold an additional fourteen million pounds of Resin 731 to outsiders had it not been obliged to meet Hercules' need for the product.[29]

On the other hand, a decline in internal consumption could stimulate outside sales. Nearly all Staybelite® resin, for example, was "sold" internally to make resins and soaps used in rubber, paper, and adhesives. A slump in the market for protective coatings in 1947 hurt Synthetics Department sales, which, in turn, led the Naval Stores Department to rechannel its rosin output to external markets.[30]

Nearly a decade after the war, then, Hercules naval stores could be viewed from two distinct perspectives. From the outside, it was still the world's leading supplier of wood rosin, turpentine, and pine oil and was steadily expanding its roster of specialty chemicals. But when internal transfers are considered, the Naval Stores Department appears as a primary producer of pale rosins for other Hercules departments.[31] The importance of the internal markets meant that, on the one hand, Hercules was adding more value by becoming more vertically integrated and specialized. On the other hand, the Naval Stores Department had less freedom to develop its own markets.

The department—like others in the early postwar era—was also concerned about raw materials shortages. It faced two alternatives: finding more tree stumps, or converting to a new source of raw materials. Although Hercules had long weighed the latter option (and would eventually pursue it—see Chapter 10), for the present, it chose to extend its search for new cutover lands.

As the company's wood camps pushed into stump lands far from the plants, the department began to look elsewhere for possible sources of supply. In 1949, after a sample shipment of ponderosa pine from Coon Creek, California, tested favorably as a source of naval stores products, the board of directors authorized more than $1.1 million for the purchase of a fifty-acre site and the building of a pilot extraction plant at Klamath Falls in southern Oregon. The plant would extract crude oleoresins and ship them to the Southern plants for refinements.[32] George Bossardet, plant superintendent at Brunswick, headed west to run the new operation, which employed approximately one hundred workers when production began late in 1950.

Although Hercules launched the Klamath Falls operation in the hope that it would result in considerable expansion of its naval stores business, the project proved ill fated for several reasons.[33] Ponderosa pines, unlike pines in the vast, continuous Southern belt, grew in scattered stands at very high elevations, making them difficult to harvest, especially after the arrival of winter snows. Transportation costs were high. Perhaps most important, the wood was not suitable for the Yaryan steam-solvent distillation process because it yielded a mixture containing considerably higher boiling terpenes and different resin acids. As a result, Hercules could neither refine the rosin above FF grade nor sufficiently raise its acid number. By the end of 1953, the experiment was over, and the Klamath Falls plant was put up for sale.[34]

In spite of this failure and declining overall performance, an impressive new product offered significant hope: the highly toxic, broad-spectrum insecticide known as Toxaphene.

Leading the Pesticide Revolution

American agriculture experienced phenomenal growth in the quarter-century following the Second World War; only two other industries—public utilities and communications—matched its gains in productivity. Several factors accounted for this spectacular performance: better feed and seed, extensive private research and government assistance, increased mechanization, and new liming materials and fertilizers.[35]

Agricultural chemicals—insecticides, pesticides, fungicides, and the like—also played a key role in the "green revolution." Prior to the war, inorganic insecticides such as calcium arsenate and copper sulfate had proved effective but tended to accumulate in the soil; supplies of natural organics (rotenone, nicotine, and pyrethrum) were limited, especially during the war. But the development of synthetic organic compounds, especially DDT, during and soon after World War II brought revolutionary changes in insect control. For the first time, highly effective insecticides could be produced on a large scale to serve a vast and rapidly growing market.[36]

In the late stages of the war, Hercules had produced DDT at Parlin; the company held high hopes for the product in peacetime agricultural markets. Amid a large number of competitors, however, Hercules was handicapped by its small and relatively inefficient operations at Parlin, and abandoned DDT production soon after the war. Its experience with the product nonetheless had impressed research personnel with the effectiveness of chlorinated insecticides. Meanwhile, the resumption of peacetime trade brought fresh shipments of African pyrethrum, sounding "the death knell to our once fluorishing Thanite business." By 1947, Thanite production was halted at Brunswick, and its impressive "fly lab" left virtually inactive. By that time, however, Hercules was prepared to begin commercial-scale production of a new insecticide that would soon dwarf its predecessors in the marketplace.[37]

Toxaphene—the brand name for chlorinated camphene—was created at the Experiment Station in 1944. Originally called simply "3956," it was one of several compounds synthesized by chemist George Buntin. The key to its effectiveness was a chlorine content of 67–69 percent, above or below which its toxicity to insects declined dramatically. Although chlorine was a relatively expensive and scarce chemical commodity and accounted for roughly two-thirds of the insecticide, the camphene needed to complete the compound was made by isomerizing Hercules *alpha*-pinene. Toxaphene was especially lethal against cotton insects such as the boll weevil, boll worm, Southern green stink bug, and flea hopper. So, in 1946, Hercules targeted cotton growers as its first market.[38]

Toxaphene proved popular among cotton growers almost immediately, but the turning point for the product came in the following year. After extensive testing, the USDA and several state agricultural authorities added Toxaphene to the list of approved and recommended insecticides against cotton pests. Requests for three million pounds of Toxaphene flooded in in 1947, and orders doubled in 1948. Hercules quickly converted its Thanite facilities to make the new product. The board allocated about $250,000 for a new facility at Hatties-

Star performers of the early postwar era included Toxaphene insecticide and sodium carbyoxymethylcellulose (CMC), a water-soluable polymer.

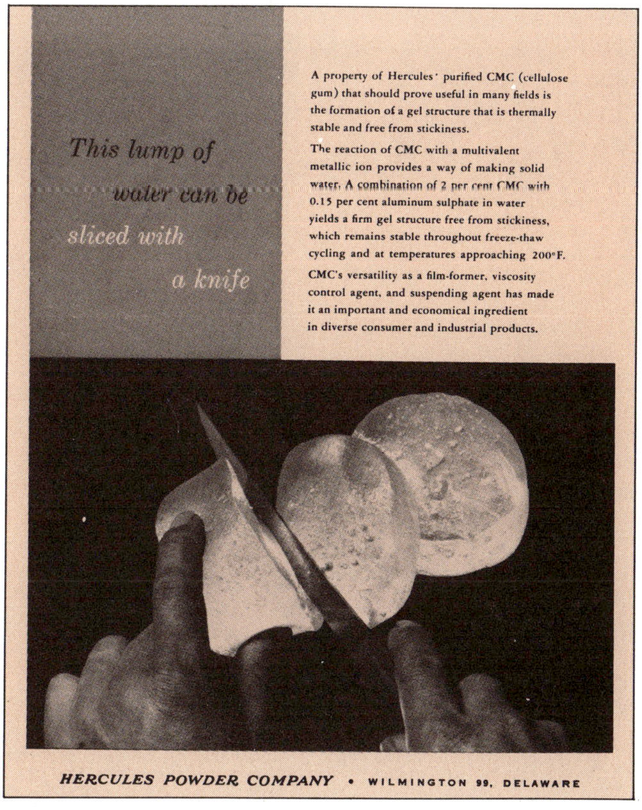

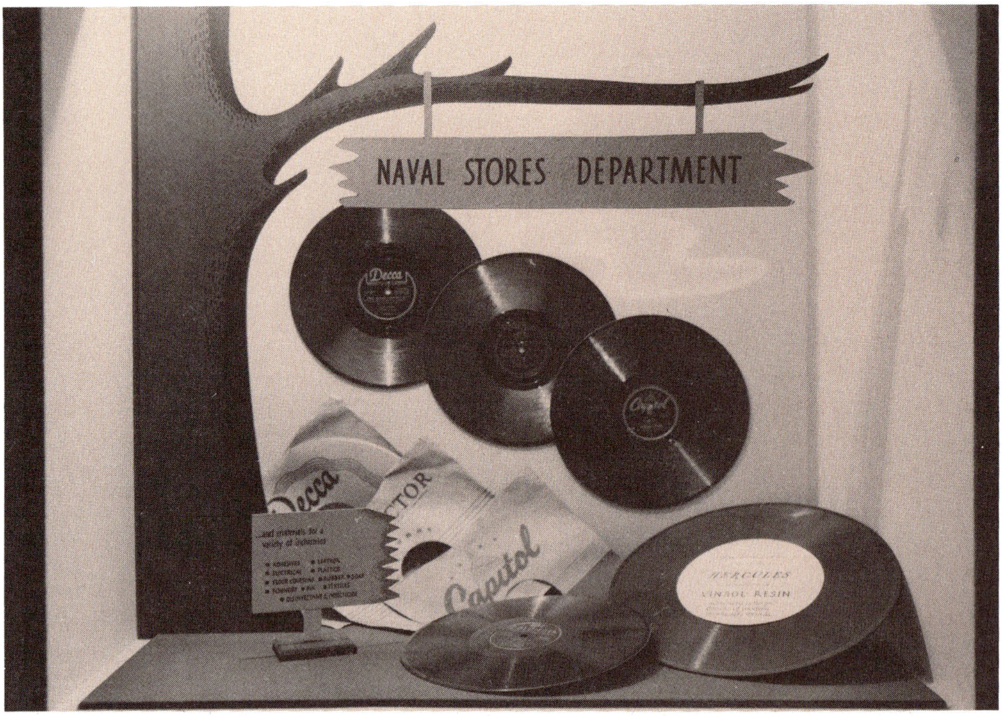

Hercules also sought peacetime markets for its chemicals. Vinsol® resin, used to stabilize aircraft runways during the war, afterward found a momentary application in phonograph records.

Paper chemicals became an enduring source of profits thanks to hands-on selling, here demonstrated in an advertisement featuring Hercules salesman Bob Leahy (right).

EXHIBIT 9.4 Hercules' Toxaphene Sales and Profits, 1947–1954

Source: Naval Stores Department, Annual Reports for 1946–1954.

burg and made plans to double the capacity of its new commercial-scale Toxaphene plant at Brunswick.[39] (See Exhibit 9.4.)

The reasons for the product's sudden, dramatic success were obvious. Government approval, of course, was a key factor; only two other poisons, benzene hexachloride–DDT mixtures and calcium arsenate–nicotine mixtures, had earned USDA endorsements. In addition, Toxaphene quickly established itself as a versatile insecticide, proving more effective across a wide range of applications than even DDT.[40]

In 1949, for example, Toxaphene was approved for use against grasshoppers to fight massive infestations in the Midwest and West. Soon thereafter, it was used against army worms and salt marsh caterpillars, as well as in sprays and dips for killing ticks, lice, horn flies, and other pests on livestock (except dairy cattle). The insecticide also provided good control of a series of cereal, forage, truck crop, tobacco, and ornamental and shade tree insects. In short, because Toxaphene was "effective against most insects that are harmful to agriculture," farmers could save both money and effort by relying on that single insecticide.[41]

For its part, Hercules saw Toxaphene not only as a very profitable outlet for naval stores products but also as a boon to its cellulose businesses, because it promised to increase cotton yields. The insecticide also generated tremendous goodwill and name recognition for the company, especially with aggressive promotion.[42] In 1948, the company abandoned proprietary rights to the trademark, a step that resulted in *toxaphene* becoming a generic name for 67–69 percent chlorinated camphene. The following year, as part of "one of the biggest single product advertising campaigns in Hercules history," the company distributed 750,000 pieces of literature and posters and sponsored extensive radio, billboard, and print advertising for its product.[43]

The program grew in scope and sophistication in the early 1950s. In 1951, Hercules spent $215,000 (or about 4¢ per pound sold) on advertising for toxaphene, produced and distributed a film called *Cotton Insects,* and supported a dozen specialized salesmen. It also cultivated close relations with rural radio and television farm directors, "who carried a lot of power in the farm states" and whose early morning broadcasts strongly influenced farmers. To persuade farmers to become "methodical" about applying the product, Hercules advocated in its radio and print advertisements regular weekly applications of toxaphene.[44]

It is worth noting that the toxaphene strategy resembled the company's successful marketing campaign for pine oil in the 1920s. Hercules was not a retailer of toxaphene. The pure substance was sold to formulators, who normally converted it into dusts, sprays, and baits of various potencies, which were then distributed to retailers. Through its advertising, Hercules helped pull the product through the distribution channels. Toxaphene and pine oil were among the few products in the company's history that created public awareness of the Hercules name.[45]

Another key marketing strategy for toxaphene was cultivating consumption overseas through export sales and a liberal licensing policy. Markets in South America seemed especially promising: "Just about the time we complete our poisoning program in the United States," predicted Forster in 1948, "the South American crop will be ready for insect control." Soon, product sales supervisor Frank Rapp was touting *toxafeno* in South America. Although sales to Brazil, Peru, and Venezuela were constrained by a lack of hard currency in those nations, by 1951 Hercules was exporting two-thirds of its toxaphene output.[46]

Overseas expansion of production followed rapidly. In 1950, Hercules opened a toxaphene plant in Oswaldtwistle, England, in cooperation with Cocker Chemical Company. A year later, it granted nonexclusive licenses to Hercules Powder Company Ltd. (HPC) and Shell Petroleum in England to make, use, sell, and grant sublicenses on toxaphene outside the United States and Canada and began forging similar agreements in other countries.[47]

By this time, despite the capacity increases at Brunswick, projections showed that demand would continue to outrace Hercules' ability to make toxaphene. The problem was exacerbated by periodic shortages of chlorine. Nonetheless, hoping to satisfy demand for the 1951 growing season, the company planned a new toxaphene facility at Hattiesburg and in 1950 began building a $2 million, two-million-pound-per-month plant at Henderson, Nevada.[48] The site was chosen largely because of its proximity to producers of key raw materials. In the next two years, Hercules negotiated guarantees with the Algonquin-Missouri Chemical Company (with facilities at Huntsville, Alabama) for chlorine and caustic soda, with Stauffer Chemical of Nevada for chlorine, and with Dow for carbon tetrachloride. Toxaphene certainly seemed to warrant the investment. In 1950, just five years after it first appeared on the market, toxaphene edged out pale rosin as the company's leading naval stores product.[49]

But this storybook performance ended abruptly in 1952. Early in the year, unusually hot and dry weather across the South devastated the market for insecticides. Sales of toxaphene "melted away like snow in the bright sunshine," recalled the director of naval stores operations. "Cotton farmers had no money for pesticides, our formulator customers found their warehouses overflowing with toxaphene products, and the most strenuous efforts of our large sales force could not find any buyers for that product." Production at Brunswick was curtailed for many months; the half-constructed projects at Hattiesburg and Henderson were halted permanently.[50]

Competitive rivalry in pesticides, moreover, intensified markedly in the early 1950s. In 1953, for example, Shell introduced Endrin, a formidably popular product. To meet such threats and help counter the effects of wide swings in demand, Hercules began offering consignment warehousing and quantity price breaks and building up its inventory as a buffer. The situation was a stark reminder that toxaphene would not escape the risks of any business closely associated with the agricultural sector.[51]

In 1954, Paul Mayfield, Forster's successor as general manager of the Naval Stores Department, gave the presidential address at the National Agricultural Chemicals Association. Pointing out the tremendous progress and promise of the field, he noted that 80–90 percent of agricultural chemicals then on the market had been unknown in 1945, and that as yet, only fifteen acres out of every hundred under cultivation in the United States were being treated against insect damage. Mayfield also discussed the special public relations and liability issues inherent in the business. The health hazards associated with DDT, he noted, were already pushing the agricultural chemicals industry into "a defensive stand."[52]

It was not many years before his remarks seemed prophetic indeed. For the present, however, Hercules remained content to enjoy the rising new star in the otherwise calm galaxy of rosin products. Although 1954 turned out to be a relatively mediocre year for toxaphene sales, that single product—scarcely a decade old—was the source of about 5 percent of the entire company's profits.[53]

Coming of Age in Synthetics and Paper Chemicals

In the first postwar decade, the Naval Stores Department generally increased its sales to other units in the company. The Synthetics Department, for example, demonstrated virtuosity in bringing new products to market, thereby deepening Hercules' involvement in specialty chemicals markets. Between 1946 and early 1948, the department began offering Cellolyn® resin (a new product for nitrocellulose lacquers), Petrex® resin, and Arbitol® hydroabietyl alcohol, the first alcohol derived from rosin. Part of the department's search for new products was motivated by a desire to lessen its dependence on paint and varnish markets—where prices were linked to unpredictable supplies of linseed, tung, and soybean oils—and to increase sales to other customers, such as manufacturers of printing inks, floor coverings, and oil additives.

But the highlight of the decade was the department's success in diversifying at the raw materials end of the production chain. Just as the Naval Stores Department was striving to reduce its role as a supplier of raw materials to the Synthetics Department, so, too, the Synthetics Department sought to diversify into businesses not dependent on Naval Stores raw materials. Pentaerythritol, Hercoflex® plasticizer, and Metalyn® distilled methyl ester of tall oil (used as an oil additive base and inexpensive modifier for plastics) were among the products leading the quest, which by 1950 took shape as an explicit "diversification program." A year later, general manager Josh Billing noted with satisfaction that roughly 40 percent of his department's sales consisted of resins, plasticizers, alcohols, and by-products based on raw materials outside naval stores.

The pattern of growth in synthetics was perhaps the least steady of any Hercules department between 1946 and 1954, but the general outline paralleled that in most other businesses and the company as a whole: sales and profits roughly doubled, to about $20 million and $3.5 million, respectively, by 1954. But the department improved its relative importance to the company only slightly; by 1954, Synthetics accounted for slightly more than one-tenth of company sales and profits (see Exhibit 9.2).

For the PMC Department, the postwar decade marked a kind of coming of age. A marginal performer during most of its early years under Hercules' control, the PMC Department emerged after the war as a significant, steady source of revenues, new products, and new customers. Sales grew from $18.8 million to $25.2 million, and profits nearly doubled to $6.3 million. Return on sales for PMC remained consistently high, never falling below 18.3 percent and once (in 1948) reaching 26.9 percent. The department became a significant component of Hercules as well, accounting for roughly 15–20 percent of total corporate sales and nearly one-quarter of total profits throughout the period (see Exhibit 9.2).

The turnaround at PMC resulted from several factors. The department continued to divest or close out many marginal operations and products that had come with the acquisition in 1931. In 1947, for example, the Atlanta and Marrero plants were sold to General Chemical Company; six years later, because of chronic milk shortages and government controls on milk prices and imports, Hercules permanently shut down its casein operations. By 1954, PMC had concentrated production in seven domestic and three foreign plants.[54]

Research breakthroughs, such as in the area of wet-strength resins (which impart strength to papers that remain in contact with water), also contributed to PMC's improved performance. The company had become interested in such products late in the war, when a competitor developed a wet-strength resin that it promoted in conjunction with its rosin size, thereby threatening Hercules' sales. In response, Hercules assembled a team of researchers, led by William Campbell and Gerald Keim, a newly minted Ph.D. chemist from Brooklyn Polytechnic Institute, to develop its own wet-strength resin.[55]

By experimenting with nitrogen compounds, Keim produced a urea formaldehyde resin that had unusual strength, even in small doses. It was duly named Kymene® 138 wet-strength resin. After a shaky start at the Mansfield synthetics plant, the product was scaled up at Savannah, spurred by large orders from National Paper Company. More important, Kymene® 138 gave Hercules an entree into the market and laid the groundwork for its development of Kymene® 557, which proved to be one of the most successful products in its history.

The researchers also investigated reactive sizes, which increased bonding strength dramatically through chemical reactions. In the late 1940s, success came in the form of Aquapel® alkylketene dimer. After extensive market trials, the company scaled up production for Aquapel® in 1954. As a sealant for milk cartons and other wet paperboards, Aquapel® would soon dominate the U.S. market.[56]

PMC also succeeded in developing new products for customers outside the paper industry. For example, the postwar era offered continuing opportunities to serve manufacturers of synthetic rubber. Dresinate® rosin soap and Dresinol® resin dispersion chemicals were commonly used as additives—emulsifiers and tackifiers—in synthetic rubber. Between 1945 and 1951, sales of Dresinate® 731 tripled and profits soared nearly tenfold, making it the second-leading product (after wet size) in the department. Sales and profits fell off thereafter as substitute products came on the market, but PMC's supply to the synthetic rubber industry remained a significant and profitable market for decades (see Exhibit 9.5).

Despite such products, the principal source of PMC's postwar success was a sound competitive strategy as a supplier to the paper industry.[57] In the first place, Hercules provided a broad range of services to its customers, including: a strong emphasis on technical sales (PMC formed a special development division for market and product development in 1947); a full line of products, rather than only the most profitable stars; and a willingness to customize products to exact consumer specifications.[58]

Second, the PMC Department specialized in understanding customer needs. Jack Starne, a regional manager, liked to assert that Hercules paper chemical technical salesmen knew more about the papermaking process than the papermakers themselves, a marketing philosophy that presaged IBM's now-famous dictum that a seller should understand customers' needs better than the customers themselves. There were practical reasons for this approach, beyond pleasing clients and anticipating their needs. The conventional wisdom in the PMC Department was that Hercules would be held responsible for customer problems, whatever their cause; therefore, why not be prepared to solve them? Indeed, many PMC technical salesmen believed that some customers occasionally feigned problems to obtain Hercules' advice.

A third reason for the company's success in paper chemicals was its geographical coverage. Shipping accounted for a large fraction of costs in the paper chemicals business. By siting plants at many locales—enough to serve most leading paper plants and manufacturing centers—Hercules could both reduce shipping costs and offer quick response times.

Finally, the culture and traditions within the PMC Department were important sources of its service excellence. Early leaders such as the charismatic,

EXHIBIT 9.5 Hercules' Dresinate® 731 Sales Volume and Profits, 1946–1954

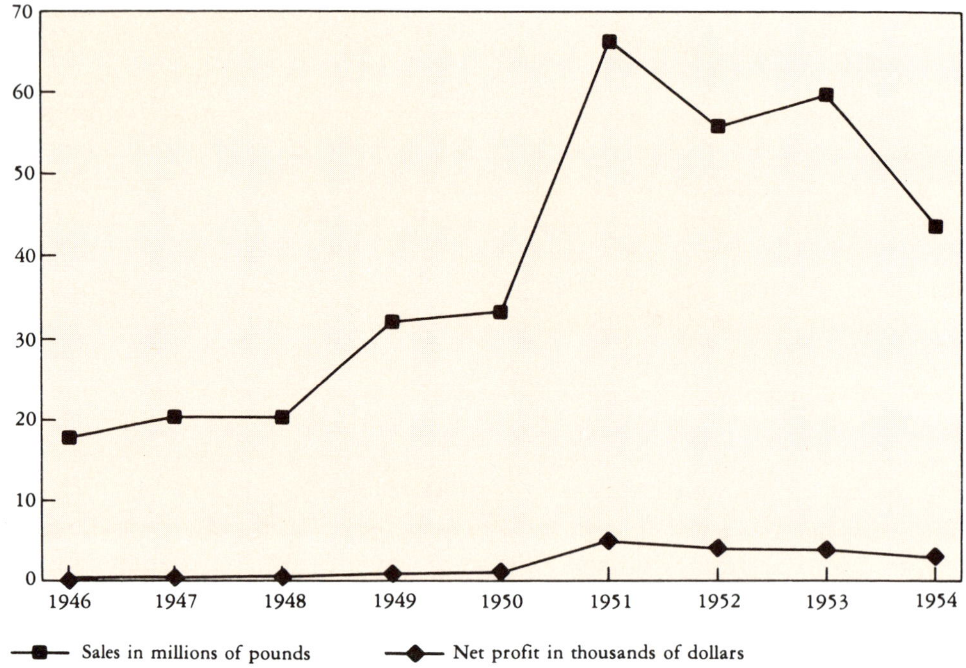

Source: PMC Department, Annual Reports for 1946–1954.

hard-driving Jimmy Foxgrover set a tone and direction that endured for decades. The typical paper chemical salesman of the postwar era spent at least half of his time on the road, often traveling to remote locations and repeatedly uprooting his family. The sense of belonging to an elite corps was vital to success in a business that demanded long hours and close working relationships with customers.

Dedication and morale reached a peak as the business became tremendously successful. It was relatively immune to cyclicity, in part because the paper industry itself took precautions to avoid cyclical downturns. Indeed, when the economy began to falter, paper companies placed added emphasis on quality in order to hold market share and actually increased consumption of paper chemicals. Moreover, paper chemicals performed important functions in the papermaking process but accounted for only a tiny fraction (typically a few pounds per ton of paper) of materials used. As a result, demand for paper chemicals was relatively inflexible and allowed high price levels to be sustained. Backward integration by the paper companies was unlikely not only because of the small quantities of chemicals used in the papermaking process, but also because the companies lacked the technical expertise required to succeed in paper chemicals. Suppliers such as Hercules invested far more in R&D than their customers were prepared to match. The paper companies were heavily reliant upon their chemical suppliers, and as a result, PMC's margins ranked among the highest in the company.

Because of its strategy and the characteristics of its industry, Hercules became the leading paper chemicals supplier in the United States after the war. The business also offered appealing prospects overseas. Despite such promise, however, Hercules soon encountered limits to growth. One problem was the relatively slow growth of the overall market. Given that the company was already the leading producer of paper chemicals, its fortunes could only improve with the health of its customers. Another problem was escalating rivalry: in 1946, the PMC Department controlled an estimated 65 percent of the U.S. market for rosin size; five years later, increasing competition, especially from American Cyanamid, cut its share to 56.2 percent.[59]

Three roads into petrochemicals

In the decade that followed World War II, Hercules attained respectable growth and impressive profitability in its main lines of business, despite periodic shortages of raw materials. The company's principal successes involved the development of new products such as CMC and toxaphene and the emergence of PMC as a very strong contributor to overall performance. Higgins's confidence notwithstanding, as time wore on, the company's traditional lines appeared less and less attractive as sources of continuing growth. Not only was the company bumping up against limits in its major product lines, but competition and substitutes were also beginning to threaten some of its strongest products.

Pressure mounted inside the company to diversify, especially into the most attractive growth area of the chemical industry: petrochemicals. The managerial and technical personnel who urged such a course, however, met with considerable resistance at the top of the company. Soon after the war, for example, Hercules was given an opportunity to enter the polystyrene plastics business. Shell Chemical Company (a subsidiary of Shell Oil) had operated a government-owned West Coast plant that made polystyrene-butadiene copolymer for synthetic rubber tires. According to Jack Martin, who moved from Explosives to Cellulose Products in 1947, the plant could have easily been converted to the manufacture of polystyrene plastic. David Wiggam, director of development in Cellulose Products, and Walter Gloor, a product development specialist, both of them very familiar with polystyrene, heartily endorsed the proposal. But a major obstacle stood in the project's way: the deal required an investment of more than $10 million, an amount that would have entailed borrowing funds and perhaps assuming a "bet-the-company" risk. Not surprisingly, the executive committee declined to bid, and the plant was later acquired by Shell.[60]

A similar case came up several years later. In the early 1950s, Jack Martin, by then assistant general manager of Cellulose Products and a kind of "czar of petrochemical-based research," got wind of a rumor that Du Pont, then the object of a major antitrust investigation, was worried that its exclusive position in polyethylene would come under scrutiny and was therefore willing to consider licensing its polyethylene process.[61] When Martin pursued the lead, he was indeed offered a license, but with the stipulation that he return quickly from Hercules with a fully approved offer. Du Pont apparently wanted a deal concluded before word spread that its technology was available.

To Martin, the opportunity seemed exceedingly attractive. Hercules already possessed expertise in ethyl alcohol and other essential know-how; the Du Pont license, he believed, could put the company into the polyethylene business. With the support of all of the company's general managers and its director of engineering, Ernie Wilson, and with positive feedback from several industry experts, Martin presented his case on the "thirteenth floor" (then the site of the executive committee meeting room). On that occasion, board member Petrus Meyeringh, although usually supportive of new products, asked pointed questions, raising doubts that silenced the project. And throughout the discussion, recalls Martin, "Higgins never said word one."[62]

The company's refusal to enter the petrochemicals industry in a big way thus partly reflected the concerns of senior management. Proponents had not only Morrow's legendary financial conservatism to contend with, but also the worries of Higgins and others about investing in areas in which the company lacked expertise. As a junior executive of the time put it, the executive committee was motivated at least partly by a "hidden fear that we shouldn't get out too far beyond what we knew best." Martin, another impatient young manager, crisply summarized the prevailing attitude as, "If the boll weevil won't eat it, if it doesn't come from a pine stump, or if it doesn't go boom, we won't touch it."[63]

The philosophically conservative cast of senior management notwithstanding, the internal debate about growth and diversification at Hercules continued under Higgins's successor as president, Al Forster. Higgins's resignation in April 1953, shortly before his sixty-fifth birthday, did not signal unrest or unhappiness at the board level with the state of the company. Indeed, Higgins, who had led Hercules through fifteen years of consistently high profitability, retained his seat on the board, along with the powerful finance committee chairman, Ed Morrow.

Forster was fifty-three when he assumed leadership of Hercules. A mining engineer by training, he had joined Hercules in 1925 as a technical service man in the San Francisco sales office. After five years on the job, he left to work on a series of projects in South America, returning in 1934 to take a position in the Explosives Department. Working his way up through assignments at plants and in sales offices, Forster got his career break in January 1940: he moved to the Naval Stores Department as assistant general manager before stepping up, eight months later, to the department's top position. During the next eleven years, he presided over the blossoming of the Naval Stores Department: after more than two decades of struggle, the business at least realized the company's goal of providing steady earnings based on a commanding market position. Elected to the board in 1940, Forster was promoted to vice president and member of the executive committee in 1951.

Although Forster was known to be concerned about Hercules' prospects for growth, there was ample reason to be circumspect about diversifying into petrochemicals. As a major supplier to plastics fabricators and rayon manufacturers, Hercules was reluctant to compete against its customers or contribute to the decline of their markets. Although this position became increasingly untenable during the 1950s, it still had powerful adherents among the senior executives.

In addition, the technology of petrochemicals was foreign to Hercules. Although the company had expertise in some petrochemical areas—production of synthetic ammonia from natural gas, for example, or supply of catalysts and additives to synthetic rubber manufacturers—its role was minor in scale and on the periphery of new developments. Entry into petrochemicals would stretch the company's technology base. Although Hercules had long worked with natural polymers, the company was slow to develop capabilities in polymer science, partly because of the nature of its existing skills in research and production. The company boasted outstanding chemists such as Ott, Cairns, Spurlin, and many others, but their work was classically chemical in nature. That is, in the simplest terms, Hercules refined and purified raw materials and then transformed them by various traditional chemical processes into useful products.

The companies leading the petrochemicals revolution, however, approached their business in a different manner. Although trained as chemists, the polymer pioneers were developing expertise in molecular engineering. Their work on molecular weights and structures resulted not so much in derivatives and modifications of natural compounds as in wholly new organic compounds that could be engineered and tailored for different purposes—a "materials," as opposed to a "chemicals," mentality. Such an outlook was alien to Hercules.[64]

And the new industry was expensive. Even at fire-sale prices, war surplus petrochemical plants cost tens of millions of dollars; building a commercial-scale greenfield plant would cost even more. (By way of comparison, Hercules' largest single investment before 1946 was just over $4 million, for a cellulose acetate plant—see Chapter 8.) During the postwar decade, moreover, new entrants included some of the largest industrial companies in the world, raising the specter of fierce competitive rivalry as the industry matured. In the event of a shakeout, it seemed, large and wealthy participants stood a better chance of survival. Not for the first time in its history—and not for the last—Hercules' medium size limited its options.

For many reasons, then, the company declined to attempt a frontal assault on petrochemicals. When Hercules made its way into this new area, it was with a characteristically cautious and prudent approach, through careful preparation and development of a proprietary position. The hard work of Hercules representatives in Europe and plain good fortune also played important roles.

The Brave New World of Petrochemicals

The petrochemicals industry—the manufacturers of synthetics commodities and fabricated materials from natural gas and petroleum feedstocks—had its roots in the chemical industries of Europe and began to flower in the United States during the 1930s.[65] Early in the decade, Union Carbide & Carbon Chemicals began making polyvinyl chloride on a small scale. In 1937, Dow commercialized Styron polystyrene plastic, and two years later, Union Carbide began producing several other synthetics from petroleum and natural gas. Du Pont pioneered the tremendously successful, petroleum-based synthetic fiber, nylon, in the late 1930s and developed other synthetic fibers, such as polyester and acrylic, during or soon after the war. During the early 1940s, Du Pont also began making high-pressure polyethylene plastic.

After the war, the petrochemical industry took off in an astonishing burst of growth. The strength, flexibility, resistance to stains, low cost of raw materials, and other characteristics of the new materials when fabricated into plastics, fibers, and other structures made them attractive and competitive in many markets. The chemical production base in the United States, moreover, was intact—unlike Europe's. American petroleum companies were ambitious in their construction programs. Technological know-how from the German chemical companies was made available to American manufacturers through the work of the Combined Intelligence Objective Subcommittee (CIOS), an agency representing the occupying powers. CIOS's Field Intelligence Agencies Technical (FIAT) Reports provided detailed documentation of German industrial facilities and processes. Finally, American companies benefited from generally open exchanges of technical information at international conferences and association meetings and from liberal licensing policies.[66]

It was, in short, an era of opportunity, and most large companies pursued the rewards avidly. Many bought chemical plants they had built for the government during the war. B.F. Goodrich, for example, acquired a large GR-S (Buna) synthetic rubber plant and converted part of it to make polyvinyl chloride. Several companies entered the field via styrene and later moved into other products. Dow bought a styrene plant in Valasco, Texas; Monsanto purchased a fifty-thousand-ton styrene plant at Texas City, Texas, and built a polystyrene plant at Springfield, Massachusetts; Koppers bought a styrene plant at Kabuta, Pennsylvania, and also erected a polystyrene facility; and Standard Oil of New Jersey and Shell began licensing production of their styrene catalysts.

In 1949, Celanese Corporation, a maker of cellulose acetate fibers and plastics, pursued another avenue of entry: it built the first multipurpose propane oxidation plant in the world at Bishop, Texas. Union Carbide became the leading ethylene producer, deftly negotiating to buy ethane from oil companies along the Gulf Coast before they fully realized the by-product's value. And several joint ventures were formed between chemical and petroleum producers to ease entry into the field. Standard Oil Development Company and Phillips Petroleum Development Company, for example, made important breakthroughs.

By the mid-1950s, the industry was already crowded with competitors. Oversupply plagued some products, such as acetaldehyde and acrylonitrile, which were made by more than a dozen manufacturers. Still, prices and profits generally remained strong. The market was simply large enough to embrace many producers, and more continued to enter. As one industry expert describes the mood of the era, it was "an unusual time when the only heinous errors were those of omission. In most companies, the annual review of the year's activities brought regrets and recriminations only for any failure to pursue an opportunity."[67] At Hercules, the early 1950s brought regrets, but also small signs of encouragement.

A Phenomenal Phenol Process

Although the Explosives Department had been involved with petrochemicals since the late 1930s, when its first synthetic ammonia operations came onstream, Hercules' first major petrochemical innovation emerged from the Naval Stores Department. Before the war, researchers at Brunswick and the Experiment Station became interested in the oxidation of terpenes (alkyl aromatic hydrocarbons), which could lead to synthetic pine oil and other organic compounds with commercial potential. During the war, a terpene-oxidation process became the foundation of a new method of producing *para*-menthane hydroperoxide, a catalyst for making GR-S synthetic rubber, and a significant contribution to the American war effort.[68]

Soon after the war, leaders of the Naval Stores Department, including Forster, Paul Mayfield, Donald Sheffield, manager of development, and R.S. George, manager of research, continued the push to discover new products that might be made from rosin and terpenes. A fruitful line of inquiry was oxidation processes. With Dr. Ott's support and encouragement, researchers George and Dr. Eugene J. Lorand investigated the oxidation of chemicals, such as cumene and cymene, that were structurally similar to terpenes. Their general goal was to explore new ways to synthesize organic compounds such as the phenol formaldehyde resins used in formica and other construction materials. In the course of his experiments, Lorand developed an entirely new process for making phenol and several related compounds that proved far more economical than traditional methods. His key innovation was a unique reaction process whereby benzene, a simple petroleum by-product, was reacted with propylene to get cumene, then air-oxidized to form cumene hydroperoxide, and finally split into phenol and acetone.[69]

The new process, demonstrated in a pilot plant in Brunswick in 1949, appeared just in time. The usual ways of making phenol started by reacting benzene with either chlorine or sulfuric acid, two products subject to frequent shortages in the late 1940s. In addition, phenol itself was in short supply. A valuable ingredient of many plastics, varnishes, enamels, and pharmaceuticals, phenol had been considered so vital to the war effort that the government had ranked it near the top of its list of essential commodities. Hercules' new process, moreover, could yield other valuable chemicals in controllable proportions and without useless by-products: *para*-cresol (for dyes, insulating compounds, electrical panel boards, and the like), acetone (still a valuable commercial solvent), and cymene alcohols (used for flotation and as wetting agents).[70]

In short, Hercules had discovered a unique path to producing a variety of valuable petrochemical commodities. But such products were alien to Hercules, and the new process touched off a major debate inside the company. The situation grew more complicated when, in the course of conducting patent research, Hercules learned that the Distillers Company in England had independently arrived at the same process. In the fall of 1950, to preempt a potential dispute, the two companies forged an agreement that granted territorial rights for the process to Hercules in North America and to Distillers in Europe.[71]

At this point, Hercules appeared to move confidently toward the development of its new process. In 1951, it assigned HPC, Ltd., the right to manufacture phenol with the "Hercules-Distillers process," granted a license to B.A. Shawinigan, Ltd. of Canada, and made plans to build a phenol plant at Montreal.[72] In addition, late in the year, the board appropriated $6.67 million for a "cresol-phenol-frother 80 plant." Soon thereafter, Hercules made its plans public and attracted considerable attention in the industry. On January 18, 1952, President Higgins (and the phenol process) were featured on the cover of *BusinessWeek*. At the end of the story, he was quoted as saying that the new investment "follows a Hercules policy of sticking close to profit producing fields of endeavor until research has had the time to forge strong links with new raw materials and wider markets."[73]

Despite outward appearances, the move hardly represented a bold new step for Hercules. Although the Naval Stores Department stood firmly behind the development, the phenol proposal had repeatedly met with "fortified resistance" in the boardroom, especially from financially minded executives such as Morrow. To say that this was a conflict between an older, more conservative generation of directors and a younger, more aggressive cadre would be only partially true. A.B. Nixon, for example, although then in his thirty-sixth year at Hercules, was "an adventuresome spirit by history and personality and approach" and favored the new investment. Forster, concerned about the maturity of existing business, and Fulenwider, intrigued with the possibilities in petrochemical plastics, also tended to favor moves toward higher growth.

Opposition to the move into phenol stemmed from two major concerns apart from financial considerations. First was a reluctance to permit Hercules to become a large-scale producer of commodity chemicals. Some directors were simply unwilling (or believed Hercules was unable) to invest in global-scale plants to compete in markets that were unfamiliar. Second, insofar as the phenol market was understood, it appeared to be highly competitive. While there were many uses for phenol in construction materials, appliances, detergents, and other products, some key executives believed that the market had been thoroughly "worked over."[74]

Given these concerns, the strategy that emerged represented a compromise of sorts: Hercules would build a phenol plant, not to become a world-class producer of that commodity, but "as a means of selling licenses on the phenol process," and as a demonstration of Hercules' ability to engineer and develop commercial processes for making related compounds. (Although the cost of the proposed new facility would be more than ten times that of an early phenol facility started at Brunswick, it was still a small amount by petrochemical standards.) However, if Hercules could develop either new commodities or less expensive commodities that were in short supply or without established market

Champions of change: Although Hercules concentrated on traditional businesses after the war, a growing cadre of managers and researchers pushed to develop business based on petrochemical feedstocks. Pictured here: The group responsible for developing a new process for making phenol: Paul Mayfield, Emil Ott, Don Sheffield, Stan George, Reg Ivett, Ernest Turk, and Eugene Lorand.

Albert E. Forster's election as president in 1953 sparked a period of furious expansion. Above: Forster presides at a meeting of the executive committee, including Mayfield, Jack Fulenwider, Ed Morrow, and John R.L. Johnson, Jr. Below: Under Forster, Hercules sought growth everywhere, including internal development. In 1956, the company poured money into an expanded research center. Peter Van Wyck explains the building program to Forster (left) and Bob Cairns.

Acquisitions of the era included Huron Milling Company, a producer of food ingredients added to the Virginia Cellulose Department. Below: New raw materials were another source of growth. At Missouri Chemical Works (a former World War II ordnance plant operated by Hercules), the company produced a wide variety of chemicals from hydrocarbons.

Headed by Henry Thouron (standing), Hercules' International Department in 1959 included (left to right) Bill Cahall, Henry Reeves, "Sam" Beasley, Harvey Taufen, John Present, and Joe Carbonara.

During Forster's tenure, Hercules also joined the space race. Above: Jack Hayes (left) and Fred Hakenjos meet with company lawyer (and future U.S. Senator) Bill Roth to discuss a missile development contract.

positions—perhaps new oxidation products from the Hercules-Distillers process—it would seriously consider becoming a large-scale producer.[75]

Early in 1952, Hercules chose a site just northwest of Gibbstown, New Jersey, for its phenol plant, which was named after President Higgins. The proposed plant would be supplied with terpenes by rail and truck, and benzene and propylene would flow in via pipelines from nearby petroleum plants.[76]

Meanwhile, licensing of the Hercules-Distillers phenol process proceeded. In 1952, Dow and the California Research Corporation each paid $150,000 plus royalties for rights, and other interested parties followed suit. Ironically, although Hercules had been a latecomer to petrochemicals, its policy of licensing phenol production to "any and all comers" represented a new trend in the industry, one that diverged sharply from prewar practices.[77]

Chemicals for Synthetic Fibers

Between 1938 and 1950, three petrochemical-based synthetic fibers—polyamide, acrylic, and polyester—had begun to transform the manufacture and use of fibrous materials throughout the industrialized world. These synthetics, along with cellulosic fibers (rayon) and, later, polyolefin fibers, would account for virtually all of the growth in fiber consumption after 1949; markets for natural fibers (cotton, wool, and silk) would remain stagnant.[78]

Polyamide fiber was developed by a group working with Wallace Carothers at Du Pont in the mid-1930s and commercialized in 1938 under the trademark Nylon. Acrylic fiber was commercialized as Orlon by Du Pont in 1950, by Union Carbide as Dynel in 1952, and by American Cyanamid and Chemstrand (a joint venture between Monsanto and American Viscose) soon after.

The most promising of the synthetic fibers for textile applications—and hence, for the mass market—was polyester. Chemists J.R. Whinfield and J.T. Dickson of Calico Printers Association (CPA), an industrial research laboratory in the United Kingdom, had first discovered polyester in the late 1930s. Pursuing a line of research that Carothers had discarded, they found that an ester of terepthalic acid (TPA), when combined with ethylene glycol, produced a polyester with interesting properties. In particular, the material, which Whinfield dubbed Terylene, shared many of nylon's positive features such as strength, elasticity, and a high melting point. Like the older miracle fiber, polyester was as resilient as wool, but it could also be "heat set" at high temperature to resume its original shape after washing (the so-called permanent-press effect). Shortly after the war, CPA licensed Imperial Chemical Industries (ICI) to make polyester, and production started in 1951. Meanwhile, Du Pont had independently developed a type of polyester during the war but agreed to purchase American rights from ICI in 1946. Du Pont began producing its first polyester fiber (called Dacron) at Seaford, Delaware, in 1950.[79]

A formidable obstacle to the commercial success of polyester was the expense involved in purifying TPA. Both ICI and Du Pont made polyester in several steps. First, *para*-xylene (produced by cracking hydrocarbons) was oxidized with nitric acid to produce TPA. Next, the TPA was esterified with methanol, a reaction that generated several by-products that were difficult to separate and remove. Once the TPA ester was purified, it was reacted with ethylene glycol to obtain the polymer. It was a costly and complex process, and the need to simplify it opened the door for Hercules and other companies to participate in the synthetic fibers revolution.[80]

Hercules arrived at its position as a major independent producer of a key intermediate of polyester fiber from a roundabout direction. The point of departure was the Synthetics Department, which, after the war, was seeking growth opportunities to supplement its rosin-based business. Department general manager Josh Billing, recalls a colleague, "was very anxious to get into some kind of different businesses" for two reasons. First, he saw limits to growth as a supplier of rosin-based derivatives—the products were too specialized to find continual new applications. Second, Billing was frustrated by his unit's dependence on the Naval Stores Department for its raw materials. Not only was the Synthetics Department constrained in terms of materials availability and prices, but some Hercules managers remained skeptical about the need for a separate department to refine naval stores products into synthetic resins. To be sure, soon after the war, the directors had shown their confidence in the future of synthetics by appropriating more than $2 million to build a new plant at Burlington, New Jersey. But a new raw materials base, reasoned Billing, might increase both growth and independence for the department.[81]

A promising lead developed from the department's work with esters of pentaerythritol (PE), which were useful constituents of plasticizers and lubricants. Shortly before the war, Hercules had built a PE pilot plant. Through this work and subsequent market development efforts, Paul Johnstone (director of development for Synthetics), Dr. Harvey J. Taufen (Synthetics Department manager of research), Eero Erkko, and others in the department realized that the growth of the business would depend upon finding new sources of straight-chain fatty acids, a key ingredient in making PE esters. Erkko's study of FIAT reports suggested that a German process for the oxidation of straight-chain paraffins (made by the Fischer-Tropsch reaction) to straight-chain fatty acids might be adaptable to Hercules' purposes. (During the war, the Germans needed long-chain fatty acids as a substitute for the palm and coconut oils that were unavailable.) Although no West German company still used the process, know-how was available.

In November 1950, Johnstone, a brilliant quick-study as a chemist, headed for Germany with the blessings of Billing and Ott. Johnstone sought either to secure sources of animal and vegetable fatty acids, compounds, and oils with which Hercules might conduct its experiments or, better still, to learn from and possibly forge an agreement with a European company more advanced in this work.[82] After visiting several companies, Johnstone arranged to purchase alcohols from Henkel & Cie.; as for know-how, he quickly focused on a small German soap and detergent company, Imhausen & Cie. Its founder, Karl-Heintz Imhausen, was "an absolute German gentleman" and a Jew who had avoided the fate of many of his coreligionists during the war because of his valuable knowledge of fatty acid esters, which were used by the Nazis in torpedo oils.[83]

The Imhausen company was a potential partner because of its experience with the paraffin oxidation process and Dr. Imhausen's "excellent contacts and relations" with East European plants still using it to make straight-chain fatty acids. By the summer of 1951, Johnstone won the approval of Hercules' directors to negotiate with Imhausen for rights to its know-how in paraffin oxidation. The company acquired rights to the technology in North America in exchange for $10,000 plus a small royalty on output. But before going ahead with its own plant, Hercules invested $90,000 in a pilot plant at the Imhausen facilities to investigate the process and produce larger test samples.[84]

This operation, which went on line early in 1952, proved economically marginal, but its performance was soon overshadowed by another opportunity that, as it turned out, was vastly more significant. During the fall of 1951, Ewald Katzschmann, an Imhausen researcher with whom Johnstone and Taufen had become friendly, developed a new process for air-oxidizing xylenes to make a variety of products, including some constituents of plasticizers and synthetic fibers. Imhausen offered the new technology to Hercules, and when Billing learned of the opportunity, he was immediately supportive. Billing and Johnstone glimpsed the vast potential of intermediates for synthetic fibers, but to persuade reluctant members of the executive committee to investigate such an alien technology, they focused on the utility of the new process in making plasticizers. In the summer of 1952, the executive committee went along with a modest proposal to erect a facility at Burlington, New Jersey, to oxidize xylenes.[85]

The Burlington plant represented an investment of $350,000, a sum far below that necessary to support commercial development of a significant new product. From the outset, however, the plant focused not on making the constituents of plasticizers, but on the air-oxidation of *para*-xylene into dimethyl terephthalate (DMT), an attractive alternative to TPA for making polyester fiber. The Imhausen process (also commonly known as the Chemische Werke Witten process) produced DMT in a highly pure form that could be reacted with ethylene glycol to make high-quality polyester. Although the Burlington facility was plagued with problems and sorely undercapitalized (one manager called it "a pile of junk"), it nonetheless succeeded in producing enough pure DMT to interest the dominant producers of polyester in the world, Du Pont and ICI.[86]

Hercules employed different strategies to woo the two chemical giants. Du Pont's appetite for DMT was so huge that a sole-source supply contract was out of the question, and Hercules proposed a license agreement. After examining the Imhausen process, however, Du Pont calculated potential savings at a level too low to justify the royalties Hercules sought.[87] Although Hercules was surprised, if not suspicious, at this response, the company eventually succeeded in gaining from Imhausen more liberal royalty terms for third-party licensees. In the spring of 1953, Du Pont finally licensed the Imhausen process through Hercules.[88]

Hercules' relationship with ICI was more straightforward. Although unimpressed with Burlington's production capabilities, ICI was very impressed with the quality of its DMT, which it sought to use at a polyester plant operated by its North American sudsidiary, Canadian Industries, Ltd. To demonstrate the reliability of the process, Hercules took six months in 1953 to build at the Experiment Station a pilot DMT plant that provided convincing data on purity and yields. The company also delayed development work on DMT to concentrate on producing *para*-xylene to ICI's specifications.[89]

By the close of 1953, Hercules DMT proved satisfactory to ICI. On December 22, with President Forster as chairman, the executive committee appropriated just over $4 million for a DMT plant at Burlington with a capacity of one hundred thousand pounds per month. The following week, two days before the end of the year, Hercules signed a long-term contract to supply ICI with DMT beginning in July 1955.[90]

Into Polyolefins

Hercules' third, and ultimately most significant, road to petrochemicals started in the Cellulose Products Department. Like his counterparts in Naval Stores and Synthetics, general manager Fulenwider was driven by a desire for new growth products, primarily to find a successor to cellulose acetate, which was then falling under increasingly heavy assault from petrochemical plastics.

Fulenwider, moreover, was keenly interested in new avenues of research. For example, he sponsored the continuing work of James Shapleigh, who had developed a nickel catalyst tubular furnace that showed promise of yielding olefins from oil at a higher rate than ordinary cracking. (Work along these lines dated back to the 1930s, when Hercules had learned how to make ammonia from natural gas and reform natural methane into synthetic gas and hydrogen.) Fulenwider also supported Jack Martin's investigations of polyethylene.[91]

The department's quest for new opportunities relied heavily on the efforts of company emissaries such as David Wiggam, director of development for Cellulose Products, and Dr. Arthur L. Glasebrook, a research chemist with a background in high-pressure reactions and a group leader at the Experiment Station. Together and separately, Wiggam and Glasebrook made frequent journeys to Europe in search of new technologies. The work required a great deal of creativity and initiative, even aggressiveness—indeed, colleagues recall Glasebrook as a man who "knew how to get into places that no one else could."[92]

The executive committee's unwillingness to invest in polystyrene and polyethylene was intensely frustrating to these two men. Pressure to find some significant entry point into petrochemical plastics continued to mount, however; the department was acutely aware that Hercules was losing ground to its competitors. By the early 1950s, many at the company believed that Hercules could not afford to pass up many more chances. When such an opportunity presented itself in connection with the work of a prominent German chemist, Prof. Dr. Karl Ziegler, managers in Cellulose Products were determined that "we were not going to miss this time."[93]

Ziegler ranks as one of the pioneers of modern polymer chemistry.[94] Born in Helsa, Germany, in 1898, he had been inspired by his first encounter with a physics textbook as a boy and earned his Ph.D. by the age of twenty-one. In 1943, he became director of the Max Planck Institut für Kohlenforschung [Coal Research] at Mülheim, an institution funded by several Ruhr coal companies to advance industrial technology in the field. A brilliant researcher and demanding manager, Ziegler made significant contributions in free radical chemistry, large-ring compounds, and other fields, but his greatest successes involved the *Aufbau* (growth) reaction in straight-chain polymers.

Most natural polymers (such as cellulose) are organic—principally a spine of carbon into which atoms of other elements may be interspersed and onto which groups of atoms may be attached. Beginning in the 1920s, the Swedish scientist The Svedberg argued that such natural polymers were not, as had been previously believed, mere physical aggregates of molecules, but actually molecules of unimagined size and weight. Hermann Staudinger, the famed German organic chemist, went further and argued that the linking mechanisms in the huge molecules were chains, and that the chains might be tens of thousands of times longer and heavier than conventional wisdom thought possible. In the following decade, researchers at Du Pont and ICI convincingly demonstrated Staudinger's theories by synthesizing polymers. In 1933, researchers at ICI discovered the first high molecular–weight polymer—high-pressure polyethylene (called Polythene in the United Kingdom). The product had obvious drawbacks—weakness, softness, and a low melting point—but nonetheless prepared the way for a revolution in materials engineering.

ICI's manufacturing process for high-pressure polyethylene revealed the importance of a catalyst and an oxidizing agent, these being peroxide and air for this polymer. The subject engaged many leading polymer chemists—in addition to Ziegler, Herman Mark, Maurice Huggins, Paul Flory, Calvin Schildknecht—who made (and at times failed to recognize) advances in straight-chain polymer chemistry.

As Ziegler progressed with his research, he formed relationships with leading chemical companies throughout the world. A former businessman as well as a scientist, Ziegler believed that he could and should benefit, financially and personally, from the institute's work under his direction. In accepting leadership of the institute, he had insisted on receiving not only research freedom but also the right to retain royalties on any patentable technology developed under his supervision not directly related to the coal industry. Even so, Ziegler did not negotiate like a seasoned businessman. He spurned the advice of lawyers and based many of his dealings on personal relationships; those who earned Ziegler's trust also gained a clear advantage in negotiations.[95]

In the 1940s, Ziegler published and lectured about his attempts to synthesize the first truly straight-chain polymers (using metal compounds such as aluminum triethyl). From many quarters, both academic and commercial, he encountered indifference and skepticism. Dr. Otto Bayer, for example, mocked Ziegler's work and rejected an early contract—a move that would later haunt him. To be sure, Ziegler had yet to demonstrate the ability to grow molecular chains of impressive and uniform length, but this did not discourage E.T. Borrows of Petrochemical Ltd. (PCL). Sometime around 1950, Borrows forged a close friendship with Ziegler and negotiated (without the help of attorneys) exclusive rights to Ziegler's discoveries in the United Kingdom.[96]

By both coincidence and design, Hercules entered the scene at this critical early stage, becoming the first American chemical company to approach Ziegler. In 1950, Wiggam and Glasebrook were on a scouting mission in Europe, accompanied by Max Riemersma from the office at The Hague. Upon reaching Mülheim, the group toured Andreas Hofer, a scientific machinery company, then visited the Max Planck Institute, which, Riemersma knew, was investigating high-pressure reactions. Although they did not meet Ziegler on that occasion, they learned of his work with ethylene polymers, an intriguing subject not only because of Martin's interest in polyethylene, but also because of ethylene's potential as a source for making pure *para*-xylene for the DMT process.[97]

Little came of that first visit, and Wiggam continued his investigations at other sites, including Farbewerke Hoechst of Frankfurt, Germany (a visit important to later developments). About eighteen months later, in May 1952, he returned to Mülheim and met with Ziegler. The two men discussed the institute's work on aluminum catalysts, and Wiggam expressed Hercules' interest in negotiating a license agreement. Ziegler was willing to consider an offer, and the two arranged to meet again.

That summer, Wiggam and Riemersma met several times with Ziegler to hammer out a deal. The terms called for Hercules to receive an exclusive one-year option (in North America) to evaluate Ziegler's process for making *para*-xylene, as well as rights to investigate Ziegler's developmental work on the preparation of aluminum organic compounds and their use in chemical reactions. In September, Paul Johnstone, who happened to be in Europe on business for the Synthetics Department, obtained Ziegler's signature. Apparently, the deal was clinched when Johnstone, a shrewd negotiator, offered to underwrite the expenses of Ziegler's impending lecture tour to the United States. Later that year, Ziegler sold Hoechst a "semi-exclusive" license in Germany covering similar research.[98]

In the early 1950s, as word of Ziegler's work began to spread throughout the chemical industry, representatives of the major competitors frequented Mülheim, crossing paths and trading stories at local hotels and restaurants, as well as at the institute. (Indeed, Glasebrook formed a luncheon club with representatives from Hoechst, PCL, and Montecatini Chemical Company [now Montedison] of Milan, Italy.) In 1952, Giulio Natta, one of Italy's leading polymer chemists, secured Italian rights to Ziegler's work relating to the "transformation of olefins" for his corporate sponsor, Montecatini.[99]

During the option year, with Ott's support, Hercules technical staff pored over information provided by Ziegler or made available to Hercules personnel in Mülheim. (Wiggam was a frequent visitor, and Glasebrook spent May through July 1953 at the Max Planck Institute.) Initial studies showed that the process for making *para*-xylene was not economical, but that the process for synthesizing aluminum alkyls seemed more promising. Accordingly, Hercules focused on this aspect of Ziegler's work.

In May 1953, Dr. Gunther Wilke, Ziegler's assistant, accidentally discovered that the addition of nickel to the reaction of aluminum triethyl and ethylene greatly extended the chain length of the polymer.[100] Subsequently, in July, E.

Holzkamp, a graduate student working with Wilke at the institute, actually produced a sample of high-density polyethylene in a reactor vessel using chromium co-catalyst with an organic aluminum compound. Glasebrook, then in Mülheim, was extremely interested—indeed, initially more interested than Holzkamp himself—in this result.[101]

As the expiration date of Hercules' option (September 30, 1953) approached, Glasebrook prepared a summary report of his work at the institute and urged that the option be continued, "provided it is realized that we are gambling on the future of an uncompleted laboratory investigation and are not buying processes which have stood the test of pilot test evaluation, cost studies, and market surveys." He estimated that "it will take several years of work to establish the commercial practicality and worthwhileness" of Ziegler's processes.[102] In September, Glasebrook also marshaled the support of key personnel in the company—Forster, Ott, Billing, and Wiggam—and even arranged for Sir Robert Robinson, a British Nobel laureate, to lend his authority to the cause. Although Fulenwider was sympathetic, from his perspective, the preponderance of opinion was not enough to convince the executive committee to spend the money. "He told me," recalls Glasebrook,

> "I have to have something. Can you tell me one thing that we could build a factory on?" He said, "I have to have something to hang my hat on, too." Well, I had to be truthful with him. I said, no, this thing needs further development. Well, he didn't want to turn it down, he wanted to go ahead, but he was reluctant to push it as hard as he could.[103]

Hercules' option expired without renewal, although Wiggam attempted throughout the fall of 1953 to structure a new arrangement. (Meanwhile, Ziegler's agreements with PCL in the United Kingdom and Montecatini in Italy covering rights to his work were extended in those countries.)

On October 26, 1953, the company's bargaining position was severely weakened. Less than a month after Hercules' exclusive option expired, chemists at Ziegler's laboratory produced a sample of linear crystalline polyethylene. While testing new metal catalysts, Ziegler's team polymerized a sample of ethylene with zirconium into polyethylene. Of the resulting specimen, Ziegler recorded that "it is impossible by my own hands to put it into two pieces." Improvements followed with new catalysts. Ziegler's laboratory was catalyzing polymer chains of enormous molecular weight at normal pressures and mild temperatures.[104]

This was a significant commercial—as well as scientific—breakthrough. It was not the first petroleum-based polymer plastic, but it promised to be one of the best: straight-chain, crystalline polyethylene possessed a combination of attributes—stiffness, flexibility, and strength—superior to those of its predecessors, polystyrene, polyvinyl chloride, and high-pressure polyethylene. Production of the last, as noted, was tightly controlled by Du Pont and ICI. Unlike Ziegler's linear polyethylene, the older variety was composed of branched molecules prepared by a dangerous and relatively expensive high-pressure process. Linear, low-pressure polyethylene's more crystalline structure gave it more strength and hardness and a higher melting point.[105]

Hercules learned of Ziegler's breakthrough in December, when he sent Wiggam a sample of the polyethylene bearing an image of the German eagle imprinted from a coin. (Honoring a gentlemen's agreement, Ziegler remained committed to Hercules and rejected entreaties from Union Carbide and perhaps other American companies who sent representatives to the institute. For his part, Wiggam "gave strict orders" against analyzing the sample for traces of the catalysts, a procedure that probably would have revealed Ziegler's secrets.)[106] The gesture rekindled the interest of Hercules senior management, who dispatched Wiggam and C.M. Rutteman, managing director of N.V. Hercules, to Mülheim to negotiate new terms. By early 1954, however, Ziegler had not only successfully produced low-pressure polyethylene but had also developed a low-cost process for synthesizing aluminum alkyls. As a result, he wanted separate licenses for each process. He also revealed that Hoechst had licensed the polyethylene process nonexclusively.[107]

In the spring and summer, "a continuous exchange of letters and proposals" between Hercules and Ziegler narrowed the gap between the two. The company moved swiftly to conclude the agreement only after learning in July that Koppers had arranged a nonexclusive license on the polymerization process. Soon thereafter, Goodrich-Gulf and Union Carbide forged agreements with Ziegler.[108]

Hercules reached its own agreement with Ziegler on September 13, 1954. Although Hercules was not the first U.S. company to obtain rights to Ziegler chemistry, its terms were the broadest and most favorable obtained by the early licensees—credit to the many months of hard work by many Hercules representatives. The company paid $350,000 for a nonexclusive one-year option on Ziegler's know-how with respect to ethylene polymers, co-polymers, and other olefin polymers, as well as for a nonexclusive license covering a new process for the manufacture of polyethylene. Hercules also obtained an exclusive agreement to sell as well as make Ziegler's catalysts (which others could produce only for their own use).[109]

By the fall of 1954, Hercules was at last positioned to take a giant step into the burgeoning petrochemicals industry. Its path to this point was twisted and full of difficulties. As we will see in the next chapter, the road from licensing to commercialization would also be riddled with sharp curves, unexpected obstacles, and occasional downhill slides.

Conclusion

The postwar decade was a major turning point in Hercules' history, marked by its growing awareness of the limits of the chemical businesses that had propelled it forward since the early 1920s, and its first forays into the rapidly expanding world of petrochemicals. The move into new areas did not come easily. Diversification was motivated neither by crisis nor by a windfall opportu-

EXHIBIT 9.6 Comparative Performance of Chemical Companies in 1954 vs. Pre–World War II Base Period (1935–1939)

	NET SALES	*OPERATING PROFIT*	*EARNINGS PER SHARE*	*CAPITAL EXPENDITURES*
Allied Chemical & Dye	+256%	+233%	+105%	+1,255%
American Cyanamid	+502	+876	+250	+1,014
Atlas Powder	+270	+316	+141	+709
Dow Chemical (year ended November 30)	+1,682	+1,457 (est.)	+369	+1,558 (est.)
Du Pont	+557	+857	+368	+286
Du Pont excluding General Motors			+492	
Eastman Kodak	+399	+479	+241	+286
Hercules Powder	+407	+557	+269	+573
Hooker Electrochemical	+750	N/A	N/A	N/A
Monsanto Chemical	+987	+741	+253	+770
Olin Mathieson Chemical	*	*	+345	N/A
Union Carbide & Carbon	+522	+310	+150	+579

*Omitted because figure is not comparable, owing to consolidations.

N/A—Not available.

Source: Hercules Powder Company, Treasurer's Department, Economic Research Division, "Comparative Performance of Chemical Companies from 1935 to 1954," June 1, 1955.

nity. To the contrary, Hercules' conservative business strategy and prudent financial policies produced impressive results: reasonable levels of growth and high levels of profitability in its major product lines (see Exhibits 9.1 and 9.2). From the shareowner's point of view, the company was very well managed.

The pressure to change came from the ranks of department managers and middle managers who were encountering ominous signs and brooded about sustaining the company's traditional high performance. Although some new products of the postwar era—CMC, toxaphene, Dresinates,® Kymene,® solid-fuel propellants—suggested that the company's main lines of business would continue to generate opportunities, other products, particularly in Cellulose Products, were underachievers.

Feeding the uneasiness were scenes of dazzling growth elsewhere in the chemical industry. A comparison of performance indicators of the leading chemical companies from the late 1930s to the early 1950s revealed that Hercules, although a profit leader, was falling far behind its peers in rates of growth and capital spending (see Exhibit 9.6). Indeed, Union Carbide, American Cyanamid, Monsanto, Du Pont, and especially Dow Chemical outpaced Hercules' growth by a wide margin. (Dow zoomed from fifteenth to second place in the industry, thanks largely to the growth of its petrochemical plastics, especially polystyrene.)[110]

At the root of its relatively low performance, of course, were the conservative policies that had served Hercules well since its founding. Addressing a group of Hercules executives in 1952, Ed Morrow confessed, "I think most of you would classify me as belonging to the 'Society for the Preservation of Conservative Capital Structures,' and I would have to admit membership." Morrow then asserted unequivocally one of the company's core beliefs: "It is never convenient to pay off borrowed money."[111]

It would have been difficult to quarrel with this way of thinking during the 1920s, when Hercules gained strength and pulled away from Atlas, or during the 1930s, when high debts ruined thousands of companies. But the post–World War II business environment posed dramatically different challenges, and Hercules was now out of step. As one industry analyst put it: "If there should be a receding tendency in business, [Hercules'] conservative financial policy will pay off. But if general business expansion continues, as I expect it will [and it did], this financial conservatism will continue to be a handicap."[112]

This analyst also noted that Hercules' research was highly regarded. The same could be said of the company's marketing and operating organizations. Hercules' strengths in these areas had enabled it to meet new challenges in its traditional businesses and would be largely responsible for the success of its new ventures into petrochemicals.

And although Hercules was dominated by conservatives after the war, there were dissenters on the board, including the new president, Forster, and throughout the middle ranks of the company. Nor was the company's prevailing conservatism without merit. A late entrant and a middle-sized competitor, Hercules had chosen its niches in petrochemicals very carefully, neither betting the company nor losing its independence. Meanwhile, the events of the decade prepared Hercules for change, perhaps in a less conservative direction.

CHAPTER

10

EXPANSION AND METAMORPHOSIS, 1955–1961

"We clearly needed a wider base for our operations," recalled Al Forster of his first two years as president of Hercules.[1] During that time, the company responded to his call: the most obvious changes involved diversification into a new raw materials base, petrochemicals, as well as increased capital spending for new or refurbished ammonia plants in California, Missouri, and Alabama. These moves were but the prologue to a period of bold expansion under Forster.

Between 1955 and 1961, Hercules' sales soared from $226.7 million to $380.1 million—an increase of 68 percent and the fastest peacetime growth rate in the company's history. Hercules also showed a new willingness to spend heavily to fund its growth: capital expenditures averaged more than $27 million annually, with particularly large investments made in areas that Forster believed would sustain the business down the road: polyolefins, DMT, and chemical propulsion. (See Exhibit 10.1.)

Investment in R&D also climbed to a new height. Under research director Bob Cairns, who succeeded Emil Ott in November 1955, Hercules boosted annual spending to above 4 percent of the chemical industry average.[2] The funds were used to double the capacity of the Research Center (as the Experiment Station was renamed in 1956), and helped launch Hercules into what promised to be a new era of discovery and growth.

As a result of such actions, between 1955 and 1961, the company grew along vertical lines by integrating backward into new raw materials and forward into new markets, and along horizontal lines by entering new, related lines of business, often through acquisitions. It increased its geographical reach by extending its capabilities to manufacture and market its products overseas. Luck—or good timing—played a part, too: the decline of the commercial explosives business was more than offset by the astonishing rise of chemical propulsion.

EXHIBIT 10.1 Hercules' Financials, 1955–1961 (Dollars in Thousands, Except per Share)

	1955	*1956*	*1957*	*1958*	*1959*	*1960*	*1961*
Net sales	$226,651	$235,903	$245,265	$236,513	$283,650	$336,905	$380,182
Earnings before income tax	42,348	38,541	37,094	36,569	49,007	55,188	59,664
Total assets	154,596	169,556	181,126	191,148	212,878	261,845	291,235
Fixed[a]	62,264	83,139	88,963	87,851	89,539	135,361	152,776
Current	87,616	80,755	85,543	96,707	110,515	110,818	122,594
Other	4,716	5,662	6,620	6,590	12,824	15,666	15,865
Stockholders' equity	100,253	114,543	127,147	137,008	151,494	213,242	188,989
Long-term liabilities	18,833	20,608	22,465	24,042	25,466	26,719	28,982
R&D Expense	7,903	10,504	10,172	10,816	11,602	14,090	15,409
Dividends/share	$3.30	$1.10	$1.10	$1.10	$1.30	$1.30	$1.30
Earnings per share (common)	$6.90	$2.13	$2.14	$2.04	$2.73	$3.05	$3.05
Number of employees	11,259	11,365	11,497	10,743	11,221	13,810	15,596
Return on sales	18.7%	16.3%	15.1%	15.5%	17.3%	16.4%	15.7%
Return on assets	27.4	22.7	20.5	19.1	23.0	21.1	20.5
Return on equity	42.2	33.6	29.2	26.7	32.3	25.9	31.6
Debt/equity	18.8	18.0	17.7	17.5	16.8	12.5	15.3
R&D/Sales	3.5	4.5	4.1	4.6	4.1	4.2	4.1

a. Gross fixed assets, including depreciation.

Source: Hercules Powder Company, Annual Reports for 1955–1961.

Simultaneous growth in all directions had a profound, liberating effect on Hercules, especially on its younger employees. Nonetheless, the new era did not represent a radical departure from the company's traditions. Indeed, when Hercules announced a realignment of its management structure in the summer of 1961, Forster took care to emphasize continuities with the past. The company stayed with decentralized departmental authority and conservative financial management. The result was exceptional growth and profitability, and industrywide acclaim for outstanding management.

PETROCHEMICALS: SCALING UP AND MOVING DOWNSTREAM

In the early 1950s, sensing limits to growth and concerned about the price and availability of raw materials in its traditional businesses, Hercules had followed three paths into petrochemicals: a new process for making phenol and related chemicals; another new process involving the air-oxidation of *para*-xylene to produce dimethyl terephthalate (DMT), a precursor for making polyester; and polymerization of polyolefins based on discoveries of Karl Ziegler.

In identifying and pursuing these opportunities, many of the younger managers and researchers displayed ingenuity and resourcefulness—abilities that would again be in demand as Hercules turned to the challenge of proving the commercial worth of its new ventures. In particular, the new processes had to work: they had to be scaled up and found economical, and markets for their products had to be cultivated.

In each case, Hercules succeeded, although not always in the expected manner. On the way to commercialization, the company encountered many obstacles and discovered that the dynamics of competition in petrochemicals were qualitatively different from those in its traditional lines of business. Indeed, Hercules found itself competing in strange circumstances characterized by the presence of many rivals from previously distinct industries, such as chemicals, oil, rubber, and metals; extreme difficulty in sustaining proprietary process advantages because of liberal licensing policies and easy diffusion of technical information across the industry; the necessity of building and operating large plants to capture economies of scale; rapid growth of demand, but even faster growth of capacity additions; and severe troubles in maintaining price levels.[3] The full measure of these problems became apparent during the 1960s and 1970s, although Hercules received a foretaste of what was to come almost immediately after its entry into petrochemicals.

Phenol and DMT: A Study in Contrasts

In 1955, Hercules' first two petrochemicals facilities came onstream: the Higgins oxychemicals plant in Gibbstown, New Jersey, started operations in January; nine months later, the new Burlington, New Jersey, plant began producing

DMT. In many respects, the fates of the two operations offer a study in contrasts, as well as a demonstration of the perils and rewards of competing in petrochemicals.

At Gibbstown, as conservatives on the board had feared, Hercules encountered fierce competition and worldwide overcapacity in the commodity petrochemical, phenol. On the other hand, the plant succeeded in demonstrating the effectiveness of the Hercules-Distillers process and provided an enduring and valuable source of license income. Hercules' success in also finding profitable niches for other oxychemicals made at Gibbstown reinforced the company's preference for making specialty chemicals.

Hercules DMT was introduced into a rapidly expanding market served by only a handful of competitors. In these enviable circumstances, the company's product commanded premium prices for its quality and paid immediate dividends. By the end of the 1950s, however, disquieting signs appeared in the company's enthusiastic campaign to develop the business. Investment capital to fund the rapid growth of DMT ate into profit margins. Worse, the lure of high growth attracted competitors from other chemical companies and oil producers, many of whom were better able than Hercules to control raw materials costs, assure demand, afford capacity expansions, and endure price competition. Worse still, the critical competitive advantage of DMT over terephthalic acid (TPA)—its higher quality—began to erode with the introduction of a new process for making TPA.

The Higgins oxychemicals plant in Gibbstown, New Jersey, was an impressive sight: the $10 million, thirty-five-million-pound-per-year facility was a showcase of state-of-the-art technology. Featuring sixty-five-foot high towers and brightly "color-conditioned" piping, the plant was one of the first in the United States to use color coding to identify the functions and contents of platforms, tanks, pipes, and structures.[4]

Despite its striking appearance, the Higgins plant was never intended to produce phenol on a large scale. Rather, as administered by the new oxychemicals division of the Naval Stores Department, it was designed first to demonstrate the Hercules-Distillers process and promote the sale of licenses; second, to develop new applications, not so much for phenol as for acetone, *para*-cresol, and related oxychemicals; and third, to supply Hercules' own needs for these commodities.

The plant quickly achieved its primary objectives. License revenues from the Hercules-Distillers process were hefty. Monsanto and Shell Development Corporation, for example, each paid an initial fee of $1 million, royalties of $1.5–2 million, and an additional royalty of 1¢ per pound on all phenol sold to third parties. As G. Fred Hogg, general manager of Naval Stores, remarked in 1958, "Shell's phenol plant will be the twelfth in the world based on cumene hydroperoxide as the intermediate. Less than fifteen years ago, this latter material was a chemical curiosity."[5]

Although Hercules immediately ran up against "extremely keen competition" and "disappointing" sales in phenol, it could not produce enough acetone and *para*-cresol "to keep our head above water." As a result, it purchased outside supplies of these commodities and, in the summer of 1956, allocated nearly $3 million to double plant capacity to six million pounds.[6]

Demand for new derivatives of *para*-cresol and other oxychemicals proved especially strong. In 1956, Hercules began converting about half of its *para*-cresol into Dalpac® (ditertiary-butyl-*para*-cresol), a product that gained wide and diverse acceptance as an antigumming agent in gasoline and an antioxidant in animal foods. That same year, the company also introduced Di-Cup® dicumyl peroxide, a compound that initially promised to be the best agent for vulcanizing rubber (replacing sulfur) since Goodyear invented the process in 1839. Di-Cup® replaced several conventional additives besides sulfur and generally improved the quality of the rubber, especially light-colored products, which it prevented from discoloring.[7] In time, Di-Cup® would become one of the most enduring contributors to profits from the oxychemicals businesses.

The Higgins plant also produced millions of pounds of various hydroperoxides—cumene hydroperoxide (CHP), diisopropylbenzene hydroperoxide (DIBHP), *para*-methane hydroperoxide (PMHP), as well as *alpha*-methylstyrene. Within two years of the plant's startup, all of its major products were profitable, and the top four—phenol, acetone, *para*-cresol, and Dalpac®—accounted for roughly 13 percent of Naval Stores' gross profits.[8]

Although Hercules earned profits on its phenol operation, this first experience with commodity petrochemicals provided insights into the difficulties the company could expect to encounter as it expanded its role. To begin with, the fragmentation of the customer base among scores of manufacturers of phenolic resins and petroleum products underscored the need to develop new channels of distribution and obviated the need for technical support. Second, Hercules' very success in licensing its technology contributed to widespread overcapacity in phenol. The company estimated that, in the late 1950s, U.S. phenol producers collectively were running at only 80 percent of capacity, and that Hercules was meeting roughly 7 percent of total demand. Despite increasing sales of derivatives, the Higgins plant seldom operated above 70 percent capacity, in part because the scale of its operations was insufficient to match the lower costs of its principal competitors.[9]

Hercules' venture into DMT offered contrasting hopes and lessons. At the outset, many managers were concerned not about the process but about its prospects, given Hercules' reliance on a sales contract with Canadian Industries Ltd. (CIL). A writer for the *Hercules Chemist* succinctly stated the issue: the company had invested millions to build a "one-product, one-customer, one-use plant based on new chemistry" (namely, DMT, CIL, polyester, and the oxidation of *para*-xylene).[10]

Nevertheless, early signs were encouraging; almost immediately, the plant ran at full tilt. And despite several expansions of capacity, demand outpaced supply for the remainder of the decade—precisely because Hercules added new customers, uses, and products.

To expand its roster of customers, Hercules looked beyond North America to Europe and Japan, where it found willing customers among textile producers.[11] These and other companies—including many in the United States—began to explore new applications for DMT and its copolymers, especially in films and tapes, wire enamels and dyeing assistants, and high temperature–resistant silicone resins and oil additives. Thanks largely to such uses, U.S. sales of DMT between 1957 and 1958 surged from 600,000 to 2.4 million pounds, with Hercules supplying much of the DMT needs for pilot plant work. By the end of 1960, after four expansions costing nearly $8 million, Hercules was the largest independent DMT producer in the world, earning about $2 million a year from the business.[12]

In the first few years of the DMT business, the high reputation of Hercules' product, especially among ICI affiliates, combined with the protected nature of worldwide polyester production, acted to temper competition.[13] By the end of the decade, however, signs of change were already apparent. First, new rivals in DMT emerged, most notably, Amoco Chemicals Corporation.[14] An enormous company that controlled its own raw materials, Amoco was a major threat. Worse, in 1956, it acquired a new, economical manufacturing process for preparing pure TPA, developed by the New York chemical engineering consulting firm, Scientific Design Company. As a result, Amoco became a formidable competitor not only in DMT but also in its ultimately cheaper substitute, TPA.[15]

Another concern for Hercules was the state of the market for polyester. In 1960, Du Pont, protected by patents, controlled 95 percent of U.S. production. But in the following year, as the patents expired, new entrants (such as Fiber Industries) appeared, and many others seemed likely to follow. (Indeed, by the late 1960s, Du Pont's market position would erode to less than 50 percent.) Hercules was initially optimistic that the new producers would mean new customers for DMT. As the 1960s wore on, however, competition in polyester fibers would hold prices down and expand the search for alternative precursors.[16]

For the present, however, Hercules was pleased to share in the fabulous growth of polyester as it replaced natural fibers throughout the industrialized world. In 1961, profits from DMT accounted for 6 percent of the company's after-tax earnings. More expansions would soon follow, and as we will see in the next chapter, DMT would play a key role in Hercules' international expansion.[17]

Ziegler Chemistry: The Race to Commercialize

Scaling up to make crystalline polyethylene, like the process of negotiating for the technology itself, was the most complex of Hercules' initial efforts in petrochemicals and the last to come to fruition.

In February 1955, several months before Hercules had to decide whether to pay the remaining sum to Ziegler for a nonexclusive license (see Chapter 9), the company began research and pilot plant work at the Experiment Station and dispatched a team of researchers, including Walter Gloor, Courtland White, and George Hulse, to Mülheim to study Ziegler's polyethylene process. The team returned with pertinent documents and apparatus, but it had found little useful operating information. Like other Ziegler licensees, Hercules learned that the professor's know-how on the polymerization of ethylene amounted to scarcely more than a "book" of laboratory data, and that access to the technology was, for those who planned to commercialize the process, essentially a hunting license. "There we were with a patent," said Elmer Hinner, general manager of Cellulose Products, "not knowing what to do with it."[18]

The primitive state of research on Ziegler chemistry raised serious obstacles to commercialization; in 1955, no one understood basic issues such as how to regulate the molecular weight of the polymer, or the process dynamics and proper sequencing of operations, much less how to manufacture the catalysts and polymer on a large scale.[19] Fortunately, Hercules chose not to tackle those problems alone. In the spring of 1955, it forged a technical exchange agreement with the German licensee of the Ziegler catalyst system, Farbewerke Hoechst of Frankfurt, to develop the new manufacturing process.

Founded in 1863 as a dyestuffs manufacturer along the Main River, Hoechst had been one of the main constituents of I.G. Farben (IGF), the giant German chemical cartel formed in 1925. After the war, IGF was broken into three groups of roughly equal size—BASF, Bayer, and Hoechst—thereby returning the industry roughly to its pre–1925 structure (much like the final decree of 1912 had attempted to turn back the clock on the explosives industry in the United States). The new Hoechst was born in the spring of 1953 with fixed assets worth 285.7 million deutsche marks (DM), 26,000 employees, annual sales of DM 1 billion, and about 9 percent of German chemical output. Pharmaceuticals and dyestuffs, including the latter's aromatic starting and intermediate products, made up the core of its business, but the company also made fertilizers, plastics, solvents, and artificial fibers and film. The smallest of the three new companies, Hoechst expanded aggressively at home and had begun buying chemical and dyestuff companies in the United States.[20]

The roots of collaboration between Hercules and Hoechst date to the fall of 1950, when Art Glasebrook met Dr. Otto Horn, director of research for the organization that would become Hoechst, at a meeting of the American Chemical Society. Later, after Hoechst and Hercules licensed Ziegler's polymer chemistry, Horn and Glasebrook rekindled their friendship, and Horn agreed to keep Glasebrook apprised of progress at Frankfurt. Through this conduit, Glasebrook learned that Hoechst, because of its research and experience with aluminum organic compounds, was ahead of Hercules in developing a process to manufacture Ziegler's catalysts and was moving quickly to complete a pilot plant to make polyethylene. Aware of the potential advantages of cooperating with Hoechst, Glasebrook persuaded Horn to help arrange for a Hoechst representative to meet with several senior managers at Hercules in the fall of 1954.[21]

Hercules had several motives for wanting a technical trade agreement with Hoechst. First, it obviously wanted to catch up. Researchers at the Experiment Station had encountered difficulties in pilot plant work, especially in making the catalysts and controlling the molecular weight (if too high, the material cannot be extruded or molded) and color of the polymer. Second, like the American chemical industry in general, Hercules maintained a high regard for the technical capabilities of the leading German chemical companies. Indeed, managers in the Cellulose Products Department recalled that Hercules had imported German know-how on ethylcellulose, cellulose acetate, and chlorinated rubber prior to the war.[22] Finally, Hercules was acutely aware of the importance of timing in the rapidly changing competitive environment of plastics, and it sensed that a research trade agreement with Hoechst might recapture some of the valuable time lost when it had wavered over its initial commitment to Ziegler. For its part, Hoechst was interested not only in sharing the expenses of development work, but also in learning more about how to market plastics.[23]

In February 1955, A.B. Nixon, who had been authorized by the board to spend as much as $3 million to secure Hoechst's polyethylene pilot plant know-how, left for Germany with Fulenwider, Hinner, and Glasebrook. The week-long bargaining was anything but easy; as Hinner recalled (with a bit of hyperbole), "There were four of us on one side of the table, and twenty-nine Germans on the other. We didn't dare *breathe* hard." Still, Hercules was pleased with the outcome: it procured Hoechst pilot plant know-how for only $1 million. (Apparently, Hoechst had been offered five times that amount by W.R. Grace, with whom Hoechst had had previous dealings, but instead chose to work with Hercules because of its experience and knowledge of markets in the plastics business.)[24]

For the next two years, a Hoechst chemist was stationed at the Research Center, and Dr. Karl Winnaker, managing director of the German company, paid annual visits to Hercules, where he became fast friends with President Forster.[25] The collaboration of the two companies worked smoothly, although progress was uneven. The manufacturing process itself quickly assumed definite shape. The first step was to feed ethylene gas into a polymerizing unit in the presence of the catalysts (an aluminum alkyl and a metal compound) and a diluent. In this batch operation, the polymer formed quickly but was contaminated by the continuing presence of the catalysts. The next step, accordingly, was to draw off the slurry mixture from the polymerizing unit and allow the catalysts to decompose. Then, in a continuous process, the polymer was filtered and purified to remove the diluent, which could be recycled. Finally, the polymer was dried and pelletized to produce resin.[26]

Straightforward in concept, the process was nevertheless complicated in practice. For a time, Hercules had to import and rely on a German diluent and experienced difficulties in making the titanium catalyst and producing aluminum alkyls on a sufficient scale. Indeed, a fire resulting from the tendency of aluminum alkyls to inflame spontaneously when exposed to oxygen caused an accident that destroyed Hercules' first pilot plant. Thereafter, research findings were tested at Hoechst's pilot plant in Germany.

Despite these difficulties, the combined resources of the two companies resulted in several significant innovations. Edwin J. Vandenberg, a mechanical engineering graduate from Stevens Institute of Technology and a sixteen-year research veteran at Hercules, made a significant discovery by demonstrating "the hydrogen effect," which showed that "hydrogen was effective in decreasing the molecular weight of an olefin polymer prepared with the use of a Ziegler catalyst and that the extent of the decrease was proportional to the amount of hydrogen added." As a result of this finding, plant engineers could predict with great accuracy the exact flow specifications of the polyolefin as it flowed through the plant and could therefore, in effect, tailor the output for different customer segments.[27]

Such innovations enabled the Hercules-Hoechst collaboration to achieve its goal quickly. By the fall of 1955, scarcely more than a year after Hercules signed its second agreement with Ziegler, the Cellulose Products Department felt confident enough to recommend the construction of a full-scale commercial operation to make crystalline polyethylene. The proposed plant would have unique mixing, agitation, and steam distillation features and, with the help of Hoechst—especially in the design of the critical catalyst inactivator—would achieve a "high productivity polymerization process yielding a product of good bulk density and color."[28]

Hercules' arrangement with Hoechst, in short, saved many months, probably years, of development time. In June 1956, the Hercules board appropriated $9.8 million for the construction of the new crystalline polyethylene plant at Parlin.[29] The site was selected primarily because of its proximity to a new ethylene plant under construction by Enjay Company (predecessor of Exxon Chemical) at nearby Bayway, New Jersey. Hercules committed itself to long-term bulk purchases, and Enjay agreed to install ethylene refining and piping facilities and to begin supplying the olefin to Hercules no later than April 1, 1957.

Although construction proceeded swiftly and the plant was ready to commence operations by the spring of 1957, Hercules suffered an embarrassing moment and a frustrating setback when Enjay failed to meet its commitment. At the plant's formal dedication on June 18, Ziegler himself appeared at Parlin to open the ethylene valve and, symbolically, to start the flow of monomer into the polymerizing unit—but there was no ethylene to flow. "It was a little grim," recalled Forster. "A beautiful new plant all ready to roll, and not a damn thing to pump into it."[30]

Parlin produced its first commercial batch of crystalline polyethylene in September, a few months after Hoechst had started manufacturing the product in Germany. Although Hercules was the first U.S. company to produce the polymer using the Ziegler process, the product was introduced into a market already crowded with competitors. Earlier in the year, four companies—Phillips, Celanese, W.R. Grace, and Allied Chemical—had begun making crystalline polyethylene using a process developed (and licensed) by Phillips. In the fall of 1957, then, Hercules was the fifth U.S. producer to enter the business. And it was far from the last: Koppers and Union Carbide opened plants by the end of the year, and another half a dozen producers were expected to join the fray the following year.[31]

The Second Horse

Amid such competition, Enjay's delay in delivering ethylene to Parlin could have been devastating but for a second propitious turn of events that resulted in the availability of an alternative feedstock, as well as a new plastic. To explain these developments, we must backtrack for a moment.

In October 1954, soon after it had accelerated development work on crystalline polyethylene, Hercules also launched a general research effort into Ziegler chemistry to increase its understanding of how the catalysts worked. A week into the assignment, a research team that included Vandenberg achieved a significant breakthrough. Attempting to polymerize propylene, Vandenberg succeeded in producing a small sample that had attractive properties: a high melting point, and the ability to be stretched into film or drawn into fiber.[32]

Hercules' enthusiasm for the new polymer was soon tempered, however. To begin with, initial investigations were disappointing: samples featured either unacceptably low molecular weights or "so much amorphous or rubber polypropylene that the overall properties were no better than polyethylene." Although researchers were confident that such problems could be overcome, they also anticipated a lengthy development process.[33] Second, it soon became apparent that Hercules was not the first to polymerize propylene using Ziegler's catalysts. Indeed, claimants for priority were legion. Through its European scouts, Hercules learned that not only Ziegler but also licensees in Italy (Montecatini, under the direction of Giulio Natta), Germany (Hoechst), and England (Petrochemicals Ltd. [PCL]) had already discovered crystalline polypropylene. Indeed, it would later turn out that at least nine independent research groups had made crystalline polypropylene between March 1953 and early 1955.[34] (After nearly three decades of litigation, appeals, and reversals, the courts awarded priority to researchers John Paul Hogan and Robert L. Banks of Phillips Petroleum in 1983. In the interval, however, most industry participants and followers credited Natta and Montecatini for the discovery; indeed, Hercules licensed Montecatini's patent.)

Despite these problems, leaders of the Cellulose Products Department recognized that if Hercules could not be first to the patent office, it might be first to the marketplace. Accordingly, Vandenberg and other researchers pushed ahead with studies of polypropylene and processes for producing it. The hydrogen effect, for example, proved as critical in making polypropylene and other isotactic polymers as in producing crystalline polyethylene. In June 1956, after months of careful experiments on polymerization reactions, the Cellulose Products Department reported making polypropylene "that had such outstanding physical properties that it would warrant a full-scale development program."[35]

Aware that any Hercules advantage was likely to be fleeting, Hinner made a highly important decision: rather than mount an intensive effort "to develop an optimum poly-propylene process," he chose instead to see whether it would be feasible to make the polymer at the polyethylene plant being designed for Parlin. "If this were possible," reasoned Hinner, "we felt we could be the first in the market with this important new plastic." Studies were encouraging, and by the end of 1956, he was "reasonably sure that we could not only obtain propylene of satisfactory quality from a nearby source at a reasonable price, but also that we could polymerize it at practical rates to give the product we wanted."[36]

The next step was to persuade senior management to go along with the investment—no easy task. As Hinner put it, the department was in the position of selling the Hercules board on "a handful of ugly gunk" on the basis of engineering and market research, rather than an established track record. Some executives were unimpressed with polypropylene, believing that it was simply too early to gamble on its commercial potential. Others were nervous either about licensing yet another foreign technology or about investing large sums in yet another unproven project. As Forster conceded, "For us, this was a pretty adventurous step to take."[37]

On the other hand, polypropylene had much in its favor. First, the plastic possessed many unique properties that seemed promising. Although at first it was seen as a substitute for crystalline polyethylene because it was harder, stronger, and higher melting, polypropylene soon found singular applications of its own, especially when drawn or stretched into strong fibers and film. Second, the ability of the Parlin plant to accommodate polymerization of either ethylene or propylene was significant: the department managers did not have to ask for funds for an entirely new facility. Third, the plant at Parlin was idle, owing to Enjay's failure to deliver ethylene monomer. Finally, the project had powerful champions among present and former Cellulose Products Department managers, including Jack Fulenwider, Jack Martin, Paul Johnstone, and Werner Brown. Johnstone was particularly influential. His dogged insistence that the project move forward—called "bullheadedness" by his colleagues—helped carry the day. Johnstone and others may also have feared repeating the decisions that had cost Hercules the exclusive opportunity to develop crystalline polyethylene in 1953.[38]

In the end, the Cellulose Products Department prevailed: the executive committee authorized the conversion of one of Parlin's two polyethylene lines to make polypropylene. Most of the modifications to the plant centered on the catalyst feed system, solvent recovery, and retention time. To test the process, Hercules built two pilot plants, one at the Research Center in Wilmington and another "plant-in-miniature" at Parlin that included "a scaled-down version incuding every unit of the big plant from which exact process and operating information was required." To avoid a repetition of the Enjay debacle, Hercules arranged to have propylene shipped in by tank truck. Exulted Hinner, "suddenly [we had] two horses in the race."[39]

In December 1957, despite problems in the slurry filtration, centrifuge, and drying stages of the operation (resulting in some residual dirt and ash), Hercules became the first producer in the world to offer crystalline polypropylene.

Markets for New Plastics

By the end of 1957, a scant three years after obtaining its nonexclusive license from Ziegler, Hercules began commercial production of crystalline polyethylene, which was marketed under the name Hi-Fax® polyethylene resin, and crystalline polypropylene, which was sold as Pro-Fax® polypropylene resin. It was an impressive achievement, and testimony to the company's resourcefulness in shortening the development cycle and dedication in overcoming technical and manufacturing obstacles.

Producing the new plastics, however, was only half the battle. The other half, finding markets for the products and establishing a secure competitive position, was also a considerable challenge. Not only did the company have many rivals to contend with, but it also faced continuing pressure to upgrade and differentiate its plastics.

Hercules' initial strategy was to market Hi-Fax® as a substitute for high-pressure polyethylene in many existing applications, such as housewares, coated wire, detergent dispensers, fibers (for seat covers, ropes, fishing nets, and similar products), bottles, toys, and chemicalware. In 1958, the product received a significant boost from the craze for Hula-Hoops, which consumed two and a half million pounds of the polymer. Hi-Fax,® however, quickly found more enduring applications, too. By investing heavily in development work and manipulating the characteristics of the polymer, Hercules was able to introduce new product viscosities and colors. For example, the company's ability to modify its polymer resin for specific uses helped it gain a secure foothold in the market for blow-molded bottles such as those used to contain liquid detergents and bleach.[40]

In the next few years, marketing efforts for Hi-Fax® became more focused. But competition, especially from Phillips, Allied Chemical, and Du Pont, was keen, and Hercules found it extremely difficult to make significant headway. From the outset, capacity increments outran demand, and every producer struggled against the downward pressure on prices. Between 1955 and 1960, total U.S. capacity for polyethylene nearly quadrupled, soaring from 450 million pounds to 1.7 billion pounds, a level about 50 percent above demand.[41] Although Hi-Fax® found niches in low-viscosity applications such as injection-molded housewares and toys, as well as in high-viscosity markets such as blow-molded bottles, it was a product under siege almost from its birth.

The Pro-Fax® picture was brighter: the product's attractive characteristics—a high melting point, stiffness, and strength—proved attractive for many applications, especially in "high quality injection molding material." Sales also showed a strong trend toward fibers applications (one-fifth of sales in 1958, one-third in 1959). Heavy denier polypropylene fibers were sold to makers of seat covers, rope, outdoor furniture webbing, scouring pads, and similar products. To broaden the market, Hercules started investigating deep-dispersion dyeing to produce bright, colorfast resins for fabricators. By 1959, output of Pro-Fax® reached fourteen million pounds, surpassing production of Hi-Fax® for the first time.[42]

Pro-Fax's® performance was encouraging, and increasing Hercules' lead in the marketplace was a compelling reason to expand the business rapidly. To maintain its advantage—and to avoid competitive problems such as those plaguing Hi-Fax®—Hercules made the largest investment in its history. In October 1959, the board appropriated $16.2 million for a sixty-million-pound-per-year Pro-Fax® plant at Lake Charles, Louisiana. Like many petrochemical producers, Hercules had gravitated toward the Gulf Coast with its complex of oil and chemical refineries connected by a "spaghetti bowl" of pipelines. At Lake Charles, executives believed, Hercules would have abundant supplies of propylene monomer nearby.[43]

At left: Hercules' business in aerospace soared after the 1958 acquisition of Young Development Laboratory, whose founder pioneered filament-wound cases for rocket motors. Below: Hercules' fiberglass-enclosed rocket motor formed the third stage of the Minuteman, the United States' first ICBM.

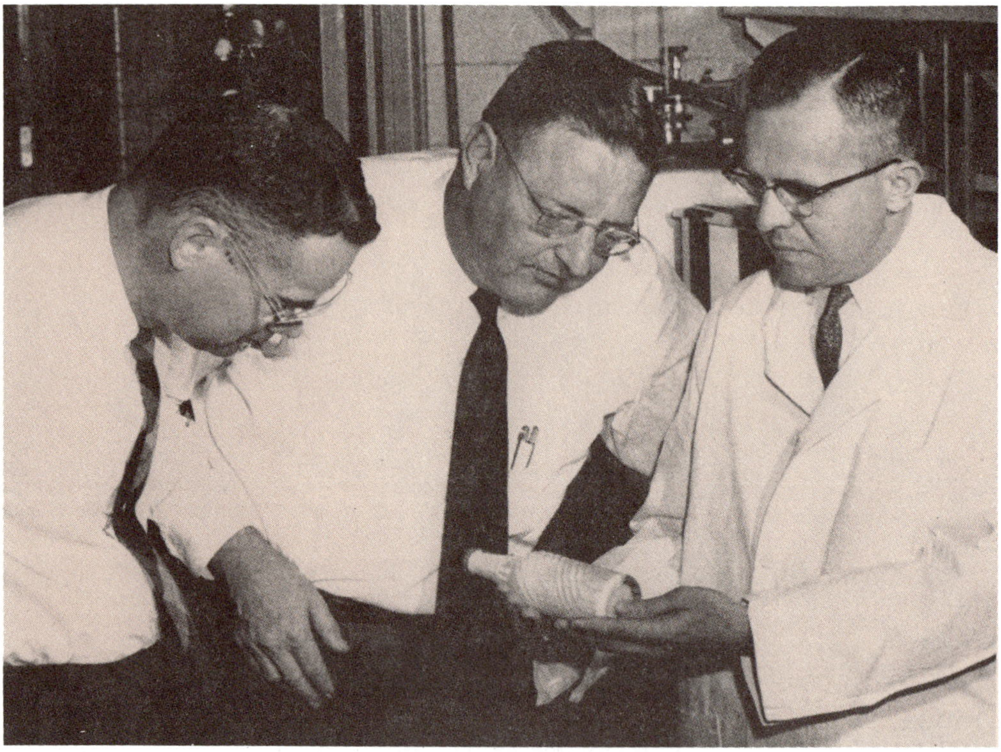

The company's principal engine of growth in the postwar era was polyolefins. Above: Edwin J. Vandenberg made and recognized Hercules' first crystalline polypropylene in 1955. Vandenberg (right) poses with fellow researchers Howard Tennent (left) and Harold Spurlin (center) around a small pressure bottle made of polypropylene. Below: In June 1957, Dr. Karl Ziegler opened a valve at Parlin to dedicate the first commercial plant to make linear crystalline polyethylene. On that day, however, the pipes were empty because of a supplier's failure to deliver ethylene monomer. Observing the scene: Elmer Hinner, Forster, Dave Bruce, and Parlin plant manager Earp Jennings.

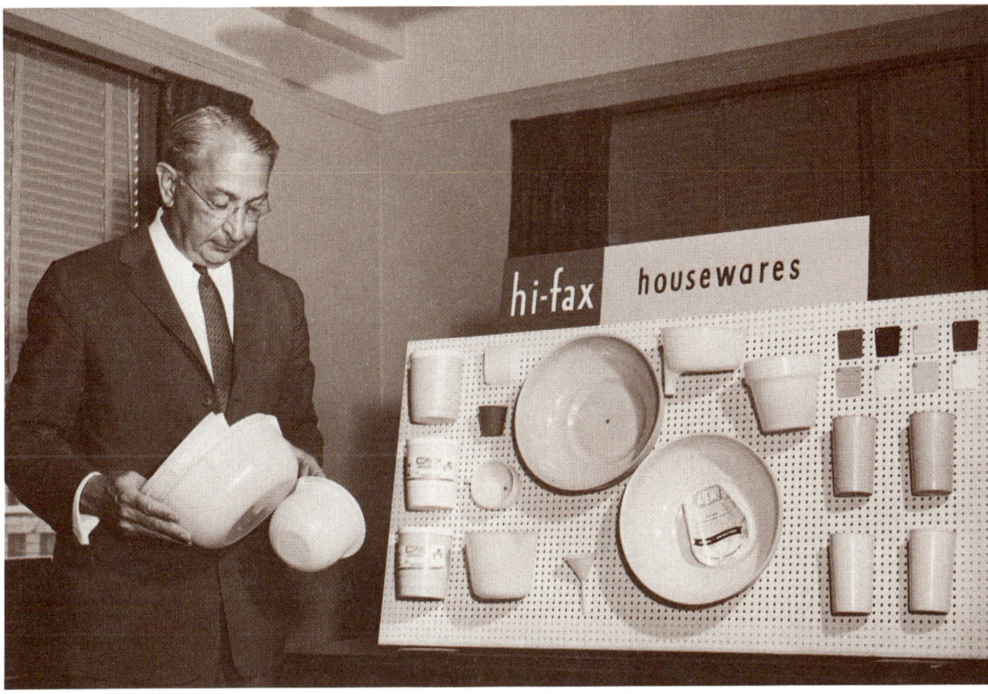

The first markets for Hi-Fax® polyethylene included kitchenware, as demonstrated by Elmer Hinner.

The first company to commercialize polypropylene, Hercules remained an industry leader for three decades, before selling the business to Montedison, S.p.A. in 1987. During that time, Hercules Pro-Fax® polypropylene resin was the industry's most prominent brand.

Above: Hercules moved quickly into downstream applications, retooling a synthetic fibers plant in 1960 to make polypropylene fibers for carpets and upholstery. Below: At Lake Charles, Louisiana, Hercules opened the world's largest polypropylene plant in 1961.

DMT, a precursor for polyester, proved a highly profitable business in the 1960s, but a disaster during the following decade.

Under President Thouron (above), Hercules poured investments into DMT in the United States and abroad.

When the Lake Charles plant came onstream in the spring of 1961 under the direction of plant manager Earp F. Jennings, it was the largest crystalline polypropylene plant in the world. Yet Hercules had little time to rest on its laurels. Almost immediately, the plant was doubled in size, bringing capacity up to 120 million pounds per year. Despite such investments, the company felt intense pressure from the competition. By 1961, its U.S. rivals included Phillips, Humble, AviSun, Tennessee Eastman, Dow, and Firestone, as well as resellers such as Du Pont, Monsanto, and Rexall. Between 1958 and 1961, the price of polypropylene plunged from 65¢ per pound to 42¢ per pound.[44]

Observing the situation, a writer for *Fortune* magazine speculated that "the only real answer to overcapacity appears to be to find new markets for upgraded products."[45] By the time these words were printed in 1961, Hercules was already immersed in a program to integrate forward into polypropylene fibers and film.

Forward into Fibers and Film

Hercules' increasing emphasis on polypropylene was accompanied by an aggressive strategy for developing new markets. Several factors strongly influenced the company's decision to integrate forward into fibers and film. First, obviously, the intense competitive rivalry in polyolefin resins made it difficult to earn profits: success in an environment of heavy capital investments and simultaneous price cuts could not be sustained indefinitely. Despite its early market leadership in crystalline polypropylene, Hercules was concerned about its long-term ability to maintain its position. Upgrading resins to fibers and film offered the prospect of increasing earnings. In addition, the company expected that developing downstream applications for polypropylene would also stimulate primary demand for its polypropylene resins.

Finally, the company's traditional reluctance to integrate forward lest it compete with customers was not an impediment in this instance. In the late 1950s, not only did few producers manufacture fibers and film from polyolefin resins, but Hercules' existing fibers and film customers that upgraded cellulose acetate were clearly suffering from the irreversible decline of their business. Between 1955 and 1956, Hercules' sales of cellulose acetate plunged more than 20 percent, resulting in a net loss of $209,000. The product was losing out in plastics applications to high-impact polystyrene; American Viscose, for example, reduced, then totally halted, its purchases for textile fibers at the end of 1957. After notifying its customers, Hercules first stopped making flake and molding powders, then ceased cellulose acetate production altogether in 1959. The company was able to salvage much of its equipment at Parlin for other operations and undertook the painful process of laying off hundreds of workers. It was an unhappy ending for a product in which Hercules had invested huge sums and great expectations.[46]

Given incentives—and lack of disincentives—to integrate forward in polyolefins, Hercules avidly pursued opportunities in fibers and film. The idea to move into polypropylene fibers apparently originated outside the company. In 1960, an executive from Burlington Mills, the textile and carpet manufacturer, called on Hinner and Johnstone soon after returning from an inspection tour of Montecatini's polypropylene operations in Italy. The executive explained that Montecatini was exploring the manufacture of polypropylene fiber, and he inquired whether Hercules might also be interested. "We said we don't make fiber," Hinner recalled, "[and that] you can't support a fiber operation on one product." The executive persisted, arguing that the market for polypropylene fiber was likely to reach one hundred million pounds in five or ten years. Moreover, he told Hercules, Industrial Rayon Corporation was willing to sell its plant at Covington, Virginia, which included equipment for making rayon and nylon fibers for industrial and heavy-duty applications and a small pilot plant for making polypropylene fiber.[47]

Hinner and Johnstone were intrigued and quickly followed up the lead. In the summer of 1960, they recommended that Hercules purchase the Industrial Rayon facility, a recommendation endorsed by the board in October, when it appropriated nearly $7.5 million for that purpose. Along with the assets, Hercules acquired technical and research personnel experienced in synthetic fibers, as well as a knowledgeable work force. Although Hercules knew that Industrial Rayon's existing equipment would be useless for making polypropylene fibers, the company also suspected that its employees would prove invaluable. The gamble paid off: within six months of the acquisition, Hercules began commercial production of Prolene® polypropylene multifilament fibers, a product targeted for textile applications.[48]

Hercules also established a Fiber Development Department with a two-year mission to "determine what are the strengths and weaknesses" of polypropylene fiber, as well as to find "outlets for it in our industrial economy." Initial studies confirmed the fiber's appealing properties, such as strength, tenacity (in which it compared favorably to nylon), light weight, and an extremely low density (which meant that lesser amounts would cover more area than fibers of higher density; for example, 70 percent more cotton would be required to make an equal amount of yardage). Subsequent research focused on methods of dyeing and pigment-coloring the fiber for use in textiles. Meanwhile, Prolene® showed early promise for applications in cordage and canvas goods, as well as industrial uses.[49]

The development of polypropylene film started earlier and culminated later than that of fiber. Hercules again resorted to a familiar tactic to speed the development process: it negotiated a technology-sharing agreement with European partners, ICI and Kalle (a film-producing subsidiary of Hoechst). In the late 1950s, Johnstone learned that the two producers were developing alternative approaches to manufacturing films; he arranged to exchange Hercules' know-how in polypropylene resins for its partners' know-how in film-forming processes. At that time, ICI employed the tubular ("bubble") process to make biaxially oriented polyester films. (The orientation of thermoplastics refers to the ordering or aligning of the long-chain molecules to impart clarity, stiffness, impact resistance, and, especially, strength. Fiber was oriented in a single direction; film in two directions—biaxially—by drawing through various processes. (See Chapter 1.)[50] ICI began by blowing hot air into a molten cylinder of plastic.

The resulting bubble was continuously expanded until it reached a specified thickness; then it was slit on two sides, flattened, and rolled into sheets. Kalle, on the other hand, relied upon the "drawing and tentering," or tenter, method: a thick sheet of plastic was extruded and then stretched in two directions to make polyester film.

After evaluating the two methods, Hercules chose to invest in ICI's tubular process, which was then capable of running at higher speeds and producing thinner films. In June 1959, at the urging of Hinner and Johnstone, the board appropriated $1.5 million to build an exact duplicate of ICI's three-hundred-ton-per-year film pilot plant at the Research Center. The company also launched a major study of the suitability of commercial-scale operations and likely applications for the product.[51]

From the outset, it was apparent that polypropylene film possessed a variety of desirable properties, including excellent transparency, gloss, and sparkle, low water and gas permeability, resistance to grease and oil, stability at high temperatures, high tensile strength, and durability, printability, and abrasion resistance. One of its most appealing features involved economics: pound for pound, polypropylene film provided a greater yield of coverage than any existing commercial film. (More than sixty-one thousand square inches of half-millimeter film could be fashioned out of a single pound of polypropylene.)[52]

Initially, the company concentrated on the market for shrink-wrapping phonograph records, toys, boxed apparel, and other products, hoping to displace films made of polyvinyl chloride. (Shrink-wrapping occurs in oriented films because of a phenomenon known as "plastic memory"—the tendency of the film's molecules to revert to their original, unoriented state when exposed to heat.) Soon, however, Hercules began pushing hard in a far more promising area: food packaging, for which it gained FDA approval in 1960.[53]

"In the year 1960," reported Werner Brown, new general manager of Cellulose Products, "our entry into the plastics business came of age." Sales of polyolefins doubled, turning losses into gains equivalent to 50¢ per share of Hercules stock before taxes. The following year, production of Pro-Fax® began at Lake Charles, and fiber and oriented-film operations began at Covington, where capacity was immediately doubled. "For the first time," reported Brown, "we enjoyed a substantial business in fine denier filament and film."[54]

In sum, less than a decade after its first steps into petrochemicals, Hercules had become a significant presence in the industry. Its phenol process was a source of steady license income and valuable oxychemicals; DMT achieved the happy twin results of high growth rates and high profitability; and Ziegler polyolefins, especially polypropylene, were growing rapidly and promising to open new markets.

VERTICAL GROWTH

Hercules' entry into petrochemicals resulted from a search for a new raw materials base to supplement what appeared to be limited opportunities for growth in the company's existing businesses. While it was exploring unfamiliar terrain, the company also sought to extend its range of products made from traditional raw materials, as well as to find new raw materials sources for its traditional products. Although the company pursued these aims across most of its product lines, its strategy of vertical integration was most evident in two areas: chemicals based on natural gas, and naval stores. In the first instance, Hercules integrated forward in the search for profitable markets for its upgraded chemicals; in the second, it integrated backward to ensure a steady source of raw materials.

New Products from Familiar Sources

During the first year of Forster's presidency, Hercules had enlarged its production of synthetic ammonia by expanding existing capacity (investments at the Hercules Works), forming a joint venture (Ketona Chemical Corporation), and acquisition (the purchase of Missouri Ordnance Works, quickly renamed the Missouri Chemical Works [MCW]). The principal motive for these actions was to secure raw materials and intermediate chemicals for the company's existing businesses: explosives, agricultural chemicals (fertilizers, fungicides, and pesticides such as Thanite and Metadelphene), paper chemicals (including Kymene® wet-strength resin), nitrocellulose, synthetic resins (such as Rosin Amine D, Rosin Amine D acetate, and Polyrad® ethylene oxide adduct), and other products.

There were other motives for these expansions. The acquisition of MCW, for example, was driven at least partly by Hercules' interest in processing hydrocarbons and broadening the range of its product offerings. MCW represented more than a feeder plant for Hercules' existing operations: it was a foundation for new business, a centrally located petrochemicals complex in its own right. Indeed, the economics of natural gas facilities such as MCW mandated nothing less than this expectation; for such a plant to operate efficiently, it had to produce far more output than Hercules could consume internally. As Forster put it, "This is the kind of situation where economy dictates that your plant must be bigger than any one user's needs for raw materials." Accordingly, Hercules sought to produce "building block" chemicals derived from natural gas, such as methanol, formaldehyde, and pentaerythritol (PE), which were consumed in a variety of industrial products.[55]

In 1956, Hercules started construction of a chemical complex adjacent to its thirty-eight-thousand-ton-per-year anhydrous ammonia plant at MCW. The expansion included new formaldehyde facilities and plants to make methanol (Hercules was the first U.S. plant to use the Swiss "Inventa" process) and PE. Most of the construction at the 548-acre site was completed by the end of 1957. Soon thereafter, the company's PE plant in Mansfield, Massachusetts (acquired in 1941), was closed and much of its equipment shipped to Missouri, where PE production began in December 1956.[56]

By the late 1950s, MCW was producing a variety of chemicals and intermediates. Ammonia, made from natural gas, steam, and air, was used primarily in fertilizers but was also consumed in explosives, acids, and refrigerants. Methanol (made from natural gas, steam, and carbon dioxide), was used in antifreeze and was a primary precursor for formaldehyde used in plastics. Formaldehyde, in turn, when reacted with acetaldehyde in the presence of lime, yielded PE. All of these products were in healthy demand both inside and outside Hercules, but the brightest star—the product MCW was oriented toward—was PE.[57]

Hercules had begun making PE in 1941 with the acquisition of the John D. Lewis plants for the Synthetics Department (see Chapter 8). Sales of Hercules PE and PE derivatives to makers of cements, chewing gum, floor coverings, explosives, paint and varnish, adhesives, plastics, printing inks, and other products grew steadily, and, after the war, rapidly. Between 1949 and 1955, PE sales grew tenfold, from 1.3 to 13.2 million pounds per year. Demand for PE-based resins in plastics was particularly strong, soaring 70 percent from 1952 to 1955 and offering "little indication of slackening." The new plant at MCW effectively doubled the company's capacity.[58]

MCW's early success encouraged further expansion. In 1961, Hercules added new facilities to make methanol and formaldehyde at the Hercules Works in California. This move not only extended the company's geographical market coverage for natural gas–based commodities, but it also helped support the development of a promising new fertilizer, UN-32, a product made from ammonium nitrate and urea formaldehyde.[59]

Although some experts viewed Hercules' expansion plans with skepticism and predicted a future of overcapacity and depressed prices, demand for most of Hercules' natural gas derivatives kept up with supply during the late 1950s and early 1960s. Equally important to Hercules, however, the new plants supplied the key chemical intermediaries consumed by many of its departments and did so with cheap and abundant raw materials.

Familiar Products from New Sources

Hercules also pursued a strategy of vertical integration in its naval stores product lines. The company was concerned about the cost and availability of pine stumps and began to integrate backward to ensure a steady supply of raw materials.

In the mid-1950s, shortages of naval stores raw materials remained a continuing concern. The failure of the pilot extraction plant in Klamath Falls, Oregon, only accentuated the problem. Hercules reconsidered an option it had always discarded before: producing rosin from tall oil, a waste by-product of the kraft (sulfate) papermaking process. This time around, however, it appeared that technological developments in tall oil distillation and the spread of sulfate papermaking could at last make the process economical.

The term *tall oil* (an American adaptation of *tallolja,* the Swedish word for pine oil) is often used to describe one of several related compounds: crude tall oil, tall oil soap, acid-refined tall oil, distilled tall oil, and tall oil rosin, fatty acids, heads, and pitch. The raw material considered by Hercules was actually tall oil soap, a product in abundant supply at paper mills. In the kraft process, pulp is cooked in a solution containing sodium hydroxide and sodium sulfide to separate the cellulose from impurities. The spent liquor is partially concentrated and on cooling the tall oil soaps separate and are skimmed off. Tall oil soap is composed of resin acids and fatty acids, of varying proportions, and a small percentage of unsaponified matter. When treated with acid, tall oil soap yields crude tall oil.[60]

Tall oil was discovered and commercialized in Sweden around the turn of the twentieth century. By the 1930s, continuous-process plants were operating in Sweden and Finland and shipments began to reach American shores, where the "liquid rosin" (as it was sometimes called) was used in inks, soaps, insecticides, and waterproof board, often replacing rosin oil or pine tar. A few American companies set up recovery units, the first being the West Virginia Pulp and Paper Company (later Westvaco).[61]

Hercules researchers had first investigated tall oil during the 1930s, but it found tall oil rosin "more or less uninteresting" because of impurities (including a foul smell) in the product.[62] After World War II, however, an important breakthrough was made: importing distillation techniques from the petroleum industry, Arizona Chemical Company (a joint venture between American Cyanamid and International Paper) "demonstrated the first successful fractionation of tall oil into high purity fatty acids and rosin on a commercial basis." During the early 1950s, tall oil became firmly established alongside gum and wood rosin as a key naval stores raw material and appeared to be a particularly attractive source for making rosin size.[63] (See Exhibit 10.2.)

At the same time, the increase of kraft papermaking in the South was making tall oil more abundant. Moreover, its supply—again, because it paralleled sulfate papermaking—grew steadily and reliably, in contrast to the yearly swings that plagued most sources of animal and vegetable oil and naval stores materials. Finally, because it contains fatty acids as well as rosin, tall oil appeared to offer new commercial possibilities. Although some observers considered fatty acids an "entirely new" business for Hercules, most common applications for both fatty acids and rosin acids were in markets the company had long served, such as inks, coatings, rubber, and detergents.[64]

In the spring of 1955, Hercules committed to the distillation of tall oil and entered negotiations with Southern paper mills for supplies of tall oil soap. The company arranged long-term supply contracts, agreeing to pay as much as $30 per ton for the by-product and to split profits on outside sales. Meanwhile, the Engineering Department was charged with the task of developing or locating tall oil distillation technology. Perhaps because the fractionation of tall oil was a difficult and delicate process (involving compounds that had high-temperature boiling points and degraded rapidly in the presence of oxygen), the department opted to follow a well-worn path: importing technology from Europe.

After an extensive search, the Engineering Department identified a new process developed by Ake Linder, a Swede regarded as one of the fathers of tall oil fractionation. Linder claimed to have developed a new column for separating the sensitive compounds under relatively mild conditions. In his "Linder tower," tall oil followed a convoluted path through a series of trays, separating into fatty acids at the top of the giant column and into rosin acids at the bottom. Hercules

EXHIBIT 10.2 U.S. Rosin Production, 1900–1965

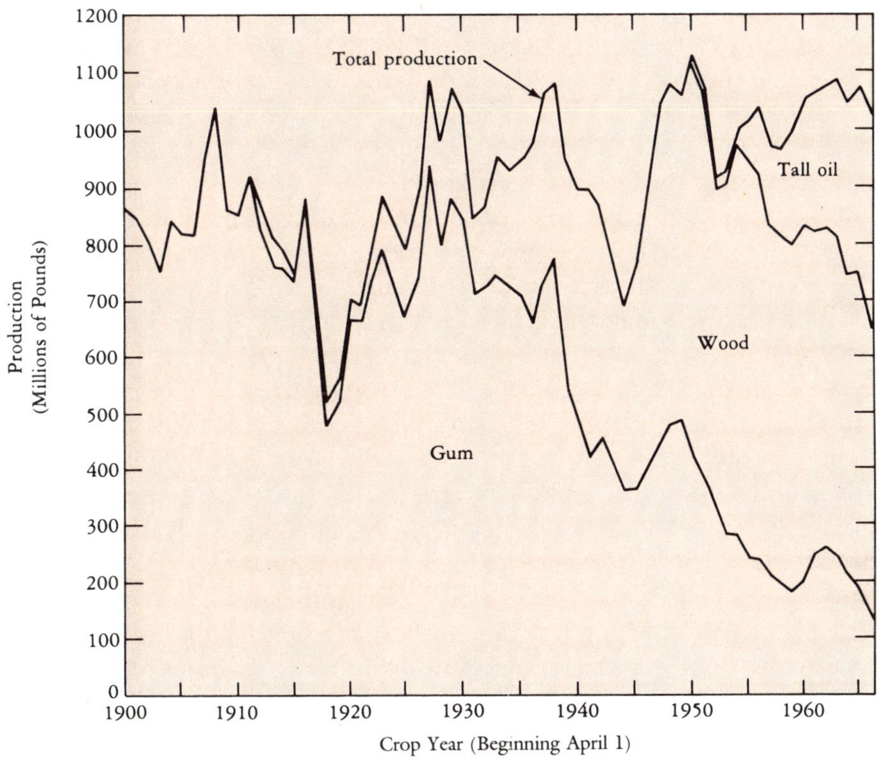

Source: Herman I. Enos, Jr., et al., "Rosin and Rosin Derivatives," in *Kirk-Othmer Encyclopedia of Chemical Technology* (New York: John Wiley & Sons, 1968), XVIII, 480.

paid 500,000 Swedish crowns (approximately $96,750) for an "irrevocable, nonexclusive, royalty-free license" to Linder's process and the consulting expertise of its renowned founder.[65]

In August 1955, the board authorized nearly $3.8 million for the construction of a tall oil refinery at its paper chemicals plant in Savannah, Georgia, an operation sited near major customers in the paper industry. At the same time, a Linder pilot column was erected at Brunswick. Early test results were discouraging: Hercules' engineers found wide pressure differentials across the pilot column, a circumstance that resulted in impure products that decomposed.[66]

In 1956, however, Hercules engineer Donald Boynton redesigned the pilot column to make it operable under low pressure across the column. Within months, Hercules had tall oil plants operating at Savannah and Franklin, Virginia (another paper chemicals facility). Each plant featured five large-diameter, stainless-steel Linder fractional-distillation towers—from which the trays were removed. (The columns were repacked to approximate distillation apparatus with which Hercules was familiar.)[67]

Hercules called its new fatty acids PAMAK and funneled most of the material through its Synthetics and PMC departments, where it was transformed into ingredients for protective coatings, printing, rubber, gasoline, detergents, and agricultural chemicals and in flotation separation processes. Most of the resin acids (Pamite® synthetic resin) were converted by the PMC Department into size and other rosin products. They were also combined with polyhydric alcohols (PE and glycol) and acids or anhydrides to make alkyd resins for many uses.[68] Although PAMAK rated near the top of the Pine and Paper Chemicals Department's product list in quantities sold by 1961, it was also the only product in the new department to lose money.

By getting involved in the distillation of tall oil, Hercules repositioned its naval stores raw materials base for the future. Not surprisingly, tall oil rosin continued to grow in importance, and Hercules became its leading processor. In 1962, tall oil production in the United States roughly equaled that of gum rosin; in three years, it was 45 percent higher than gum rosin production. Tall oil was already well on the way to replacing tree stumps as the preeminent source of naval stores raw materials in the United States, just as stumps had supplanted living pines in the 1930s.[69]

Despite its new tall oil plants, Hercules continued to experience difficulty in meeting its own raw materials needs for naval stores derivatives. In the late 1950s and early 1960s, for example, the company was forced to buy large quantities of gum rosin, steam-distilled rosin, and pinene.[70] Subsequently, the company took several steps to ensure a steady supply of naval stores raw materials: it formed a joint venture (Pine Gum Production Company) to investigate production of gum rosin; it undertook a major expansion of the Brunswick plant and resumed production of FF rosin; and it built or participated in the construction of several naval stores extraction plants overseas.[71]

These actions helped ease critical raw materials shortages for the company's naval stores, paper chemicals, and synthetic resins businesses. The strategy of backward vertical integration proved effective, just as forward integration had helped expand the company's lines based on natural gas. In other businesses, Hercules sought to extend its product lines and augment sales by acquisition and joint venture.

Partners and Purchases

For a quarter-century after its purchase of Paper Makers Chemical Corporation in 1931, Hercules made no acquisitions of note.[72] Under Forster, Hercules usually sought every opportunity to grow, including partnerships in and purchases of new businesses. Between 1956 and 1961, Hercules formed three joint ventures and completed five acquisitions.

In making these moves, Hercules was not driven by the imperative to diversify. On the contrary, as Forster put it, "Any new endeavor must be logical."[73] However much the transactions differed in scale, scope, and intention, all shared a common characteristic: they were in fields closely related to Hercules' existing lines of business. Along the way, Hercules learned several key lessons about its capabilities, as well as its obligations in bringing new partners into the fold.

Of the company's joint ventures, one proved insignificant, another stable and enduring, and the third a failure that taught a valuable lesson. The insignificant transaction was Pine Gum Production Company (1960), a brief exploration of opportunities to produce gum rosin as an alternative source for several Hercules departments. The venture was soon terminated as other raw materials sources became available. The stable and enduring joint venture was Texas Alkyls Corporation, a fifty-fifty partnership between Hercules and Stauffer Chemical. Formed in 1958 to produce aluminum trialkyls—catalysts used in the Ziegler polymerization reaction—Texas Alkyls successfully provided catalysts to Hercules and other plastics-makers that relied on Ziegler's processes.[74]

The third joint venture, Hawthorn Chemical Company, owned in equal shares by Hercules and ICI, was the most ambitious and the briefest. Organized in 1956, Hawthorn was established to manufacture methyl methacrylate (the ester of methyacrylic acid), which was used to make clear, glasslike sheets of plastic.

In the late 1920s and early 1930s, the monomer of methyl methacrylate had been discovered by ICI, and the polymer by Rohm and Haas and Du Pont. The polymer had attractive properties. In particular, it could be fashioned into clear, hard, flexible sheets that substituted for glass in many (especially hazardous) applications. Marketed as Plexiglas by Rohm and Haas and as Lucite by Du Pont, the plastic generated considerable demand during World War II as it replaced glass in military aircraft (when pierced by a bullet, the plastic simply left a hole instead of shattering). To Du Pont, Lucite was one of many plastics; to Rohm and Haas, however, Plexiglas was a major new business and a means of growing from a small specialty chemicals company into a leader in acrylics. With its future at stake, the smaller company outraced the chemical giant in product and process improvements, as well as in plant and equipment investments. With these tactics, as well as its willingness to price aggressively, Rohm and Haas effectively prevented Du Pont from earning profits in the business.[75]

Hercules and ICI might well have taken a lesson from Du Pont's experience. In the spring of 1956, the two companies began discussing a possible joint venture to make methyl methacrylate. Their motives were straightforward: on ICI's part, it had expertise in the monomer and no longer (after 1950) had a cooperative agreement with Du Pont to stay out of the U.S. market; Hercules saw another opportunity to sell the natural gas–based derivatives used in making the product and to generally expand its plastics business in partnership with one of the world's largest producers. By summer, the deal was concluded, and the partnership soon announced plans to construct an $11 million, thirty-five-million-pound-per-year plant adjacent to MCW. In January 1957, Richard T. Yates, formerly of the Naval Stores Department, was named president of the new corporation.[76]

Ten months later, however, Hercules pulled out of the venture and Hawthorne Chemical Company was disbanded.[77] The reason was partly related to the market, which displayed signs of maturity rather than of growth, as well as to technical difficulties. But the closure may also have been influenced by competitors' responses: price signaling and capacity expansion announcements helped drive Hercules and ICI from the field.[78]

Hercules had better luck in fields closer to home, such as the Texas Alkyls joint venture and a series of acquisitions related to the company's existing operations. Two small transactions, for example, helped extend the company's growing business in agricultural chemicals. Nitroform Agricultural Chemical Company of Woonsocket, Rhode Island, purchased in 1960 for about $500,000, was in the fertilizer business. As a maker of urea formaldehyde fertilizer compounds (known particularly for its Powder Blue and Blue Chip brands), Nitroform subsequently drew on the commodities made at the expanded MCW, while serving markets through long-established ammonium nitrate fertilizer channels.[79]

A year later, similar reasoning lay behind Hercules' acquisition of Reasor-Hill Corporation of Jacksonville, Arkansas, for approximately $100,000. Reasor-Hill operated a plant in Jacksonville that manufactured 2,4-dichlorophenoxyacetic acid (2,4-D), 2,4,5-tichlorophenoxyacetic acid (2,4,5-T), and other herbicides. Just as DDT, toxaphene, and other agricultural chemicals had transformed the science of insect control after the war, 2,4-D was the first profoundly successful herbicide; it initiated a "revolution in chemical weed control" in the 1940s and 1950s. Developed secretly in World War II and commercialized by several companies thereafter, 2,4-D achieved sales of thirty-six million pounds by 1960 and spurred the development and use of other weed control chemicals. Citing the value of its existing channels of distribution to formulators, Hercules managers saw the acquisition of Reasor-Hill as a logical step: "As long as we're going to call on [the same customers] anyway," explained Donald Sheffield, "we might as well sell this, too."[80]

Similar reasoning lay behind larger acquisitions of the era: Huron Milling Company (1956) and Imperial Color Chemical and Paper Company (1960). In each case, the company used the purchase to broaden its existing position in major markets.

Food Starches and Proteins

The first large acquisition of the period—Huron Milling Company, a wheat flour processor based in Harbor Beach, Michigan—was made for several reasons. First, Hercules sought to supplement its sales of cellulose and rosin derivatives in traditional markets such as food additives, paper, and adhesives. Second, as Ed Crum, general manager of the Virginia Cellulose Department reasoned, "Starch has the same molecular configuration as cellulose, and we wanted to apply some of our chemistry to it." With Hercules' know-how, the argument ran, new products and low-cost organic chemicals could be produced from new raw materials. Finally, Huron Milling's manufacturing process paralleled that of Hercules' CMC operations, especially in the mixing, handling, and packing stages. In short, Huron Milling appeared to be a close fit in terms of technology, manufacturing, and markets.[81]

Huron Milling had started as a flour mill in 1876. Two decades later, it set a new course when it hired a graduate in chemistry from Michigan State named William L. Rossman and established a small research lab. In the next half-century, the company became a leader in the art and science of separating wheat into its major components—gluten and starch—and from these, formulating a variety of edible and nonedible additives useful in several industries.[82]

By 1956, Huron Milling employed five hundred people and had annual sales of $12 million. (Among its employees were the mayor of Harbor Beach, its school board president, and other town officials.) The company's main starch products included Red Stave laundry starch, edible food starch, wallpaper paste, and oil well–drilling mud starch. On the gluten side, Huron Milling marketed food flavorings and enhancers such as monosodium glutamate (MSG) and hydrolyzed vegetable protein (HVP), medicinal glutamic acid, and enriched baking proteins. These intermediary products were sold to makers of food flavorings, breakfast cereals, baked goods, starch, and adhesives.

Huron Milling first came to the attention of Hercules through Leland Burt, supervisor of product development in the Virginia Cellulose Department, and interest soon grew among Ed Crum, Dr. Richard E. Chaddock (director of development in the department), and Paul Mayfield (general manager of Naval Stores). In August 1956, these men convinced the Hercules board to spend as much as $5 million to buy the company, issuing one hundred thousand new shares of common stock for the purpose. On December 3, 1956, Hercules exchanged the shares (then valued at approximately $3.8 million) for Huron Milling's assets.[83]

Once in control, Hercules sought (as it had in earlier acquisitions) to maintain the operating integrity of the plant while integrating it into the Hercules "family" of plants. Before the deal was consummated, Hercules' board voted generous retirement pensions for Huron Milling's top executives and retained its former president, Robert Farr, as a consultant. Carl Smith, a vice president of manufacturing and research, was promoted to plant manager. Remaining senior management positions at Huron Milling were filled by personnel from the Virginia Cellulose Department, including Charles Grant as head of sales. At the end of the year, Hercules extended its benefits package to employees at Harbor Beach.[84]

Despite the logic of the transaction and Hercules' care in assuming control, the new unit's financial performance was mixed at best. Hercules had held high hopes for MSG, for example, but discovered shortly after the acquisition that growth of the market was slowing, and that Huron Milling was vulnerable to competition. In 1958, Hercules began selling a new vital gluten product for bakeries, Vicrum, which proved popular in the marketplace and partly offset the slump in MSG. On the whole, during Huron's first four years under Hercules, the proteins proved more profitable than the higher bulk starches; in 1961, proteins earned slightly more than $250,000, and starches lost more than $600,000.[85]

Pigments and Wallpaper

Hercules applied similar reasoning, and achieved similar mixed results, when it acquired Imperial Color Chemical & Paper Corporation of Glens Falls, New

York, in 1960. At that time, Imperial was the world's largest producer of chemical pigment colors, with annual sales exceeding $27 million and net income of nearly $2 million. To acquire its assets, Hercules exchanged 158,807 shares of convertible stock and 315,722 of common stock for the assets of the company. The deal was closed on April Fools' Day of 1960.[86]

Imperial brought with it three plants as well as sales offices and warehouses throughout North America.[87] The largest factory—70 acres, 1,000 workers, and 150 laboratory professionals at Glens Falls—constituted the unit's core business, its Pigment Color and Chemical Division. The facility supplied color pigments for makers of a wide range of high- and low-value products: paint, printing ink, paper, floor covering, carbon paper, rubber, plastics, leather, textiles, crayons, and roofing granules. Imperial also produced the same products for distribution throughout Canada at a wholly owned subsidiary in St. Johns, Quebec. (Many by-products and commonly used raw materials were also manufactured at the two plants: sodium sulfate, vanadium oxides, sodium bichromate, dichlorobenzidine, sodium ferrocyanide, and tobias acid by the chemical division; lead chemicals, basic acetate, and carbonate by the color division.)

Imperial's second line of business was wallpaper, including branded lines such as Imperial washable wallpaper, Glendura fabric wallcovering, and E-Z DU, a fully trimmed, prepasted wallpaper targeted at the do-it-yourself market. These products were manufactured at a fully integrated paper plant at Plattsburgh, New York. Located along the Saranac River, which provided power and carried in logs, the plant employed about five hundred people and possessed somewhat antiquated technology.

In considering the acquisition, Hercules was interested primarily in the Pigment Color and Chemical Division, whose products shared distribution channels with many Hercules products. Both companies had cultivated long, close relationships with makers of rubber, plastic, paint, ink, and textiles and were known for their strong technical support and service. In addition, Hercules sought to employ Imperial pigment technology in coloring its own polypropylene fibers. Finally, Hercules was impressed by Imperial's apparently sound financial condition, as well as by the investment opportunity. With its "good solid group of northern New York state people running it (our type of folks), solid balance sheet, good earnings, maybe some growth," Imperial appealed to even the most conservative members of the board.[88]

Although Imperial appeared to fit well with Hercules, there were few commonalities on the manufacturing side. Indeed, Imperial's technology was virtually new to Hercules—and a bit mystifying. The Glens Falls plant, for example, required eight hundred raw materials to create a thousand distinct colors. The process of precipitating, washing, filter pressing, drying, and grinding the pigments to achieve a perfect color match, more akin to an ancient handicraft than a modern chemical process, was "an art which cannot be trusted to the cold, impersonal mechanism of an instrument control board." And although Hercules was familiar with the papermaking side of the business, from the outset it planned to sell the Plattsburgh wallpaper operation.

Before splitting up Imperial, Hercules took steps to integrate it into its new parent. Although Hercules recognized the increasing difficulties of maintaining a family spirit in a period of increasing diversity and growth via acquisition, it nonetheless made the attempt. Richard B. Douglas of Hercules' Public Relations Department described the general approach:

> One of the first things we would do when we consummated an acquisition, would be to go up and talk to plant people . . . put on a little show for them, show them what we were, "This is Hercules." We would take a movie [and] maybe two or three vice presidents; hold a big meeting; go around and shake hands with everybody; let them know that we were a real company; put them on the distribution list for things such as the *Mixer*; try to tie them in as rapidly as we could to the company and make them a part of the company.[89]

Such gestures were meant to soften the inevitable difficulties of the transition. Indeed, soon after the transaction, Imperial was visited by a delegation from the Home Office, including President Forster, who spoke of "a strong feeling of close family relationship" that had long characterized the two companies.

Given Imperial's size, the new unit was not folded into an existing organization but was made a separate operating department. Key top managers of Imperial, including its president, Arthur F. Brown, and vice presidents Albert L. Emerson and Alfred E. Van Wirt, were retained, with Brown serving as general manager and as a member of Hercules' board of directors.

Once again, however, the integration of a new acquisition proved difficult. Imperial's first full year of operation under Hercules, according to Van Wirt, was "rather turbulent."[90] Some of the strain, he continued, was caused by steps taken to "fit our fiscal and functional policies into the Hercules mold." In particular, Van Wirt believed that "records, systems and procedures," even employee benefits, were converted "to the Hercules pattern" too quickly.

Much of the stress resulted from personnel cutbacks in the Wallpaper Division (25 percent in 1961) and, beginning in July, highly visible efforts to sell the business. Demand for wallpaper had declined drastically after the war (sales were at 25 percent of their postwar level), when painting became fashionable. "Certainly," admitted Van Wirt, "our wallpaper business does not present a promising growth or profit picture as a diversification for a chemical company." The business was finally sold in 1962, when investors from Cleveland purchased the Plattsburgh operation through a leveraged buyout.

Hercules soon learned, moreover, that many of Imperial's pigments were "showing signs of . . . maturing." New cadmiums and mercadiums were needed to replace lead chromates and other pigments, and while Imperial's strong products were based upon older chrome yellows, molybate oranges, and iron blues, the department was finding it "difficult to develop a significant quality advantage any more." Its work on the coloring of polypropylene fibers also proved complex and arduous: "We are really sweating this one out," admitted Van Wirt.[91]

In 1961, it was too early to pass judgment on the joint ventures and acquisitions of the Forster era. In the best cases, some product lines of the new partners generated immediate profits that allowed Hercules the luxury of gaining its bearings without drawing down its own resources; in other instances, Hercules either saw quickly that its acquisitions were in poorer condition than expected—reminiscent of its early experiences with Yaryan, Aetna, and PMC—or simply foundered for a while as it learned the new business.

But in each case, Hercules attempted to soften the shock of acquisition on both sides before methodically instituting its own reporting systems, engineering practices, employee benefits, and other policies, hoping to both understand and improve the new business. The addition of new businesses made Hercules larger and more complex, and therefore more difficult to manage, a potential problem tempered by Forster's insistence on pursuing related diversification.

Entering the Space Race

One of the most striking areas of growth under President Forster involved rocketry. Indeed, in the late 1950s, Hercules made great strides in large rocket motors for ballistic missiles and space exploration. The achievements of this era formed the foundation of what would become a major business for Hercules in its third quarter-century.

The sudden and dramatic surge in Hercules' involvement in rocketry came about for several reasons. One was defensive: chemical propulsion grew out of explosives, a business that declined sharply after the mid-1950s. Hercules needed to replace the lost revenues, if possible by utilizing its existing plants and personnel. Second, important changes in the geopolitical setting—intensification of the Cold War, a more aggressive, technology-based foreign policy under President John F. Kennedy, and the race for the moon—wrought an unprecedented demand for large military and space rockets, a demand that Hercules was well positioned to satisfy.

Finally, the rise of large-scale rocketry at Hercules was a result of technological breakthroughs, managerial initiative, and no small amount of bravado and luck. That Hercules had a firm foundation in solid propellants and small rocketry hardly preordained its ultimate role as a leading government contractor for the large rocket motors that power space vehicles and intermediate-range and intercontinental ballistic missiles.

Twilight of Commercial Explosives

In the mid-1950s, Hercules' Explosives Department faced a crisis: the rapid substitution of ammonium nitrate–fuel oil (ANFO) mixtures for traditional nitroglycerin explosives such as dynamite (see Exhibit 10.3). The problem was especially acute in open-pit mining and quarrying operations, where ANFO mixtures provided ample blasting force at a bare fraction of the cost of dynamite. Although Hercules manufactured ammonium nitrate and continued to supply detonators for ANFO explosives, low prices and fierce competition for ammonium nitrate and low sales volume for detonators made for sharply declining sales and falling profits.[92]

EXHIBIT 10.3 U.S. Industrial Explosives—Market Distribution by Explosive Type, 1915–1978

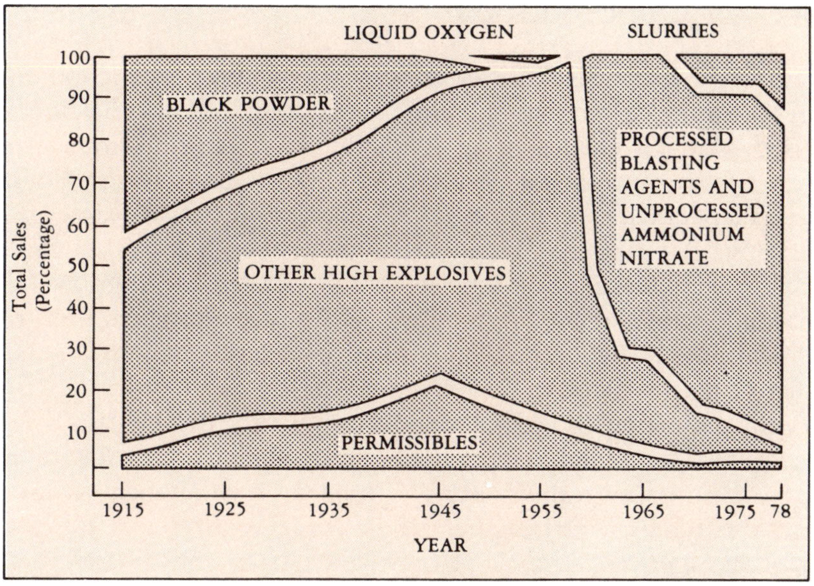

Source: Robert B. Hopler, "The History of Explosives," *Explosives Engineer,* Supplement (January 1980), 3.

Other factors threatened to worsen the crisis. In the late 1950s, Standard Oil, U.S. Steel, and Standard Pacific built ammonium nitrate plants in the West, contributing to "very severe" competition in the region. And as other explosives-makers, such as Atlas, were hit by the impact of do-it-yourself explosives, they reduced their purchases of explosives ingredients from Hercules.[93]

Such problems produced inevitable results. On Friday, January 6, 1956, the dynamite plant at the Hercules Works, after churning out some 561 million pounds in 43 years, permanently closed its doors. Some personnel and equipment were transferred to Bacchus, Utah, site of a concurrent renovation. The encroachment of residential neighborhoods was partly responsible for the Hercules Works shutdown. Nonetheless, a trend was forming: in 1961, Hercules' dynamite line at Ishpeming, Michigan, was closed. When Kennecott Copper announced its conversion to ANFO in the same year, the Bacchus Works lost 60 percent of its market and began the painful process of closing its dynamite lines. Eight explosives sales districts were consolidated into four regional offices. Between 1955 and 1960, Hercules' profits in explosives plummeted by nearly half, and the blasting cap business fell seriously into the red. Although it would continue to manufacture commercial explosives for several more decades, Hercules was playing out an endgame in its original business.[94]

The waning of commercial explosives, however, was more than offset by exciting new applications of explosives technology. At the very moment that sharp decline set in, changes in national defense policy provided enormous opportunities in large-scale rocketry.

Skyrocketing Opportunities

On Friday, October 4, 1957, one of the pivotal events of the second half of the twentieth century occurred: the launching of *Sputnik I,* the world's first satellite, into orbit by the Soviet Union. In the United States, the significance of the launch was not so much strategic—although such factors were a concern—as psychological: *Sputnik* represented a crisis of confidence, a shattering of the assumption that only a free and democratic society could achieve such a technological milestone. The resulting acceleration of U.S. space programs, including the formation of the National Aeronautics and Space Administration (NASA), was soon accompanied by rising investment in defense programs generally, as well as increasing public expenditures on education and technology.[95]

As both a cause and an effect of these trends, "the middle 1950s," observes historian Walter McDougall, "were the most dynamic and imaginative years in the history of American military R&D." Indeed, as McDougall demonstrates, the era was extremely productive: "Every space booster and every strategic missile in the American arsenal, prior to the Space Shuttle and the Trident submarine-launched ballistic missile (SLBM) of the 1970s, date from these years."[96] It was in this context that Hercules made its first large commitments to long-range rocketry.

Hercules already possessed more than a decade of experience as a producer of solid-fuel rocket propellant. At Allegany Ballistics Laboratory (ABL), which Hercules operated for the U.S. Navy, the company had developed a process for casting the large grains of double-based propellant used to power tactical rockets and missiles such as the Nike and Honest John (see Chapter 9). ABL was continually searching for ways to improve other characteristics of propellants; to increase impulse (power) and modulus (stiffness), for example, ingredients such as aluminum and ammonium perchlorate were added to the double-based mixture. Hercules also pioneered methods of bonding propellant to the rocket motor case using energetic material, thereby substantially improving the motor's efficiency.[97]

At the same time, Hercules worked closely at ABL with other contractors on advanced types of motor cases. Among these specialists was Richard Young, a veteran test pilot and missile and aircraft designer who was also founder and owner of Young Development Laboratories of Rocky Hill, New Jersey. In the late 1940s, Young was obsessed with the problem of strength-to-weight ratios in the constant search for stronger and lighter materials for rocket motor cases. In the course of his work, he began advocating the use of pressure vessels made of fiberglass because they offered superior strength to conventional steel cases and provided significant savings in weight.[98]

The essential challenge in building better rocket motor cases, as Young saw it, was only partly dependent upon the selection of construction materials. It also involved determining a shape for the pressure vessel that would take advantage of the enormous linear tensile strength of glass filaments (roughly 300,000 psi), as well as finding a weaving pattern that would create a vessel of uniform strength and integrity. In particular, Young learned from earlier work on pressure vessel design that "one of the major problems had been the retention of end closures." This led him to conclude that "the ends [of a pressure vessel] should if possible be retained by fibers in pure tension." After discarding a number of ideas, he later explained, it became evident that "winding fibers along helical paths along a cylindrical surface and reversing these along a spherical end zone would retain end closure while stressing the filament in tension only." Indeed, he concluded, "the load in each fiber could be constant whether generated by end retention loads or radial pressure within the cylinder."

The system that emerged involved rotating horizontally (as if on a giant rotisserie) a large cylindrical mandrel with rounded ends while layering on, back and forth, a narrow strip of glass filaments coated with a liquid resin. By adjusting the tension around the curves of the end domes, a crosshatched pattern would create continuous, uniform strength throughout the vessel. Once wound, the structure was cured to bond the filaments with the thermosetting plastic resin, and the mandrel was removed from inside the rigid new case.[99]

Soon Young was fabricating vessels that could withstand well over one hundred thousand pounds of pressure per square inch. But more impressive was their strength-to-weight ratio, or "mass fraction." Approximately 80 percent glass by weight and 60 percent by volume, the cases were as light as magnesium and as strong as steel; in short, they boasted the highest strength-to-weight ratio of any contemporary material.[100]

In the mid-1950s, cases made of Young's filament-wound Spiralloy materials proved successful in a series of tests with small rockets and missiles designed and developed at ABL.

The Birth of Hercules Chemical Propulsion

In March 1958, well aware of the crisis in commercial explosives and new opportunities in rocketry, Hercules announced a major change in the strategy and organization of its Explosives Department: the company would move beyond its traditional position as a GOCO contractor and assume "a new role as an independent producer" for military and civilian space programs. Accordingly, the Explosives Department established a new chemical propulsion division under Fred M. Hakenjos, with ABL's Dr. Lyman Bonner as director of development. The new unit's mission was to "design, develop, produce, and sell propulsion units for applications ranging from missiles and space vehicles to small, compact, packaged power units," and to manage smokeless powder production and sales. That same month, the company speeded expansion of ballistic facilities at Kenvil and undertook a $12 million chemical propulsion construction program at Bacchus. With ample acreage, an underutilized dynamite plant, and modern technology for making nitroglycerin, the site seemed ideal.[101]

These steps were the prelude to a calculated gamble. According to those who attended the company's two-day chemical propulsion conference that

March, Hercules was preparing itself to make long-range rockets—indeed, any missile or space vehicle "now contemplated," including intercontinental ballistic missiles (ICBMs) and moon rockets. Jack Martin, head of the Explosives Department, succinctly summarized the prevailing view when he instructed everyone at the conference to "think big."[102]

Later that year, Hercules learned that Aerojet General, a competitor that had recently won the contract to build the submarine-launched Polaris ICBM, was interested in acquiring Young Development Laboratories. By then, other manufacturers, including Kellogg, Goodyear, and Brunswick Corporation, had also shown interest in lightweight pressure vehicles. Although the competitive arena suddenly seemed crowded, several Hercules technical experts, including Dick Winer of ABL, saw the availability of Young Development as an opportunity not to be missed. Accordingly, Winer and sympathetic colleagues lobbied hard for Hercules to link its fortunes in large rocketry to filament-winding and to acquire Young's company. Late in the year, the executive committtee agreed, appropriating $300,000 for the purchase, which, once decided on, moved swiftly. "They called me down one morning at 10 o'clock," recollects Dick Young, "and the deal was done by the end of the day."[103]

Young Development continued under its founder's direction as a division within the Explosives Department. Although nearly all of its filament-wound structures were used by the military, Hercules hoped also to develop commercial business for Spiralloy. Theoretically, virtually any symmetrical shape without concavities—from tubes and sheets to vessels of many shapes—could be wound and cured for uses in which strength and lightness were required. Telephone poles and cherry-picker booms, automobile and boat parts, sporting goods and furniture, chemical equipment and containers, and structural building materials were among the many applications first considered.[104]

Excited by these possibilities, as well as the potential in rocketry (for not only cases but also nozzles, stabilizers, and tailfins), Hercules soon launched an ambitious expansion program. Within a year, the number of employees at Rocky Hill jumped from about seventy-five to three hundred, and square footage from eleven thousand to nearly thirty-four thousand.[105]

Although the Young Development acquisition was relatively modest for Hercules, the transaction was a key turning point in the history of the company. Filament-wound rocket cases proved to be an indispensable component in Hercules' successful bids to supply stages of the Minuteman and Polaris missiles, programs that helped launch Hercules into the world of large-scale rocketry.

Minuteman and Polaris

As a quantum leap in rocket technology and one of America's original ICBM programs, the Minuteman had a genesis as complex as any capital-intensive, politically sensitive government project. To understand Hercules' place in the program—how it was earned and what it represented—it is necessary to trace the roots of the Minuteman program as a technological and political phenomenon.[106]

In 1954, the year after the Soviet Union first detonated a hydrogen bomb, the United States initiated a program to develop missiles that could carry an atomic warhead from its western plains to the heart of the Soviet Union. At that time, the only rockets capable of such a feat were powered by liquid-fuel motors, and most development work focused on huge rockets such as the Titan and Atlas. Most experts believed that solid-fuel rocket motors were simply too small for the task. But some military personnel, including Col. Edward N. Hall of the U.S. Air Force's Western Development Division, were impressed with the research at ABL and other propulsion laboratories on solid-fuel motors that had high power and efficiency and were encased in lightweight materials. Hall believed that a solid-propellant ICBM might have significant advantages as a strategic weapon. In particular, liquid-fuel rockets required "huge trucks, filled with fuming, ultra-cold oxidants and volatile compounds of fuel," making for slow and dangerous fueling and preventing concealment in underground silos or easy mobility by truck or rail (which would permit a "shell game" defensive strategy).

In December 1954, Hall invited in Hercules, Thiokol, Aerojet, Phillips, Grand Central, and Atlantic Research (leading makers of solid rockets) to discuss his ideas. In February, Hall and Dr. Adolph Thiel were authorized to start a program to develop an ICBM with solid boosters parallel to the liquid-rocket program. As Hall saw it, great improvements would be needed in the nozzle, propellants, and case if his scheme was to work.[107] The ICBM programs gained momentum in early 1956, when the new Ballistic Missiles Committee (set up to advise the secretary of defense) approved two large solid-rocket research programs—one under the Navy and one under the Air Force—and the solid-propellant program was transferred to Wright Air Development Center.

In March 1958—at the same time that Hercules formed its chemical propulsion division and expanded Bacchus—a new solid-fuel missile program called Minuteman, the brainchild of Colonel Hall, was approved. The salient features of Minuteman were its constant readiness—with solid-fuel motors, the rocket could be stored for long periods in underground silos and fired with sixty seconds' notice (thus its name)—and its cost-effectiveness. Minuteman was designed to fire in three stages of varying sizes that could be used in other programs.

The Air Force invited bids on the program from all major producers of solid-fuel propellant, including Hercules. When news of the program reached ABL, Duard H. "Tex" Little and Alexander F. Giacco went to Jack Hayes, assistant general manager of the Explosives Department, to suggest that Hercules make an aggressive proposal. As Little recalls, Hayes said, "Hell, that's what we're in business for, let's bid for the whole damn thing." Given the green light, a team led by Giacco produced a proposal within three weeks for Hercules to develop all three stages of the missile.[108]

In July 1958, the Air Force selected its contractors.[109] Relying on earlier feasibility contracts, as well as proven technology in systems integration and large rocket motors, the Air Force designated Boeing as prime contractor for the entire missile, Thiokol as prime contractor on the first stage, a competition between Thiokol and Aerojet on the second, and Aerojet as prime contractor on the third, with Hercules and Thiokol as backups.

Of the three companies vying for the third-stage contract, according to the historian of the Minuteman project, Hercules, which proposed using double-based propellant and a filament-wound case, offered "the boldest approach." The Air Force Ballistic Missiles Committee staff was intrigued: double-based fuel promised to yield appreciably higher efficiency, and the fiberglass case would represent significant savings in weight. But they were also skeptical. "Because the Hercules approach was more advanced," recalled Maj. Ralph Harned, chief of Minuteman propulsion,

> there was even less known about many of its features. The other contractor approaches were more conservative. We had less assurance that Hercules could deliver a high-quality engine in time to satisfy Minuteman's needs. The additional performance promised by the Hercules approach was too attractive to pass up, but the risk was too great to have the complete weapon system dependent on that approach alone.[110]

The solution was to support both Hercules and another contractor until Hercules either succeeded or failed. But the company's bold approach succeeded, earning it a $15.3 million R&D contract from the Air Force on June 8, 1959, and a complete construction contract for the third stage in late spring of the following year.[111]

In September 1960, Minuteman was unveiled to the public at the annual Air Force Association Convention in San Francisco. It stood fifty-nine feet tall, weighed less than seventy thousand pounds, and could travel as many as sixty-three hundred statute miles. USAF chief of staff Gen. Thomas D. White explained its significance: "Because it is simpler to manufacture, maintain, and operate, Minuteman can be provided in great numbers at much lower cost than *any* other strategic missile system."[112] The missile was successfully test-launched on February 1, 1961. Minuteman represented a clear triumph of solid over liquid fuel for balistic missiles: solid rockets were simpler, more compact, more reliable, and safer.

The same reasoning applied to the other major long-range ballistic missile program of the era, the U.S. Navy's submarine-launched Polaris.[113] At the end of 1956, for reasons similar to those of Col. Hall, Rear Adm. William F. Raborn, Jr., head of the Navy's Special Projects (SP) Office, committed to solid-fuel rockets as the basis of the fleet ballistic missile (FBM) program. In the following months, an SP task force developed the basic concept of the new missile: its size, weight, center of gravity, power requirements, and other key factors, including details of its launch platform inside a nuclear submarine. The limitations of this environment dictated that the FBM would fire in two stages.

The flight of *Sputnik I* accelerated the FBM program, convincing SP to "leap frog some steps in development," accept a phased approach to meeting long-term objectives, and take chances on unproven technology. In April 1958, the Navy awarded the contract for both stages on a scaled-down A-1 rocket motor to Aerojet. The A-1, which used steel cases, achieved a range of twelve hundred nautical miles, rather than fifteen hundred as originally proposed. Although Hercules had been only peripherally involved in the program, the company was chosen as a development contractor for the second stage on the next-generation A-2 rocket motor in 1959. A series of explosions and delays at Aerojet had led the Navy to consider other sources; Hercules was chosen because of its expertise in double-based propellants and filament-wound cases, as well as its early success on Minuteman. Developed at ABL, the Polaris A-2 motor made its first successful test-flight on November 10, 1960, covering a range of fourteen hundred nautical miles.[114]

The success of Minuteman and Polaris marked a major shift in the U.S. strategic defense program. The two missile systems joined the B-52 to form America's basic triad defense system in the early 1960s and made liquid-rocket programs such as Titan, Atlas, and Thor obsolete.[115]

For Hercules, Minuteman and Polaris marked the company's entrance into major ballistic missile systems; subsequently, it would participate in virtually all major programs. In the late 1950s and early 1960s, moreover, the ABL-designed Altair and Antares rockets using Hercules' propellant and filament-wound cases fired successfully in hundreds of civilian space research programs. In 1958, for example, the Atlas-Able rocket used the reliable X248 Altair motor built by Hercules to drive its third stage while launching satellites, and the following year, the same rocket motor powered the third stage of the Thor-Able III rocket that boosted the *Explorer VI* paddlewheel satellite into orbit.[116] (See Exhibit 10.4 for a list of rocket and missile programs of the era in which Hercules participated.)

The company's achievements in rocketry produced tangible results: in the two years following the formation of the chemical propulsion division, employment at Bacchus soared from about one hundred to nearly seven hundred and would surpass six thousand in the mid-1960s, when production of Minuteman and Polaris rocket motors hit full stride. By 1962, the company had more than ten thousand people engaged in chemical propulsion at Bacchus, Kenvil, ABL, Rocky Hill, Port Ewen, and the GOCO plants, and revenues from the business reached about 15 percent of the corporate total.[117]

Such rapid growth was an obvious source of concern to some Hercules directors, who recalled the booms and busts of defense contracting from the world wars. In 1961, Forster was worried enough to admit, "We've almost reached the limit of putting money into rocket fuel development."[118] Nonetheless, under Jack Hayes (Martin's successor as general manager of Explosives) and Pat Butler (head of the chemical propulsion division after 1960), the business became a major contributor to Hercules' sales and earnings, and junior executives and technical experts engaged in the buildup, such as Giacco, Little, John Greer, Ernest A. Mettenet, Henry A. Schowengerdt, and Rudy Steinberger, would subsequently assume major roles in the corporation.

The International Department

In addition to growth via diversification, acquisition, and vertical integration, under Forster's leadership Hercules rethought its approach to international

EXHIBIT 10.4 Hercules' Participation in U.S. Rocket, Space Vehicle, and Missile Programs, 1960

	TYPE	*USED BY*	*STATUS OR FIRING DATE*	*DEVELOPED BY HERCULES*
Research rockets and space vehicles				
Deacon	A high-altitude sounding projectile used in a multitude of scientific programs	Army, Navy, NASA, ARPA	Operational	Complete rocket
X248 Altair	High-performance rocket motor designed for high altitudes	NASA, Air Force	Operational	Complete rocket
X254 Antares	High-altitude motor developed for Scout third stage	NASA	Development	Complete rocket
Thor-Able Pioneer	Space probe	NASA	October 11, 1958	Third stage
Atlas-Able	Radio satellite	NASA	December 18, 1958	Third stage
Thor-Able III	Satellite (*Explorer VI* paddlewheel)	NASA	August 7, 1959	Third stage
Vanguard III	Satellite	Navy-NASA	September 18, 1959	Third stage
Shotput	Inflated sphere	NASA	October 28, 1959 January 16, 1960	Second stage
Strongarm	Electronic density probe	Army-NASA	November 10, 1959	First, second, third stages
Sodium Cloud	Wind current study	NASA	November 18, 1959	First stage
Javelin	Galactic noises and radiation phenomena study	Air Force-NASA	December 22, 1959 January 14, 1960	All four stages

Journeyman	High-altitude phenomena study	Air Force-NASA	Scheduled 1960	Fourth stage
Thor Delta	Deep space probe	NASA	Scheduled 1960	Third stage
Tiros	Meteorological satellite	NASA	Schedued 1960	Third stage
Transit	Navigational-aid satellite	Navy	Scheduled 1960	Third stage
Argo	High-altitude research	NASA	Scheduled 1960	All four stages
Scout	Satellite vehicle	NASA	Development	Third and fourth stages
Military missiles and rockets				
Nike Ajax	Antiaircraft—surface to air	Army	Operational	Booster
Nike Hercules	Antiaircraft—surface to air	Army	Operational	Booster
Terrier	Antiaircraft—surface to air	Navy, USMC	Operational	Booster and sustainer
Talos	Antiaircraft—surface to air	Navy	Operational	Booster
Tartar	Antiaircraft—surface to air	Navy, USMC	Development	Sustainer
Honest John	Surface to surface	Army	Operational	Complete rocket
Little John	Surface to surface	Army	Operational	Propellant
Snark	Surface to surface	Air Force	Development	Booster
Minuteman	Surface to surface	Air Force	Development	Third-stage rocket
Weapon A	Surface to underwater	Navy	Operational	Propellant
Polaris	Underwater to surface	Navy	Development	Second-stage rocket

Note: ARPA = Advanced Research Projects Agency; NASA = National Aeronautics and Space Administration; USMC = U.S. Marine Corps.
Source: Hercules Mixer 42 (February 15–16, 1960).

expansion. The key step was the formation of the International Department in 1959, with Henry A. Thouron as general manager. Before becoming head of the Synthetics Department in 1955, where he helped make the company's DMT operations successful, Thouron had spent a year in Europe working on export sales. There, he became interested in opportunities for Hercules. Although he gained little support from the corporation, for years he "kept hammering away at it at the Board level and with Forster," until his request for an international department was granted.[119]

Thouron's enthusiasm for opportunities overseas and the board's eventual acquiescence came from several sources: Hercules' recognition of broad changes in the dynamics of international chemical competition; changes in trade and monetary practices in Europe during the late 1950s; and changes in the company itself. Hercules' traditional, export-led international strategy seemed at risk, especially if the company hoped to find new sources of growth.

By the mid-1950s, Hercules was well aware of an unprecedented level of competition between American and European chemical companies. Prior to the war, such competition had been minimal, primarily because of the "grand alliance" under which Du Pont and ICI (with the occasional participation of I.G. Farben) had allocated world markets. The alliance began to unravel during the 1930s, when new managers at Du Pont and ICI altered the terms of their cooperation, and the trend accelerated under scrutiny from the U.S. Department of Justice. Indeed, patent exchanges and joint ventures between the two companies were outlawed in the United States in 1950. Thereafter, the United States became an open market to European producers.[120]

A second broad change in the international competitive environment was the miraculous recovery of the wartorn economies of Europe. By the early 1950s, Germany, France, and Italy joined the United Kingdom both as attractive markets and as homes to strong rivals to U.S. companies.

Third, specific changes in European political and economic policies in the late 1950s made U.S. direct foreign investment in Europe more attractive. At the end of 1958, fourteen European nations, led by Great Britain, made their currencies convertible to dollars for the first time since the end of the war. At about the same time, two European trade communities formed to encourage the free flow of goods and services among participants: the European Economic Community (EEC), created on March 25, 1957, included Belgium, the Netherlands, Luxembourg, France, Germany, and Italy; the European Free Trade Area (EFTA) was formed in 1959 between Britain, Norway, Sweden, Denmark, Switzerland, Portugal, and Austria. By lowering tariff barriers among themselves—and leaving those barriers in place for outsiders—the two organizations increased trade opportunities for members and simultaneously penalized importers.[121]

As Thouron saw it, Hercules had little choice but to respond to these changes, especially given the increasingly competitive nature of the U.S. domestic market. There was also a growing sense in the company that it had "a big backlog of processes and know-how that were just waiting to be exploited."[122] As a writer for the *Hercules Mixer* explained:

During the 1960s and 1970s, Hercules moved aggressively to invest overseas. Two new ventures of this era: Dawood Hercules Chemicals, Ltd., a joint venture to produce fertilizers in Pakistan (above); and harvesting seaweed for carrageenan in the Philippines, part of A.S. Kobenhavns Pektinfabrik's business in food additives (below).

In 1967, at the dedication of the Middelburg plant, Managing Director of Hercules B.V. Jan Wagemaker shows His Royal Highness Prince Claus of the Netherlands the latest in computer-controlled processes.

> With the rise of the European Common Market and the European Free Trade Association, it became evident that it would be expedient for Hercules to modify its method of handling foreign sales to these areas. National interests were making it more difficult to sell products manufactured in the United States to these areas. Tariff barriers, the need for local customer service, and the necessity to cut transportation costs all indicated a need for the establishment of manufacturing facilities in foreign countries.[123]

Thouron pointed out that, along with the need to "extend our market coverage abroad," especially in developed areas with rising standards of living, the company also faced a clear defensive challenge. Unless Hercules built plants overseas "to protect our present markets," he warned, it would suffer. Indeed, large consumers of Hercules products in Europe had warned Thouron and other company representatives that they preferred to buy locally so as not to risk a plant shutdown if shipping problems delayed imports.

The International Department made its debut in 1959 as the seventh operating (as opposed to auxiliary) department in the company, and the first to be created internally since 1936. The new department's responsibilities were broader and more ambitious than those of its predecessor, the Export Department. Thouron later distilled International's strategy into four components: develop new markets; supply growing markets such as the EEC; protect established markets from tariff barriers and other restrictions; and increase the value of research.[124]

The new department was divided into two groups: one assumed responsibility for exports, and the other took charge of foreign investment and the management of overseas operations.[125] The latter group began modestly. Apart from Thouron and his support staff, the group consisted of three key players, each with one assistant: Warren F. "Sam" Beasley, a naval stores expert assigned to the Western Hemisphere; Dr. Harvey J. Taufen from the Research Center and the Synthetics Department, who would scour the Far East for opportunities; and Henry Reeves, Jr., the overseas PMC manager whose territory was Europe. (David M. Houston, director of the Export Department since 1945, was retained as a consultant.)

Although these men were skilled and experienced—collectively, they had served the company for seventy years—they were severely outgunned by several other U.S. competitors. As Taufen put it, "There were only six people that were going to get Hercules organized in manufacturing worldwide. Considering Du Pont and Hoechst had hundreds, it was an interesting way to go about it."[126]

The first foreign plants built after the International Department's formation produced pine and paper chemicals in Europe and Australia. In 1960, the company formed Hercules Kemiska Aktiebolag, a wholly owned subsidiary in Gothenburg, Sweden, to manufacture rosin size and, eventually, a complete line of paper chemicals. The locale was chosen, as Thouron explained, because of "the unusually rapid and increasing importance of the Swedish paper industry." Indeed, Sweden was then the eighth-largest producer of paper and paperboard in the world and the largest per capita consumer of paper in Europe. The plant at Lilla Edet, under the management of Prosper F. Neumann, was dedicated at the end of September 1961. That same month, N.V. Hercules Powder Company managed the startup of a thirty-million-pound-per-year rosin derivative plant at Zwijndrecht, Holland, and another rosin derivatives plant was under construction at Tampere, Finland, under the auspices of the wholly owned Oy Hercofinn Ab and its general manager, Donald Kane.[127]

The pattern was slightly different in Australia, where Hercules Powder of Australia formed a joint venture with A.C. Hatrick Chemicals Pty. Ltd., a Hercules distributor for three decades. Its plant at Springvale near Melbourne (managed by Courtland White and started in late 1961) made rosin derivatives and industrial chemicals such as paper chemicals and synthetic rubber emulsifiers.[128]

Under the International Department, Hercules exports reached approximately 15 percent of total sales in 1961, thereby recovering to the level that had been reached a few years prior to World War II. Greater gains were made in foreign manufacturing operations; the number of Hercules foreign plants operating or under construction doubled in 1960 and 1961 from seven to fourteen.[129]

In retrospect, the limited resources and loose mandate accorded the fledgling department worked in its favor in the early years, thanks largely to the judgment of its overseas scouts. Concludes Taufen:

> I know of very few companies that would have given us, at our level . . . when we first got started . . . the free hand that we had. . . . Nobody questioned where we went, nobody questioned what we were doing at all, it's just that they wanted some results. That turned out to be a very successful way to go.[130]

In 1959, Hercules took its first steps toward becoming a multinational competitor based on foreign investments. The challenge soon shifted to a higher level, however, as international competition intensified and the company was obliged to manage its transformation into a truly global corporation.

REALIGNING THE ORGANIZATION

On July 1, 1961, Hercules revamped its departmental structure for the first time since 1928, and for only the second time in its forty-nine-year history. Although the changing mix of business in some departments was a key factor, the immediate reason was to accommodate the new units formed in 1959 and 1960: International, Imperial Color and Chemical, and Fiber Development. To settle jurisdictional and status issues in these new operating departments and in several others, Hercules realigned its structure into eight operating departments.

Under the new plan, the operating departments were Explosives (commercial explosives and chemical propulsion); Pine and Paper Chemicals (a combination of the old Naval Stores and PMC departments, with some synthetic resins from the old Synthetics Department); Cellulose and Protein Products (Virginia Cellulose and Huron Milling); Polymers (primarily polyolefins, but also cellulose products made at Parlin); Synthetics (which took over agricultural chemicals and oxychemicals from Naval Stores and ammonia and DMT from Explosives); International; Imperial Color and Chemical; and Fiber Development (see Exhibit 10.5).

The realignment achieved several important results: it clarified responsibilities, shifted power from old areas such as Cellulose Products and Naval Stores to new areas such as Polymers and Synthetics, and generally acknowledged the broad changes in the company's business since the war. In these respects, at least, Hercules' leaders recognized the need for change.

Also notable is what the restructuring did not accomplish—or attempt. It did not change the fundamental principle of Hercules' organization, which (in most instances) remained chemical processes rather than end markets. As Forster put it, "A thorough study of the Company's organization as it existed at the start of 1961 confirmed the soundness of the structure adopted in 1928." Indeed, he added, "the new alignment . . . is essentially a clearer delineation of product lines."[131]

Nor did the realignment alter how the company was governed: the internal board consisting of corporate officers and department heads, active executive and finance committees, and vice presidents with specialized "spheres of influence," rather than formal or functional lines of responsibility. This mode of governance was increasingly rare; a 1962 survey by the National Industrial Conference Board revealed that only 4.5 percent of industrial firms were governed by exclusively internal boards. Nonetheless, Hercules wanted to retain both the close-knit interactions among the vice presidents and the autonomy of the operating departments.[132]

The realignment did not solve every organizational problem: indeed, a perennial dilemma for diversified chemical companies such as Hercules was that no organizational scheme could completely address the complex nature of the business, with its overlapping technological bases and markets. Many of the company's leading markets—plastics, coatings, adhesives, construction, and synthetic fibers—were simultaneously served by many departments, and the realignment offered no ultimate solutions to this recurring issue. Indeed, as the company pushed into still more new areas in the 1960s, it would continuously tinker with its structure.

Conclusion

The period 1955–1961 brought a new vitality to Hercules that, by most measures, had been lacking in the first decade after the war. Under President Forster, Hercules garnered the praise of industry analysts, who spoke of a

EXHIBIT 10.5 Hercules' Eight Operating Departments, July 1, 1961

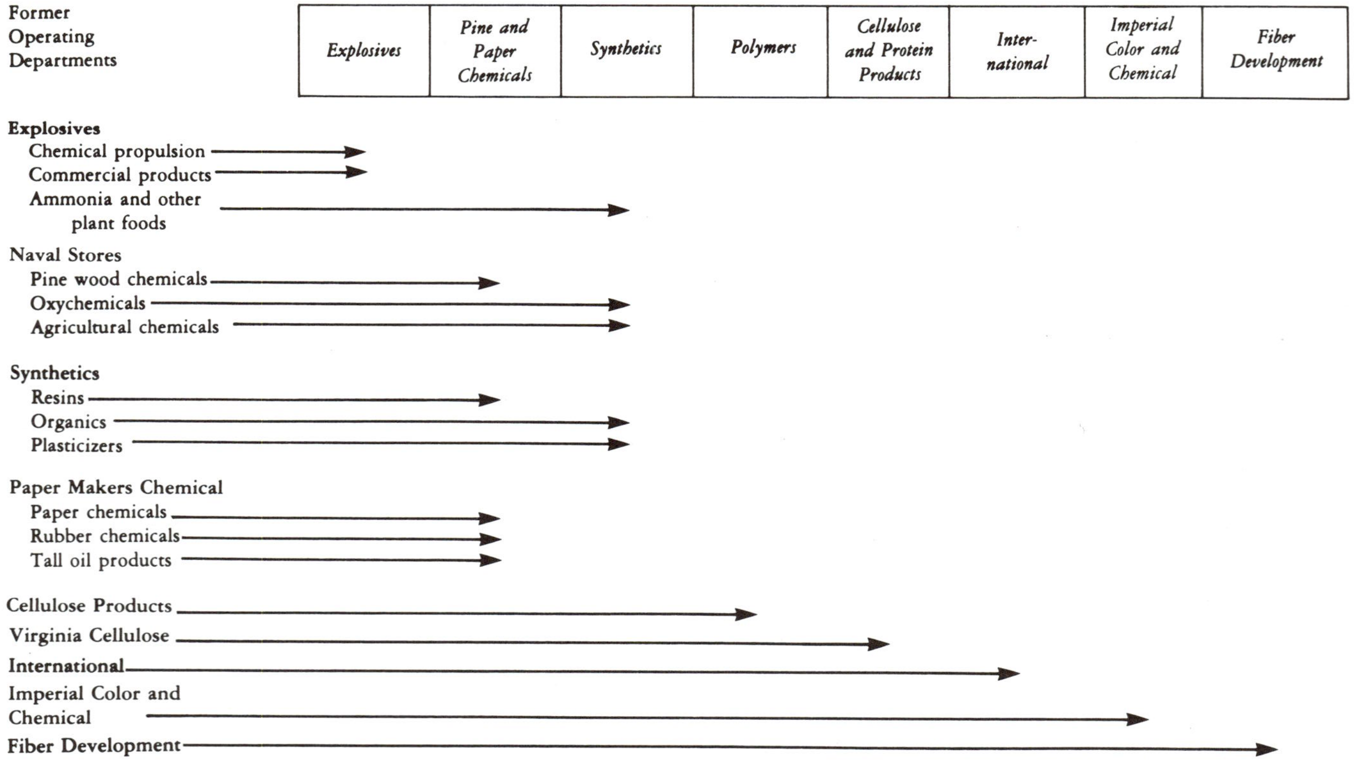

Plants	Explosives	Pine and Paper Chemicals	Synthetics	Polymers	Cellulose and Protein Products	International	Imperial Color and Chemical	Fiber Development
	ABL[a] Bacchus Radford[a] Rocky Hill Sunflower[a] Bessemer Carthage Kenvil Port Ewen Virginia (Gilbert)	Brunswick Burlington (Ont.)[b] Franklin Hattiesburg Holyoke Kalamazoo Milwaukee Portland Savannah	Burlington (N.J.) Gibbstown (Higgins) Hercules (Calif.) Louisiana (MCW) Ketona Chemical Corp.[c]	Lake Charles Parlin	Harbor Beach Hopewell		Glens Falls Plattsburgh St. Johns (Que.)	Covington

a. Operated for U.S. government
b. Wholly owned subsidiary
c. Affiliated company

Source: Hercules Mixer 43 (July–August 1961), 27–28.

"metamorphosis" within the company and ranked it among the industry's leaders. *Forbes* magazine, which devoted its cover story to Forster and Hercules in June 1961, noted that between 1956 and 1960, while average net profits for the chemical industry declined from 8.1 to 7.4 percent, those at Hercules moved in the opposite direction, rising from 7.4 to 8.3 percent. Contrasting this period with the preceding decade (when Hercules stood "pat with its traditional product lines" while others raced ahead), *Forbes* concluded that, "against the receding tide of the industry's fortunes, this onetime tagalong has pushed its way [into the vanguard] of chemical profitmakers."[133]

Similarly, *Investor's Future* lauded Hercules management, especially Forster, for doing "an outstanding job in orienting its efforts towards those fields in which the firm can grow and prosper"; *Chemical Week* spoke of a "new force at Hercules" and honored the company with its Management Award in 1962. Hercules deserved the acclaim, explained the leading industry journal, for (among other things) increasing earnings per share 43 percent between 1956 to 1961, thereby outperforming Du Pont, Union Carbide, Dow, Monsanto, Allied, Stauffer, and American Cyanamid during the same period (see Exhibit 10.6).[134]

The renaissance at Hercules was also reflected in outward appearances: the new twenty-two-story Hercules Tower became the company's new Home Office in the fall of 1960. Constructed between the thirteen-story wings of the Delaware Trust Building at 910 Market Street (Hercules' corporate address since 1921), Hercules Tower was the tallest structure in Wilmington. At its dedication, President Forster proudly cut a ribbon made of Pro-Fax® polypropylene.[135]

What lay behind Hercules' success, and what did it portend? Increased sales and profits had been driven by high rates of investment in plant construction and R&D (see Exhibits 10.7, 10.8, and 10.9). Clearly, Hercules under Forster's leadership was willing to take substantial risks—such as spending $40 million, virtually its entire cash reserves, to expand and integrate forward in polyolefin plastics between 1954 and 1957. "For us," Forster remarked, "this was a pretty adventurous step to take."[136] An additional indication of a new, aggressive spirit was the company's willingness to resort to debt to help pay for its growth. Hercules borrowed $500,000 in 1960 and another $2 million a year later. This debt represented less than 1 percent of total assets, and Forster was quick to emphasize that the company was still governed by "sound" financial practice. Nonetheless, the actions had a powerful symbolic impact.[137]

Moreover, the new financial measures reflected an even deeper metamorphosis: the new degree of balanced, diversified growth. With the new petrochemicals lines and aerospace accounting for approximately 15 percent of total profits, and natural gas–based commodities, starches and proteins, and color pigments making up small but growing portions, Hercules was propping up what had threatened (with the decline of explosives) to become a two-legged stool. At the same time, petrochemicals, synthetic ammonia, tall oil, and overseas naval stores operations had reduced the company's dependence on dwindling raw materials.

Moreover, Hercules had not pursued indiscriminate expansion. Its major

EXHIBIT 10.6 Hercules' Sales and Profits Compared with the Chemical Industry Average, 1952–1961

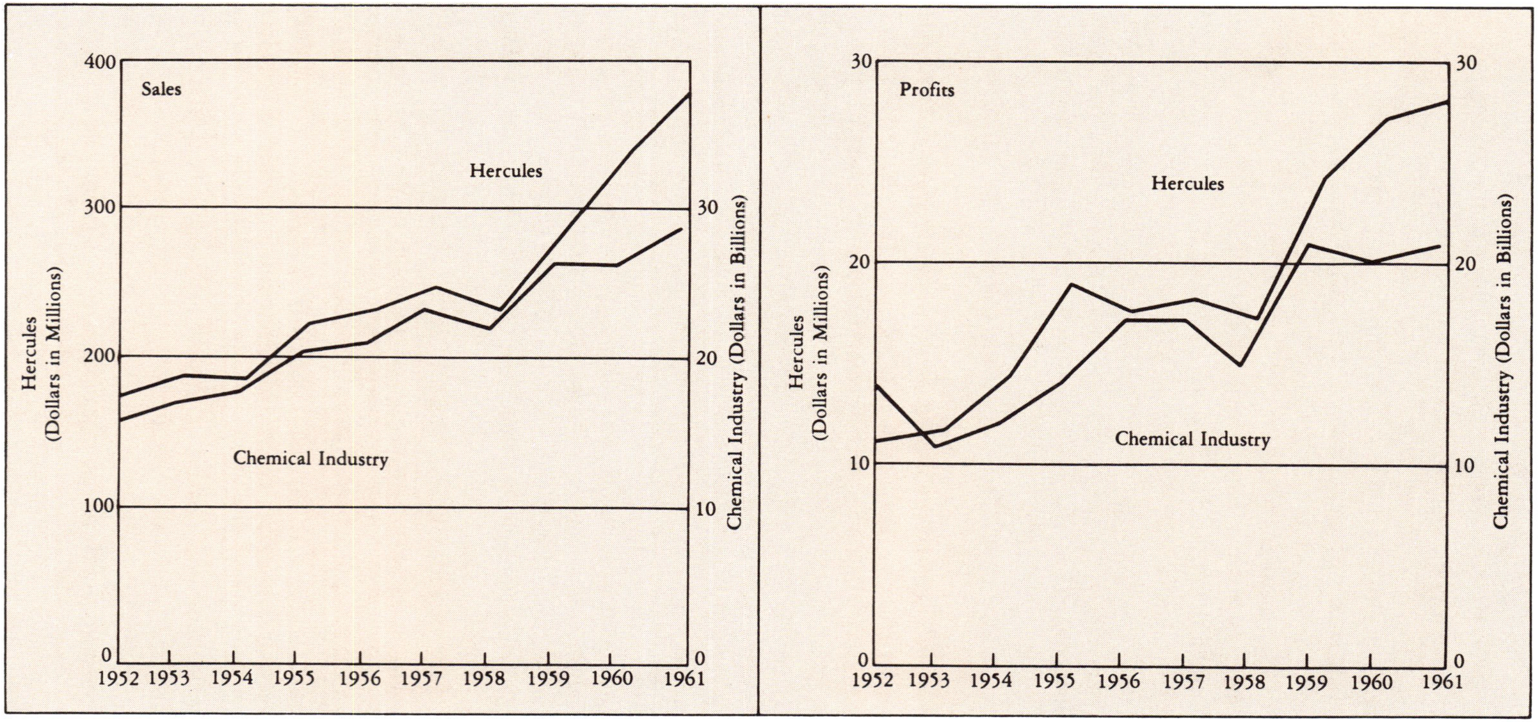

Source: Burke, "Planning for the Next Half Century," 55.

EXHIBIT 10.7 Hercules' Investment in Plant Construction, 1945–1961

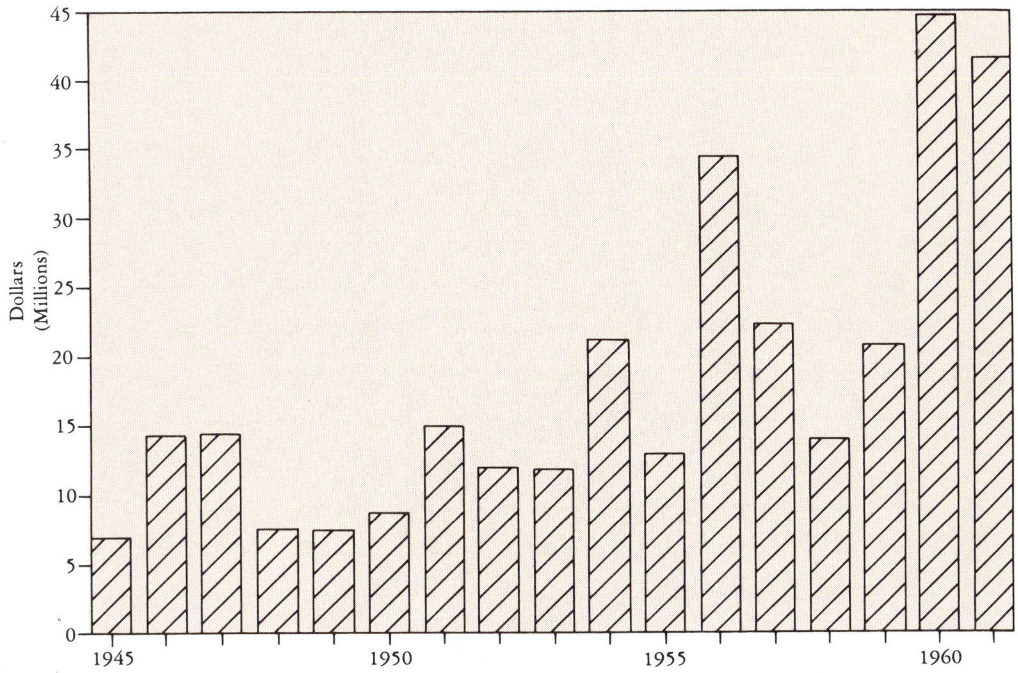

Source: Hercules Powder Co., Annual Reports for 1945–1961.

acquisitions were in areas closely related to its existing businesses in terms of processes and markets, and the company continued to spurn commodities businesses. As Forster explained, "We have tended to shy away from bulk chemicals, which require heavy capital investments, where the net return on investment is often lower than in further upgraded chemicals. In any case, the best areas in basics were already overcrowded."[138]

As for the future, the arms and space races, the wave of American investment overseas, and world demand for DMT seemed poised on the brink of explosive growth. But the picture was more troubling in petrochemical plastics. To its credit, Hercules had responded quickly to gluts in polyethylene and polypropylene; but where would it turn if fibers and film were also flooded with competitors? Also to its credit, Hercules had plunged ahead in large-scale rocketry when explosives declined, and in polyolefin fibers and film when cellulose products declined. The timing of these opportunities had been fortunate, but the company might not be so lucky again.

Some analysts were critical of Hercules' heavy dependence on petrochemicals, its internal board, and its growing tendency to license or acquire rather than develop new technology—and indeed, these issues proved increasingly controversial in the coming years. For the time being, however, its strategies were working.

EXHIBIT 10.8 Hercules' Research Costs as a Percentage of Net Sales, 1941–1961

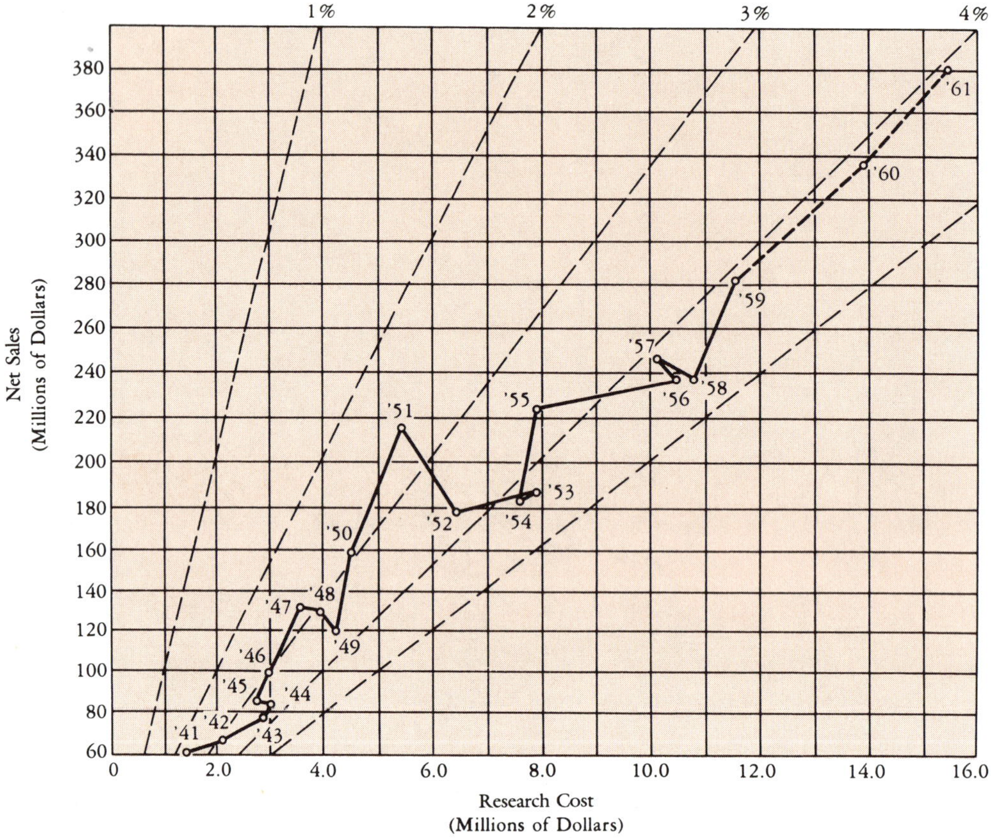

Source: Research Department Report, Fourth Quarter 1961.

EXHIBIT 10.9 Hercules' Research Costs, by Types of Work, 1944–1961

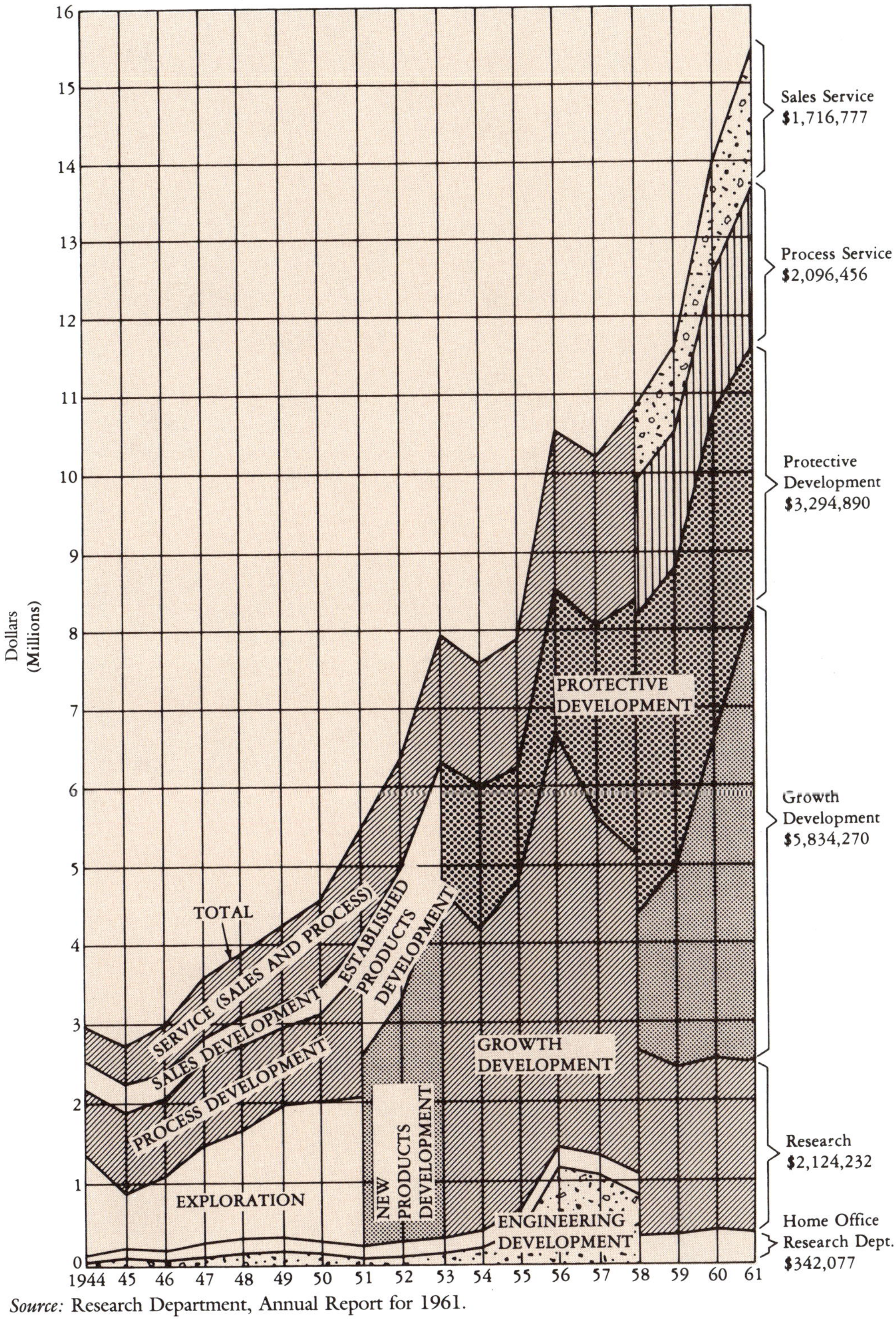

Source: Research Department, Annual Report for 1961.

CHAPTER 11

MANAGING NEW CONSTRAINTS, 1962–1973

In October 1962, Hercules paused to celebrate its fiftieth anniversary. Spirits were high, and deservedly so. Growth in the previous half-century had been impressive, particularly in the last half a dozen years. The company was settling into new overseas operations and domestic businesses (aerospace, polyolefin film and fibers), while relying on steady revenues from its healthier traditional lines (especially pine and paper chemicals and synthetics).

Still, recent times had taught Hercules top managers that the company faced several great challenges: overcoming a "profit squeeze" caused by increasing raw materials costs, competition, and taxes; moving into still more areas of potential growth in order to keep pace with the rapid rate of product substitution in chemicals; and managing an immensely complex, global, diversified company.

From 1962 to 1973—the last years of U.S. postwar economic expansion—Hercules attempted to grow in three principal ways. First, under President Forster's internationally oriented successor, Henry Thouron, it expanded its foreign operations well beyond the naval stores extraction and rosin derivatives plants of the late 1950s by building or joint-venturing operations in Germany, Italy, Japan, Australia, India, Pakistan, and other nations. Second, it made several large-scale investments to increase its capacity for making petrochemical commodities and fabricated materials—specifically, DMT and polypropylene resins, fibers, and film. Third, Hercules attempted to move into new high-growth areas, including several virtually unrelated to its traditional chemical lines. This strategy, first manifest in the acquisition of Haveg Enterprises, was later systematically carried out by the New Enterprise Department (NED). Thus, like many leading American corporations of the day, Hercules adopted the diversification strategy of the conglomerate movement.

Meanwhile, Hercules adapted to the rise of environmentalism in the

United States, a broad-based movement that included changes in public opinion toward the chemical industry as well as numerous legislative measures designed to regulate product safety, air and water pollution, working conditions, and other aspects of the environment. Although Hercules found it increasingly costly to work within the new constraints, it also cultivated the inherent opportunities by developing new products to purify wastes and protect the environment.

Few of Hercules' growth tactics proved successful in this period, so its financial performance was mixed (see Exhibit 11.1). By the time Thouron's successor, Werner C. Brown, began to confront the problems wrought by the world energy crisis of the early 1970s, Hercules was already facing many constraints. Heavy investments in environmental equipment had become a permanent requirement of doing business; heavy investments in petrochemicals had unbalanced the company's portfolio and, to some, seemed increasingly precarious; and few of the investments in the NED businesses had grown as anticipated. During the 1960s and into the 1970s, Hercules' central challenge was to make its way through an increasingly hostile business environment.

HERCULES AT FIFTY

It is likely that the several retirees joining in the anniversary festivities who had worked for the company since its beginnings were amazed at the changes in the company. With twenty-seven thousand employees (about half of them at chemical propulsion plants), twenty-seven domestic plants, seventeen plants of foreign affiliates and subsidiaries, three government plants under its operation, thirty-one domestic and nine foreign sales offices, eight hundred products, and annual sales of nearly half a billion dollars, Hercules, true to its name, was a giant. (It was ranked eighty-fifth in net profits among the *Fortune* 500.)[1]

On this occasion, many reflected on the unusual circumstances of the company's birth and compared its performance with that of its parent, Du Pont, and its sibling, Atlas. A quick comparison revealed two basic facts: sales had grown much faster than profits for all three companies, and the larger the company, the faster its relative rate of growth. Hercules had grown nearly nineteenfold in assets and profits, and—fittingly enough—fiftyfold in sales since 1913 (see Exhibit 11.2).[2]

Hercules marked its golden anniversary by releasing a new logo (chosen from three hundred sketches) that depicted a bold, "modernized" mascot, as well as a promotional film called *What in the World Does Hercules Do?*[3] Given the company's broad spectrum of activities, it was a reasonable question even for Herculites to contemplate. Indeed, in the fall of the anniversary year, the company's top managers did so at Spring Lake Beach, New Jersey, at the third major management conference in the company's history.

The Management Forum, which met September 9–16, was attended by 456 executives from around the globe. Because its purpose was to take stock, improve communications and morale, and generally discuss new directions

EXHIBIT 11.1 Hercules' Financials, 1962–1973 (Dollars in Thousands, Except per Share)

	1962	*1963*	*1964*	*1965*	*1966*	*1967*	*1968*	*1969*	*1970*	*1971*	*1972*	*1973*
Net sales	$454,829	$476,462	$530,975	$532,373	$609,975	$642,625	$718,307	$745,991	$798,608	$811,884	$972,267	$1,154,775
Earnings before interest and taxes	67,349	65,671	73,191	72,936	98,418	85,382	97,489	81,185	94,883	93,967	132,227	157,355
Total assets	318,587	345,591	369,628	423,620	529,948	655,631	798,689	787,341	824,793	781,764	911,569	1,036,304
Fixed[a]	159,265	162,356	163,541	197,898	230,256	334,000	394,201	404,977	426,097	411,465	449,199	522,456
Current	137,525	160,013	180,313	182,966	240,493	235,907	257,609	274,800	302,356	275,929	359,189	401,052
Other	21,798	23,222	25,774	42,756	59,199	85,723	146,879	107,565	96,340	94,370	103,181	112,796
Stockholders' equity	232,124	256,271	273,758	289,178	324,795	355,961	384,761	393,166	414,753	457,649	508,829	597,517
Long-term debt				35,000	60,000	159,500	213,337	184,645	190,267	129,742	125,198	177,222
R&D Expense	17,624	19,042	16,472	18,678	20,684	21,469	23,336	23,231	21,587	22,179	23,628	24,542
Dividends/share	$0.75	$0.75	$1.00	$1.00	$1.10	$1.20	$1.20	$1.20	$1.20	$1.20	$0.63	$0.71
Earnings per share (common)	$1.69	$1.72	$1.98	$2.21	$2.67	$2.37	$2.69	$2.21	$2.61	$2.78	$1.77	$2.18
Number of employees	18,441	21,042	21,418	20,790	20,961	21,821	22,474	22,797	22,112	21,125	22,395	24,063
Return on sales	14.8%	13.8%	13.8%	13.7%	16.1%	13.3%	13.6%	10.9%	11.9%	11.6%	13.6%	13.6%
Return on assets	21.1	19.0	19.3	17.2	18.6	13.0	12.2	10.3	11.5	12.0	14.5	15.2
Return on equity	29.0	25.6	26.7	25.2	34.0	26.3	25.3	20.6	22.9	20.5	26.0	26.3
Debt/equity				12.1	12.1	18.5	55.4	47.0	45.9	28.3	24.6	29.7
Debt/total capitalization				8.3	6.6	9.2	26.7	23.5	23.1	16.6	13.7	17.1
R&D/Sales	3.9	4.0	3.1	3.5	3.4	3.3	3.2	3.1	2.7	2.7	2.4	2.1

a. Gross fixed assets, including depreciation.

Source: Hercules Incorporated, Annual Reports for 1952–1973.

EXHIBIT 11.2 Fifty Years of Growth: Du Pont, HPC, Atlas Percentage Increase of Sales, Profits, and Assets between 1913 and 1961

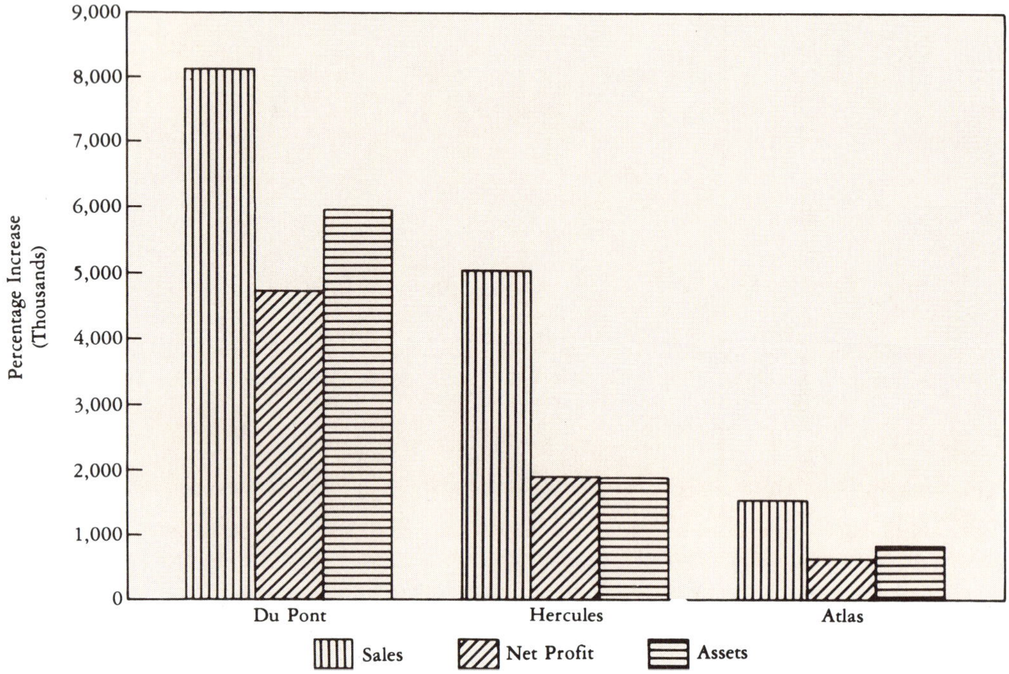

Source: Wilmington Evening Journal, July 17, 1962, 31.

rather than fashion a specific plan of action, the forum was more akin to the conference at Virginia Beach in 1946 than to the one at Atlantic City in 1919. In the first of sixty-nine presentations, Ed Morrow, chairman of the finance committee, focused on the acute need to sustain profits in order to attract investment and finance new plants, given the company's plans to invest nearly $500 million in new plants over the next decade. President Forster emphasized the fact that "people are and always will be Hercules' most valuable asset," and he foresaw a day "in the not-too-distant future [when] you younger men will be running this great company of ours."[4]

Shortly thereafter, on January 14, 1963, Forster resigned as president (although he remained as chairman), and Henry A. Thouron was elected president and chairman of the executive committee. A graduate of Princeton, Thouron joined the company as a naval stores salesman in 1934, held various sales and administrative posts in the Virginia Cellulose and Synthetics departments and at N.V. Hercules before taking the helm at Synthetics, and was elected to the board shortly before launching the International Department. Along with his proven management abilities, Thouron was described as "a serious amateur agriculturalist" and "a believer in the traditions of gracious living which he preserves in his perfectly appointed 19th century ancestral home."[5]

The Search for Growth

Fifteen months prior to Thouron's election, *Fortune* magazine ran a feature story with the ominous title, "Chemicals: The Ball Is Over." The article described how the industry had "come of age in a new and besieged market." Among the forces it identified as threatening profits and growth opportunities were rampant overcapacity on the domestic scene, especially in petrochemicals; impending overcapacity overseas due to rapid international expansion; invasion of markets by newcomers from other fields (such as chemical companies invading one another's turf and oil companies integrating forward into petrochemicals); and the difficulty of finding and controlling new products long enough to recoup R&D costs.[6]

As described in Chapter 10, most chemical companies had struggled in the late 1950s, but Hercules had prospered. By 1962, however, Hercules faced the same fundamental challenges as its peers; its response to them was straightforward in most cases. It pushed hard on the international front in those settings where it perceived an opportunity to exploit a technological advantage in a fertile market. And it tried to moderate the effects of increasing competitive rivalry by carving out market niches and, for the most part, spurning commodity businesses that required huge capital investments. But what of the dilemma of developing new products that might take years to commercialize, only to be promptly displaced by substitutes?

Henry Thouron came to have strong feelings about this issue. An internal study, recalls Chief Economist William W. Bewley, revealed that it took a decade or more before a new product discovered at the Research Center would yield $1 million in gross profits.[7] This remarkable finding, as well as the attitude emerging throughout the industry (as reflected in the *Fortune* article), convinced Thouron that ways to shorten the development process were sorely needed. (Because Thouron was not a chemical engineer, Bewley also speculates, he may not have fully appreciated "the nature of chemistry and its ability to invent new products all the time.")[8] The growing sense that traditional chemical lines were reaching maturity, in turn, fostered a new openness to nonchemical fields throughout the industry. At Du Pont, for example, President Crawford H. Greenewalt committed his company to a large program of diversification, "beyond existing fields of interest and beyond chemistry," in 1959.[9]

Indeed, the conglomerate movement in American business in the 1960s also made the time seem ripe for diversification into new, perhaps unrelated, fields. Conglomerates were, by definition, collections of unrelated businesses under a single managerial umbrella. With relatively small corporate staffs and little or no integration among operating units, conglomerates entered and exited from businesses for largely financial reasons, acquiring and divesting companies much like an individual investor juggles his stock portfolio. Between 1961 and 1968, for example, eleven conglomerates, often posting spectacular profits, gobbled up more than five hundred companies. Although the movement would falter by the late 1960s, conglomerate builders such as Royal Little of Textron, Tex Thornton of Litton, Charles Bluhdorn of Gulf + Western, and Harold

Geneen of ITT were the most celebrated business leaders in America during Wall Street's "go-go" decade.[10]

As we will see, Hercules' push into uncharted high-growth fields would intensify throughout the 1960s, culminating in the activities of the NED between 1968 and 1976. But the first significant move in that direction—and the first under Thouron's leadership—was the acquisition of Haveg Industries in 1964.

The Haveg Debacle

Hercules first became interested in Haveg through a chance encounter on an airplane between one of its managers, R.S. George, and Haveg's president, Dr. John H. Lux. Through this and subsequent meetings, Hercules learned that Haveg, a manufacturer of specialty plastic products, had grown remarkably since its founding in 1955. Employment had surged from 416 to 1,280 in 1963, and net sales from $1.3 million to more than $30 million. In the process, Haveg had acquired new operating divisions and subsidiaries: Haveg-Reinhold of Santa Fe Springs, California, in 1957; American Super Temperature Wires of Winooski, Vermont, in 1958; and a factory in Taunton, Massachusetts, in 1959. Together with Haveg's R&D laboratories in Wilmington, Delaware, these facilities manufactured a wide range of "engineered plastics" for the aerospace, atomic, and electronic industries, from tiny injection moldings to wire and cable coatings and giant tanks, pipes, and canopies. Like Hercules, Haveg also made filament-wound structures.[11]

Hercules was drawn to Haveg by a variety of hopes and expectations: that Haveg would bring high growth and profits through both new products and the licensing of new proprietary technologies; that it could infuse Hercules with some of its high-growth methodologies; and that the new company might serve as a "nucleus" for Hercules' endeavors in plastics fabrication. Most important, Hercules became interested in a Haveg process, promoted heavily by Lux, for foaming plastics (much like that later used by Dow to make Styrofoam from styrene). By using this process to foam polypropylene, Hercules hoped to produce a versatile and inexpensive (because largely air-filled) new product.[12]

On April Fools' Day of 1964, the Hercules board of directors mulled over a report on Haveg filed by its committee on mergers, acquisitions, and forward integration. Against the wishes of Ed Morrow (who would retire that summer after forty-eight years with Hercules) and John Goodman, the board voted to acquire Haveg for an undisclosed (yet, by its own admission, premium) price by exchanging stock.[13] The deal was finalized in July. The following month, Lux was elected to the (now eighteen-member) Hercules board. He retained his position as president of Haveg, which was made a Hercules department but operated as a separate subsidiary corporation.[14]

Early reports from Haveg were mixed. About half of its businesses were quite profitable, but the others were making little or losing heavily. In the closing months of 1964, Haveg earned profits of only $1.1 million on sales of $25.2 million. Its managers expressed optimism about each of the weaker lines, claiming that new processes or products were on the way. And optimism reigned for a time as Haveg acquired a new subsidiary (Glascote Products, a maker of glass-lined equipment for industry) and earned more than 10 percent on its 1966 sales of $41.6 million.[15]

But red ink began to flow in 1968. Hercules installed its own top managers (John Ryan, then Paul Graybeal), but the following year Haveg lost more than $1 million. In addition, environmental problems associated with some of the Haveg facilities were beginning to plague Hercules, as they would for many years to come. In short, the Haveg venture was seen as "a colossal mistake." The department was disbanded in 1970; some of its assets were transferred to other departments, and others were sold off piecemeal throughout the 1970s, often at bargain prices.[16]

What went wrong? Hercules executives involved in the acquisition offer several explanations. Werner Brown points out that, from the beginning, Haveg was more of a promise than a reality: "At least three-fourths of the sale was based on what they *could* do for us, not what they *were* doing." In particular, the critical plastic-foaming process turned out to be superficial. Brown recalls how, during a sales call to Owens-Illinois, he secretly stole from John Lux a sample cup made of foamed styrene. "I took it home and . . . put water in it, and the damn thing leaked like a sieve. And that's when I realized we were in trouble." Indeed, says Brown, Haveg's technology for blown polystyrene containers was "almost fraudulent."[17]

Haveg's high growth rate was attributable mainly to its acquisitions program rather than to internally generated earnings, recalls Bill Bewley, who also argues that Thouron may have pushed too hard, too fast, because he felt the need to make a decisive move early, as do many new presidents.[18] Still, the central motive behind the Haveg acquisition was finding new sources of growth outside the increasingly constrained chemical industry, a concern that haunted Henry Thouron throughout his presidency.

From a more recent perspective, the Haveg episode offers a lesson about the perils of unrelated diversification. Many Hercules executives, including Thouron, apparently concluded that Haveg's central flaw was its surprisingly weak technologies and somewhat deceptive rate of growth, rather than its being "a hodgepodge of acquired companies" only tangentially related to Hercules businesses. Such analysis would soon resurface on a grander scale during the "New Enterprise era."

Indeed, Thouron's conviction that Hercules needed new high-growth businesses was reinforced with each passing year. Consider the overall profit performance of the company's traditional lines between 1962 and 1970, the year Thouron stepped down as president (see Exhibit 11.3). The fastest growing businesses were Cellulose and Protein Products (C&PP) and Synthetics, which each nearly doubled their level of net profits during the 1960s; thus, their annual rate of growth was a respectable, but not outstanding, 11 percent.[19] At C&PP, the continuing strength of CMC and the growing popularity of new water-soluble polymers Natrosol® hydroxyethylcellulose, Klucel® hydroxyethylcellulose, and Reten® retention aid accounted for much of the growth. At Synthetics,

EXHIBIT 11.3 Hercules' Profits by Department, 1963–1970

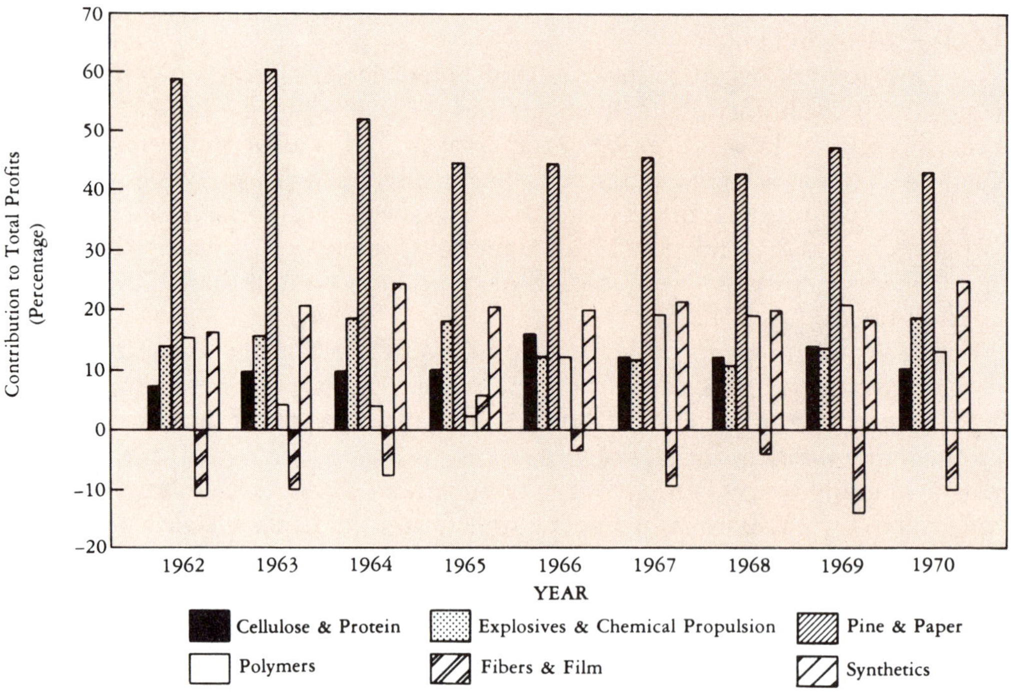

Source: Departmental Reports to executive committee, 1962–1973.

the star performer remained DMT. The other specialty chemicals businesses, however, struggled for growth during the period. Imperial Color and Chemical increased its profits by 43 percent between 1962 and 1970. The source of most of the company's profits, Pine and Paper Chemicals (P&PC), failed to increase its level of profitability, despite the enduring success of its traditional products and the introduction of new ones such as Aquapel® internal sizing agent. P&PC contributed $35.4 million in 1962, and $36.1 million in 1970, with little variation in the interim.

Explosives and Chemical Propulsion (E&CP) grew slightly in the 1960s, although the continuing decline of commercial explosives nearly offset breakthroughs in chemical propulsion. The department's greatest successes occurred in the fleet ballistic missile (FBM) program, in which Hercules supplied the second-stage rocket motors for the Polaris A-2 and A-3 missiles. In the mid-1960s, its experience gained on the Polaris helped the company increase its participation on the next-generation Poseidon FBM program. For the Poseidon, Hercules and Thiokol Chemical Corporation formed a 50-50 joint venture to develop the first-stage motor using high-energy propellant and a filament-wound case. Hercules alone was responsible for developing the second-stage rocket motor and case.

In sum, despite some important achievements, none of Hercules' departments exceeded the rate of GNP growth during this period (57 percent), nor do any of these figures reflect the impact of inflation, which eroded 20 percent of the dollar's worth between 1962 and 1970. In real terms, then, E&CP and P&PC actually lost ground.[20]

These departments were, at worst, declining slightly in absolute profits and, at best, nearly keeping pace with general economic growth. What became of the company's remaining great hopes for growth: international and petrochemicals?

THE MAKING OF A GLOBAL CORPORATION

President Thouron's lack of confidence in the future profitability of the American chemical industry may have wounded Hercules in the Haveg debacle, but it brought great benefits to the company when it came to foreign expansion. Thouron was a strong internationalist. Building on the foundation for foreign expansion that it had established during the 1950s, Hercules grew as a multinational competitor at the highest rate in its history during the 1960s and early 1970s. By 1973, with forty-four joint ventures and wholly owned subsidiaries beyond American borders, Hercules had become a truly global corporation (see Exhibit 11.4).

Space does not permit us to discuss the salient features of each foreign operation. Therefore, we will present (1) a case study of the establishment of a key foreign operation, Teijin-Hercules of Japan, and (2) a general discussion of Hercules' experiences and strategies in its three international territories, in the context of multinational American business in the 1960s and early 1970s.

Teijin-Hercules

The formation of Teijin-Hercules in 1963 was significant in several respects. The joint venture proved to be one of Hercules' most enduring and successful. Its formation also illustrates particularly well the problems the International Department faced in attempting to set up operations overseas, especially outside Europe. Finally, Teijin-Hercules manufactured DMT, one of the company's leading growth products both at home and abroad during this period.

The modern Japanese chemical industry was established under the Meiji Restoration (1868–1912) and grew rapidly between the two world wars; when measured against the chemical industries of Germany, the United States, and other industrial leaders, however, it was relatively underdeveloped by the mid-1950s. Two attitudes shaped the emerging industry—as well as the nation's overall economy—throughout the nineteenth and twentieth centuries: an aggressive search for advanced foreign technology, and strong resistance to foreign interference in Japanese political and economic affairs. As a result of its somewhat contradictory aims, Japan became one of the most restrictive nations for

EXHIBIT 11.4 Hercules' Subsidiaries and Affiliates, 1967

		Plants	*Products*
Australia	Australia Chemical Holdings Limited—Sydney	Springvale	Paper Chemicals • Dresinated Emulsifier
		Botany	Industrial Chemicals • Synthetic Resins • Pigments
		Alexandria	Electrical Components • Hardware
		Sydney	Engineered Products
		Gladesville	Vegetable Oil Crushing
		Collingwood	Die Castings • Plastic Extrusions and Moldings
Belgium	Hercules Europe S.A.—Brussels		
	Jean Saels S.A.—Brussels	Malea	Flexible Packaging
Canada	Hercules Canada Limited—Montreal	Burlington	Paper Chemicals • Pamites Tall Oil Fatty Acids
			Pamites Tall Oil Rosin • Synthetic Resins
		St. Johns	Pigments
England	Hercules Powder Company Limited—London	Erith	Paper Chemicals—Plasticizers • Resins
		Pendleton	Paper Chemicals
	Fred H. Wrigley Limited—Taunton	Taunton	Ground Products
	Nelsons Acetate Limited—Lancaster	Lancaster	Cellulose Acetate
	The Holden Vale Manufacturing Company Limited—Hasingdon	Hasingdon	Chemical Cotton
Finland	Oy Hercofinn AB—Helsinki	Tampere	Paper Chemicals
Germany	Abieta Chemie GmbH—Gerstholen	Gerstholen	Dresinate Emulsifier
	Hercules Papierchemie GmbH—Dachau	Dachau	Paper Chemicals

India	Herdillia Chemicals Limited—Bombay	Thana	Phenol • Acetone • Phthalic Anhydride
			Plasticizers • Diacetone Alcohol
Italy	Bewoid Italiano & Callegaro S.p.A.—Milan	Milan	Paper Chemicals
	Bewoid Italiano & Callegaro del Sud S.p.A.—Sora (Frosinone)	Sora	Paper Chemicals
Japan	Hercules Far East Limited—Tokyo		
	Teijin Hercules Chemical Company Limited—Tokyo	Matsuyama	DMT
Mexico	Quimica Hercules S.A.—Mexico D.F.		
	Lerma Industrial S.A.—Mexico D.F.	Salamanca	Toxaphene
	Montrose Mexicana S.A.—Mexico D.F.	Salamanca	DDT
The Netherlands	Hercules N.V.—The Hague	Zwijndrecht	Paper Chemicals • Synthetic Resins
		Middleburg	Dresinate Emulsifier • Water-Soluble Cellulosics
	N.V. Chemische Vertstoffenfabriek V[illegible]HL, TH. Ten Horn—Masstricht	Masstricht	Pigments
	Technical Packaging N.V.—The Hague		DMT
New Zealand	A.C. Hatrick (NZ) Limited—Wellington	Auckland	Industrial Chemicals
		Wellington	Industrial Chemicals • Synthetic Resins
Nicaragua	Hercules de Centroamerica S.A.—Managua	Managua	Toxaphene
Portugal	Resiquimica-Resinas Quimicas Ltda.—Sintra	Mein Martins	Synthetic Resins
Spain	Macaya Agricola S.A.—Barcelona	Barcelona	Agricultural Formulations
Sweden	Hercules Kemiska Akliebolag—Goteborg	Lilla Edet	Paper Chemicals
		Sandarne	Paper Chemicals

Source: Joseph O. Bradford, "Hercules Circles the Globe," *Hercules Chemist* 55 (September 1967), 12.

foreign manufacturers in the post–World War II period, but also a leading licensee of foreign technology.[21]

It is therefore not surprising that when the Ministry of International Trade and Industry (MITI), which controlled foreign technology and investment in Japan, formed a petrochemical technology committee (made up of twenty executives from leading petrochemical firms) in November 1954, the committee's recommendation after three months of deliberations was that Japan join the international petrochemicals race by importing, rather than developing, all technical know-how in the field. Accordingly, Mitsui Group, Mitsubishi Petrochemical, Sumitomo Chemical, and Nippon Petrochemical were soon authorized to begin constructing petrochemical complexes.[22]

Such was the state of affairs Harvey Taufen found in 1956, when Henry Thouron sent him to Japan to investigate petrochemical opportunities.[23] Extensive analysis convinced Taufen that the Japanese petrochemicals industry was "conceived but not yet born"; in the wake of the MITI committee report, there was great interest in petrochemicals on the part of scores of companies, but no concrete progress. Indeed, Japanese eagerness to move ahead in petrochemicals was reflected in the terms of a bold offer from Mitsui to Hercules: to trade technical know-how for half interest in a heavily leveraged, $25 million chemical complex to be built by Mitsui to produce phenol, terephthalic acid (TPA), low-density polyolefins, and ethylene oxide. But because of the deal's rather speculative financial structure, and the long-term demands it would have placed on the company's technicians, Hercules declined.

Still, the Japanese pressed on and made great strides in petrochemicals in the late 1950s. They secured technologies from U.S. engineering firms such as Universal Oil Products, Stone and Webster, and Scientific Design, as well as from leading U.S. and European operating companies (including the technologies for producing Du Pont's polyethylene, Montecatini's polypropylene, and Sohio's acrylonitrile). Whereas the Japanese spent $4 million for licenses and invested $62 million in petrochemical plants in 1957, these figures jumped to $12 million and $225 million, respectively, by the end of the decade.[24]

After setting up a paper and rubber chemicals joint venture in Australia with A.C. Hatrick Pty. Ltd. (Hercules Powder Company Pty. Ltd. of Australia) in 1960, Taufen returned to Japan.[25] He found that Mitsui had erected the four plants as promised four years earlier, and that many of the Japanese petrochemical companies were competing fiercely to integrate forward into fibers. MITI continued to arbitrate, allocating new processes to various competitors and, according to some rumors, taking bribes for attractive licenses.

Hercules was now prepared to act. Taufen narrowed the field to two potential DMT licensees—Mitsui and Teijin—and negotiated with both. Teijin, a leading producer of polyester fiber, was seeking to integrate backward and held an edge in several respects. As a long-time consumer of Hercules cotton linters (when it was Tekuku Rayon), Teijin was both familiar with Hercules and convinced that DMT was the best chemical route to polyester fiber. Teijin was also the only Japanese company with captive capacity to make DMT.

Nevertheless, negotiations dragged on for two years—in part because

Teijin was not sure MITI would grant it permission to manufacture DMT, even if it secured a license from Hercules, and in part because of the Japanese bargaining style, which was in sharp contrast with American expediency and directness. Taufen and his department colleague John Present finally nudged the deal to a conclusion in 1963. The result was a $2 million joint venture called Teijin-Hercules, Inc., with Hercules holding a 30 percent interest. The company constructed a thirty-million-pound-per-year DMT plant at Matsuyama (on the island of Shikoku), which began operating in the spring of 1964.[26]

The enterprise was an undisputed success for both companies. Thanks to Hercules' know-how, Teijin's polyester fibers quickly became the best in Japan. In spite of periodic gluts in the polyester market, and even with the expansion of the DMT facility at Matsuyama and the building of new plants at Tokuyama and Ehime (increasing the company's total annual capacity to nearly 250 million pounds by the end of the decade), Teijin-Hercules continued to run at or above rated capacity.[27]

Hercules, meanwhile, increased its equity participation from 30 to 49 percent in the capacity expansions and enjoyed healthy revenues—more than $1 million per year in pretax profits after 1964, and well over $2 million per year by the 1970s.[28]

Hercules' Strategy in Comparative Perspective

The story of Teijin-Hercules illustrates some of the intricacies involved in setting up foreign ventures. Indeed, within Taufen's territory, Teijin was more easily established than several other joint ventures. In India, considerable "red tape with the Indian government" thwarted until 1965 the establishment of a joint venture (Herdillia Chemicals Ltd.) with Distillers of England and E.I.D. Parry Ltd. of India to manufacture several basic organic chemicals (phenol, acetone, diacetone alcohol, phthalate esters, and phthalic anhydride) near Bombay.[29]

But the outcome most feared by multinationals came true for Hercules in Pakistan after it established an ammonia fertilizer joint venture, Dawood Hercules Chemical Ltd., in 1968.[30] "Once the plant started . . . to make money," recalls Taufen, "the Pakistani government abrogated all [of their] solemn treaties. They took away the right to sell; they took over the product; they let us have only our depreciation money to service the debt"—even though the project generated substantial taxes and employment in Pakistan, significantly increased the nation's agricultural productivity, and helped its trade balance.[31]

Did such problems plague Hercules operations in the Americas and Europe? Generally, only in the former. "In Latin America," recalls Sam Beasley, whose territory was the non-U.S. Americas, "nothing was done normally. . . . Arbitrary decisions by centralized governments and graft were the paramount things we had to fight." Heads of state such as the Somozas of Nicaragua controlled foreign investment with an iron hand. In Mexico, for example, Hercules opted to sell its interest in a successful pesticide plant at Lerma because it feared that pressure from the Mexican president's son-in-law, who operated a potentially competitive fertilizer business, would lead to the revocation of critical permits.[32]

Also within Beasley's territory, Canada more closely followed the European pattern (see below). But Hercules made little headway setting up plants there because it was more efficient to export and pay tariffs, especially since most Canadians lived near the U.S. border. Attempts to form joint ventures to produce hydrogen cyanide (with Sherritt Gordon Mining Ltd.) and polypropylene (with Canadian Industries Ltd.) never came to fruition; in Canada, Hercules only operated the plants it had acquired along with Imperial (pigments) and PMC.[33]

Not surprisingly, Europe was the easiest arena in which to operate. Henry Reeves found that, in addition to having similar cultural traditions, most European businessmen spoke English. (Still, adaptability was as desirable there as elsewhere; Reeves, for instance, took a freezing cold swim with Finnish businessmen to gain their confidence.) Moreover, trade and financial ties between the United States and Europe were well established. As a result, in its European ventures, unlike those elsewhere, Hercules normally held total ownership.[34]

What can be said about Hercules' overall strategy of foreign expansion? How did it compare with those of other multinational corporations?

When contemplating a joint venture of plant construction beyond American borders, Hercules faced several critical questions. Which nation would make the best host? Developed or less developed? European or non-European? When was the best time to build? Which products were the best to manufacture? Should foreign direct investment be funded by Hercules, through a joint venture, or with local capital? There were other important considerations: Would new foreign plants hurt Hercules exports? Would they hurt the U.S. balance of payments?

Hercules decided to build foreign plants under one or more of the following conditions: markets for Hercules products existed that could not be satisfied by its U.S. exports (because of high tariffs, high shipping costs, and so forth); competitors were interested in invading a Hercules sales territory with imports or a new plant; or special incentives were offered by a foreign corporation, government, or international organization (such as the World Bank). In addition, changes in international trade relations—for example, rise of European trade communities (see Chapter 10)—sometimes motivated U.S. investment abroad.[35]

Hercules found conditions for foreign manufacturing more favorable in industrialized nations than in less developed countries (LDCs). Factors that commonly made LDCs less attractive than developed economies (except Japan) included: low purchasing power and high inflation rates; political instability; government interference; resistance to foreign intervention (or specific anti-Americanism); and the difficulties of dealing with non-Western languages and cultural norms. Thus, West Germany, Belgium, Italy, Australia, and, to a lesser degree, most of the other West European countries permitted foreign investment during the 1960s, but India, Pakistan, Spain, Ceylon, and Japan tried to hold the line by making joint ventures a "virtual requirement."[36] Given the

International Department's limited resources, its strategy was to target a few key countries in each territory, favoring those that had more developed economies and welcomed foreign investment.

The choice of products can be summarized easily: Hercules exported its stars. Paper chemicals, DMT, toxaphene, and other very profitable products in which Hercules held strong proprietary positions were exported; explosives, cellulose products, pigments, and other widely available commodities generally were not. Moreover, when building capacity to make more competitive products, Hercules sought out markets in which its rivals had not yet established a presence.[37]

The nature of local markets mattered as well. As noted, the Scandinavian countries were among the largest per capita consumers of paper in the world, and Japan had become a leading producer of polyester fiber. The nature of Hercules' expertise also made a difference, although it was generally less important than local demand. In Japan, for example, Hercules was the first DMT supplier because its process was the best available (and because its negotiations were astute and well timed); in Pakistan, on the other hand, the company relied in part on an ammonia process developed by an American competitor. In every country, local demand set an agenda that investing companies had to meet on many levels, from personal relations between the key negotiators to financial arrangements, raw materials, and product prices.[38]

Hercules favored foreign sources of capital over its own retained earnings, debt, or equity to finance foreign operations; the norm was nine-tenths foreign capital. As noted, a high degree of local financial participation was often required by the host country, especially by LDCs. Even so, Hercules did not possess the resources to fund the level of foreign expansion it sought. Local financial participation not only benefited host nations (which Hercules tried to do in all cases, for both altruistic and practical reasons) but also tended to expedite projects, because the company paid high debt service on unfinished facilities. Finally, Hercules encouraged host nation investment to help alleviate a growing U.S. trade imbalance (see below).[39]

The impact of Hercules' overseas production on its own exports was generally very positive, not only because the company built foreign plants in markets it could not otherwise meet with exports, but also because local production of some Hercules products often spurred demand for other Hercules-produced raw materials, semifinished products, intermediates, and related products.[40] (See Exhibit 11.5.)

Similarly, the company's overseas plants helped redress the U.S. international trade imbalance in the 1960s. In 1966, soon after President Lyndon Johnson called for a "volunteer balance-of-payments" program among American multinationals, Hercules posted a favorable trade balance of $55 million. The following year, it formed (and invested $6 million in) the Hercules International Finance Corporation, which offered $25 million worth of five-year debt to foreigners in order to obtain more funds from outside the United States. (The funds were purchased in Eurodollars at the rate of 6⅝ percent interest.) In short, because Hercules' foreign manufacturing operations greatly stimulated U.S. ex-

EXHIBIT 11.5 **Hercules' International Sales, 1964–1973 (Dollars in Millions)**

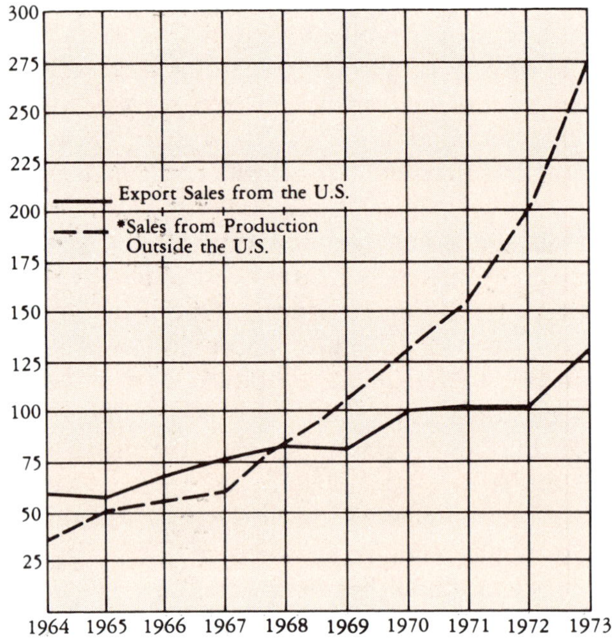

*Replaces sales from overseas production.

Source: Hercules Powder Company, Annual Report for 1973.

ports and consumed much foreign investment, they contributed positively to U.S. international trade flows.[41]

Did Hercules' experiences in direct foreign investment in the 1960s and early 1970s generally parallel those of other American multinationals? Mira Wilkins, a leading historian of U.S. corporate foreign investment, notes that most direct foreign investment after World War II was made by "manufacturing companies in market-oriented investments," that is, companies that had long histories of foreign involvement and strong positions in U.S. industries and sought to exploit advantages in technology or unique products. In addition, most investment was focused in Europe and other industrialized areas, despite U.S. government policies to encourage investment in LDCs, because "the environment abroad is more important than U.S. policy in stimulating investment." Wilkins concludes that direct foreign investment by U.S. multinationals on balance benefited both sides.[42]

Hercules' experience reflected these trends. In 1973, the company sold $170 million worth of goods produced in its European plants, compared with $57 million sold from its other foreign manufacturing sites. (Accordingly, that year the company formed a new department called Hercules Europe, headed by Al Giacco.) In Pakistan, Japan, and other difficult settings, Hercules learned that local markets and local controls mattered the most. On balance, it benefited tremendously from its foreign manufacturing operations. Just as U.S. direct foreign investment abroad grew faster than GNP during the postwar years, Hercules' profits from direct foreign investment on a percentage basis outpaced those from its domestic operations between 1962 and 1973.[43]

ENTRENCHMENT IN PETROCHEMICALS

The third area in which Hercules had anticipated high growth—petrochemicals—was less successful than the company's foreign ventures, although not for lack of commitment. The company poured hundreds of millions of dollars into its young DMT and polypropylene resins, fibers, and film businesses in the 1960s and early 1970s. The risk seemed justified: DMT became one of the company's most successful products, and its polypropylene lines seemed poised on the verge of explosive growth. Altogether, the investments in new petrochemical plants and plant expansions represented one of the greatest channelings of capital in the company's history; certainly, they represented the strategic centerpiece of the Thouron-Brown years. As such, one way or another, they would alter the shape of Hercules for decades.

DMT

Dimethyl terephthalate (DMT) was the most profitable of the company's petrochemicals at this time, and the one about which the company was most optimistic (see Exhibit 11.6). Consider the following expansions:

In the late spring of 1965, Hercules began construction of a $15 million, sixty-million-pound-per-year plant—named for President Forster—near Spartanburg, South Carolina, a community of fifty thousand chosen for its proximity to textile producers. The plant was completed in 1966.[44]

That year, the board approved the construction of another DMT plant and purchased a site at Wilmington, North Carolina. This $9.3 million "Hanover" plant went into production in 1968. By 1972, its capacity was being expanded from six hundred million to nine hundred million pounds, at an additional cost of $6 million. Adjacent to this plant, Hercules in 1973 also began construction of a forty-million-pound-per-year plant to manufacture Terate aromatic resin (derived from terephthalate) for sale to producers of adhesives, sealants, composition board, and polyols for urethane foams.[45]

The company's original DMT plant at Burlington, New Jersey, was expanded as well. Its capacity reached 80 million pounds by 1971, when the board approved another $4.5 million expansion (completed in 1973), raising total plant capacity to 150 million pounds per year.[46]

The company's overseas DMT capacity also expanded. Hercules B.V., the wholly owned subsidiary in the Netherlands, built a 110-million-pound-per-year facility at Middelburg in 1967 and expanded it to 275 million pounds per year

EXHIBIT 11.6 Sales and Profits from Hercules' DMT, 1962–1972

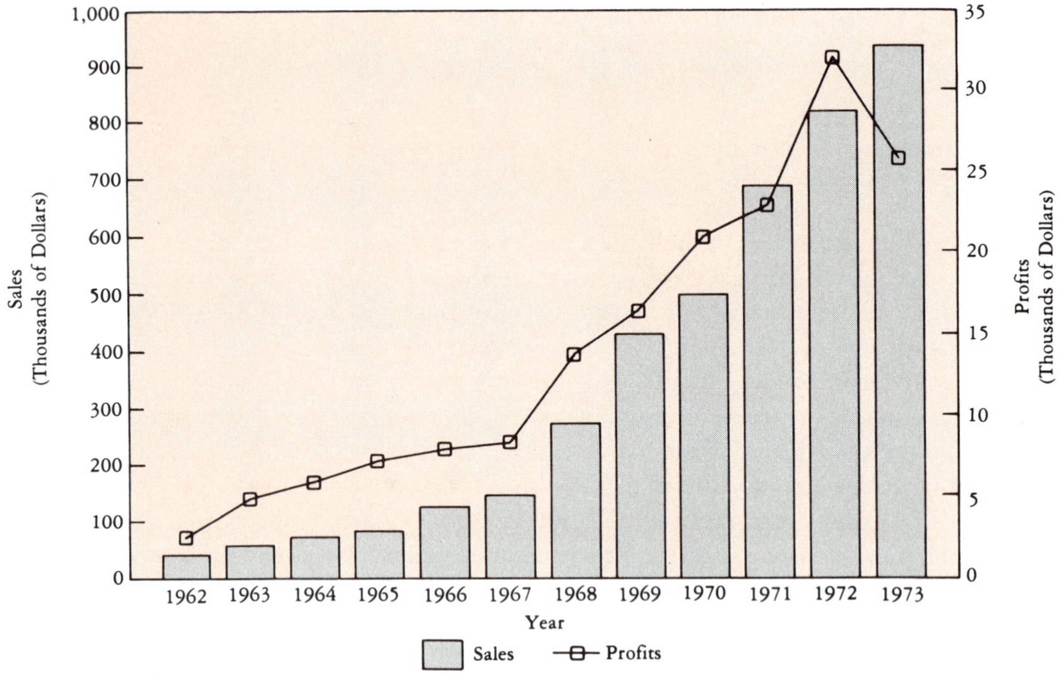

Source: Synthetics Department, Annual Reports for 1962–1973.

by 1973. By then, Hercules also owned 49 percent of Teijin-Hercules, whose three Japanese plants could produce 400 million pounds of DMT per year.[47]

Even more important than plant expansions was Hercules' attempt to integrate forward into polyester fibers as it had done in polypropylene. The move into a second synthetic "miracle" fiber was, as Thouron put it, "one of [Hercules'] most important milestones"—for its risks as much as for its opportunities. Demand for polyester—for applications in clothing, home furnishings, and tire cord and for various industrial and military uses—had been growing rapidly and seemed almost insatiable; indeed, unlike polypropylene fiber, there had recently been shortages of polyester fiber. Still, Hinner and others had to acknowledge that Hercules faced "stiff competition" in the new business, especially from the "big four" who dominated the U.S. market: Du Pont (Dacron), Celanese (Fortrel), Eastman (Kodel), and Beaunit (Vycron).[48]

To limit its risk, Hercules entered the business in 1966 through a joint venture with Hoechst called Hystron Fibers. There were several reasons for the move, apart from the proven compatibility of the two companies. Hoechst had been in the business since 1954, when it licensed polyester fiber from ICI and began producing Trevira. As the ICI patents expired (between 1963 and 1968, depending on the country), the German company began to expand aggressively in polyester fibers in Europe, South America, and Africa. By the mid-1960s, Hoechst was the leading producer in continental Europe; when it learned that

American companies planned to invade its territory, it decided to "jump into the American lion's den."[49]

Still, Hoechst proceeded cautiously, knowing that any serious foray into the American market would require at least $100 million ("an engagement" that Karl Winnaker of Hoechst described as being "on a scale that was quite new to us as far as foreign projects were concerned"). Hercules held several attractions: it possessed years of experience marketing polyolefin fibers in the United States; it was the largest U.S. producer of DMT; and its new DMT plant at Spartanburg, South Carolina, was situated in the "heart of the American textile industry" and only fifty miles from Hoechst's dyestuffs laboratory at Charlotte, North Carolina. Moreover, Hercules had "declared itself ready to incorporate its modern DMT plant into the new company." Ross O. Watson (plant manager at Covington after 1963) was named president of the new corporation, and Gunther I.O. Ruebcke of Hoechst, its vice president.[50]

Hystron proved successful—indeed, too successful as far as Hercules was concerned. The first thirty-million-pound-per-year section of the plant was dedicated (with a polyester clothing fashion show) on April 9, 1968, after delays and cost overruns. Thereafter, the plant operated smoothly, and construction of a second section of equal capacity was going well. The problem was that, finding its highest rates of growth in the United States, Hoechst wanted to expand the business quickly; Hercules wanted slower growth.[51] As Werner Brown recalls,

> [Hoechst] had the global mentality of wanting Trevira all over the world. [Hystron] represented just one leg of that many faceted marketing approach. And they were quite ready to spend another $100 million before the first $100 million had started making real money. That wasn't our cup of tea; we wanted to see that real money was made first and then go. They said you can't wait that long, it's too competitive.[52]

The companies parted ways amicably in February 1970, when the Hercules board voted to sell its portion of Hystron, including the Forster plant, to Hoechst for $32 million.[53] Brown considered it a fair price and valued the experience gained through the venture.[54]

The Hystron episode can be seen as a case of diverging strategies: Hercules wanting to "strengthen [its] position in the manmade fibers industry," and Hoechst pursuing a more ambitious global strategy. But the episode also reflects a Hercules characteristic during this period: its unwillingness and inability to commit the resources necessary to break into a major new field—the result of its deep commitments to other petrochemical products.[55]

Demand for Hercules DMT, spurred by periodic worldwide shortages and double-digit growth of polyester sales, continued to grow vigorously. At the end of 1972, the executive committee expected demand for Hercules DMT in 1974 to exceed 1.3 billion pounds. Thanks to an aggressive expansion program, Hercules was positioned to satisfy such a level of demand.[56]

There was only one troubling sign on the horizon: terephthalic acid. As noted in Chapter 9, DMT had long prevailed as the primary polyester raw material because of the difficulty of purifying TPA. DMT's quality was also higher, most customers were equipped to utilize it rather than TPA, and it commanded a slightly higher price because of its higher molecular weight. Over time, however, Amoco improved its process for manufacturing TPA, bringing its quality in line with DMT's and therefore making the price differential significant.[57]

In response to the challenge, Hercules began to build its own TPA capacity. In 1972, it started constructing a $31.6 million, 250-million-pound-per-year TPA plant adjacent to the Hanover DMT facility. It was completed the following year, when the company also formed a joint venture with Petrocel to build a 300-million-pound-per-year DMT/TPA plant at Tampico, Mexico (completed in 1975).[58]

But Hercules remained committed primarily to DMT. Its process for making TPA (not the same as the one used by its competitors) was a more expensive adaptation of its DMT technology, and one from which the company made little or no profit. The Hercules TPA plants were part defensive—"Little by little," Werner Brown recalls, "we indeed found ourselves either going out to supply TPA or losing business"—and part posturing ("Don't get the notion that we're about to curl up and die.") To become a serious competitor in TPA, Hercules would have had to rebuild from scratch.[59]

For this reason, the directors were prepared to listen to the assurances of the company's leading DMT advocate, Harvey Taufen, that TPA would never capture more than about 10 percent of the market. To protect its market share, as we will see in Chapter 12, the company would continue to expand its DMT capacity.[60]

Polypropylene: Resins, Fibers, and Film

With the startup of its Lake Charles plant in the spring of 1961, Hercules became the world's leading producer of polypropylene resins. It was a stumbling start: in the first winter, pipes and vessels throughout the plant burst because Northern engineers had failed to provide insulation. But the more serious problem was industry overcapacity, which continued to drive down prices and force plants to operate below capacity. Yet consumption of the plastic was growing rapidly.

Thus, wanting to remain dominant in the field, but to avoid overbuilding, Hercules was forced into a difficult guessing game. For the present, it opted to forge ahead. Before the first plant had been completed, the board of directors (with only Morrow dissenting) had authorized construction of a second $16 million, sixty-million-pound-per-year facility (process 2). Altogether, there were now four batch slurry lines in processes 1 and 2, and seven "cold cut" Pro-Fax® lines in the molding powder plant, boosting the company's total capacity to make polypropylene flake to some two hundred million pounds by the end of 1962.

But overcapacity ensured that process 2 was idled within months; it would run only sporadically in subsequent years. Indeed, it was several years before capacity at Lake Charles was expanded (when process 3 was added in 1968–1969), and even then the addition was motivated as much by a desire to upgrade the technology as to add capacity (to 340 million pounds). Because it relied on continuous process rather than batch technology, process 3 was dramatically more efficient than its predecessors.

Hercules' early realization that it would need to create an internal market for the commodity plastic by integrating forward into fibers and films seemed more valid than ever. As in the late 1950s, the first expansion came in fibers. In 1966, Hercules began constructing a twenty-five-million-pound-per-year Herculon fibers plant near Oxford, Georgia, a declining textile community east of Atlanta. The plant started operating in October 1967.[61]

As with all new materials, it took Hercules years to discover the best applications for Herculon fiber. One obvious possibility, thanks to the nylon revolution, was apparel and women's hosiery. But it was clear from the outset that Herculon would capture a share of these huge markets only with great difficulty. Polypropylene's Achilles' heel, its relatively low melting temperature, precluded ironing and machine drying, and thus most apparel applications. The hosiery it was tested in bagged at the knee and retained the shape of the wearer's leg. One promising avenue was large-diameter monofilaments for lawn furniture, ropes, and (because of the chemical resistance of Herculon) industrial uses. For the most part, however, it appeared as if Hercules would need to "invent" new applications for the product, rather than substitute other materials in existing applications.[62]

This attitude changed with the coming of Herculon's first major market: carpet fiber. The breakthrough account was E.T. Barwick Company in Chamblee, Georgia, in the heart of the nation's carpet-manufacturing region. Fortuitously, American carpet-makers were in the midst of a technological revolution that would transform the marketplace in the 1960s. Carpeting hitherto had been considered a luxury for most Americans. But manufacturers had begun to apply to carpet-making the same mechanized tufting techniques used for bedspreads and other fabrics, thus lowering costs and prices to within the reach of the average consumer.[63]

Hercules managers at first expressed concern that such market appeal might create an image problem for the new product and urged distributors to create a "more luxurious" image. (The same concern would resurface in the late 1970s, prompting Hercules to create the Signature Collection of upholstery fabrics.) But by mid-decade, the carpet-driven demand for Herculon was proving the product's worth, and Hercules began to promote the fiber among general consumers (see below). Sales of Herculon jumped 30 percent to 21.6 million pounds in 1965, and nearly doubled the following year. Herculon was now available in sixty varieties, including tufted, woven, nonwoven, random-shear, level-loop, Moresques, bold sculpture, lip shear, and plush. In addition, Barwick began offering a piece-dyeable class.[64]

Hercules responded promptly to the growing popularity of Herculon by adding new capacity. Under the auspices of the new Fibers and Film Department (created in late 1965 out of the film division of Polymers and the Fibers Development Department), the company expanded capacity at Covington to sixty-five million pounds per year by 1966. At the same time, the company launched an intense advertising campaign—one of the few in its history—to promote Herculon.

Hercules marketed the fiber directly to retailers and consumers for several reasons. As a unique product, Herculon had no competitors that stood to gain indirectly from Hercules' advertising. Final consumers, retailers, and intermediate producers (yarn spinners, carpet weavers, upholstery makers) needed to be educated about the product if they were to place orders. In touting the fiber to these groups, Hercules followed the example set by Du Pont, which had heavily promoted nylon—also a new synthetic fiber produced from a proprietary position—beginning in the late 1930s.[65]

To reach the mills, Hercules advertised extensively in trade journals. It also subsidized fabricators' advertising costs, often by sharing the expense of the "hang tags" attached to carpets or furniture (they carried both the name "Herculon" and the mill's brand). To foster consumer demand, Hercules hired an advertising agency and launched an expensive multimedia campaign. In the mid-1960s, millions of Americans each week read about the unique attributes of Herculon in newspapers and magazines, heard about them on local radio programs, and watched celebrities such as Frank Blair and Jack Lescoulie on the "Today," "Sunday," and "Tonight" television shows soil and clean carpet samples made of Herculon.[66]

Although a few products had previously brought the Hercules name before the general public and there were sound reasons for promoting Herculon to retailers and consumers, forward integration into consumer products was still awkward for a company that envisioned itself as a value-added developer of chemical intermediates. Two department managers recall that convincing the company's senior managers to undertake the sales campaign for Herculon was a "real cultural problem."

> [Hercules] made products that got delivered in tank cars and bags and drums, and all of a sudden here we are not only selling our product, but also going to our customers' customers—carpet distributors, retailers—and saying, "We will help you support your advertising on behalf of Herculon if you sell that product."[67]

Indeed, Herculon consumed more advertising dollars ($1.5 million in 1969) than any other product or department in this period, even though general consumer advertising of Herculon was curtailed in 1971. The program seemed successful, the growing market for synthetic-fiber carpeting in the 1960s was clearly a boon for Herculon, and demand from carpet-makers brought several capacity expansions. But profits from Herculon were far from outstanding.[68]

Moreover, carpeting proved to be neither an enduring nor the best application for the product. Its light weight, color-fastness, and stain resistance were offset in the carpeting market by its poor resilience compared with nylon. High-pile shag carpets for residential use, for example, stayed matted down from foot traffic. As a result, Herculon remained strong only in the relatively limited market for commercial and industrial carpeting, which had a lower pile and was denser. Fortunately, the carpeting revolution was spreading into these settings as well.[69]

Even more fortunately, Herculon fiber soon found a new application. Beginning in the late 1960s, furniture upholstery manufacturers (Park Silk Company of New York was the first) began to show an interest in the material. A writer for *Upholstery Industry* magazine explained:

> Herculon is one of the most stain resistant and easily cleaned of all fibers currently used in home furnishings. It does not require "post treatments" for stain release; the quality is intrinsic. In addition, olefin has excellent abrasion resistance and outstanding color fastness when offered in pigment colors.

Again, Hercules emphasized the stain resistance and durability of the product; in the new market, the company was no longer handicapped by Herculon's poor resilience.[70]

While the leading furniture upholstery makers such as Burlington, Stevens, and Milliken were loath to buy large amounts of Herculon, smaller firms (especially the Quaker Fabrics Company of Fall River, Massachusetts) were not. Many yarn makers were obtaining good results in applying a texturizing process (Taslan) developed by Du Pont in 1954. Brighter colors were coming into vogue as well.[71]

Supplying Herculon to upholstery businesses was more profitable for Hercules than supplying the carpeting industry. As we will see, a still more important breakthrough came later: bulk continuous fibers.

Hercules' first decade in polypropylene film would bear many similarities to its experience in polypropylene fibers, especially with regard to searching for the best markets and scaling up productive capacity to meet those markets. In 1963, the year after Hercules created a film division in the Polymers Department, it put into operation a new film plant at the Covington, Virginia, facility.[72] Again, the company faced critical marketing issues: Should it attempt to create new use for the film? Or should it seek to replace other materials in already existing applications? If so, which applications? The search for answers to these questions would take time and would reflect Hercules' abilities to modify its polypropylene films to possess specific properties.

Initially, the film division made a "small but aggressive" attempt to dislodge polyvinyl chloride (PVC) film from the shrink-wrap market. But efforts to sell "shrink polypropylene" never gained momentum—in part for technical reasons, but mainly because the total market (for wrapping records, playing cards, greeting cards, and the like) was relatively small.[73]

The Polymers Department had its eye on a much larger prize: to "challenge the heart of the cellophane business." For this, improvements in the heat-sealing and coating of the plastic were needed. Hercules not only drew on expertise from ICI and Kalle and licensed a high-speed coating process, but also installed a new fifteen-million-pound-per-year coating machine at Covington in 1964. Increasingly, it seemed as if the properties of Hercules' polypropylene film—lower price (because of molecular weight factors), stiffness (to run through packaging machines), clarity (for a see-through package), and long-term flexible strength (to prevent cracking from age and dehydration, like cellophane)—would make it a formidable challenger to the older cellulose-based film.[74]

By the late 1960s, urged on by Polymers managers George Taylor and Charles Grant, Hercules was ready to raise its bets on polypropylene film. The company also secured two key accounts: Philip Morris, which decided to use the film to wrap cigarette packages; and Frito-Lay, the snack-food giant, which purchased the film from converters that had printed and laminated it. In this new field, Hercules was again building a reputation on strong technical service to key accounts. As Paul Johnstone, general manager of Fibers and Film, reported in 1967,

> Packaging engineers have been moved into the districts and gained confidence as Hercules took complete responsibility for all packing lines with large volume customers. The plant has learned the quality required to keep such customers happy, and Research has responded promptly to solve unforeseen problems when they become key issues for moving ahead.[75]

To better serve Midwestern markets (and to not leave its film business vulnerable to the shutdown of a single operation), Hercules built a new polypropylene film plant at Terre Haute, Indiana, which began operating in December 1968. The facility looked more like an office building than a factory; its superclean environment was designed to prevent small particles or insects from bursting the bubbles in its four units.[76]

Although it would be three years before Terre Haute reached full capacity, and before Hercules worked out the technical issues (especially printability) involved in producing a satisfactory cigarette wrap, the greatest barrier to success was the fact that customers' machinery was designed to handle cellophane, not polypropylene film. As Fred Buckner recalls, "You couldn't walk in to a guy and say, 'Hey look, I can save you 5¢ a pound on your purchases but [you need] to replace $100,000 worth of equipment.' " To overcome the problem, Hercules often redesigned and modified customers' equipment to handle its thinner, lighter film, or it simply waited for attrition to take its toll. Indeed, by the early 1970s, new installations of equipment designed to handle cellophane had virtually ceased.[77]

At the end of its first decade, the polypropylene film business was a mixed success. Hercules controlled slightly more than half of the U.S. market, competing primarily against Mobil, but also against Diamond, Cryovac, and Du Pont (which would soon exit the field).[78] Cellophane had been soundly defeated.

Moreover, the "value-added objective was being accomplished"—that is, the film business was boosting demand for Hercules resins through direct purchases, and indirectly by increasing general acceptance of polypropylene film.[79]

But Hercules polypropylene fibers and film were hardly financially sound in their own right. Both lost several million dollars each year throughout the 1960s. When Werner Brown took office, he concluded that the infancy period for the two products was over.

MANAGEMENT BY THE NUMBERS

By the end of the 1960s, it was clear that the problem of growth that had long occupied Thouron's administration was not being solved by the company's domestic businesses, old or new. Petrochemicals, in particular, were not yielding high returns, in part because of overcapacity throughout the industry, and in part because Hercules plowed most of its earnings back into investment in plants and equipment. Thus, Hercules was continuing to search for new sources of growth when its presidency changed hands for the fourth time.

Enter Werner Brown

On May 27, 1970, the fifty-eight-year-old Thouron resigned as president and chairman of the executive committee because of ill health. He became chairman of the board, replacing the retiring Elmer Hinner, but left the company altogether at the end of the year. (Those close to Thouron knew of his failing health; indeed, he died of a heart attack five years later.) In a closed ballot election, the board voted fifty-one-year-old Werner C. Brown as the company's fifth president.[80]

Even then, it was apparent that Brown's election would not represent a sharp discontinuity: he was steeped in Hercules' traditions and perspectives, having spent his entire career with the company and risen through its ranks in a "very classical straight line." Brown had joined Hercules soon after graduating from Duke University in 1942. After stints at the Research Center and as an ordnance manager at Sunflower, he began to cultivate his skills in marketing for the C&PP and Polymers departments. He advanced steadily—to general manager of Polymers in 1960, corporate director in 1963, and three years later, vice president and member of the executive committee.[81]

Apart from his achievements, Brown was chosen because he was expected to carry on the company's basic strategic plan, which Brown himself had helped shape as an executive and as Thouron's close confidant. The new president understood his mandate to be, in the simplest terms, continued expansion internationally and in petrochemicals, the latter in order to wean Hercules further away from its dependence on mature businesses such as cellulose, rosin, and nitrogen derivatives.[82]

Diversified companies such as Hercules had long categorized their various lines of business in terms of current profit margins, perceived growth potential, and so on. But at Hercules, the practice was gaining increased emphasis because of the tightening constraints of the period. Soon after the leadership transition, for example, Brown and Thouron often described the company's lines of business as: (1) "turnaround" (static, perhaps mature, businesses that were candidates for divestiture if they did not soon show signs of a turnaround); (2) "cyclical" (those whose fortunes would parallel changes in the GNP); (3) "growth" (businesses that would thrive independently); and (4) "new enterprise" (startup businesses expected to achieve very rapid growth).[83]

Thouron had begun to translate this typology into action during his last months in office. The two most notable manifestations were the creation of the New Enterprise Department (NED) in 1968 and the initiation of a "housecleaning" program the following summer. Brown's administration carried on those programs. (The activities of the NED are discussed in detail below.) In the first two and a half years under Brown's leadership, Hercules sold off businesses at Kenvil and Parlin and divested or closed operations at Rocky Hill, Gibbstown, Spartanburg, Eugene, Tacoma, Glens Falls, Jacksonville (Arkansas), and Cleveland. Beginning in 1971, it also instituted "organizational realignments," which decreased indirect expenses to their lowest level in fifteen years.[84]

As the months passed, however, the new president was forced to intensify and modify the strategies he had inherited. U.S. economic conditions were worsening: years of heavy deficit spending to fuel the Vietnam conflict were taking their toll in the form of recession and spiraling inflation, prompting even conservative Republican President Nixon to institute wage and price controls.

First, as we will see, Brown stepped up the company's NED diversification efforts. Second, he scaled back the level of capital investment. In the last four years of the 1960s, Hercules had increased its investment in gross fixed assets more than 56 percent (from $533 million to $834 million), effectively expanding the company's sales capacity from $700 million to $1 billion. Brown pulled in the reins on new construction not only because he believed that debt was climbing out of control, but also because he was optimistic that the company could now reap the benefits of the heavy investments made under Thouron.[85]

Finally, the growing constraints of the period compelled an examination—perhaps unequaled since the postwar debate about entering petrochemicals—of the relative value of the company's various businesses. As a result, the early 1970s became a time of intense reflection about divestiture and diversification. To sharpen their thinking on the issues, the company's top managers sequestered themselves at Seaview, New Jersey, for a series of strategic planning sessions.[86]

At Seaview, they pored over reports prepared by the Boston Consulting Group (BCG). Along with McKinsey & Company and the Strategic Planning Institute, BCG had gained wide recognition in the late 1960s by developing a "portfolio approach" to strategic planning. It is useful to review two of BCG's key concepts—the "experience curve" and the "growth share matrix"—to understand the impact of its reports on Hercules.[87]

According to the BCG experience curve concept, companies become more efficient as their workers and managers gain experience—broadly defined—in a business. Improvements might result not only from greater worker efficiency (the simple "learning curve"), but also from economies of scale, product standardization, process improvements, and so on. So critical was the experience curve in lowering costs of all kinds, according to BCG, that it urged companies to invest early and heavily in new businesses to secure first-mover advantages.

Building on this notion, and to inform strategy making further, BCG also developed the "growth share matrix." One axis represented a continuum representing the relative market share of a business. The other axis reflected a business's rate of growth. By plotting a business on this matrix, managers could gauge its relative appetite for funds (based on the assumptions that greater market share resulted in lower costs, and that greater market growth rates had large appetites for capital).

BCG's growth share matrix won a wide following in part because each quadrant was conceptually easy to grasp and the quadrants were assigned catchy names: "stars" had high market share and high growth; "cash cows," high market share but low growth; "question marks," low market share but high growth; and "dogs" were low in both categories. The growth share matrix, moreover, carried strategic implications: to invest heavily in stars, minimally in cash cows, and heavily in question marks (to make them into stars) and to divest dogs.

Unfortunately, the BCG reports prepared for Hercules in the early 1970s—which included learning curve data and plotted each of the company's businesses on matrices—apparently have not survived. Still, the findings can be deduced from other sources and subsequent actions. Rosin, cellulose, phenol, and methanol, which together were responsible for most of the company's revenues, were classified as cash cows. Dogs included foundering units already divested (or about to be), such as those under the Haveg umbrella. Petrochemicals were the stars, although one startling finding was that Hercules was beginning to experience a reverse experience curve in DMT (total costs had fallen, as the theory predicted, but, unaccountably, were rising again). Finally, there were many young question marks: in modular housing, health care, waste treatment and reclamation, and electronic and printed communications. For the most part, these were the recent finds of the New Enterprise Department.[88]

Unrelated Diversification: The New Enterprise Department

As before, when Hercules sought growth through diversification, the company formed a new institutional body to administer the effort. Like the Development Department of 1926–1936, but unlike the Industrial Research Department of 1919, the New Enterprise Department was to be an ongoing operation.

The NED was organized in 1968. Its first general manager was Dr. Eugene D. Crittenden, whose sharp analytical skills and comprehensive knowledge of Hercules operations (he had held several research and management posts in the P&PC, Synthetics, and International departments since joining Hercules in 1951) made him well suited for the assignment.[89] The NED began with ten staff members and a full agenda of opportunities already under investigation within Hercules. These were divided into five groups (see Exhibit 11.7).

EXHIBIT 11.7 Starting Agenda for the New Enterprise Department, 1968

I. Research

"New products or processes from Central Research which are outside established interest of any operating department."

First leads: Polyolefin open-web film with Smith & Nephew of Great Britain; oriented polypropylene bottles; polyurethane building sealants; synthetic diamonds.

II. New Technology

"New business opportunities resulting from purchase of inventions or businesses which are based on technology outside Hercules' normal fields."

First leads: Technology Investment Program: Program to "invest in small businesses based on new technology"; will invest \$200,000 in Procedyne, Inc. (fluidized bed instrumentation and related products). Also considering two clinical laboratories; companies that make animal food additives, can openers, building installation materials, fiber-stressed beams, soil-testing equipment and valves; and processes for making protein from cellulose and polyamides.

III. New Commercial Fields

"Economic and business investigation of major lines of industry or commerce where Hercules does not have a position at present."

First leads: Leisure time and recreation: Will present a report in January about how to enter this \$100 billion field now being pursued by conglomerates.

Farming: Will present options in field of large-scale "corporate farming" or farm management in January.

Environmental systems: "Systems" approach to pollution control and water management problems seems promising.

IV. Established Diversification Goals

"Tactical planning and plan implementation to enter new commercial fields resulting from II or III."

First leads: Solid Waste Program: Will submit proposal in January to form a two-year, \$300,000 joint venture with Eriez Manufacturing to help develop their system of converting solid waste into fuel.

Medical and hospital supplies: Includes trays, monitors, broad-use commodities (syringes, scalpels, gloves, etc.) and specialties (dressings, implantable plastics, refuse disposal, etc.). Previous efforts to enter this field via acquisition have failed; must decide whether to offer "selected, proprietary products" through established distributors or commodity products as well; Smith & Nephew, a leader in the field in

EXHIBIT 11.7 (Continued)

the United Kingdom, is interested in joining with us to enter the U.S. market.

Oil: Previous internal studies have concluded that a chemical refinery and fuel chemical refinery at Allemania or merger with an integrated oil company would not serve Hercules well. Great Yellowstone Corporation plans to sell out, including Hercules' gas and oil reserves owned through the Yellowstone Oil Co., Ltd. partnership.

V. Merger/Acquisition Leads

"Search, evaluation, and approach to merger and acquisition candidates selected in IV or arising from outside sources."

Mergers: Investigations of merger with an oil company and of a "major merger" with one of "eight large companies" (at least one-third the size of Hercules).

Acquisitions: Detailed analysis, often with outside help, of twenty-one acquisition candidates.

Source: Hercules Powder Company, New Enterprise Department Report, December 1968.

Endeavors of the first three types were seen as having significant "long-term potential" and were allocated about half the department's resources; those in the last two categories were viewed as short to medium term. At the outset, it was expected that the NED would place increasing emphasis on short- and medium-term projects—especially in nonchemical fields—that would yield higher than normal returns.[90]

As Crittenden later explained, the NED screened potential new products and companies for investment or acquisition according to rather specific criteria. The overriding goal was to boost Hercules' "inherent growth rate" (roughly, its return on total operating assets) to between 7 and 8 percent by adding new growth products, which had relatively high turnover ratios (sales revenue to capital employed of greater than 1.0) and high profit margins (greater than 10 percent net after-tax profit to sales revenues).

As for acquisitions, promising candidates were defined as having a "reasonable" financial record, good prospects for success (even without the intervention of Hercules), and reasonable price "relative to Hercules" (some candidates were eliminated because their price-earnings ratios were much higher than Hercules'). Specifically, the NED would seek companies with sales in the $14–150 million range and inherent growth rates of at least 10 percent.[91]

In short, the NED was searching for new products and companies offering higher than normal growth. In that way, Hercules hoped to shore up its overall performance and identify new growth areas, while bypassing the long and expensive developmental process—even if doing so carried the company beyond the chemical realm. As Crittenden put it,

> [New growth products will be found] primarily in new or emerging markets where chemistry plays an important, but not dominant role in offering a system to the customer. . . . It is probable that any significant impact on Hercules performance within the next five years must be achieved through sound acquisitions because of lead time necessary to build sales and profits from research or venture projects.[92]

In 1969, its first full year of operation, the NED moved ahead aggressively. In addition to its initial agenda, the department opened several new fields of investigation (high-income retirement and nursing homes, oceanography, furniture, and modular housing); considered acquisitions in the flavors and fragrances field; and, most important, purchased minority interests in four companies and completed negotiations with a fifth. By 1970, under the NED's technology investment program, Hercules had invested more than $2 million in promising new companies. Although by this time the thirty-five-member department (the staff would swell to sixty-two in 1971) was investigating hundreds of new technologies, ventures, mergers, and acquisitions, most of its activities were focused in a handful of key areas: electro-optical information systems, biomedicine and health, new materials and processes, and flavors and fragrances.[93]

Hercules sought to profit from what came to be known as the "information revolution" by utilizing new electronic and graphics technologies. In 1971, for example, the company negotiated a worldwide license from Electronics Reading Systems that was developing an optical character recognition (OCR) system capable of reading text produced on an ordinary typewriter. Hercules also invested in the International Components Corporation, which manufactured telecommunications components, but sold that equity by the end of the following year.[94]

On a somewhat larger scale, Hercules bought into a Wilmington manufacturer of magnetic detection devices called Infinetics, Inc. (Hercules' 44 percent share of the company was valued at $450,000 in 1970). Malcolm M. Schwartz, the inventor of sensors capable of detecting small metal objects at a short distance, founded Infinetics in 1961, with the idea that the devices would be used primarily to prevent pilfering at industrial plants. Soon, however, airline hijackings and a growing national concern about "law and order" spawned new applications: by 1970, twenty airlines throughout the world had installed Infinetics' Friskem metal detectors at their gates, and many law enforcement officers used Friskem nightsticks and concealed sensors to scan suspects for hidden weapons. Schwartz also envisioned many industrial applications for his technology, such as aerospace magnetometers for measuring the direction and intensity of magnetic fields.[95]

Nevertheless, Infinetics failed to hold its technological lead. After the metal detectors became required equipment at all U.S. airline terminals, compet-

The Fleet Ballistic Missile (FBM) program helped sustain Hercules' growth in chemical propulsion during the 1960s and 1970s. Above (left to right): Poseidon and Polaris missiles. Below: Applying its skills in process engineering, Hercules repeatedly found ways to lower the cost of graphite composites. Pictured: The carbon fiber line at Bacchus in 1974.

The New Enterprise Department launched the company into highly diverse businesses such as credit card readers, electronic security systems, and commercialized housing. Above: A housing unit designed and built by Hercoform in Sugarbush Valley, Vermont. Below: During the 1970s, the team of Jack Martin (left), chairman of the board, and Werner Brown, president, guided Hercules into additional diversification.

itors managed to garner a large share of the market. A minority owner, Hercules watched from the sidelines as Schwartz marketed in a style quite different from its own. Hercules' profits from Infinetics fell from $63,000 in 1972 to $15,000 in 1973, and Hercules later divested most of its equity in the once-promising upstart.[96]

A larger Hercules foray into the field of information systems—broadly defined—was in photocomposition printing (mainly for newspapers) with thermoplastic relief plates. Despite its own expertise in plastics, Hercules entered the business by obtaining an exclusive North American license for know-how from a Japanese firm, Asahi, in July 1971. The Asahi liquid photopolymer printing plate system process (trademarked Merigraph® in the United States by Hercules) seemed superior to competing stereo letterpress and etched zinc-plate processes and promised a rate of return of 20–30 percent.

After several successful trials and a victory over W.R. Grace's Letterflex system at a Booth newspaper chain competition in Michigan in 1972, Merigraph® garnered a few orders from medium-sized newspapers. But liquid polymer printing plate systems remained beyond the financial reach of smaller newspapers (which accounted for approximately one-third of the market). Only ten U.S. papers had switched to Merigraph® by 1973. The new business, transferred to the Organics Department in midyear, lost money for several years before becoming a steady earner later in the decade.[97]

Hercules' largest NED-inspired investment in the information field was a Los Angeles company called Data Source, which manufactured electronic credit card–reading terminals. Hercules initially invested $590,000 for a 62 percent share in the company, then purchased its remaining equity in 1971. The Data Source system (invented by John H. Humphrey) used fiber-optic heads to read embossed letters and numbers on bank, department store, oil company, and other credit cards at the point of purchase. It relayed data via telephone lines to central computers, which verified credit and billing information and adjusted customers' accounts.

The prime selling points of the system were its power, compatibility, speed, and safeguards against fraud: by reading the embossed figures rather than magnetic strips or encoded dots, Data Source could handle about 90 percent of the three hundred million or so credit cards then in use; with customers' accounts verified in a few seconds, merchants no longer had to check long lists of stolen card numbers.

By 1973, about fifteen hundred terminal systems were in operation, but few merchants had actually purchased them. Price may have slowed the adoption of the new technology, especially among smaller merchants; the complete Data Source System 800 (including minicomputer and software) sold for $350–750. But the main problem for Hercules, recalls Frank Wetzel (who spent a short time in the NED), was that Data Source needed to be "part of an overall system" that included cash registers, telephone communications terminals, and the like. Data Source also became a heavy drain on Hercules: by 1973, it was losing more than $2.5 million a year.[98]

Health care, another glamour industry in the 1960s and early 1970s, presented the NED with scores of opportunities to consider. For a time, filament-wound structures made at Bacchus had potential for a variety of surgical applications, including catheters, heart bypasses, and other surgical tubes, but Hercules found entry into the tightly controlled medical supply field difficult.[99]

The situation was similar in pharmaceuticals. In 1971, the Research Center developed a product called Divema (a 2:1 copolymer of maleic anhydride: divinyl ester), which seemed promising as a therapeutic drug to combat certain viruses and tumors. Hercules also became interested in blood fractionation and carried out a number of tests using blood from a prostitute who apparently had a remarkably strong immune system. Nothing came of this work, nor of the NED's search for a suitable venture or acquisition candidate in the field.[100]

Hercules ultimately entered the field in limited ways. One of its new materials, Delnet (see below), found limited application in the medical supply field. In pharmaceuticals, Hercules later forged an agreement with Montedison to test and market several of the Italian company's drugs in the United States (see Chapter 12).[101]

By 1970, the NED's established diversification goals had evolved into venture projects—businesses for which Hercules erected a new plant or "acquired 100% of a minority investment company." At the time, the NED was sponsoring only two new businesses in the venture category, both of which were marketing new materials and processes.[102]

Delnet, the trademark for a kind of polyethylene netting, was developed by a British firm, Smith & Nephew. A versatile product that "breathed," came in a variety of colors, and could be printed on, Delnet was constantly spawning new applications in clothing, cosmetics, food packaging (including boilable bags and cheese netting), home decoration, medical (from bandages to blood filtration), and other fields. Hercules sold imports from Smith & Nephew before completing its own plant to make Delnet at Marshallton, Delaware, in October 1970.

Hercules sold four million square yards of Delnet in 1971, and sales more than doubled in each of the two succeeding years. While not a smashing success, Delnet was one of the few profitable projects to emerge from the NED. Because it was essentially Hi-Fax® polyethylene in netting form produced by a process akin to that for polyolefin fibers and colored by Imperial pigments, Delnet was more closely allied to products in the company's existing lines than any other new product commercialized through the NED.[103]

The same could hardly be said of Hercoform, an experiment in the production of prefabricated modular housing that became the NED's most infamous venture. In 1969, Hercules spent $520,000 to acquire a 10 percent interest (with an option on another 10 percent) in Modular Structures Inc., a small real estate development company that specialized in modular housing. The investment served its purpose well, namely, to educate Hercules "immeasurably in understanding the housing market and defining a role for Hercules." So encouraged was Hercules by the prospects in modular housing that it began drawing up plans to erect a factory to produce at least one thousand units (apartments and houses were formed at the residential site by connecting a few of the units).[104]

Why was modular housing so appealing to Hercules? The NED spelled out compelling arguments in terms of both markets and production. On the demand side, the $25 billion housing industry was expected to triple during the 1970s, with multifamily dwellings (for which modular units were ideally suited) growing at a much faster rate than new housing in general.[105] At the same time, the federal government was offering additional incentives to builders of low-cost housing. The Department of Housing and Urban Development (HUD), created in 1965, had announced a new program called Operation Breakthrough, designed to help alleviate the nation's housing shortage. Hercules believed that participation in the program would be "relatively less speculative" than regular commercial construction because HUD promised to subsidize rents for qualified low-income applicants at the rate of 25–35 percent, and to establish uniform building materials codes (through the National Bureau of Standards) to supersede local codes. Only 22 of the 236 companies that submitted proposals to Operation Breakthrough were approved, among them Hercules. The only chemical company to participate in the program, Hercules made plans to construct about a hundred Breakthrough units for placement in Macon, Georgia, and Kalamazoo, Michigan.[106]

On the marketing side, the appeal for Hercules was akin to Henry Ford's motivation when he developed the first automobile assembly line: mass production could dramatically lower costs. Indeed, just as Ford recognized the importance of "bringing the work to the workers," Hercules calculated that the centralization of materials and workers (so that only completed housing units instead of the whole means of producing them had to be transported) would result in tremendous savings. In particular, labor costs, one of the fastest rising components of construction costs, could be drastically reduced.

Hercules chose to build its own factory so that it could apply a "systems approach" to the manufacture of housing. When the company looked at the highly decentralized nature of the industry—in 1969, there were some forty thousand construction companies, most of them small, and the largest ten controlled less than 3 percent of the market—it saw millions of disorganized, unsophisticated builders needlessly duplicating their efforts. What was needed, reasoned the NED, was the "industrialization of the housing process" through uniform standards and mass-production techniques. Economies of scale would permit lower costs as well as lower prices; Hercules would become a formidable competitor while helping to solve a national housing problem.[107]

Hercules parlayed its budding Hercoform operation into a department on April Fools' Day of 1970. Under the direction of John Present, a first-rate chemist with extensive international experience but no background in housing, the new department controlled two corporate entities: Hercoform, Inc. (HI), responsible for the manufacturing and construction divisions, as well as the overall management of the business; and Hercoform Marketing, Inc. (HMI), charged with land development and marketing of the modular units (it was decided from the outset that Hercules would sell rather than lease the units). HMI, in turn, owned a half interest in a firm called Instant Modular Construction (IMC), formed two years earlier "to develop and market resort condominiums in New England." Since that time, IMC had built several ski barns and condominiums in Vermont, including a few modular units at the Middle Earth ski resort at Sugarbush Valley. Now bereft of capital and stalled by a municipal tangle, IMC sold its equity share to HMI and folded.[108]

HMI made rapid progress without its former partner. By September 1970, it had secured eight additional development sites with capacity for a total of 1,450 prefabricated townhouse units. The following month, the Sugarbush units (outfitted with carpeting made of Herculon) were completed; they received positive reviews from the local community. But Hercoform ended the year with a loss of $1,295,000. Beginning to have second thoughts about its dependency on HUD projects, Hercules began to design models for nonsubsidized modular townhouses.[109]

Still, an early loss (although not quite so large) had been anticipated, since Hercules had not finished its own modular housing factory at Bloomsburg, Pennsylvania. The factory went into production in February 1971, under the direction of Harold E. Roth, who was hired by Hercules for his two decades of experience in mobile home construction. In its first year, Bloomsburg employed 19 payroll workers and turned out 386 units. Expecting to hire about 200 factory operatives when the plant swung into full production, Roth began interviewing local job seekers.[110]

Actually, the end was near. By the summer of 1972, Hercules informed the workers at Bloomsburg of its intention to curtail operations, then put the plant on the selling block. Since the beginning of 1971, Hercoform had lost an additional $2 million. In the end, it had built 270 units at Springfield, Massachusetts, and completed or nearly finished between 150 and 200 each at Glens Falls and Auburn, New York, Greenfield, Massachusetts, Bath, Maine, and North Lebanon, Pennsylvania. Hercules began divesting itself of a business it had spent approximately $10 million to enter. The modular units it still owned and the plant equipment were sold, and Bloomsburg was peddled to Hauserman, Inc. for $850,000 on May 1, 1973—three years and one month after Hercoform was created.[111]

How had it happened? Certainly, Hercoform's deplorable performance did not bode well for the new business, although Hercules had endured such losses before in order to assist businesses it believed in through their infancy. Three years of experience in modular housing, however, had convinced Hercules that Hercoform was at a dead end.

Many of the units developed severe water leaks because Hercules used untried materials and designs. But a more intractable problem was the unexpectedly long period it took to site and sell completed units. FHA approval took nine months at best. At the local level, unionized builders and contractors, threatened by the outside business, wielded codes, restrictions, and local influence to stifle progress, while the communities, also skeptical of outside developers, held town meetings and assembled city councils to debate zoning laws and building permits.[112]

In short, although Hercules may have ultimately succeeded at churning out dozens or hundreds of prefabricated housing units a year at relatively low cost, the prospects for placing and selling them in less than two years on average were grim. The company learned a lesson that had been well known since Henry Ford's time: to reap the benefits of mass-production, a manufacturer needs mass-distribution. Indeed, the need to maintain high throughput in demand as well as supply compelled America's first mass-producers (of canned goods, cigarettes, soap, and the like) to integrate forward during the late nineteenth century.[113]

From its creation, the NED devoted considerable effort to identifying an acquisition candidate in flavorings and fragrances. Such a move promised to supplement existing Hercules lines in food additives and cosmetic ingredients and to expand the company's reach in these huge markets. On October 1, 1973, the NED ended its four-year search when Hercules purchased (by exchanging stock) a well-established flavor and fragrance maker called Polak's Frutal Works (PFW). Based in Middletown, New York, PFW employed six hundred workers at plants in Middletown, Canada, Holland, England, Germany, and Australia and distributed its products worldwide. One of the last acquisitions directed by the NED, and one of the most closely related to Hercules' existing operations, PFW eventually became the most successful new venture of the New Enterprise era.[114]

Not all of the new businesses Hercules entered between 1968 and 1973 came out of the NED. The company's operating departments continued to cultivate new products and markets, including some that either relied on technologies foreign to Hercules or reached forward into new markets.

An example of the former (and another incursion into construction materials) was Snowden lightweight aggregate building material, used to make strong but relatively light concrete beams, cinder blocks, panels, and poured molds. In 1969, Hercules began building an eighteen-hundred-ton-per-day plant at Snowden, Virginia, which relied on a process that was developed decades earlier by Stephen J. Hayde and was subsequently used throughout the United States and Canada. The Snowden operation became part of the newly formed Industrial Systems Department in 1970. Suffering from raw materials, technology, and marketing problems, the business lost between one and two million dollars per year through 1973, when the oil crisis dealt a heavy blow by sharply raising gas prices, a major element of the unit's cost structure.[115]

As for forward integration into new markets, the Polymers Department did so in 1973 when it formed a home furnishings division and introduced the line of Vantage plastic furniture frames. The polypropylene frames, blow-molded at a plant in Oxford, Georgia, and distributed to large furniture manufacturers, were strong, light, durable, and almost indistinguishable from wood once upholstered. Nevertheless, the business did not blossom and was curtailed after a few years.[116]

Thus, while the overall financial performance of NED-inspired new businesses was quite dismal, Snowden, Vantage, and similar cases indicate that Hercules fared little or no better when it entered unrelated businesses through more traditional channels. Indeed, the NED hardly deserves the scornful place it came to occupy in the minds of many Hercules managers; the department carried out its mission ambitiously and as directed, reviewing an almost staggering number of new technologies and companies during its brief existence.

Most managers, however, understood that the problem lay at a more fundamental level, namely, the predominant beliefs of the company's top managers: that "a good management team can run anything," that "Hercules management is great," and "that management is management." The Hercoform venture, for example, was launched on the assumptions that homebuilding in the United States was an antiquated and backward process, and that Hercules' systems approach to management would more than compensate for the company's lack of experience.[117]

The notion was simple, confident, and easy to convey, as demonstrated by this passage from a Data Source advertisement explaining why a chemical company was selling credit card readers: "We already have the management, technical and marketing techniques that have made us big in the chemical industry. And there's no reason why the systems approach that has worked so well in one field can't be applied equally in others."[118] As it turned out, there were many reasons why the strategy was doomed, not the least being Hercules' lack of the technical and marketing expertise to manage most of the new businesses into which it plunged.

Hercules was hardly alone in learning this difficult lesson. In 1964, because it was moving into nonchemical fields, Monsanto dropped the word *Chemical* from its name; three years later, it created the New Enterprise Division for the purpose of "searching for and creating new business ideas inside and outside the company, for testing and evaluating them, and bringing them to predetermined profit levels." New fields in which it invested, with limited success, included electronics, graphic systems, educational toys, engineered composite systems, and protein foods.[119]

Du Pont's adventures in unrelated diversification bore remarkable parallels to those of Hercules'. In the early 1960s, Du Pont executives determined that the chemical giant needed to speed the development of new businesses by acquiring "small, highly technical companies" and aggressively developing the fruits of its own R&D. It formed the Development Department, which invested about $100 million in a variety of generally unprofitable new ventures, including a $50 million building materials debacle! According to the historians of Du Pont R&D, the company's new venture program was a strategic and financial failure.[120]

As scholars explore the histories of other leading American chemical enterprises, similar parallels are likely to emerge. At a time of diminishing expectations about profitability, it was not surprising that large companies in the nation's most diversified industry would use further diversification as a strategy, especially when the trend was gaining momentum throughout the economy.

But was Hercules trying to transform itself into a conglomerate? The answer is a qualified no. The company never explicitly adopted such a policy, although it was aware of and influenced by the growing phenomenon and adopted its basic tactic.[121] But there were limits to how far afield the company would venture. As Werner Brown recalled, Hercules "talked about, laughed about, and set aside" offers to buy the Dr. Pepper soft drink line, the Philadelphia Eagles football team, and similar unconventional acquisitions. Nor did it invest more than a small percentage of total assets in any one venture.[122]

Probably the closest Hercules came to a large-scale acquisition in a new industry was in oil. The NED considered numerous oil services companies and reported a near miss with the strongest candidate (BKO) in 1971. Brown recalled that Hercules also came very close to striking a deal with one of the largest oil producers in Texas. Still, Hercules was interested in oil because of its heavy dependence on the raw material and its synergy with chemicals; the risk would have been much larger than any carried out by the NED, but a closer fit with existing businesses than some of the acquisitions of the period.[123] It was fitting that the major lesson of the New Enterprise era played a role in quashing the project. As Brown recalls, "The real reason we got cold feet was . . . [that] we were scared that if we did get in we would be asked to make big money decisions without really understanding the business."[124]

On the one hand, Hercules so strongly sought wellsprings of new growth that it considered virtually any field. On the other hand, as Brown's words indicate, the company's leaders felt an often prudent fear of the unknown. The result was a hesitancy to make heavy, sustained investments in many new fields at once. "You needed to have a portfolio of ten or more positions," Crittenden explains, "that you could monitor, manage . . . and yet leave alone enough to be successful." This requirement gave the New Enterprise Department an almost impossible mission—to identify and generate new fields without the managerial and financial support to do so. Thus, a lack of full commitment, along with a lack of expertise in research, production, and marketing, undermined the NED more than the weaknesses of its individual projects.[125]

HERCULES AND THE RISE OF ENVIRONMENTALISM

Whereas big business in the nineteenth century was uninhibited in its pursuit of profit, giant corporations in the twentieth century have increasingly confronted the countervailing power of government regulatory agencies, labor unions, consumers, special interest groups, and public opinion.[126] The change was never more profound or apparent than with the rise of environmentalism in the 1960s and 1970s. Chemical companies, perhaps more than any other American industry, became a target of the movement. As a result, the government intervened in the affairs of Hercules and its peers to an unprecedented extent, permanently and significantly altering many of the ways they conducted business.

The environmental movement had deep roots in the American past, from early naturalist thinkers such as Thoreau and Emerson to the conservationist and public health campaigns of the Progressive Era. But modern environmentalism took hold faster, gained wider support, and focused on broader issues. For example, within a decade of the publication of Rachel Carson's *Silent Spring* in 1962—an event that conveniently marks the beginning of the environmental movement—a wave of regulatory measures and organized advocacy for the protection of the environment swept the country. Important new federal agencies were created—the Council on Environmental Quality (1969), the Environmental Protection Agency (1970), the Occupational Safety and Health Administration (1970), and the Consumer Product Safety Commission (1972)—that implemented scores of new laws and supplemented the work of state agencies. In addition, membership in long-standing, private nature groups such as the Sierra Club (1892) and the Audubon Society (1905) grew explosively, and new associations of activists—Friends of the Earth (splintered from the Sierra Club in 1969), the Environmental Defense Fund (1967), and the National Resources Defense Council (1970), among others—made their first appearance. Indicative of the times, tens of millions of Americans celebrated Earth Day on April 22, 1970, an event endorsed by President Nixon, corporate and educational leaders, and activists alike.[127]

The environmental movement affected Hercules in three important ways. First, new state and federal regulations prompted the company to devote more resources to toxicological testing of new products, as well as to retesting (and, in some cases, legal defense) of existing products. Second, Hercules had to institute procedures and hire new professionals to control polluting emissions from and safety conditions within its domestic plants. Finally, the company perceived in the environmental movement not only compliance challenges but also business opportunities to develop products and technologies for the treatment of solid wastes.

Toxaphene under Fire

Rachel Carson's important book is an especially apt starting point from which to explore the impact of environmentalism on Hercules. By increasing public awareness about the interrelatedness in the ecosystem and the potential dangers posed by the misuse of chemical pesticides, *Silent Spring* spurred the enactment of a variety of laws that affected Hercules plants and products. As one key government official put it, "There is no question that *Silent Spring* prompted the federal government to take action against water and air pollution—as well as against persistent insecticides—several years before it otherwise would have moved."[128]

More specifically, Carson's book focused on the widespread, often reckless use of chlorinated hydrocarbon pesticides such as endrin, dieldrin, heptachlor, and toxaphene. Although *Silent Spring* did not discuss Hercules in particular, it did contain dramatic passages about the dangers of toxaphene, thereby drawing

Hercules into a prolonged and expensive legal and public relations battle to defend one of its leading products.[129]

Carson considered toxaphene "one of the most destructive to fishes" of the chlorinated hydrocarbon pesticides. In characteristically vivid prose, she described a series of fish kills in fifteen tributaries of the Tennessee River in Alabama adjacent to fields treated with toxaphene. In the summer of 1950, heavy infestations of weevils prompted cotton farmers in the region to spray their fields with insecticides, especially toxaphene, and to repeat the treatments several times following heavy rains. Near Flint Creek, according to Carson, the average acre of cotton received sixty-three pounds of toxaphene, some as much as two hundred pounds, and one farmer, "in an extraordinary excess of zeal," applied more than a quarter of a ton to the acre! On August 1, Carson continued,

> Torrents of rain descended on the Flint Creek watershed. In trickles, in rivulets, and finally in floods the water poured off the land into the streams. . . . By the next morning it was obvious that a great deal more than rain had been carried into the stream. Fish swam about in aimless circles near the surface.[130]

Popular game fish (white crappies, bass, sunfish) and rough fish (carp, catfish, buffalo, drum) alike suffered. Those removed and placed in spring water survived, those left behind died. Other storms—and other fish kills—followed the next two weeks, and test goldfish in cages confirmed the presence of the lethal level of toxaphene and other pesticides in the river.[131]

Silent Spring's impact on Hercules was felt first and most acutely by the Medical and Legal departments. The company's medical director, Dr. Lemuel C. McGee, reported in 1962 that Carson's book (and the concurrent thalidomide tragedy) had already "influenced legislation as well as administrative decisions on matters of safety." The following year, chief counsel Charles S. Maddock began to carefully follow related legal developments: a proposed amendment to the Federal Insecticide, Fungicide, and Rodenticide Act that would shift the burden of proof of safety from government to industry; the tightening of Department of Agriculture (USDA) standards for pesticides; and the introduction of a proposed ban on all "hard" (persistent) pesticides in Rhode Island.[132]

As such legislation gained scope and momentum, Hercules stepped up its already extensive pesticide-testing procedures. Prior to 1962, the company had elaborate and costly procedures for testing new pesticides. They included preliminary and secondary screening, preliminary and secondary field-testing, persistence studies, and large-scale field-testing that included long-term toxicity studies to "establish beyond any reasonable doubt" the safe level of ingestion of the pesticide as a residue on edible crops. Hercules estimated that only one out of approximately three thousand pesticides cleared all those screening stages. As a result, each marketable product required an average investment of more than two years and nearly $2 million (not including subsequent marketing and plant construction).[133]

Soon after *Silent Spring* was published, Hercules ordered new reproduction studies on food additives, pesticides, and drug ingredients under development, which added as much as 50 percent to toxicology costs. Meanwhile, older products, including toxaphene, were "reevaluated" at a cost nearly equal to that of the original tests. Most of the work was done in advance of legal requirements. Soon, however, the Medical Department found its resources taxed to capacity, forcing McGee to contemplate the postponement of more additions to overhead "until and when legislative action forces us (and our competitors) into this needed activity." In order to gain greater control over the discovery and testing of new products, the company established three field research stations in toxicology (instead of farming out much of the work to the USDA, universities, and small cooperators) in 1964.[134]

But by this time, Hercules' emphasis was shifting toward the defense of existing products, such as toxaphene and Delnav® insecticide. *Silent Spring*'s attack on DDT had ensured that "all chlorinated pesticides were being regarded as prime candidates for cancellation," recalls Charles Dunn, who became manager of agricultural chemical development in 1965. Dunn devoted his energies to vindicating toxaphene. He presented evidence at many state forums and testified before House and Senate committees. "The message we carried to all of these state legislators, and ultimately to the federal people," he recalls, was that "before you throw out all the chlorinated hydrocarbons, look up details, look up the facts, and then determine whether [the products are] worth saving."[135]

The legal battle made it increasingly clear that while toxaphene might survive on its own merits, it was suffering from its close, long-time association with DDT. Thus, a novel approach to marketing toxaphene was needed, one that would dissociate it from DDT both legally and in the public mind. That campaign would represent a complete reversal of the approach used since toxaphene had first been marketed, when it was discovered that DDT-toxaphene mixtures possessed a synergy that made them more effective than the sum total of the two insecticides used separately.[136]

A new campaign "to separate toxaphene from DDT and other hard pesticides" was initiated in the mid-1960s. As summarized in a later annual report, "toxaphene is not biologically persistent; does not magnify to any significant extent in the food chain; and remains near the site of application, where it is biologically degraded." In 1962, the "primary emphasis" of the company's toxaphene advertising had been "to make clear the visual symbol of the DDT-toxaphene combination." By the end of the decade, the Advertising and Public Relations Department under Carl Eurenius (indicative of the times, the department's name was changed from simply "Advertising" in 1970) employed a well-received "public service approach, stressing the point that toxaphene is not a contributor to environmental pollution and is relatively safe to use."[137]

Controlling Air and Water Pollution

While Hercules was confronting regulatory challenges to the safety of products such as toxaphene, it also began to reevaluate the nature and impact of air and water pollution generated by its domestic plants. Again, much of the effort was carried out in anticipation of new regulatory measures. In 1964, McGee warned the executive committee that Hercules probably would soon be required to clean up all plant wastes (unless it could prove they were not harmful, which was unlikely because Hercules had "no facts on which to argue"). He recommended that Hercules conduct an extensive two-year study of all its plants, a study that "should include analytical measurements of the chemical nature and quantity of our liquid wastes, their persistence, dilution, biodegradability and toxicity."[138]

Hercules' top managers, especially Thouron and Ed Crum, were sympathetic to the recommendations. In 1967, the company created the Environmental Health Committee (EHC) to "coordinate the companywide pollution-abatement programs." The EHC's chairman and executive secretary was Dr. Richard E. Chaddock, a Ph.D. in chemical engineering who had served the company in sales, research, development, and planning since 1945. Its other nine members represented the Legal, Medical, Advertising and Public Relations, C&PP, and Engineering departments. Although the EHC met every few weeks, Chaddock performed his important new function essentially alone, reporting directly to Vice President Ed Crum and, in turn, Henry Thouron, who strongly endorsed the program.[139]

Chaddock's first order of business was to take stock of the situation. He compiled files on every Hercules plant, spending most of his time on the road. Since on-site liaisons were needed, Chaddock established "pollution abatement coordinators" at the major Hercules facilities, requiring them to submit periodic reports on water effluent and air emissions, as well as financial information on yearly cleanup expenses and projected capital costs. (At smaller plants, these tasks were performed by superintendents or managers.) From the start, Chaddock believed that since "pollution starts at the plant, the plant has to clean it up."

Not surprisingly, Chaddock encountered resistance on the operational level. His agenda—to institute new controls and to call for (often significant) expenditures for process modifications or waste treatment facilities—often diverged sharply with the goals of directors of plant operations, who were responsible for holding down costs. In the short run, this potential obstacle was overcome by the strong support the EHC received from Crum, Thouron, and the board. In the longer term, the company's pollution control efforts earned the support of plant operators precisely because they often lowered costs by reducing waste. (Waste reduction through process improvements not only saved expenditures on purification equipment but also helped recover profits, in the form of finished product, that had formerly washed or floated away.)

The EHC's initial task was to ensure compliance with the panoply of state environmental laws (the EPA did not institute federal controls until 1970), which presented a complicated set of criteria; Hercules operated plants in more than two dozen states, each with a unique set of environmental regulations. New Jersey, for example, placed strong emphasis on air quality, while the Western states were more concerned with water purity. In Chaddock's view, Hercules enjoyed positive relations with state regulators, and the company earned a reputation for keeping its promises.[140]

The work had a decidedly public relations component as well. Often the first step was to follow the Chinese proverb, that is, eliminating "what people can see, feel, and smell." Open burning was halted; noxious odors were filtered. The Glens Falls pigments plant, for example, stopped discharging about a million gallons of colored water into the Hudson River each day, a minor—but glaringly obvious—safety hazard. Moreover, unlike many companies, Hercules openly discussed its cleanup gains, thereby admitting its problems. The Public Relations Department helped produce a film called *Our Priceless Heritage,* which contained dramatic "before and after" footage at several Hercules plants. In 1972, the company held a press conference, attended by *Chemical Week* reporters, to discuss its cleanup efforts at Brunswick and Parlin. The results of such openness were generally positive; for example, Hercules won the Sports Foundation's National Gold Medal Award for "outstanding achievement in the fight against water pollution" at Hattiesburg in 1973.[141]

Hattiesburg was one of about a dozen plants that received the most attention during the late 1960s and early 1970s. Water-purification equipment was also installed at Burlington and Gibbstown, New Jersey, Franklin and Hopewell, Virginia, Wilmington, North Carolina, Savannah and Brunswick, Georgia, Hattiesburg, Mississippi, Hercules, California, Lake Charles, Louisiana, and Glens Falls, New York, and air filtration apparatus was installed at Hopewell, Glens Falls, and Parlin.[142]

Although Hercules' plant cleanup efforts were supported and successful, they were clearly altering the way the company conducted business. Many of the trouble spots were at recently acquired facilities, so Hercules began incorporating an environmental review into its acquisitions process. The costs were also significant. The Parlin installation alone consumed $4 million in 1971, or nearly 8 percent of the company's profits for the year. Typically, waste treatment facilities added 3–5 percent to total plant investment. Apart from new administrative costs, which added more than $7 million per year to operating expenses by 1973, Hercules spent more than $37 million on pollution abatement equipment between 1965 and 1973.[143]

New Opportunities in Waste Treatment

The passage of the Federal Water Quality Act of 1965 represented more than a new set of constraints for Hercules. It also signaled an opportunity for the company to apply its chemical expertise to the solution of waste treatment problems. In 1966, U.S. companies invested $2.2 billion in water and waste treatment, and the market was growing rapidly. They sought not only to treat polluted water leaving plants but also to speed the purification of incoming water (used in manufacturing processes, to feed boilers, and for cooling) to prevent algae buildup, erosion, and scaling.[144]

In 1967, Hercules established the Environmental Services Division (ESD) in the C&PP Department to carry out this work. Although the ESD was primarily a sales-oriented profit center, it also featured a waste treatment R&D group housed at Marshallton, Delaware. The company hoped to capitalize on several products developed by the C&PP Department—especially polyelectrolytic flocculents such as Reten® retention aid, used to settle waste materials suspended in water. Because Hercules lacked marketing expertise in the field, it acquired a Houston water management services company, Aquatrol, in July 1967. Although Aquatrol was both small ($1 million in sales) and regional (it served Louisiana, Texas, and Oklahoma), the company featured technical service teams knowledgeable about influent, effluent, and in-plant water problems. Building on this base, the ESD expanded rapidly (into Wilmington and Milwaukee) but retained its decentralized structure; each regional office had a separate technical service staff, laboratory, small plant, and territory. By 1969, the network included eighty-seven employees, about half of them dedicated to direct sales.[145]

Hercules' expansion in the water treatment field gained momentum in the early 1970s. In 1970, the company acquired a small waste treatment engineering firm in Gainesville, Florida, called Black, Crow & Eidsness (BC&E, named after Charles Black, William Crow, and Fred Eidsness, whose company earlier had done contract work for Hercules). From its offices in Florida, Georgia, Ohio, and California, BC&E's 150 employees provided consulting and engineering services for the design of more than 100 industrial and municipal plants. The following year, Hercules formed a joint venture called Advanced Waste Treatment Systems (AWT) with Procedyne (an NED discovery) to utilize that company's fluidized bed technology.[146]

For new housing developments that lacked adequate waste disposal facilities, AWT offered an ingenious solution: it would install small, totally self-contained waste treatment plants, capable of processing the waste from hundreds or thousands of dwellings, inside inconspicuous, houselike structures. The AWT plants promised several advantages over conventional plants: lower construction costs per gallon of capacity; unattended operation with little noise or odor; greater flexibility in the range of treatable wastes; and reduced final waste. Using a series of treatment steps—screening, settling, chemicals, magnetic separation, and carbon purification—the plants could convert 95–99 percent of suspended solids, phosphates, and biochemical oxygen demand from an entire neighborhood into a few bags of ashes, a "wisp" of carbon dioxide, and water suitable for irrigation. The heart of the system was the company's chemical flocculents, which reduced the time needed to settle suspended wastes from several days to several hours.[147]

Another ambitious waste disposal project was the reclamation of city garbage for Newcastle County, Delaware, under a contract with the state. Hercules proposed to build a plant capable of recovering about 95 percent of the county's refuse (sewage sludge from the Wilmington treatment plant as well as hundreds of tons of metals, glass, plastics, paper, organic matter, and so forth, per day). Hercules hoped to profit from the sale of both recyclable materials and organic humus, useful for growing mushrooms and other crops. In short, the company was attracted to the project for familiar strategic reasons: the chance to "add value" to a waste product while operating in a comfortable niche. (Indeed, Hercules viewed the business as directly analogous to naval stores.)[148]

The 1960s had heightened the nation's awareness of the problems associated with pollution and waste disposal; the early 1970s found Hercules actively working toward solutions, hoping to profit in the process. Its activities ranged from consulting engineering services to the production and marketing of flocculent chemicals, to the construction of "miniaturized treatment plants" and large waste-recycling facilities. In addition, the company had other experimental projects under development: Haveg was working on processes for neutralizing steelmaking by-product acids and degrading chlorinated hydrocarbons, and the Explosives and Chemical Propulsion Department (which carried out the Newcastle reclamation work) was attempting to desalinize seawater.[149]

Unfortunately, all but one of these businesses were short-lived. The Newcastle plant made it through the design stage, then stalled permanently for several reasons: federal funds were delayed for years; the community demanded profits from the recovered materials; and Hercules feared entanglements in local politics. AWT completed a demonstration unit at a 130-home development in Freehold, New Jersey, but the corporation was dissolved in 1974. BC&E was sold to CH^2M Hill in 1977.[150]

The projects that foundered generally suffered from mediocre to poor returns—at a time when Hercules was placing greater emphasis on strict financial performance criteria. But it is also noteworthy that only the chemical businesses survived in the new environmental field. By attempting to build purification and reclamation plants or to sell engineering services, Hercules was diversifying into unrelated areas—primarily engineering rather than chemical endeavors. As discussed earlier, the problem also plagued most of the ventures cultivated by the New Enterprise Department in this period.

The environmental movement indisputably gave rise to additional constraints for American companies. New worker safety and pollution standards raised costs and narrowed options. Companies required new organizational skills to operate—and compete—effectively in an era of increased government intervention. For chemical companies, the most important constraint was the broadening definition of product safety. Whereas before 1962 premarket testing was reserved for obviously toxic substances, by the 1970s (especially with the passage of the Toxic Substances Control Act in 1977), virtually all chemical products had to be proved safe before being marketed.

What generalizations can be made about Hercules' responses to the new constraints? Administratively, the process of building organizational capabilities to deal with the changed regulatory environment took several years, as was typical throughout the industry. Initially, much of the responsibility fell to the Medical Department, which soon became overtaxed. In 1965 alone, that department's seven professionals and five secretaries completed twenty-eight toxicology studies, answered 450 inquiries, submitted fifteen petitions (six in the United States), worked on three proposals, visited sixteen plants, gave nine addresses, and served on eight trade association committees and four professional society committees.[151]

By the end of the decade, however, the burden had been spread around: the Medical Department's resources had been increased; other auxiliary departments—especially Legal, Advertising and Public Relations, Research, and

Safety—had assumed greater responsibilities in environmental matters; and specialized internal groups (such as the EHC) had been formed. In addition, operating departments had assumed greater roles in the defense of their own products. By 1970, for example, the Synthetics Department had four full-time professionals to defend toxaphene—a reasonable investment, since the product was setting sales records in excess of eighty-four million pounds per year.

As the company developed its organizational response to the rise of environmentalism, it became increasingly proactive (rather than reactive). By 1970, Dr. Howard L. Reed (the company's new medical director) had become convinced that the chemical industry was fighting a "hopeless" battle—because it was an "entirely defensive battle to maintain the status quo"—and called for greater leadership from industry in establishing standards and procedures.

Nowhere was the company's growing activism more apparent than in the international arena. Inspired by the environmental movement in the United States and Europe, several nations, international organizations—such as the World Health Organization—and the EEC set new standards of acceptability, some of which applied to Hercules exports. The company responded to some of these standards individually, such as when chief toxicologist John Frawley flew to Egypt to head off a toxaphene boycott. But in general, Hercules sought coordinated action through Codex Alimentarius, a kind of "international EPA" concerned mostly with pesticides and food additives. According to Dr. Frawley, Hercules helped found and was active in Codex because it was easier to effect change in Washington through an established international organization than through direct lobbying. At the same time, Hercules protected its exports through active participation in several international standard-setting organizations.[152]

Hercules accepted the validity and permanence of environmentalism from the beginning; in this sense, its response to the "crisis" was somewhat atypical for the chemical industry. "Henry Thouron, who really got things moving, knew [that environmentalism] was here to stay and we had to clean up," recalls Chaddock. Thouron's outspoken support of Chaddock's EHC established and reflected a tone in the company that persisted through subsequent presidencies.[153]

To be sure, the new regulations—and their implementation—were not without their critics in the company, especially those under direct fire. Medical Department directors described controls over some chemicals as "unreasonably restrictive" and considered the imminent passage of the Toxic Substances Control Act as "almost unbelievable." But much of the dissatisfaction stemmed from two sources: first, relations between federal regulators and the industry had become unnecessarily adversarial; and second, scientists were frustrated with the nonscientific components of the debate. As Reed put it in 1969: "The national and international pesticide situation has become almost ridiculous. The controversy is so charged with emotion and politics, that almost anything can happen." The company's toxicologists acknowledged "the philosophy of the environmentalists . . . that chemical technology has outstripped the capability of the biologist to evaluate the effect on man and the environment," but rejected "an over-concentration on the negative aspects of chemical innovation." In short, they called for a "benefit-risk ratio . . . to guide national policy" so as to avoid "undesirable social, medical and economic consequences."[154]

The company's legal experts, meanwhile, accepted the inevitability of the new controls. The imminent passage of the Air Quality Act, wrote the chief counsel in 1967, "in effect [marked] a new era . . . in environmental control." Now "we are going to have clean air and clean water—*period*." The top management of the company concurred. Unlike Monsanto and Dow, for example, Hercules issued no scathing attacks on Rachel Carson's book or the environmental movement as a whole.[155]

Moreover, Hercules experienced no serious environmental tragedies during this period, as did many of its peers. Indeed, the company probably learned from the adversities of others. For example, many cases of cancer were diagnosed among bis(chloromethyl)ether workers at nearby Rohm and Haas beginning in 1962, a tragedy ultimately resulting in at least fifty-one related deaths. The incident drew widespread coverage from the media; Rohm and Haas, with no public relations department and no resident toxicologists or epidemiologists, was "poorly equipped and ill-prepared" to handle the crisis. Along with Hercules, Du Pont also began to conduct extensive testing beyond minimal requirements in the wake of this disaster and the release of *Silent Spring*.[156]

Hercules probably escaped such calamities for other reasons as well. It operated only a handful of plants that used or produced as by-products highly toxic chemicals such as chlorine. More important, the company's concern for safety, dating back to its early days as an explosives-maker, made the adoption of new controls and safety measures easy to administer and to accept.

Hercules became a different company in the decade after the publication of *Silent Spring*. It created an organized body of professionals who carried out coordinated efforts to ensure the safety of the company's products, people, plants, and environs.[157] The new testing, monitoring, reporting, cleanup, and public relations functions became a permanent feature of the company. Indeed, in 1972 at least ten Hercules products faced serious challenges from regulators: toxaphene, carrageenan, locust bean gum, amidated pectin, guar gum, lead pigments, diarylide pigments, cadmium pigments, chrome pigments, and Parlon® chlorinated rubber.[158]

While Hercules accepted the urgency and importance of environmental protection, it was stung by the movement's economic consequences. Still, economics did not dictate that the answer was to abandon heavily regulated products: in 1969, one-quarter of the company's profits came from FDA-cleared products, which averaged twice the profit rate of other Hercules products. Rather, environmentalism was seen as an inescapable yet essential constraint in an already profit-constricted era. Dr. Reed stated it well at the dawn of the 1970s: "Hercules and other members of the chemical industry face a severe challenge in the seventies to maintain strong profit positions in spite of increasing government regulation [environmentally related] of business practices."[159]

Conclusion

Between 1962 and 1973, Hercules continued to grapple with a problem that some of its managers had identified as early as the 1930s: the company's overdependence on maturing businesses, namely, rosin- and cellulose-derived products. The issue had become the animating force behind Hercules' postwar strategy. At first reluctantly, then aggressively, the company sought solutions through new aerospace and natural gas facilities, foreign expansion, related acquisitions, and, most important, petrochemicals.

The 1960s and early 1970s were a critical time because the outcomes of earlier investments were becoming apparent. In addition, the need to find sources of growth was intensified by new, industrywide constraints on profitability, especially increasing competition in petrochemicals and other commodity fields (leading to overcapacity and price cutting) and a wave of environmental regulations.

Only the international component of Hercules' postwar strategy was a resounding success in the immediate term. Many of the acquisitions of the late 1950s and early 1960s (Huron Milling, Imperial Color & Chemical) grew only moderately, as did the chemical propulsion field. The outcome of petrochemicals was more difficult to diagnose: DMT became a star, but in spite of growing sales, profits in polypropylene resins were eroded by competition. "Unfortunately," complained Werner Brown, while head of the Polymers Department in 1964, "the history of this product has shown that every increase in operating efficiency is obliterated by deterioration in price."[160] Fibers and film, meanwhile, still struggled through their infancies and began to evolve from promises to profits only in the mid-1970s.

To seek growth opportunities, Hercules hired consultants and applied new, increasingly quantitative performance criteria to guide investments in existing lines as well as divestitures of business units. In keeping with the general business climate, including the strategies of competitors such as Du Pont, it simultaneously diversified farther afield than ever before, often into businesses unrelated to its own expertise or, in some cases, to the chemical industry.

But most of these efforts–first Haveg and then the NED's projects–miscarried. The reasons were complex; some lay at the level of the individual project—from management to miscalculations or even deception. On the whole, however, the company's diversification projects reflected the consequences of two broader trends. The first was the conglomerate movement and the rise of portfolio management consulting. By the early 1970s, conglomerates had fallen into disfavor, as had more simplistic varieties of portfolio analysis of the late 1960s. Perhaps Hercules can be faulted to the extent that it followed these trends uncritically.

The second factor that undermined Hercules' diversification efforts was its own lack of commitment. As a Hercules representative told the financial community in 1972: "Don't be surprised, then, if a favorite project from yesterday is scrapped. . . . Don't get excited if we abandon a project of which earlier we had high hopes. A high fatality rate is inherent in the field of new ventures and advanced technology."[161] Werner Brown put it more succinctly: "Those that don't develop properly we won't hesitate to drop."[162]

We can never know, of course, whether greater commitments of managerial and financial resources would have sustained any of these projects through their infancy, or if their success would have been good for Hercules in the long run. (PFW, one of the few surviving NED ventures, later became one of the company's star performers.) But there is an essential irony to the New Enterprise era: although the department was created out of a recognition by company leaders that the time and resources needed to develop a new business had become unacceptably costly, the NED itself was unequipped to carry out its assignment for more than a few years.

Hercules' second response to the outcomes of its earlier investments and the growing constraints of the period was to invest heavily in petrochemicals. Only a portion of these investments—those in DMT—were substantiated by immediate financial returns. Those in polypropylene sprang from both great hopes for the business and a determination to maintain the company's hard-won lead in the field.

Quite apart from the financial soundness of the strategy, which it was still too early to judge, heavy investments in petrochemicals changed Hercules in important ways. Because of its middling size, Hercules could not afford to make simultaneously investments of a similar scale in other fields, such as those under development by the NED. As Brown notes: "DMT, polyethylene, polypropylene . . . that was a pretty big bite, and we didn't need to do a whole lot more than be sure that those were taken care of."[163] Thus, incalculable opportunities were lost in aggressive petrochemical expansion.

Entrenchment in petrochemicals and the failure of unrelated diversification also reshaped Hercules by focusing its portfolio of businesses. By 1973, the company's sales were relatively more concentrated in a few markets than they had been in 1962, its destiny tied much more closely to demand for certain petrochemicals. And as it was, the worldwide petrochemicals market was about to experience its greatest shock.

CHAPTER 12

RESTRUCTURING AND REORIENTATION, 1974–1987

In 1973, Yom Kippur, the holy day of atonement for Jews, fell on a Monday, the twenty-second of October. On that day, as the Jews of Israel attended to prayers and observed traditional rites, Egyptian tanks crossed the Suez Canal into Israeli-occupied territory. At the same time, Syrian forces bombarded the north of Israel in an effort to reclaim territory lost during the six-day war of 1967. The fighting in the Yom Kippur War of 1973 ended in a stalemate within a matter of months. Although the conflict was brief and directly engaged only the immediate participants, it was to have momentous consequences for the Western world.

The American government's support for Israel during the conflict led to a boycott of trade with the United States by the Arab nations. At the same time, Arab oil producers banded together with leaders of other oil-producing nations to transform the ineffectual Organization of Petroleum Exporting Countries (OPEC) into a militant cartel. In the spring of 1974, the embargo was lifted; in its place, however, came a sequence of OPEC-led hikes in the price of crude oil. A barrel of crude, which had traded for $3.70 at the end of 1972, sold for more than three times that amount eighteen months later. In the United States, federal price controls on gasoline and fuel oil helped shield consumers from the full impact of the change. For producers of petrochemicals and petrochemical intermediates such as Hercules, however, there was no such protection.

Twice in its lifetime, Hercules had been transformed by its experiences in foreign wars. Once again, in the mid-1970s, a foreign war stimulated its transformation. Although Hercules played no direct role in the Middle East conflict, the company nonetheless had to come to terms with its consequences. And the problem of adapting to new circumstances was made more difficult by other worrisome trends already present in the business environment: the breakdown of the postwar international monetary system; the rise of environmentalism and consumerism; the escalation of government regulation and intervention in the economy; nearly a decade of high inflation; and the globalization of competition.

In 1974, few at Hercules saw the magnitude of these trends or foresaw the need to reexamine the company's basic assumptions about growth and competition. It was not long, however, before change became imperative, resulting in the most turbulent era for Hercules since the Great Depression (see Exhibit 12.1).

Time of trial

In 1974, Hercules posted its fifteenth consecutive year of record sales, and its fifth straight year of improved profits. The company's total revenues of $1.53 billion and net income of $93.7 million seemed to validate its recent efforts to concentrate on developing its core chemical and petrochemical businesses and pruning some of the more exotic operations of the New Enterprise era. Under the leadership team of Werner Brown, the fifty-five-year-old president, and board chairman Jack Martin, Hercules had expanded vigorously: in polypropylene resins—a successful "debottlenecking" operation at Lake Charles, the completion of the second capacity increment of two hundred million pounds per year at Bayport, the beginnings of new facilities at Varennes, Quebec, and Paal, Belgium, and a fifty-fifty joint venture in Taiwan; in DMT—major capacity expansions at Burlington, New Jersey, and Wilmington, North Carolina, the startup of a joint venture plant in Tampico, Mexico (Petrocel), and the beginning of work on an eight-hundred-million-pound-per-year plant at Eastover, South Carolina (Wateree); in other lines—expansion of facilities to make polypropylene film, CMC and pectin, and agricultural chemicals.

Hercules, moreover, had just completed three important acquisitions: Pennsylvania Industrial Chemicals Company (PICCO), a manufacturer of hydrocarbon resins; Copenhagen Pektinfabrik (CPF); and, as mentioned in Chapter 11, Polak's Frutal Works (PFW), a leading maker of aroma and flavor chemicals. In contrast to many deals of the previous decade, these were related moves to support and extend the company's existing businesses in organics and food additives and flavorings. Hercules had not abandoned diversification, however. In 1974, at the prompting of Paul Johnstone, vice president and member of the executive committee, the company entered into a partnership with the Italian chemical giant Montedison. The fifty-fifty joint venture, Adria Laboratories, provided Hercules with a long-sought toehold in biomedical products: the company was formed to seek Food and Drug Administration (FDA) approval for—and subsequently, to distribute—Montedison's pharmaceuticals in the United States. In 1974, Adria Labs marketed its first product, an anticancer drug called Adriamycin.

Although the company noted an economic downturn during the fourth quarter of 1974, Brown pointed out that, "historically, this type of unexpected

EXHIBIT 12.1 Hercules' Financials, 1974–1987 (Dollars in Thousands, Except per Share)

	1974	1975	1976	1977	1978	1979	1980	1981	1982	1983	1984	1985	1986	1987
Net sales	$1,525,489	$1,413,111	$1,595,956	$1,697,737	$1,946,477	$2,345,425	$2,485,226	$2,718,366	$2,621,859	$2,628,954	$2,570,965	$2,587,213	$2,615,110	$2,693,003
Earnings before interest and taxes	144,599	40,650	200,931	106,079	177,755	238,388	114,738	180,444	103,338	222,224	190,863	121,583	217,319	1,286,818
Total assets	1,327,212	1,316,252	1,430,282	1,477,543	1,596,598	1,761,177	1,889,679	1,997,144	2,101,723	2,175,173	2,388,489	2,658,822	2,914,319	3,492,077
Fixed[a]	654,607	712,178	699,383	721,292	714,286	772,889	872,656	907,733	948,455	793,313	906,031	1,000,212	1,152,141	1,075,389
Current	541,431	485,072	567,808	599,494	707,055	784,485	791,723	854,210	851,029	826,462	942,523	1,045,771	1,079,821	1,793,467
Other	131,174	119,002	163,091	156,737	175,257	203,803	225,300	235,201	302,239	555,398	539,935	612,839	682,357	623,221
Stockholders' equity	659,159	664,910	742,020	757,570	818,451	945,422	1,009,746	1,051,357	1,078,911	1,288,118	1,366,911	1,474,201	1,703,151	2,189,867
Long-term debt	348,551	334,224	326,368	329,443	295,969	280,619	334,530	454,356	431,919	350,494	421,015	530,779	546,113	488,538
R&D Expense	30,021	30,025	35,389	37,361	40,081	46,701	53,462	61,410	74,173	73,925	71,690	76,137	71,142	73,819
Dividends/share	$0.80	$0.80	$0.85	$1.00	$1.00	$1.08	$1.20	$1.26	$1.32	$1.38	$1.48	$1.60	$1.72	$ 1.85
Earnings per share (common)	$2.17	$0.77	$2.44	$1.36	$2.36	$3.89	$2.60	$3.09	$2.10	$3.17	$3.54	$2.40	$4.02	$14.74
Number of employees	25,335	23,476	23,957	24,002	24,453	24,409	22,928	22,777	24,450	24,221	26,262	25,448	25,120	23,152
Return on sales	9.5%	2.9%	12.6%	6.2%	9.1%	10.2%	4.6%	6.6%	3.9%	8.5%	7.4%	4.7%	8.3%	47.8%
Return on assets	10.9	3.1	14.0	7.2	11.1	13.5	6.1	9.0	4.9	10.2	8.0	4.6	7.5	36.8
Return on equity	21.9	6.1	27.1	14.0	21.7	25.2	11.4	17.2	9.6	17.3	14.0	8.2	12.8	58.8
Debt/equity	52.9	50.3	44.0	43.5	36.2	29.7	33.1	43.2	40.0	27.2	30.8	36.0	32.1	22.3
Debt/total capitalization	26.3	25.4	22.8	22.3	18.5	15.9	17.7	22.8	20.6	16.1	17.6	20.0	18.7	14.0
R&D/Sales	2.0	2.1	2.2	2.2	2.1	2.0	2.2	2.3	2.8	2.8	2.8	2.9	2.7	2.7

a. Gross fixed assets, including depreciation.

Source: Hercules Incorporated, Annual Reports for 1974–1987.

contraction has been sharp and self-correcting in a relatively short period of time," and he showed little concern. "Throughout its history," he noted, Hercules "has been able to maintain a posture of flexibility and adaptability in response to the current and anticipated future needs of the marketplace." On the surface, then, Hercules had reason for optimism. As Brown summed it up, "Record profits, new plants, momentum generated for continued future growth—all these made 1974 a good year for the company."[1]

When Brown wrote these words in January 1975, however, it was fast becoming apparent that the future would not be as bright as the past. The most immediate problems were related to the oil crisis and associated rumblings in the international economy. More than half (56 percent) of the company's asset base was involved in petrochemicals, with the DMT investments most seriously exposed. But the company's other lines of business suffered as well. As the Vietnam War wound down to its conclusion, for example, production at the GOCO plants fell off, and opportunities in chemical propulsion seemed less than promising.

At the same time, a crisis loomed in naval stores, which had long been one of Hercules' strongest cash contributors. The first problem was raw materials shortages. In 1973, unusually wet weather in the Southeast led to a poor harvest of stumps. The shortage was compounded by federal wage and price controls from 1972 to 1974, which led many of Hercules' rosin (especially tall oil) suppliers to break contracts and export to obtain much higher prices outside the United States. Although the company searched diligently for new sources—expeditiously, it arranged to import gum rosin from the People's Republic of China (newly open to Western trade)—it was unable to find enough. As a result, Hercules watched many customers for rosin-based chemicals shift to substitute products. The situation grew still worse after price controls were lifted in the spring of 1974. The price of rosin "essentially doubled overnight," recalls Keith Smith, then a planner in the Organics Department. "A lot of people who were using rosin derivatives . . . reformulate[d] their products and eventually switch[ed] to things like hydrocarbon resins. We lost volume in those years that we've never regained." Production of pale rosin fell from 224 million pounds in 1973 to 95.5 million pounds in 1975. Such declines, coupled with the falloff of general business activity, rippled throughout Hercules' resins and paper chemicals businesses.[2]

Nor did Hercules' international business provide relief: total sales overseas (exports plus foreign production) dropped 15 percent between 1974 and 1975, with markets in Europe particularly hard hit. In past recessions, says Arden B. Engebretsen (then treasurer and now vice chairman and chief financial officer), Hercules had been protected by its diversification: "When one business had gone down, another business had picked up the slack. But this was the first time the whole world went down, and all the businesses went down with it. For the first time, diversification didn't work very well for us."[3] The company's financial performance in 1975 was disastrous: net sales declined 7 percent to $1.4 billion, and net income plunged 65 percent to $32.5 million. To help maintain investor confidence, the company paid out more in dividends than it actually took in.

Nonetheless, the price of Hercules stock plummeted from a high of $44 in the second quarter of 1974 to $21 at year end, and it hovered in the low twenties throughout 1975.[4]

Brown and the executive committee took a series of extraordinary measures: annual bonuses were canceled, and four corporate officers took early retirement. In August 1975, Hercules redefined the roles of its management team, appointing several new vice presidents with broad responsibilities to work with the operating departments on marketing and planning, manufacturing, engineering and construction, finance, and R&D. The company also suffered "a loss of enchantment" with the activities of the New Enterprise Department. "The battle cry that good managers can manage anything began to sound a little hollow," recalls David S. Hollingsworth, then the newly installed head of the department (and now chairman and CEO).

> It hit me one day because I went into the executive committee with a scheme to minimize the losses in our credit card–reading business. . . . [My report] wasn't greeted with a great deal of enthusiasm. I remember going back to my office and Brown came wheeling down the hall and came in. He said, "You don't understand. I don't want to know how to cut my losses in that business. I don't want to *be* in that business." "Okay, partner. If you don't want to be in the business, we'll get the hell out of it." So we got in the position of winding up that area when I went into it.

Thereafter, the NED became a planning and analysis function, until it was disbanded in 1976.[5]

Disaster in DMT

Hercules also undertook a searching examination of its basic strategy, a project headed by Al Giacco, then vice president for marketing and planning. The major issue, apparent to everyone, was the company's DMT business. There were several key problems. To begin with, Hercules had recently chosen to double its capacity by building the new Wateree plant in South Carolina. As explained by Brown, the decision had been based upon careful analysis. The company's biggest customer, Du Pont, had demanded assurance from Hercules that it would continue to supply Du Pont's fast-growing needs for DMT and "stated unequivocally that its needs would far outstrip Hercules' ability to supply from existing capacity." Hercules believed that with its greater production volume, its plant and indirect costs would fall, making the business "highly competitive and offering excellent profit margins." Finally, Brown acknowledged, "the whole world was caught up in the polyester growth syndrome, and while we knew that growth could not go on forever at the current rate, one more major plant seemed hardly a bad risk marketwise, and a new location offered extra protection from accidents and shutdowns."[6]

The timing of the Wateree decision could hardly have been worse. "You see," says Giacco, "DMT was a product in which the raw material accounted for 30 percent of the cost. With the energy crisis, it became 80 percent of the cost."[7] In other words, Hercules' opportunity to add value to the product—and hence to achieve attractive returns—was sorely constrained.

Market conditions were turning sharply for the worse. In the late 1960s and 1970s, DMT production was suffering at the hands of a cheaper, competitive alternative, terephthalic acid (TPA), a chemical that Hercules was not well positioned to supply in bulk (see Chapter 11, and Exhibit 12.2). Although some executives perceived the threat of TPA, especially as developed by Amoco, opinions about its seriousness were divided at the top. Although it did indeed have a molecular cost advantage over DMT—it took less TPA to make a pound of polyester—Brown, Harvey Taufen, and others believed that Hercules' "volume and low-cost process made up for a portion of that disadvantage." At the same time, says Brown, "we were under the impression that with some modification of our process we could add a step and also make enough TPA to satisfy those customers that demanded it." Finally, Hercules' leaders believed that Du Pont, regarded as "the biggest and the best" in the polyester business, would stick with DMT "because its fiber process was geared advantageously to that feedstock."[8]

Worse still, Hercules' faith in the "polyester growth syndrome" proved unfounded as demand for the fiber slowed down in the mid-1970s. Long one of the most successful of the miracle fibers, it was increasingly regarded as artificial in the worst sense. As consumers shifted back to natural fibers or other, more attractive synthetics for apparel garments, demand for polyester (and hence for DMT) flattened. After growing at 23 percent annually from 1963 to 1973, polyester sales rose less than 7 percent annually during the next five years.[9] (See Exhibit 12.3.)

In short, Hercules found itself heavily invested in a declining product in a slow-growth market, and dependent on a single large customer. Disaster struck in 1975, when Du Pont unexpectedly announced its decision to become a self-sufficient producer of polyester intermediates. (According to Brown, Du Pont arrived at this decision after one of its suppliers in another product line suddenly raised prices. Because Du Pont relied heavily on the supplier, it had little choice but to pay. To avoid what they regarded as extortion—or the threat of it—Du Pont's top executives resolved that the company would no longer tolerate dependence upon any single supplier.) Because Hercules had started construction at Wateree on the basis of requirement letters from the Du Pont Purchasing Department, as well as "total assurance from departmental managers at Du Pont that increased orders would take place," the reversal came as "an enormous shock." Construction was halted immediately, resulting in a write-off of about $14 million, while the company contemplated additional measures. According to Brown, "We talked seriously about suing the Du Pont company for leading us on in this manner but decided there was no profit in attempting to sue one of our most important customers. We already had a penalty clause in our contract, and indeed, Du Pont paid a modest penalty for materials it did not take."[10]

Hercules realized that its smartest option was to find a way out. After an extended series of executive committee meetings, Brown, Taufen, Giacco, and

EXHIBIT 12.2 U.S. Production of DMT and TPA, 1960–1987

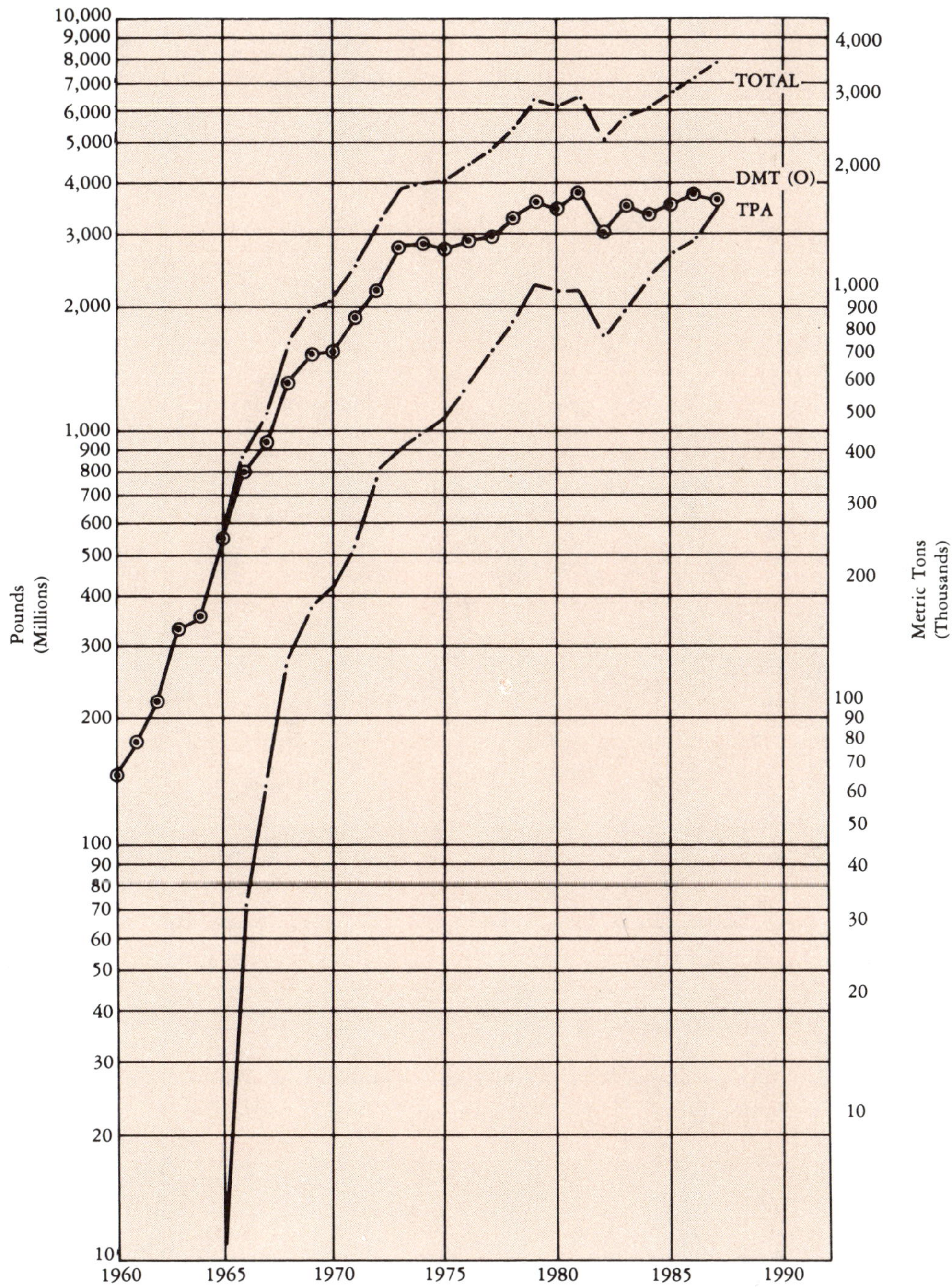

Source: SRI International, *Chemical Economics Handbook* (November 1988), 695.40210.

EXHIBIT 12.3 U.S. Production of Man-Made Fibers, 1965–1982

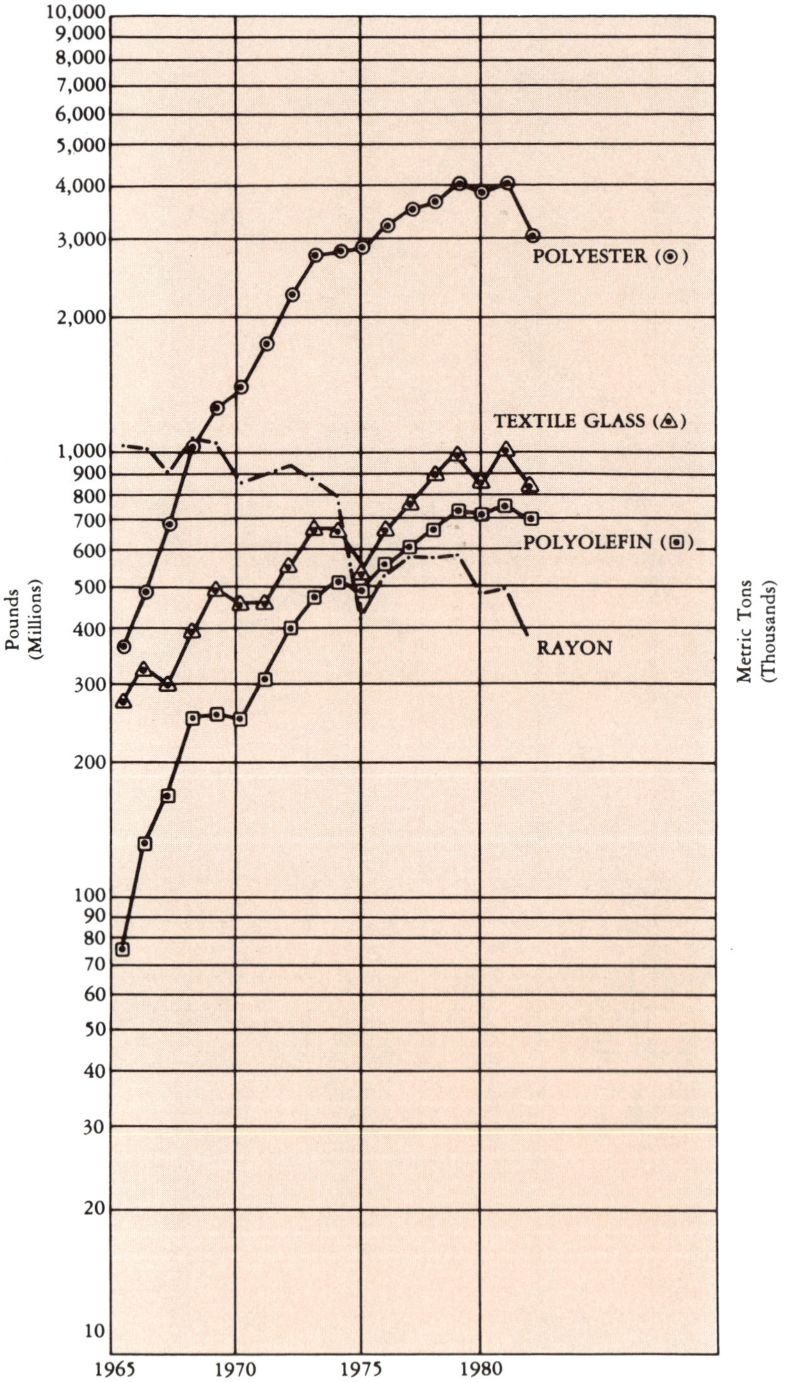

Source: SRI International, *Chemical Economics Handbook* (August 1983), 543.1000.

other officers were dispatched to find a buyer or joint-venture partner for the DMT business. Given a business in which the raw materials accounted for 80 percent of the cost, and only a single market existed, the most logical candidates were oil companies or fiber producers. After approaching Exxon, Texaco, and other companies, in the summer of 1976, Hercules at last found a taker: American Petrofina (the U.S. subsidiary of the Belgian oil giant, Petrofina), which saw the opportunity as a good long-term investment. Negotiations for a joint venture, led by Giacco, were concluded two weeks before OPEC announced a major price increase for crude oil. Under the agreement, Hercules' U.S.-based assets for making DMT, TPA, *para*-xylene, and methanol were placed under the management of a new company, Hercofina. Hercules retained a 75 percent interest in the venture and sold the remaining 25 percent to American Petrofina for $87 million in cash, notes, and other considerations. The agreement also included provisions for Petrofina to acquire an additional 25 percent interest over time as it made capital investments in the business. The unit's first president, Tex Little, was a long-time Hercules veteran.[11]

Hercules' complete withdrawal from DMT would take nearly a decade. The formation of Hercofina nonetheless represented a frank recognition among top executives that the company's future was not in commodity petrochemicals but in specialized niches where it could wield greater control over its own fate.

From Plastics to Applications

In Hercules' other major petrochemical business, polypropylene, the picture was far brighter, principally because of the diverse nature of its markets: whereas DMT was consumed primarily in a single application—textile fibers—polypropylene had a multitude of uses in products for many industries, including transportation, appliances, fibers, packaging, housewares, housing, and medical supplies.[12] And although the oil crisis sent prices of propylene soaring and Hercules' sales of polypropylene dropped 14 percent during the recession, the situation quickly improved. In part, polypropylene benefited from the even greater price hikes in other petrochemical plastics, as well as from huge increases in the price of steel, aluminum, and other energy-intensive materials. But the advantages of polypropylene were also coming to the fore. The major automakers, for example, began to make more fuel-efficient, lighter weight cars and used polypropylene plastics and fibers for interior trim and carpeting.

Moreover, Hercules was better able to defend its position in polypropylene than was possible in DMT. As the U.S. market leader in polypropylene by a healthy margin, the company had gained the benefit of know-how exchanges with Hoechst and ICI and was able to secure favorable terms on licenses of key process innovations made by Mitsui and Solvay. Finally, Hercules could rely on its growing business in fiber and film applications, especially the latter; the company acquired a steady stream of major new accounts in packaging and bottling.

The pickup of sales and profits in the film business came just in time. In the late 1960s and early 1970s, Hercules had made major capacity investments in polypropylene film at Covington, Virginia, and Terre Haute, Indiana, to serve key accounts in the tobacco and snack-food industries. By the middle of the 1970s, top management was anxious to see the payback from these investments. Fred Buckner, director of sales for film products in 1974 (and currently president and chief operating officer), recalls a pointed discussion on the matter at a management retreat.

> Werner [Brown] cornered me and told me that the film business would have to show some profit, and show it soon. He said it's fine to build these businesses and put money into them and have great hope and faith in the future, but after ten years they've got to start to pay off and we just about have had all this that we can take. He was nice about it but he made the point: time had run out! Well, profitability obviously started to get better.[13]

According to James E. Knox, manager of business development for the Polymers Department in the mid-1970s (and now corporate vice president of marketing), the cost breakthrough helped clinch the victory of polypropylene film over cellophane for many packaging applications. Polypropylene had a very low density, and for a given amount of weight, it covered a greater area than cellophane. Polypropylene also had great clarity and excellent moisture resistance; cellophane was neither as strong nor as clear and needed to be coated to prevent moisture from passing through it. "So," concludes Knox, "we had property advantages, and as we got our cost down and plant scale up, the density began to play to our advantage, too."[14]

Changing of the Guard

Despite progress in film and other lines—one of the strongest contributors in this period was CMC, millions of pounds of which were consumed by the oil industry for drilling mud—by the end of 1976, Hercules was not out of the woods. Although it reported profits of $107 million for the year, more than one-third of the total reflected returns from the partial sale of the DMT business as well as the sale of other assets. Such measures improved Hercules' balance sheet and income statement but did little to impress the stock market. At year end, Hercules' shares continued to trade in the low twenties, a precarious level that could (and, in fact, did) drop as OPEC raised oil prices again in 1977.

As Brown puts it, Hercules' growth strategy in the preceding two decades had depended upon "cheap and abundant raw materials, major technological breakthroughs, and a seemingly limitless demand for chemicals"—three characteristics, it now appeared, of a vanishing era. At the same time, there was

> a growing realization that major changes are taking place in the industrialized world, especially in the manufacture of and markets for chemical products. New and formidable worldwide competition, increasingly expensive raw materials, lower economic growth, and increasing

> Government intervention in the private sector—all indicate a more cost-sensitive environment and a need to find new ways of doing business more efficiently and economically.[15]

Although Brown saw the need for fundamental changes at Hercules and approved plans drawn up to that end, he decided that the new strategy needed another leader to preside over its implementation. Accordingly, he gradually withdrew from the running of the company. In December 1976, Al Giacco was promoted to a new position, executive vice president, and designated "senior operating officer of the company, supervising all aspects of corporate manufacturing and marketing operations and related staff functions."[16] In the process, he became Brown's heir-apparent. Ten months later, Giacco was elected president and chief executive officer of the corporation, and Brown moved up to succeed the retiring chairman of the board, Jack Martin.

STRONG MEDICINE

In many respects, the new president was an anomalous figure at Hercules. When he took office, Giacco was fifty-eight—the same age as his predecessor. A chemical engineer trained at Virginia Polytechnic Institute, he had spent much of his career in sales positions, where he displayed remarkable ingenuity and a knack for getting results. Giacco had joined the company in 1942 and worked throughout the war as a technical supervisor at the Radford Ordance Works. After the war, he held a series of assignments in Explosives and Chemical Propulsion, where he attracted notice in the early 1960s by leading successful bids on strategic missile programs. One of several promising young managers with a background in chemical propulsion promoted to work in other areas of the company, he moved into the Polymers Department in 1965, rising to become general manager three years later. In 1973, he became the first general manager of Hercules Europe and was elected a corporate vice president the following year. Although he worked his way to the top by serving in many parts of the company, Giacco was Hercules' first president since Dunham who lacked firsthand experience in the company's traditional chemical businesses. He was also something of a loner—not aligned with any camp in senior management and standing apart from the debates on strategy and growth that had raged through the company in the 1960s. A brilliant, charismatic leader who could appear alternately charming and ruthless, Giacco did not shrink in the face of challenges.

When he assumed the presidency of Hercules, Giacco had definite aims in mind.[17] Since 1974, in fact, when he joined the executive committee, he had also chaired the corporate long-range planning committee. From this position, he had participated in the major decisions of the period: the formation of the Adria Labs and Hercofina joint ventures, new investments in fiber and film products, and divestitures of marginal operations. At the close of 1977, he was eager to reshape and restructure the entire company. And with the continuing plunge of the stock price—to $14.75 in the fourth quarter—Giacco believed he had a strong mandate for change.

Moving to a Matrix

The first step was to shake up the organization, a goal Giacco apparently had mulled over for some time.[18] From the outset, he sought a radical break with the past and considered provocative gestures such as relocating the company to new headquarters, perhaps even out of Wilmington (see Chapter 1). More immediately, he acted swiftly and dramatically to alter the culture and style of management. Reversing the customary order of strategic change, Giacco believed that Hercules would have to change its structure in order to reformulate its strategy.

During his ten months as executive vice president, Giacco had traveled extensively to visit the company's operations. At every location, he had met with sales and operating people to get a sense of the problems and opportunities in each business. But, he added, the meetings had a second objective as well:

> What I was really interested in finding out was the feeling of the rank and file of Hercules marketing and operating people. How did they feel about the company, what did they think it needed, and so forth. And what I basically got out of this thing was the feeling that because of the departmental structure, people didn't identify with Hercules, but rather with each different business. The guys didn't feel close enough to the company because our lines of communications were too long and too fragmented.

In the fall of 1977, Giacco summoned the department heads to a management retreat at Hershey, Pennsylvania, to discuss and plan "a rather radical change" in the company's organization. What came out of the meeting was the outline of the first major corporate reorganization in fifty years. The new structure certainly fulfilled Giacco's hope for dramatic change. The executive committee of the board, which had governed Hercules through a process of consensual decision making since 1928, was rendered powerless. In its place came the "office of the president," consisting of the president (Giacco), chairman (Brown), and two senior vice presidents (Harvey Taufen and S. Raymond Clarke III). Two other senior vice presidents (Spencer H. Hellekson and John R. Ryan) were made the heads of domestic and international operations, respectively. The remaining former department heads were reassigned as vice presidents with responsibilities for coordinating major functions—operations, marketing, sales, production, development, administration and public affairs, control, and treasury—across the whole company.

Hercules also abolished the positions of department general manager and assistant general manager and reorganized operations into sixteen "business centers" for the major product areas (plastic resins, water-soluble polymers, organics [resins], food and flavor ingredients, and so on). To help offset the disorienting effects of the reorganization and the centrifugal effects of decentralization, the entire management team—members of the office of the president, the functional vice presidents, and the business center heads—gathered periodically to discuss major issues affecting the corporation as a whole. (See Exhibit 12.4 for an organization chart of the matrix as it stood in 1980.)

In making these changes, which took effect on January 1, 1978, Hercules in effect adopted a project-oriented or matrix structure in which key decisions were negotiated between two bosses: the functional vice presidents and the business center directors. The business center directors had final decision-making authority (subject to Giacco's review), but their staff support people—for example, the directors of sales or production at a business center—also reported to the functional vice presidents at headquarters. Not surprisingly, the reorganization created massive confusion. Giacco quipped, "What we did was send everybody out in the halls and close the doors and blow the whistles, and nobody knew where to go." At the Research Center, recalls Ted Bednarski (currently vice president, Science and Technology), the joke went around that the company should enter a rapidly growing new line: printing business cards.[19]

Although Giacco has been criticized for using the reorganization to seize greater control of the company, he responds by citing the need to change systems and procedures that had blocked critical information from reaching the top and had led the company, he believed, to make a series of poor decisions. (His criticism of the departmental structure, incidentally, is shared by many of today's senior managers.)[20] He admits that by breaking the power of the department heads, the new structure represented "a real centralization of the company"; on the other hand, he denies that he centralized because of his own management style: "The answer to that is no. We centralized because what we had to do as a number-one priority was change the portfolio. And the only way we were going to change the portfolio was to put [all the businesses] on an equal basis."

Giacco also viewed the change as an opportunity to develop a new generation of management, and quickly: "The limitation of the Hercules management system in the past," he says,

> was that people stayed in their functions for too long. You could grow up having started out as a salesman . . . until you became a general manager. So now you're a generalist, but you have no generalist training. What I was trying to create in these business center positions was a way to give people training in a generalist way. Before they have to take over, have them learn about operations, have them learn about development, have them learn about all these things that have to be put together. That's what I did [performing assignments in different functions and different departments]. I learned by doing it. Well, by 1977, we didn't have time to do that.
>
> I never contemplated the structure being described as a matrix organization or anything like that. I grew up in Aerospace. There we used to call it project organization. But it was just a means to neutral-

EXHIBIT 12.4 Hercules Incorporated Organization, 1980

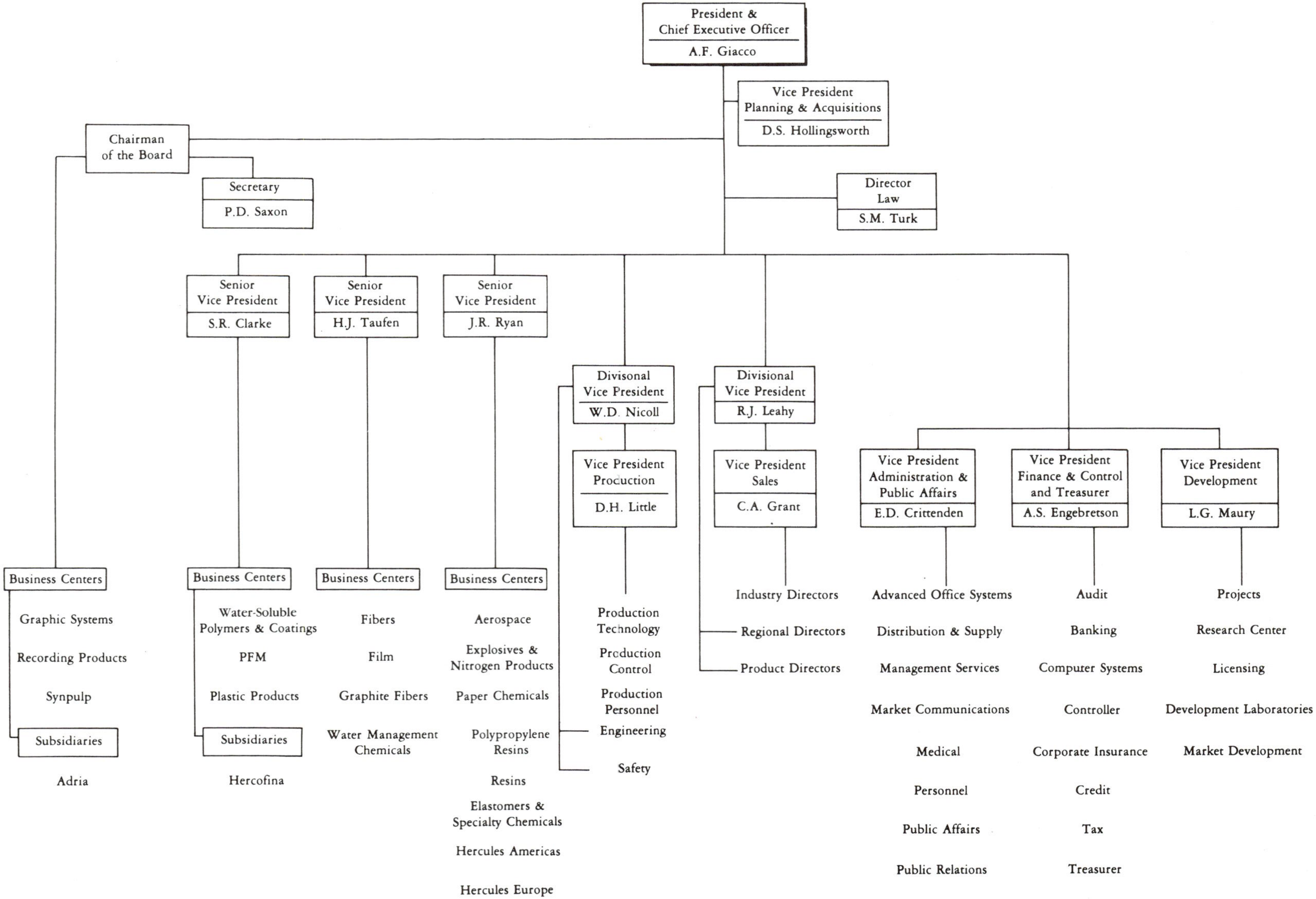

> ize the field by putting the senior people in the functions and giving the young people [at the business centers] freedom to run with their businesses. I didn't have to worry very much about them because the old guys would keep them in line. The point was, we had to create a very massive change in organization.[21]

The result of the new organization and associated policies was a fundamental shift in how Hercules operated. "Al's objective was to change the culture," recalls Robert J. Leahy, who became the functional vice president for marketing in the new structure. "Believe me, he succeeded eminently." For example:

> In business, there are lots of ways to look at things. Take profitability by product line. You [a business center manager] can look at that one way and I [a functional vice president] might look at it another. My object was to sell everything we had profitably. Your object might have been to sell the one, single most profitable product, and if you don't know the market opportunity, you might end up with a hell of a lot less business, but profitable. Well, what I realized was, in the matrix system, you were free to go ahead and do that even though I as vice president in marketing say, that's a damn dumb idea. Al said when the thing was formed, he wanted to create conflict. The matrix does a beautiful job of creating conflict. Everything was negotiated. There wasn't any longer the seat-of-the-pants decision and go take the hill. No more of that. It was strictly a negotiated thing.[22]

To Leahy, the results of the reorganization were evident: "We did pretty well." Frank H. Wetzel, who had worked for Giacco in the Polymers Department, became a business center director for water-soluble polymers in 1978. In his view, "The whole focus of the matrix was Giacco's desire to totally destroy the old departmental structure. Just totally remove any vestiges of the departmental culture that had existed in Hercules for fifty years." From Wetzel's perspective, the change was clearly for the better: "I thought the [new] system was fine. I liked the matrix system a lot. For the first time, you could really run a business internationally. In the old department setup, that was never possible."[23]

The matrix also improved the operation of the auxiliary or staff departments. According to Dr. Eugene Crittenden, who became vice president for administration after the reorganization, the new structure succeeded in "aligning staff services much more with the businesses." As a result, lines of accountability and responsibility became sharper and the staff units more clearly supported the functioning and profitability of line operations.[24]

The change was especially marked at the Research Center, where Giacco appointed his former chemical propulsion colleagues, Jack Main and Dick Winer, as directors. These appointments signaled a new thrust in R&D: Giacco wanted to change the direction of the research policy developed by Emil Ott and continued by his successors: conducting basic chemical research. Henceforth,

Hercules would focus instead on projects of immediate interest to the business centers. "At the risk of making a lot of people mad at me," says Giacco, by way of explanation, "Hercules is not a research company. It's an applications company. We're very good at applications, but we don't have the critical mass to afford long-range research."[25]

In choosing to emphasize applications-oriented research in the mid-1970s, Hercules was accompanied by other chemical companies frustrated by heavy investments, long development periods, and uncertain paybacks from basic research.[26] Under Main and Winer, "the very traditional research organization with six operating divisions, each one having its own defined R&D group," says Bednarski, was transformed by the matrix, in which "everything got mixed up." Project managers represented the businesses and worried about budgets, scheduling, deadlines, and results, whereas functional managers were responsible for maintaining technical expertise (in areas or disciplines such as analytical chemistry, engineering sciences, polymer science, and so on), as well as for training, facilities, and equipment.

Tensions between the two sides occasionally erupted into conflicts, recalls Bednarski, but the overall impact of the matrix was positive. "For one thing, the scientists learned to communicate with the business people, meeting on a quarterly basis to discuss sins or accomplishments. That communication loop wasn't there before the 1970s, but it was important." Bednarski points out another benefit of the matrix: young technical people were forced to work with plant managers and marketing personnel to achieve specific results within a given time frame. "These people were given a lot of responsibility and received high visibility early in their careers. Project management was a tremendous training ground."[27]

Although some managers adapted well to the matrix at Hercules, many others did not. When he announced the reorganization, Giacco had predicted that there would be casualties because of both the consolidation of six departments into one and an early retirement program established during the summer of 1977. In fact, he anticipated annual savings of about $45 million from salary reductions. In retrospect, this total seems a likely, perhaps even conservative, estimate in view of the actual turnover at the top. Three years into the system, by Giacco's own reckoning, less than a fifth of the top twenty-five functional managers remained, while ten of the original business center directors had moved on.[28]

Fire Sale

Organizational upheaval was but a preview of more changes. The new structure achieved the goal of clarifying the performance of the business units—an accomplishment, according to Wetzel, that was long overdue. Under the previous arrangements,

> A department like Coatings and Specialty Products not only had water-solubles in it, but it also had food ingredients, nitrocellulose

> business, and other things. These businesses inevitably were all rolled together on a departmental basis, and a business per se was never really looked at from the standpoint of growth, risk, and investment. If C&SP had a good year, all the businesses were allocated collectively. Water-solubles should have been run as a business all by itself for a very long time before that. The same with coatings, because nitrocellulose is dramatically different from the other cellulosics. The food ingredients business should have been broken out that way, too.[29]

The matrix allowed Hercules to reorganize its operations based on logically defined strategic business units. Under the new scheme, poor performance in one business could no longer be disguised by consolidation with the results of other businesses—especially given Giacco's energetic attention to the company's affairs. (In the early 1980s, he had installed in his office a computer terminal that provided him—at least in theory—with the means to monitor operating results across the company worldwide, supposedly updated every twenty-four hours. The symbolism of his terminal was apparent to everyone in the company.)[30]

With the new organization in place, each business standing on its own feet, and a new generation of managers on the scene, Hercules was in a position to implement a new strategy. As Giacco saw the company's predicament, "the radical and unexpected change" in energy prices had altered the economics of many of its operations. The situation was made worse by the debt that Hercules had arranged in the late 1960s and early 1970s to fund its expansion in petrochemicals, as well as by an increasingly intolerable burden of overhead. (Between 1974 and 1978, general indirect expenses had soared by 38 percent.) With the slowdown of growth, Giacco believed, the immediate tasks were to reduce the company's dependence on petrochemical commodities, lower its long-term debt, and restructure and reduce overhead costs.[31]

Although the Hercofina joint venture reduced Hercules' exposure in petrochemical commodities, the slumping DMT business remained a serious problem, and the company was eager to liquidate its stake. Doing so, however, proved to be a slow process of piecemeal transactions. In 1979, Ashland Oil acquired Hercofina's methanol plant in Plaquemine, Louisiana. The following year, Hercofina arranged to sell the plant in Burlington, New Jersey, to the Bombay Dyeing and Manufacturing Company of India. Hercofina also closed its DMT plant in Middelburg, the Netherlands, writing off $10.7 million in the process.

To reduce its debt burden, Hercules acted quickly to place other units on the block. "The situation was urgent," recalls Brown, "so we looked at selling a lot of businesses, including some of our recent acquisitions such as the flavors and fragrances businesses." The company also relied heavily on the techniques of portfolio analysis, which categorized units on the basis of market attractiveness and competitive position. Such analysis allowed top management to see which businesses would be profitable contributors to Hercules over the long run, and which would require too much scarce investment capital to sustain. Additional criteria, developed as the restructuring proceeded, reflected assessments of different kinds of "unacceptable risk," economic or social. Unacceptable economic risks, according to Giacco, involved assets or businesses that were heavily dependent upon hydrocarbon feedstocks, in mature or "sunset" industries, or exposed to escalating global competition. Unacceptable social risks were those encountered in businesses that generated environmental pollution, were heavily regulated, or exposed the company to potential court battles in "our increasingly litigious society."[32]

Operations that failed to meet the company's criteria for performance were soon divested. In the summer of 1978, the pigments business, classified as both an economic and social risk, was sold to the Swiss chemical and pharmaceuticals giant, Ciba-Geigy. That fall, the declining agricultural chemicals business (with toxaphene under siege from the EPA) was spun off as Boots-Hercules, a forty-sixty joint venture with the Boots Company of Nottingham in the United Kingdom. At year end, the lightweight aggregate facility in Snowden, Virginia, which had been leased for two years, was sold to Amlite Corporation. Soon thereafter, MCW ceased production of urea formaldehyde. By 1985, more than twenty businesses, representing about $650 million in assets—roughly 40 percent of the company's asset base at the start of the period—were sold or joint-ventured (see Exhibit 12.5).

By these and other measures, Hercules succeeded in shifting the balance of its portfolio away from petrochemical commodities toward more profitable specialty chemicals and aerospace products. By 1982, only 28 percent of the company's gross fixed assets were employed in commodity petrochemicals (down from 43 percent in 1975); two years later, that figure had dropped to 19 percent. Proceeds from the divestments were used to reduce the company's long-term debt: between 1977 and 1983, Hercules' debt-to-equity ratio dropped from 55 percent to 40 percent.[33]

Giacco's measures to contain costs took effect quickly. The company curtailed expenses across its operations and made several strategic moves to improve relative cost position. In polypropylene, for example, Hercules formed a joint venture with Enterprise Petroleum Corporation, a Texas energy trader, to split propylene from propane and deliver the monomer via pipeline to Hercules' Lake Charles and Bayport facilities. This partial step toward backward integration helped ensure the low-cost supply of a critical raw material.[34]

The most visible savings, however, resulted from the loss of about seven hundred corporate jobs between 1977 and 1980. About half resulted from early retirements, and half through attrition. The only outright layoffs came in 1980, when fourteen salaried employees were let go after Hercules abolished its corporate Advertising Department. A study revealed that the advertising function was best carried out by each separate business, and that, in any case, expecting a small corporate department to serve the needs of sixteen business centers was (and partly proved to be) a recipe for conflict. As a partial result of such downsizing initiatives, between 1978 and 1982 corporate revenues rose 27 percent, but general indirect expenses increased only 11 percent.[35]

Giacco also relied heavily on information technology to help thin the ranks of management. Hercules was one of the first large companies to make intensive

EXHIBIT 12.5 Restructuring the Portfolio: Divestments, 1975–1985

DMT—Burlington, New Jersey
DMT—Wateree, South Carolina
Nitrogen products—Hercules Works
—Missouri Chemical Works
Torak—Plaquemine, Louisiana (insecticide)
Methanol—Plaquemine, Louisiana
Toxaphene—Brunswick, Georgia
—Managua, Nicaragua

Pigments and colors—Imperial
—Ten Horn
—B.F. Drakenfeld
Haveg products—Santa Fe, New Mexico
—Marshallton, Delaware
Cesalpinia—Italy (guar and natural gums)
Fibers—Covington, Virginia

Lightweight aggregates—Snowden, Virginia
TPA—Hanover, North Carolina
Hydrocarbon resins—Baton Rouge, Louisiana
Polypropylene resins—Partial (HIMONT)
Hercor—Puerto Rico
Black, Crow & Eidsness—Gainesville, Florida
Hercofina—xylenes and terephthalates
Commercial explosives
Boots-Hercules—agricultural chemicals

Source: A.F. Giacco, speech to annual stockholders' meeting, March 26, 1985; Hercules Incorporated, Annual Reports for 1975–1985.

use of personal computers, satellite communications, voice messaging, videoconferencing, and other marvels of the information age. Starting in the late 1970s, and accelerating after headquarters' move to Hercules Plaza in 1983, the impact of the new technology was felt on staffing levels and productivity. In a typical company, for example, top and middle management account for 65 percent of all management; in 1985, in the fibers and film operations, those ranks accounted for only 37 percent of all management employees (see Exhibit 12.6).[36]

Signs of Hope

The proceeds from the sale of the DMT business and other assets, cost-containment measures, and strong earnings generated by CMC, naval stores, and paper chemicals helped provide resources to invest in the company's more promising businesses. The positive side of Giacco's early strategy involved balancing the portfolio according to several criteria and moving the company toward higher value-added products and services.

EXHIBIT 12.6 Impact of Information Technology on Hercules' Management Ranks, Fibers and Film Group, 1985

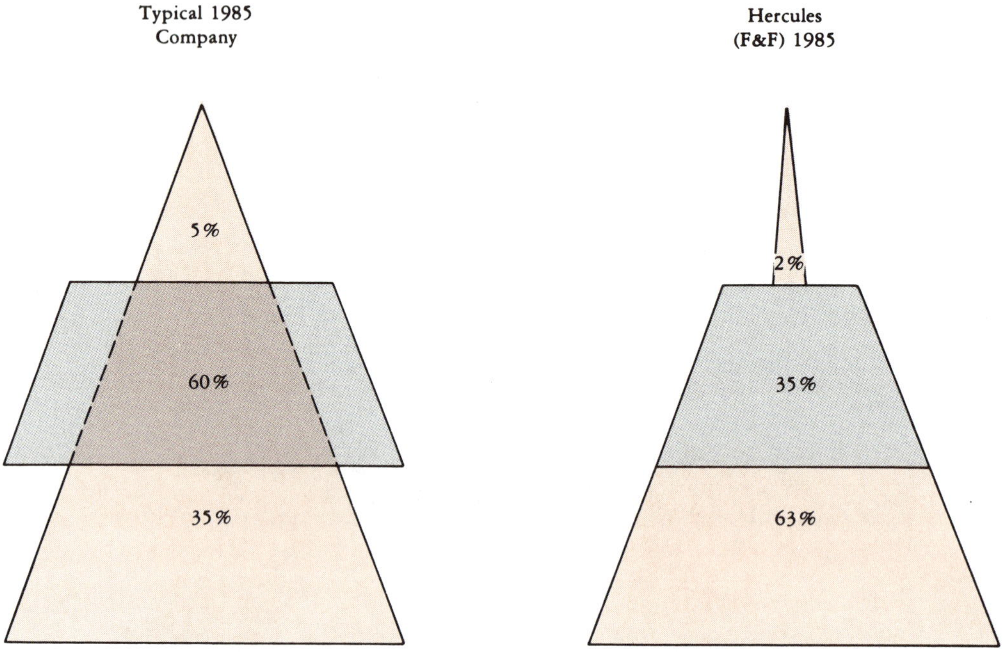

Source: Robert G. Eccles, Jr., and Jane Linder, "Hercules Incorporated: Anatomy of a Vision," Case 186–305. Boston: Harvard Business School, 1986, p. 18.

Giacco's concept of balance was multifaceted. Seeking to insulate the company from a downturn in any of its major businesses, Giacco aimed to strike balances in raw materials sources, end-use markets, and geographical market coverage. Accordingly, Hercules reduced not only its dependence upon petroleum and hydrocarbon feedstocks but also its sales to cyclical businesses (such as housing and textiles) relative to noncyclical businesses (such as food and materials) and its production base and sales mix in the United States and overseas.[37]

Hercules' emphasis on value-adding meant several things at once. First was the popular view of the company's history held by senior management. The most successful businesses, it was believed, entailed taking waste or low-value products such as tree stumps and cotton linters and adding value to transform them into chemical specialties. It followed that Hercules' best hopes for the future lay in developing similar new businesses. Second, according to Giacco, the key to the company's success in naval stores and cellulose products was not control of raw materials costs, as conventional wisdom dictated, but rather "relative cost minimization—that is, minimizing raw material cost relative to final selling price." This goal was best accomplished not by backward integration, but by "continued forward integration to better, more valuable products for the marketplace." In Giacco's view, "the basic corporate strategy for all our major businesses is built on this simple principle—minimize the raw material cost component relative to sales price—relative cost minimization. Or it can be expressed as—maximize the added value." In 1979, the company's "average value-added factor" was three times its cost of raw materials. At that point, Giacco directed capital spending to increase the factor to four times, by increasing investment in such areas as polypropylene fibers and film and aerospace. In support of the strategy, Hercules pushed its annual spending on R&D from about 2 percent of sales to nearly 3 percent.[38]

In polypropylene fibers, for example, Giacco approved plans to build a one-hundred-million-pound-per-year staple fibers plant at Oxford, Georgia, as the first step toward developing markets for nonwoven applications such as cigarette filters and disposable diapers. At the same time, Hercules poured investment capital into Terre Haute and Covington to install a new process technology and increase film capacity; it also sought to expand the business overseas through three new projects: a plant at Varennes in Canada, a fifty-fifty joint-venture plant in Brazil, and acquisition of a small film producer in the United Kingdom.

Hercules also launched an aggressive strategy to develop its aerospace business. Since the end of the Vietnam War, the company's sales to the government had been flat, hovering at $75–100 million per year. In 1978, however, two steps were taken to improve this situation. First, Hercules replaced Rocketdyne as management contractor at the Navy's solid-fuel rocket facility at McGregor, Texas, thereby strengthening its ability to make tactical rocket motors and acquiring expertise in metal cases. Second, Giacco ordered the Aerospace Division (Business Center) to relocate from Wilmington to Bacchus and told its managers to build the business. Recalls Henry A. Schowengerdt, one of the managers dispatched to Bacchus, "Every time I saw him after that, he would ask, 'What are you doing to grow it?' "

At first, Schowengerdt says, he was skeptical of such interest, given Hercules' traditional reluctance to increase its dependence on government contracts. He sensed a changed attitude, however, when he appeared at a board meeting to discuss Hercules' participation in the largest peacetime military program in the company's history: development of the Trident I C-4 submarine-launched ballistic missile (SLBM). If all went well with the program, scheduled to move into production in the early to mid-1980s, Hercules' sales to the Navy could soar into the hundreds of millions of dollars per year. Schowengerdt remembers, "One of the older members [of the board] said that by his calculations the program could raise Hercules' sales in aerospace to something like 30 percent of the total corporation's, and he asked whether they shouldn't discuss that a little bit." The concern was that aerospace might become too big a segment in the corporate portfolio. "After a few minutes of discussion," Schowengerdt recalls, "Al said, 'That is not Hank's problem. Hank's problem is to grow aerospace, and it is our problem to grow the chemical business as fast as he is growing aerospace.' [Thereafter,] we had excellent support."[39]

The aerospace bet proved to be a good one: sales doubled between 1978 and 1980, and profits surged even faster. The Trident I C-4, moreover, became a breakthrough program for Hercules. Not only was it a larger missile than the

EXHIBIT 12.7 Hercules' Restructuring: Gross Fixed Assets Distribution, 1975–1985 (Dollars in Millions)

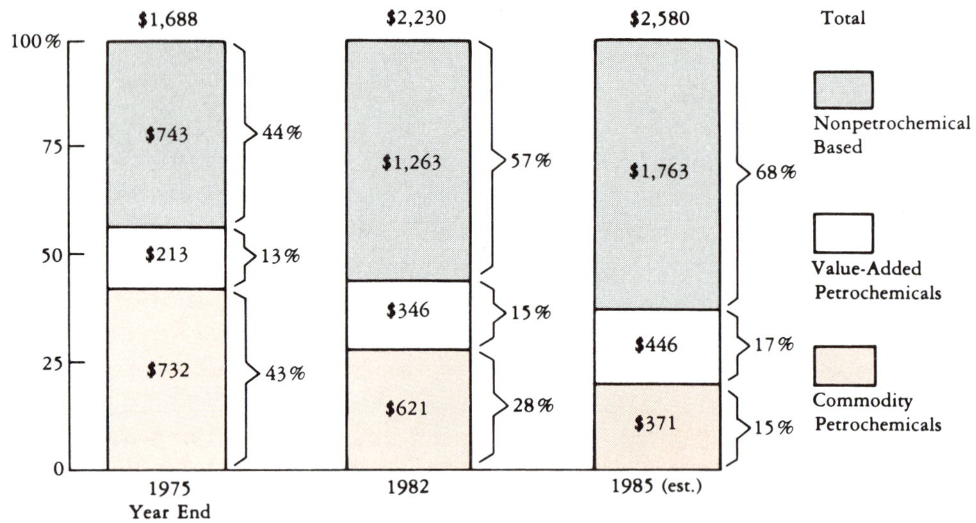

Source: A.F. Giacco, speech to annual stockholders' meeting, March 26, 1985.

company had ever produced, but it also used for the first time a new high-energy propellant that has since become standard for other ballistic programs. As the missile moved to production in the 1980s, Hercules increased its participation—and its earnings—by building composite cases for the missile's second stage.[40]

By 1980, Hercules' restructuring showed clear signs of paying off. Despite an increasingly difficult economic environment—extraordinarily high levels of inflation in most developed countries, a second energy crisis, and a worldwide recession—the value of Hercules' stock inched upward into the low twenties. Although restructuring and the sale of unwanted assets would continue into the mid-1980s (the last piece of Hercofina was sold off in 1985), the company turned to its next challenge: finding and developing, once again, new areas of growth (see Exhibit 12.7). Its focus on the future, however, was distracted by a celebrated lawsuit.

Interlude: Agent Orange Defendant

As a defendant (with six other chemical companies) in a class-action suit filed on behalf of veterans exposed to the defoliant Agent Orange during the Vietnam War, Hercules faced enormous potential liabilities in the early 1980s. Indeed, when the original suit was filed against Dow Chemical and several other chemical companies (not including Hercules) in 1979, attorneys for the veterans sought damages "in the range of $4 billion to $40 billion."[41]

Although few observers expected that such extraordinary sums would ever be paid, Hercules attorneys were nonetheless concerned. In part, their anxiety reflected the times: the dark side of the modern industrial economy was coming under severe scrutiny. Hazardous wastes found in the Love Canal near Niagara Falls and in the James River near Richmond had raised public outcry to new levels. "That was a very sensitive time," recalls Dick Douglas, retired director of public relations, "and a lot of people were very nasty and very angry toward the chemical industry."[42]

The 1970s, moreover, witnessed frequent and increasing mass tort litigation in cases involving transportation accidents, hazardous drugs, materials and consumer products, hotel fires and building collapses, and many other such issues and incidents. The stakes in such cases were high, and juries were likely to be sympathetic to individual plaintiffs in suits against giant corporations. In the most notorious example, Johns-Manville Corporation, a processor of asbestos, filed for bankruptcy protection in 1982 in the face of nearly 16,500 separate suits and billions of dollars in potential damages. The specter of a similar outcome hovered over the Agent Orange defendants throughout the suit. "For a time it was an all-consuming subject matter," recalls S. Maynard Turk, Hercules vice president and general counsel. "There was serious consideration among all the chemical companies that were involved that the Agent Orange case might lead to a Johns-Manville fate."[43]

Hercules' involvement in the case stemmed from its role as a producer of Agent Orange (a fifty-fifty mixture of two herbicides, 2,4-D and 2,4,5-T) for the U.S. military between 1965 and 1968. A manufacturer of herbicides since its 1961 acquisition of Reasor-Hill Corporation's plant at Jacksonville, Arkansas, Hercules was the third-largest producer of Agent Orange used in Vietnam, supplying 19.7 percent of the total. In 1968, the peak year of production, Hercules' sales of herbicides amounted to just over $12 million, with profits of about $1.8 million. This performance was anomalous, however, and the company lost money on herbicides in eight of the ten years after it acquired Reasor-Hill, and on balance for the decade. Indeed, the dismal economic prospects for the business led Hercules to cease operations at Jacksonville in 1970 and to begin negotiations for the sale of the plant.[44]

The veterans' suit against the Agent Orange producers raised many complex and troubling issues: the state of knowledge in the 1960s about the harmful effects of the defoliant; establishing a causal link between exposure to Agent Orange and medical problems suffered years later by the veterans and their families; the quality of the product as manufactured by different companies; the mixing, handling, and application of the defoliant by the military; the extent of liability on the part of military contractors that manufacture products at the government's request and to the government's specifications; and a host of legal questions, such as the definition and certification of a class, the applicability of federal common law to the case, and the role of individual judges in resolving these and other matters. Complicating the case further were the motives and beliefs of different stakeholders, including the veterans, their lawyers, the defendants (separately as well as collectively), the government, the judges, and the public, especially as opinions of the case were shaped by the media, usually on behalf of the plaintiffs.[45]

Among many central issues, the purity of the defoliant proved significant. Agent Orange was suspected to cause harm to humans in proportion to the amount of dioxin (short for 2,3,7,8 tetrachlorodibenzo-*para*-dioxin) contained in the product. A by-product of an intermediate stage in the manufacture of the Agent Orange constituent 2,4,5-T, dioxin was known in the early 1960s to cause chloracne ("a severe but usually reversible skin disease") in humans. More harmful effects, such as possible links to cancer and birth defects, were not publicly known until the late 1960s, and widespread use of Agent Orange ceased soon thereafter.[46] An essential part of the plaintiffs' case, then, was establishing the presence of significant amounts of dioxin in 2,4,5-T, and further, proving the connection between these amounts and the ill effects suffered by the veterans.

In this respect, Hercules had a strong defense: its 2,4,5-T contained extremely low, virtually undetectable levels of dioxin. (According to Dr. John Frawley, by then general manager of Hercules' Health and Environment Department, equipment sensitive to the presence of 0.2 parts per million of contaminant found no traces of dioxin in Hercules' products. Other manufacturers were found to have as much as 47 parts per million in their products. (Prolonged or repeated exposure to one part per million was believed to be dangerous.) The relative purity of Hercules' 2,4,5-T resulted from an unusual feature of the manufacturing process, whereby the chemical intermediate was washed to remove another impurity, a step that also removed dioxin. A measure of the effectiveness of this process was that no outbreaks of chloracne had occurred among employees at Jacksonville.[47]

On the basis of such findings, in May 1983, Hercules and another manufacturer with low dioxin levels were dismissed from the case by Judge George C. Pratt of the U.S. District Court for the Eastern District of New York, in Uniondale. Before the company's lawyers could relax, however, Hercules was again besieged. Pratt, who had recently been promoted to the federal circuit court, relinquished control of the case before signing the order of dismissal. His replacement, Judge Jack B. Weinstein, refused to accept the dismissal, stating that a jury ought to resolve the issues and that "everybody stays in until the litigation is over, unless there is a very good reason to let them out." On the other hand, Judge Weinstein also believed the case would be "better settled than tried" and actively encouraged the parties to fashion a settlement before the scheduled start of the trial the following May. Through frequent and aggressive interventions by Weinstein and his deputies, culminating in an intensive, round-the-clock period of negotiations as the trial date approached, the parties agreed to settle the case hours before jury selection was scheduled to commence on May 7, 1984.[48]

Under the terms of the settlement, the manufacturers agreed to pay fines totaling $180 million into a fund from which claims would be paid to eligible veterans. (This amount was essentially imposed on the parties by Judge Weinstein, who was "very sympathetic to the veterans and . . . the injuries they and their families had suffered," indignant at the chemical companies' "cavalier attitudes," and deeply resentful of the government's "stonewalling and detachment" from the case. Nonetheless, the veterans' case seemed to him "shaky on the merits," and he felt "an obligation to the legal system as a whole" to ensure that the total fine did not "signal that the case was stronger than it actually was," thereby encouraging "groundless mass toxic tort litigation in the future." For their part, the companies agreed to pay the total to avoid a prolonged and uncertain trial that might have resulted in a higher liability.) Individual manufacturers' contributions were apportioned according to two criteria: their share of production of Agent Orange used during the war, and the dioxin content of their product. Hercules, which had manufactured about one-fifth of the defoliant used, paid one-tenth of the fine, $18 million. (The heaviest fines were incurred by Monsanto, Dow, and Diamond Shamrock.)[49]

The cost to Hercules was considerable in terms of money, management time, legal expenses, and a tarnished public image. As noted, such considerations were subsequently factored into the company's strategic thinking about the social and environmental, as well as financial, risks of its businesses. On the other hand, Douglas believes that Hercules emerged from the case in better shape than most of the other defendants, thanks to the purity of its product, the excellent work of its lawyers—including Thomas Hunt, who devoted virtually all of his time to the case for years—and the company's candor and moderation in responding to journalists' inquiries.[50]

To the company, however, the case itself was more significant than its resolution. Although Hercules had paid the first big fine in its history, it remained troubled by other, larger issues: the indeterminate nature of causation in the veterans' claims; the political nature and populist overtones of the suit; the arbitrary nature of the settlement; the nature of a judicial system in which settlements are reached out of fear of costly delays or popular impressions rather than on the merits of a case; and especially, the government's unwillingness to support its contractors or to share in their liability. As Brown put it, succinctly, what happened to Hercules seemed "just so unfair."[51]

The Business of Creating Wealth

In undertaking a massive restructuring during the late 1970s and early 1980s, Hercules was not alone among large industrial companies. Indeed, the same forces that swirled around it—excess capacity, raw materials shortages, technological change, global competition, government intervention, and stagnant markets—affected many industries, especially those based in the United States and Europe. Automakers, primary metals producers, heavy goods manufacturers, and industrial suppliers and equipment manufacturers of all sorts struggled to make major adjustments in their businesses.

In the chemical industry, Hercules competitors and peers contended with the same global forces. In 1976, for example, Rohm and Haas reported the first loss in its history. The following year, it abandoned production of polyester, selling its major U.S. plant to Monsanto and closing other fibers-related assets. At about the same time, Union Carbide began divesting its commodity plastics operations and eventually sold its petrochemical business in Europe to British Petroleum. Monsanto and Dow also retreated from petrochemical investments in Europe; in addition, Dow reversed its long-standing strategy of backward integration into petrochemical feedstocks by selling off most of its oil and gas ventures and pulling back from its riskier efforts to diversify. Du Pont, in contrast, attempted to solve the feedstock problem by acquiring a major petroleum company, Conoco, in 1980 for the enormous sum of $7.6 billion.

In Europe, similar events played out: British Petroleum, an oil company, and ICI, a chemicals giant, swapped key parts of their portfolios in an effort to rationalize operations, as did the Italian chemical company Montedison and oil company ENI. In Germany, BASF closed polyethylene and polyvinyl chloride plants, and Hoechst slashed its commodity petrochemical plastics capacity by roughly one-quarter.[52]

The major chemical companies also sought new avenues of diversification. Allied Chemical acquired Bendix Corporation, thereby obtaining major positions in the aerospace and automotive supply industries. Monsanto bought G.D. Searle's unit that had developed the new artificial sweetener Nutrasweet and made several small investments in biotechnology startup operations. Du Pont sought to develop products and technologies that offered high rates of return with modest capital investment: medical instruments, agricultural chemicals, and specialty plastics and fibers.[53]

Observing these and other trends in the global petrochemical industry, Harvard Business School professor Joseph L. Bower discerns a common pattern: in most companies, successful restructuring followed the appearance of a new leader and included a major reorganization, a fresh emphasis on portfolio analysis, and the use of "credible outside sources of information"—consultants, academics, industry panels, governments—to justify crisis actions and to "make unpalatable decisions legitimate."[54] In Giacco, the matrix, and the refashioned portfolio of businesses, Hercules appeared to be almost a textbook case. Only in the last element of Bower's pattern—the use of outside experts—was Hercules an exception, although the severity of the crisis and the widespread publicity it received provided its own justification.

Indeed, looking back on the early 1980s, says Giacco, "Restructuring was the easy part. Rebuilding was the hard part. The first couple of years, I wasn't really working on it. I was working on restructuring. And it wasn't until, say, the third or fourth year, that we really started on the business of creating wealth for the shareholders."[55]

At the end of 1979, Giacco reorganized the office of the president: two officers responsible for growth areas—Hollingsworth as vice president for planning and acquisitions, and Dr. Lucien "Rusty" Maury as vice president for development, including research—replaced Brown and Taufen, who retired. (At the public announcement of these changes, an analyst asked Giacco: "Your newly created office of the president puts you into the strongest position of any president in all the time I've been following Hercules. Does this mean no one can say 'no' to Al Giacco?" Giacco's reply was that Hollingsworth often had been a critic of his policies.)[56]

The challenge these leaders faced was to establish a balanced position in the chemical industry's major growth markets of the future. Given its size and assets, Hercules could not expect to pursue every opportunity. For example, although the company had established a modest position in pharmaceuticals through its interest in Adria Labs, it was unlikely that it could compete successfully in the long run against industry giants like Merck or Pfizer. On the other hand, Hercules could build on its strong foundations in aerospace (including composites) and materials (primarily plastics and applications derived from polypropylene). The rapid growth of these areas, Giacco believed, would complement Hercules' steadier business in specialty chemicals and help carry the company through the 1980s and 1990s. And so began, he believed, the second phase of Hercules' restructuring.[57]

Rockets, Composites, and Systems

In aerospace, the momentum of the late 1970s accelerated into the 1980s. By itself or in combination with its joint-venture partner, Morton Thiokol, Hercules participated in development or production programs for virtually every strategic and tactical program in the U.S. missile triad: sea-based Trident missiles; the land-based Peacekeeper, MX, and Pershing II missiles, small ICBMs, and a host of tactical weapons such as the Stinger; and the air-based Sidewinder missile and AMRAAM (advanced medium-range air-to-air missile). For the largest of these weapons, Hercules produced not only the rocket motors but also the filament-wound composite cases.

The growth of this business was spurred in part by the defense buildup of the Reagan administration. But Hercules adroitly positioned itself to exploit opportunities by steadily upgrading its capabilities to develop propellants and composite cases. Fred Buckner (a native of Magna, Utah) points out that

> Hercules got market share a lot faster than the market grew, and the reason is because we gained credibility during our twenty years' experience to that point. During those years, we just kept building on our success: we kept expanding our capability to make bigger missiles, to develop higher energy propellants and make glass-fiber and composite graphite fiber cases to replace steel.[58]

According to Ruth L. Novak, vice president (strategic programs and marketing) of Hercules' Aerospace Products Group, the company's reputation was established during its participation in programs for the third stage of missiles, in which the need to keep weight to a minimum stimulated a series of important innovations. On the Poseidon program, for example, Hercules introduced a "flex seal" that allowed a single nozzle to replace four nozzles on the previous-generation Polaris missiles. The high-energy propellants and composite cases used on the upper stages of the Trident I doubled the missile's effective range.[59]

For the land-based Pershing II, recalls Ernest A. Mettenet, retired vice chairman of Hercules' aerospace operations, "we achieved a fantastic win over Morton Thiokol by introducing a safer propellant and a composite case suitable for a wide range of temperatures for tactical army applications."

Hercules' sales in rocket motors and composite cases soared to $200 million in the early 1980s. "That was big for propulsion," says Mettenet, "but then we said, hey, we've got the technology now to go after the big one": solid-rocket boosters for the Space Shuttle. This became Hercules' first opportunity to win a large graphite composite case program. With Giacco's blessing, early in 1981, the aerospace division began the long, slow process of persuading NASA to consider a second-source alternative to the boosters made by Morton Thiokol. The essence of Hercules' pitch to NASA was better performance derived from the greater strength-to-weight ratios of graphite composite cases, compared with those made from other materials.[60]

The opportunity to make large rocket cases provided an obvious boost to Hercules' graphite composites business, which was already coming into its own as a supplier to the aircraft industry. The company had entered the business in 1969 by licensing a manufacturing process from Courtaulds. At the outset, technical and managerial personnel in chemical propulsion, including Dick Winer and John Greer, believed that composite materials made from graphite fibers would eventually lead to revolutionary changes in the aircraft and space industries. The initial market for graphite fiber composites, however, developed in consumer sporting goods, such as golf clubs and tennis rackets. In the mid-1970s, those products showed signs of becoming commodity items, and Hercules abandoned consumer applications, focusing instead on aerospace customers that valued performance characteristics more than price. At the time, graphite fiber traded for several hundred dollars per pound, and somewhat more in the form of "prepreg" (a woven fabric or tape impregnated with polyester or epoxy resins that fabricators could mold into parts or structures).

Hercules' strategy was to lower this cost by improving the manufacturing process and increasing the scale of operations. The business built up slowly; the company managed to place a small amount on the tail of the F-16 fighter, about 2 percent of the surface area of the aircraft. By the late 1970s, next-generation aircraft used significantly larger amounts of composites: 10 percent for the structure of the F-18 fighter, and 28 percent for the AV8B.

In 1981—a decade after the first plant came onstream at Bacchus—Hercules produced its one-millionth pound of carbon fiber. "It took us ten years to make the first million pounds," recalls Jon DeVault, group president for composites products. "In the next two years we made another million pounds, and then the third year after that we made a million pounds that year. So we kind of sat there for ten years getting this to be an accepted product, and then it really took off."[61] By the early 1980s, the price of graphite fiber had fallen to roughly $40 per pound, and Hercules was the largest producer in the world.

In addition to its successful marketing program for high-performance aircraft, two other key factors spurred the growth of the business. In the early 1970s, the company developed a method for the continuous processing of fibers, from the precursor spool through oxidation and carbonization. Not only did this result in significant cost savings, but the tensile strength of the finished fibers actually increased by 20 percent over the fibers made by the original Courtauld's process.[62]

The second key factor came in 1978, when Hercules severed its relationship with Courtaulds and entered into a joint venture with Sumitomo Chemical of Japan (Sumika-Hercules Company, Ltd.) to make its own precursor, polyacrylonitrile (PAN) fibers. Thereafter, Hercules not only controlled the key raw material for making graphite fiber, but it also became the world's only fully integrated producer. Before the deal with Sumitomo, says DeVault, "We were good, but we didn't have a technical advantage. When we teamed up with the Japanese, that is when we really jumped up with a superior product and captured a dominant market position." DeVault believes that

> the real key over our competition was that we were totally integrated. We were competing with guys that just sold fiber or just sold prepreg. We had fiber, prepreg, and design and manufacturing capabilities. When we talked to General Dynamics about getting our material on the F-16, we could talk about fiber, prepreg, and we could talk about the design and fabrication of parts. The integration of those three skills is really what gave us our lead, not having a superior product.[63]

Hercules' success in rocket motors and graphite fibers led to a major acquisition in 1983: Simmonds Precision Products, a privately held company that designed and manufactured instrumentation and sensors for aircraft and missiles, with annual sales of roughly $150 million. The motive for the acquisition, explains Giacco, was the desire to capture greater value on aerospace programs—the same motive, for example, that had led to the acquisition of Young Development Laboratories a quarter-century earlier. Hercules had gradually expanded its participation in major rocket programs. As Giacco saw it, the problem was what to make next.

> Only 4 percent of the value of a rocket involves propulsion and propulsion components. The rest of it is electronics and everything else. So the trick was to expand our ability in other areas. Simmonds had certain electronic capabilities and mechanisms that would do things that could ultimately be translated into rockets, like pushing nozzles, like force and separation.[64]

Although Simmonds' historic levels of profitability were below those of Hercules, thereby diluting the parent company's earnings per share, Giacco felt the investment was worth it: "You don't always get everything you want; sometimes you have to take the pieces you can get when you can get them."[65]

Film, Fiber, and a Big Deal

The second growth area that Giacco identified in the early 1980s was materials, a designation he preferred to "plastics" or "polypropylene." This thinking reflected his 1960s' experiences in the Polymers Department, as well as his conviction during the 1970s and 1980s that Hercules' future lay with value-added products. Although the company maintained its leading position in polypropylene, Giacco believed that the business derived its strength from vertical integration: resins into film, fiber, and other applications.

In the 1960s, Hercules film and fiber products consumed about 40 percent of the company's output of polypropylene. That left the Polymers Department's sales force with the problem of finding markets for the other 60 percent. "So," recalls Giacco, "we started looking for things which were structural, like the under fender for the Oldsmobile. We kept that contract for years, and we got a good price premium for the business." The goal of selling polypropylene at a higher price than the industry average remained an enduring objective. As Giacco puts it, "That's where you make your money." To get there, he believed, required a fresh approach to marketing:

> We learned how to sell properties. We said we were not in the polypropylene business, we were in the business of providing properties. When we built a fiber plant later, we built it because you look at the stuff and it feels like cotton and you say, hey, I can make cotton cheaper than cotton. So the property we were selling was soft covering. We sold properties. [We concentrated on] what is it the consumer [end-user] really wants, and started thinking of the consumer and not our immediate customer.[66]

In the film business, despite some rocky moments during the 1980s, this strategy played out well. By the late 1970s, the conquest of cellophane was complete and Hercules faced its next challenge: to replace glassine, a specialty paper used for bottle labels and candy bar wrappers. For soda bottle labels, polypropylene film offered superior wet strength and flexibility to glassine, which was affected by moisture during the bottling process and later tended to crack. For candy bar wrappers, film offered the advantages of thinness and strength as well as lower cost.

To dislodge glassine, however, Hercules needed to make white opaque films that could be printed. This, in turn, entailed developing a different film-forming process. Hercules had built its business using the tubular ("bubble") process licensed from ICI. For making white opaque film, however, the bubble process was unsatisfactory. Mixing calcium carbonate with the polymer, to achieve opacity, tended to abrade the tubular process machinery, and, according to Buckner, "It was just not economical to think in terms of abrading away the parts and replacing them and abrading them away." The alternative was to use the tenter process for forming film, a technology widely used by Hercules' competitors.

In 1981, Hercules licensed the tenter process from a Japanese manufacturer, Honshu Paper Company. Although installing the new process at Terre Haute "wasn't as easy as we thought it would be," recalls Buckner, it was nonetheless a success: in 1983, Hercules secured a promising future for opaque films by winning major contracts with Coca-Cola and Mars Candy Corporation.[67]

Hercules' focus on selling properties also paid off in the fibers business. During the 1960s and 1970s, polypropylene fibers were consumed principally in two markets: industrial carpeting and upholstery fabrics. In the late 1970s, Hercules explored new applications such as nonwovens to expand the business and help offset its cyclicality. After the second energy crisis, which created a relative cost advantage for propylene versus other petrochemical monomers, Giacco believed that polypropylene staple fibers might supplant polyester staple fibers in certain markets. ("Staple" refers to short, squiggly fibers made to resemble natural cotton or wool fibers. Synthetic staple fibers can substitute directly for natural fibers in many textile applications.) He reasoned that inexpensive raw materials, coupled with a world-scale manufacturing plant that featured the latest technology, could quickly open new markets.[68]

In 1979, Hercules took a substantial risk by committing $30 million to build a one-hundred-million-pound-per-year plant in Oxford, Georgia, to make ultrafine-denier staple fiber. When this investment was made, the technology was uncertain and Hercules had no assured customers. As Buckner tells the story,

> That plant was built on the basis that polypropylene fibers were cheap, cheaper than any of its competitors. Giacco simply said, "By God, it's cheap, and therefore there's a market for it." We built it, and as it turns out, he was right, but he was just barely right. Virtually all of it goes into a single application.

That application turned out to be one of the hottest markets of the 1980s: disposable diapers. By the end of 1983, after debottlenecking, production at the Oxford plant was approaching capacity, and Hercules found itself the world's leading producer of polypropylene staple fiber.[69]

By 1983, virtually everyone at Hercules acknowledged the growth potential of the polypropylene film and fibers businesses, but opinion was far from unanimous about the prospects for polypropylene itself. On the one hand, Giacco was a devout believer in the promise of polypropylene as a material, "because nobody's going to build another steel mill, an aluminum mill, or things like that. We're going to be able to replace these materials with plastics." And he was convinced that polypropylene, in which Hercules held a leading position, was a material of the future, based on its properties of lightness and toughness as well as its low cost.

But many other Hercules executives voiced skepticism: twenty-five years of investment in the polypropylene business had produced significant growth but not significant earnings (see Exhibit 12.8). The dispute between Giacco and his lieutenants reached a critical point at a planning retreat in 1980 led by the two new members of the office of the president, Hollingsworth and Maury. The top forty officers of the company were divided into study teams and asked to

EXHIBIT 12.8 U.S. Capacity and Domestic Sales for Polypropylene Resins, 1958–1975

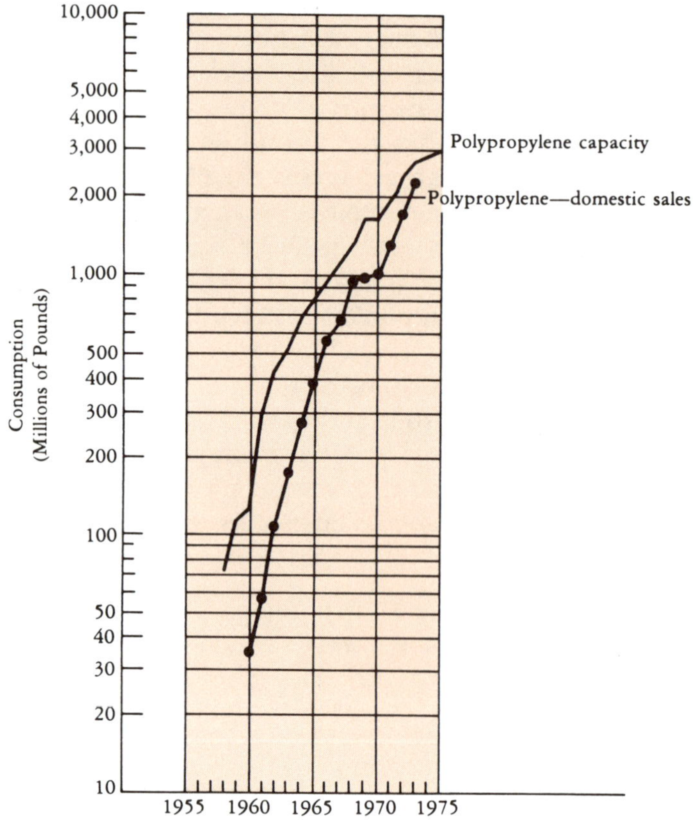

Source: Peter H. Spitz, *Petrochemicals: The Rise of an Industry* (New York: John Wiley & Sons, 1988), 518.

report on how the company could achieve stated targets for growth and earnings, with recommendations on which businesses to invest in and which to consider divesting. According to Giacco:

> They all came up with the recommendation that we ought to divest the polypropylene business. All of them! Well, it's the one time I lost my temper in public. I get irritated a lot of times, but not in front of a whole group of people. [In this case], I said what the hell do you propose we replace it with! Tell me. Well, the general consensus was specialty chemicals, which in those days would have been the greatest disaster we ever had because that market went to hell in a basket.[70]

"Al was absolutely livid," recalls Hollingsworth. "God, he was mad. And yet, I think, down in his heart, he probably knew where they were coming from. The polypropylene business had a history of lousy returns. You couldn't justify it on the basis of return on investment."[71]

The polypropylene business survived the confrontation, but another problem soon surfaced: Hercules found it was falling behind in developing the next-generation process technology. Although Hercules had invested heavily to develop a new catalyst system of its own, it had encountered a series of obstacles that placed it well behind Montedison and Mitsui of Japan. In the early 1980s, those companies developed a revolutionary new process (called Spheripol) for making polypropylene. The Spheripol process, which used catalysts ten thousand times more efficient than those in the traditional slurry process, offered tremendous advantages: fewer steps in the manufacturing process; siting on half the acreage; compatibility with cheaper feedstocks; 50 percent savings in capital investment; 50–90 percent savings in energy consumption; no by-products, wastes, or pollution; and low maintenance. Not only could it make polypropylene at lower cost, but it could also produce resins and copolymers with a higher degree of purity and a significantly broader range of properties than conventional methods. These resins, moreover, could be tailored for specific markets. In other words, the Spheripol process represented a fundamental breakthrough in both cost and quality.[72]

At Montedison, the development of the Spheripol process coincided with two other concerns: its wish to reenter the U.S. market for polypropylene, which it had abandoned in 1979, and its desire to establish "a viable, self-sufficient multinational presence in pharmaceuticals."[73] In both respects, Hercules appeared to present problems for Montedison. For one thing, although it had a process technology advantage over Hercules, Montedison faced a long, uphill fight to gain market share in the United States, given Hercules' entrenched position. For another, the two companies were still partners in pharmaceuticals. And although Adria Labs had grown to become a nearly $100 million business by 1982, according to Montedison's then-chairman, Dr. Mario Schimberni, the joint venture "was not without problems." In particular, he said, "there was an imbalance in the type of inputs that the partners brought to the venture. We contributed our products and technology, while Hercules contributed only finance."[74]

The two companies, moreover, diverged in their strategic aims for the joint venture. Montedison hoped to grow Adria Labs into a $500 million business by increasing expenditures on R&D to as much as 10 percent of sales, an expensive route that Hercules was reluctant to follow. The partners also differed on other objectives: Montedison hoped to use Adria Labs as the U.S. sales arm of its global pharmaceutical business; Hercules wanted to make Adria the basis of a new, self-sufficient company capable of its own product development and manufacturing.[75]

These events and circumstances threatened the relationship between the two companies but also fostered creative thinking about a possible rearrangement. Montedison first asked Hercules about its interest in selling its polypropylene business. Receiving a cool response, Montedison next asked about Hercules' interest in licensing the Spheripol process. At this point, talks between Giacco and Schimberni veered off in another direction entirely: a worldwide joint venture to manufacture and market polypropylene. In the fall of 1982, working groups from both companies began meeting regularly to plan the creation of "a free-standing company capable of attracting financing in the public and private markets, primarily in the United States."[76]

By the following spring, the terms of the deal were settled. Control of the venture would be shared equally between the partners. Because Hercules contributed substantially more assets to the venture, Montedison agreed to pay $55 million in cash, notes, and a note convertible to stock in its new pharmaceutical subsidiary, Erbamont, in return for a 50 percent share. Hercules' film and fiber businesses were not included in the deal, nor was its interest in joint-venture resin plants in Taiwan and Brazil. Hercules' interest in Adria Labs was exchanged for a 13.5 percent share in Erbamont, an investment valued at $87.5 million.[77]

The new company, HIMONT Incorporated, was based in Wilmington, with Al Giacco designated as chairman for its first five years. Although the transaction made obvious strategic sense, in retrospect, Hollingsworth believes yet another reason lay behind the deal:

> I think Al, being an entrepreneur and wanting to do something big, was another factor. See, he had a golden opportunity to do all this at one time, perhaps improve the profitability of a chronically unprofitable business, update our technology, knowing full well that we were probably going to go under unless we had that technology and produced something splashy and big. And, by God, it worked![78]

RESTRUCTURING, PART II

Hercules' recovery under Giacco earned it and him the admiration of Wall Street. By 1983, the company's stock was trading in the mid-thirties and was viewed as one of the chemical industry's best investments. For presiding over this performance, *Financial World* hailed Giacco as one of three outstanding leaders in the chemical industry in 1984. The *Wall Street Transcript* followed suit, naming Giacco the outstanding chief executive in the chemical industry three consecutive times between 1983 and 1985.

In the fall of 1983, when Giacco was one year shy of Hercules' mandatory retirement age, the board extended his contract until early 1987. The board stated that, "since Mr. Giacco was the architect of the corporate strategy, it would be prudent to ask him to remain until these strategic moves were implemented." At the same time, however, several outside directors were concerned about developing an eventual successor, as well as about the longevity of the matrix organization and the number of Giacco's direct reports.[79]

The following summer, the board's concerns dovetailed with Giacco's strategy of shifting the balance of the portfolio toward aerospace and materials and prompted another corporate reorganization. In July 1984, Hercules regrouped its operations into three new operating units: Hercules Aerospace

Company (HAC), which included rocket motors, composites, Simmonds, the GOCO plants, and the residual explosives business; Hercules Engineered and Fabricated Products Company (HEFPC), which was the film and fibers business, including advanced materials in development; and Hercules Specialty Chemicals Company (HSCC), which included the corporation's traditional businesses in organics, water-soluble polymers, paper chemicals, and coatings and additives, as well as newer lines such as foods and fragrances (PFW) and electronics and printing products (chiefly the Merigraph® photopolymer printing plates and other small businesses). In addition, Hercules continued to participate in equity companies such as HIMONT, Hercofina, Boots-Hercules, and many other joint ventures. (In 1984, net sales of Hercules' affiliates and nonconsolidated subsidiaries surpassed $1.8 billion, the bulk of which came from HIMONT.)[80]

One purpose of the reorganization was to groom candidates for senior corporate management by placing greater responsibilities in the hands of the operating company presidents and their staffs. As presidents of these new units, Giacco named Schowengerdt (HAC), Buckner (HEFPC), and Hollingsworth (HSCC). The reorganization did not, however, alter Hercules' basic strategy of focusing on value-added applications and extensions of existing operations. Although reporting channels changed, the business centers in each operating company continue to function as before.

In HAC, for example, the unit pursued a systems strategy for rockets and missiles—acquiring a second defense electronics company (Sperry Microwave) in 1985—and pushed the fabrication of structures from graphite composites. Most of these structures were aimed at the aerospace market. Late in 1986, Hercules received priceless publicity from a spectacular public demonstration of this product when pilots Dick Rutan and Jeana Yeager flew the *Voyager*—an experimental plane made almost entirely from Magnamite® carbon composites from Hercules—nonstop around the world. The company also contributed structural parts to race car teams and formed a joint venture with Montedison (Intermarine) to market composite structures for advanced ship designs and the marine industry.

The company also continued to develop bigger and more powerful rocket motors, such as the Space Shuttle boosters. In 1983, Hercules began construction of a massive, highly automated plant at Bacchus for the casting of propellants for large rocket motors. The initial investment of $150 million was a gamble somewhat reminiscent of the building of the Oxford staple fibers plant. When funds were committed for the new plant (known as Bacchus West), Hercules had little guaranteed business, although it seemed likely to win the production contract for the Trident II D-5 rocket motor. On the other hand, that program alone would not fill the plant's capacity, and it was far from certain that Hercules would be awarded the second-source contract for the Space Shuttle boosters. When the *Challenger* exploded in January 1986, NASA still had not made a formal commitment to Hercules. In the aftermath of the tragedy, however, Hercules opportunely benefited from the rebirth of expendable launch vehicle programs, which had been dormant during the heyday of the shuttle. In

1986, the company secured major production contracts for solid propulsion systems for the Delta II and Titan IV rockets.[81]

In HEFPC, Buckner described his portfolio as "two investments operating in the red and two divisions scaling up from 1980 to 1982."[82] The latter, of course, were the film and fiber businesses, as they cultivated new markets for opaque films and staple fibers. In the mid-1980s, Hercules developed several new film products for packaging applications based on techniques of layering or bonding polypropylene film to metallic or other plastic materials. In fibers, at Giacco's urging, the company launched a determined effort to replace cellulose acetate with polypropylene staple fiber in cigarette filters, a market whose size the company estimated at $1 billion annually.[83]

The two businesses "operating in the red" in the mid-1980s were synthetic pulp (a blend of polyethylene and polypropylene) and METTON® high-performance structural plastic made from liquid molding resins. Hercules' interest in synthetic pulp (also known as "synpulp," or by Hercules' registered trademark Pulpex®) started in the late 1970s as a joint venture with Solvay to synthesize an alternative to wood pulp for making very fine papers. In 1982, Hercules bought out Solvay's interest, and although the paper market failed to materialize, synpulp found applications as the material in teabags and as a replacement for asbestos in certain kinds of flooring tiles. According to Jim Knox (who succeeded Buckner as head of the renamed Hercules Engineered Polymers Company [HEPC] in 1986),

> We are on the brink of a couple of exploding new applications areas. The move to microwave—75 percent of American homes have microwaves—has created a demand for dual-ovenable containers and trays, which can go into a regular oven or a microwave. You can't do that with aluminum foils, of course, but you can with synthetic pulp in combination with paperboard.

The total demand for such products, predicted Knox, will fall in the multibillion-dollar range.

The high-performance structural plastic, METTON,® appeared to hold even greater promise, with annual sales forecast at $200 million or more by the early 1990s.[84] The material, protected by nearly thirty product and process patents held by Hercules, was manufactured by reaction-injecting two liquids—a rubberized monomer called polydicyclopentadiene and a catalyst—under pressure into a mold. The result, moments later, was a tough, light thermoset plastic with several key advantages over most engineered polymers or sheet-molding compounds. METTON® plastic was lighter, tougher, and less expensive to produce and tool than many competitive products. It was particularly suited to large structural shapes such as automobile body panels and bumpers, as well as the contoured parts of recreational vehicles such as golf carts and skimobiles. In 1985, Hercules formed a joint venture with Industries PPD of Canada to mold structures made from METTON.® Hercules also joined with Teijin of Japan to produce and market the product in Asia.

The corporate push to achieve growth via value-added products was also manifested on a global scale in HSCC. Although Giacco was not bullish about the prospects for rapid growth in many of the company's older businesses, he nonetheless encouraged the unit to continue its efforts to develop new products and augmented services.

In the paper chemicals area, for example, Hercules attempted to build on its strong ties with its customers by providing new services in process technology. In the early 1980s, as the paper industry adopted a new, less expensive alkaline process to replace the traditional acid-alum sizing process, Hercules developed an efficient method of applying additives, using foams and emulsions. Later, Hercules extended its efforts with a systems approach to serving the paper industry, including involvement with automated equipment and new process technologies. One of the publicly stated motives for the Simmonds acquisition, for example, was the potential synergy between electromechanical and electronic systems and Hercules' knowledge of paper chemicals.[85] In 1984, the company acquired the Ross Pulp & Paper Division of Midland-Ross Corporation, a leading maker of specialty drying and process control equipment for the paper industry. A year later, Hercules followed this acquisition by establishing a joint venture with Devron Engineering of Vancouver, Canada, to supply state-of-the-art actuators and controls to the pulp and paper industry.

Many of the best opportunities for paper chemicals were to be found overseas, says Thomas L. Gossage, who was recruited from Monsanto to become president of HSCC in 1988:

> Hercules is certainly a leader, not only in the United States; we're a leader in Europe, we're a leader in Canada, we're strong in Australia, and we're strong in Japan. From a strategic standpoint, we think that we can maintain that leadership position in these developed countries. And we also see considerable opportunity for growth in the developing countries. We've identified five or six developing countries where we believe either the paper industry is or will be an important industry, and we want to position our paper chemicals business in those countries.[86]

In the organics area, the sixty-year-old refrain of concern about the eventual depletion of the pine stumps continued to echo. Although the company developed a promising process to stimulate the formation of rosin and turpentine by inoculating the root collars and stumps of pine trees, it proved unworkable in practice. The process achieved the goal of producing more rosin per stump; however, it proved too labor-intensive and expensive to justify commercialization.[87] On the demand side, Hercules continued to cultivate its traditional customers: producers of adhesives, printing inks, and synthetic rubber.

In the mid-1980s, Hercules was optimistic about what seemed to be the most promising area of growth in specialty chemicals: food additives and flavorings. Its optimism reflected the nature of the market, which was large and growing steadily, as well as the structure of the industry, which was fragmented.

The combined market for food and flavor ingredients exceeded $10 billion and was growing at a rate of 5–8 percent per year. Certain segments of the market—packaged items such as breakfast cereals and convenience foods—were expanding at a much faster clip, as much as 20 percent per year, as a steady stream of new products bumped older products from supermarket shelves. In addition, said Gossage, the market was tending toward "natural products in food ingredients, especially in the United States, but we think also in Europe," where the company's Danish subsidiary, CPF, contributed strong performance. He predicted that Hercules' strength in pectin and carrageenan will serve the company well, because "the American public wants more and more labels on their food that say no chemical ingredients."[88]

No single producer, however, dominated the market. Indeed, the largest maker of flavorings controlled less than 7 percent of the market, while the top 100 competitors accounted for only 50 percent. A producer that could break out of the pack seemed likely to enjoy a prosperous future. In 1984, Hercules consolidated its natural gums (CMC and carrageenan) business with its flavors and fragrances business, forming a single unit, the Food and Flavor Ingredients Group (FFIG). In the process, Hercules became the first large competitor that could promote all aspects of taste—appearance, mouth feel, aroma, and flavor—on a global scale. Indeed, D. James MacArthur, vice president of FFIG, believed that global-scale advantages in research and marketing would determine the long-term winners in the industry, and that Hercules is well positioned to become a leader.[89]

Unfortunately, the same could not be said for nonfood uses of CMC, long the cornerstone of Hercules business in water-soluble polymers. In the late 1970s and early 1980s, CMC and other water-soluble polymers had been one of the corporation's most successful businesses. "CMC used to make a 16 percent return on assets—after taxes," recalls Giacco.

> Everybody used to call it the crown jewel. But then, all of a sudden, it turned out to be paste. The reason is that it was living on the oil business. Single-market product, the whole production base in the United States. When the energy picture changed in the early 1980s, this whole business changed.[90]

Frank Wetzel, who ran the CMC business in the early 1980s, says these problems were compounded by an inability to attract investment capital and difficulties in defending its position in Europe. Although growth of the business was funded out of its own earnings, Wetzel had "a strong suspicion that the corporate strategy was such that they were pulling cash out of our unit to put in polypropylene, which was an enormous cash consumer."[91] This policy had two adverse effects on the business: investment capital was not available for modernizing plant and equipment, and business center managers were frustrated by an inability to expand capacity to match demand. In the early 1980s, for example, Wetzel requested a major appropriation to expand a small facility in Alizay, France, to make highly purified CMC for the petroleum industry. The request was turned down, and from Wetzel's point of view, the decision was "an absolutely dramatic, near fatal stroke" that allowed several competitors and new entrants to add capacity and led to depressed conditions throughout the industry for most of the 1980s.[92]

Whether the misfortunes of CMC were attributable (as Wetzel believed) to "a lack of understanding in the top management of the corporation," or (as Giacco stated) to a precipitous decline in its principal market, or to both, soon became a moot point. By the mid-1980s, the water-soluble polymers business had reached a crossroad. Hercules' choices were to continue to harvest the business until an ultimate write-down or exit, to make major investments to rebuild it, or to exit immediately through a joint venture or an outright sale. In late 1986, Hercules opted for the joint venture: at that time, Hercules and Düsseldorf-based Henkel KGaA banded together to form a new unit to manufacture and market water-soluble polymers on a worldwide basis. The parent corporations held equal stakes in the new company, the Aqualon Group, which was based in Wilmington, with Eugene Crittenden as its first president. Hercules contributed its CMC, Natrosol® hydroxyethylcellulose, and Klucel® hydroxypropylcellulose businesses, and Henkel added its capacity to make methylcellulose, a key ingredient of detergents, wallpaper paste, and other consumer products, and hydroxypropyl guar, used in the petroleum industry.[93]

Hercules' decision to place its water-soluble polymers in a joint venture was one of several signs that the restructuring begun in the mid-1970s was an ongoing process. Other evidence included the final sale of the company's interests in Boots-Hercules (1984) and Hercofina (1985), as well as its remaining business in commercial high explosives (1985). By 1986, in fact, Hercules had abandoned many of the businesses that had sustained it during much of its corporate lifetime: explosives, cellulose derivatives, insecticides, and DMT. The divestment of these product lines in the space of a decade represented a fundamental transformation not only of the corporate portfolio but also of the corporate culture. And in the company's seventy-fifth year, there was yet one more act to come.

THE END OF PETROCHEMICAL COMMODITIES

Late in 1986, the board, which had recently expanded to include three new outside directors, elected Hollingsworth to succeed Giacco as chairman and CEO of Hercules after the company's annual meeting the following spring. Although he had worked closely with Giacco since the late 1970s, Hollingsworth had been regarded as a dark horse candidate for the top job. His election suggested a renewed emphasis on the company's traditional chemical businesses.

A chemical engineer who had joined Hercules in 1948, Hollingsworth had spent his early career in a series of sales, marketing, and general management positions in paper chemicals and naval stores. In 1974, he served briefly as general manager of the New Enterprise Department before moving on to head the Food and Fragrance Development Department. Under the matrix structure, Hollingsworth directed the Organics Worldwide Business Center and was promoted to vice president (planning) and business group director in 1979. Elected to the board three years later, he moved rapidly through a sequence of senior positions: group vice president (polypropylene); divisional vice president (marketing); president of HSCC, and, in 1986, vice chairman of the board.

Dissimilar in background to Giacco, the fifty-eight-year-old Hollingsworth also represented a sharp contrast in personality and style. A keenly intelligent, soft-spoken, and unassuming man, he was perceived as a healer who would help soothe the wounds of restructuring and guide Hercules on a steady course toward renewed growth. Hollingsworth also showed a greater willingness to share power than his predecessor. When he assumed office as chairman and CEO, he was joined in a top-management team by Fred Buckner as president and chief operating officer and Arden Engebretsen as vice chairman and chief financial officer.

When Hollingsworth's election was announced, it was expected that Giacco would remain on Hercules' board and continue to serve as chairman of HIMONT. In one of his last actions, however, Giacco set in motion a chain of events that would fundamentally alter Hercules' strategy, as well as his relationship to the company. Early in 1987, HIMONT announced an initial public offering to trade 22.6 percent of its stock on the New York Stock Exchange at $28 per share. That fall—less than a month before the coincidental seventy-fifth anniversary of Hercules' incorporation and the crash of the stock market on Black Monday—Hollingsworth announced the sale of Hercules' interest in HIMONT to its partner Montedison for a pretax gain of $1.51 billion. Giacco, who was not in favor of the sale, soon resigned from the Hercules board, thereby ending a forty-five-year association with the company.

By far the largest transaction in Hercules' history, the sale of its stake in HIMONT was even more significant than the size of the sale suggests. In a single, breathtaking stroke, Hercules divested its major engine of growth (though not of earnings) since the early 1960s and placed one of its major operating units, HEPC, in the discomfiting position of no longer having control over its principal raw material. On the other hand, says Hollingsworth, the sale also represented the culmination of Hercules' strategy of more than a decade to divest itself of commodity petrochemicals. "Once and for all, we've declared [ourselves] out of the game of petrochemical and chemical commodities."

HIMONT's decision to go public accounted for the timing of the sale: according to several senior Hercules officers, the step became inevitable once Hercules lost its 50 percent share, and hence its control, of the joint venture. At the same time, the parent companies found themselves in disagreement about the disposition of HIMONT's earnings. Hercules sought a greater cash dividend to show the overall impact of the equity investment on its own earnings per share, whereas Montedison wanted to reinvest in the business. By the summer of 1987, the instability inherent in many joint ventures was starting to show in HIMONT, and to Hercules senior management, the moment seemed ripe to sell.[94]

Hercules realized a significant premium for its HIMONT stock, which had risen to nearly $60 per share when the transaction took place on September 25.[95] The company now had a substantial war chest to finance its future growth. The question was, at the end of its seventy-fifth year, where would its opportunities lie?

EPILOGUE: ENTERING THE 1990s

In January 1989, the top fifty officers of Hercules gathered near Princeton, New Jersey, for a management retreat. In some respects, the event recalled meetings that had been held after the world wars, when the company, flush with cash, had reflected on its past and pondered its future. The sale of its interest in HIMONT, as well as the recent change of leadership, marked a similar occasion. As Hollingsworth put it, the HIMONT transaction "completed the strategy of the late 70s and 80s," in which Hercules had restructured to escape its dependence on petrochemical feedstocks. Looking forward, he had prepared a corporate vision statement for the 1990s that the assembled executives reviewed, discussed, and ultimately endorsed. According to Hollingsworth's vision,

> Hercules will build a major position in advanced materials, and related systems and structures. It will be a leader in key materials for markets which include the space, defense and transportation industries. Chemical technology and materials engineering will be the foundation on which Hercules will build a future that includes high-performance composites, adhesives, sealants, ceramics, coatings, structural polymers and additives. Hercules will aggressively expand its position in flavors and food ingredients. Hercules will maintain leadership positions in its core businesses.[96]

In essence, the new strategy marked a return to Hercules' historical foundations, as well as a renewed concern with growth. The statement's focus on materials, systems, and structures was a natural extension of Hercules' longstanding involvement with plastics and composites, especially in aerospace, and its stress on industrial markets represented a continuation of the company's traditional strength in specialty chemicals. As for growth, Hollingsworth pointed out that real sales volume increased annually by 8 percent from 1949 to 1974 but had remained flat ever since (see Exhibits 12.1 and 12.9). "The challenge to management in the immediate years ahead is clear," he stated. "It is to get us back onto a solid growth path—growth in volumes, earnings, and employment levels."[97]

Hercules appears to be following the vision statement closely. The proceeds from the sale of its interest in HIMONT were used to buy back stock, fund investments in R&D and in the aerospace and film businesses, and pay for the upgrading of facilities across the company. Hercules also maintained a reserve capacity to help pay for acquisitions to augment its core businesses, if and when suitable opportunities appear.

EXHIBIT 12.9 Hercules' Growth Rates: Sales and Employment, 1949–1987

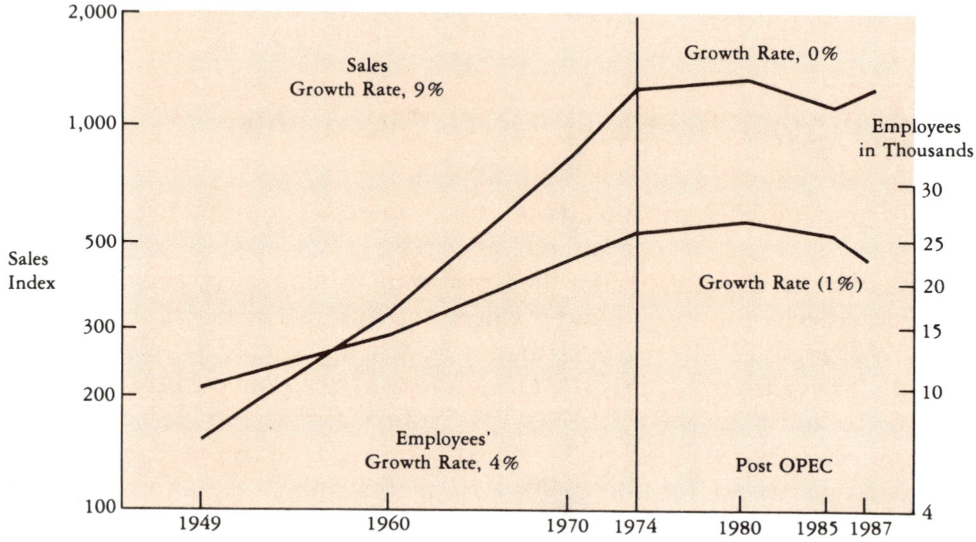

Source: D.S. Hollingsworth, speech to annual stockholders' meeting, March 22, 1988.

Hollingsworth was determined to position Hercules for the future. The first step involved purging marginal operations such as Ross Pulp & Paper, Inc., Caribbean Lumber Company, Champlain Cable Corporation (the last remnant of Haveg), and Industries PPD (the METTON®-fabrication joint venture). In 1988 and 1989, Hercules also sold its polypropylene carpet fiber business to Du Pont, closed its chemical cotton operation in Hopewell, Virginia, and ceased production of Parlon® chlorinated rubber at Parlin, New Jersey. Finally, the company announced plans to sell its Electronics and Printing Products, Aircraft and Electronic Products (Simmonds), and Intermarine (the joint venture to manufacture composite structures for marine applications) businesses.

In some respects, these moves suggested a retreat from the strategy of Giacco's last years: using small acquisitions to pursue incremental growth opportunities in existing businesses. Hollingsworth, however, was more inclined to accentuate the positives. To him, Hercules' key objectives were "to build, grow, and maintain," that is, "to build a new business in advanced materials . . . to grow vigorously in our flavors and food ingredients businesses . . . and to maintain our existing leadership position in all our core businesses."[98]

Accordingly, in the spring of 1989, Hercules acquired Henkel's 50 percent stake in the Aqualon Group, a move that reestablished Hercules as the clear global leader in cellulose derivatives and water-soluble polymers. (In 1988, Aqualon had posted sales of $359 million.) Hercules also declared itself in the market for an acquisition to augment its FFIG business, although Hollingsworth indicated that the company would not make such a deal if it involved diluting corporate earnings.[99] In the summer, the company created a new operating unit, Hercules Advanced Materials and Systems Company (HAMSC), to combine reactive polymers (chiefly METTON®) from HEPC with advanced fiber products and composite structures from HAC. The new unit includes new joint ventures with Sumitomo Chemical to manufacture PAN fiber (the precursor to carbon fiber) in Decatur, Alabama, and with Italy-based BAT International to design, manufacture, and market composites in Europe.

Concerned about the prominence of its aerospace operations in the era of glasnost, Hercules is seeking to limit its reliance upon U.S. military contracts, focusing instead on commercial and international opportunities, as well as smaller programs. In November 1988, for example, the company acquired an equity stake in Orbital Sciences Corporation (OSC), a commercial space technology company in Fairfax, Virginia. Hercules and OSC also signed a joint venture to develop, produce, and operate the Pegasus air-launched space booster. Powered by three Hercules solid-fuel rocket motors, Pegasus was designed to be launched from a high-flying B-52. The vehicle was designed to carry small commercial satellites into orbit. Pegasus completed its first mission successfully in April 1990.

Beyond these changes in the portfolio, Hollingsworth made changes in the company's management and organization. He hired two senior managers from outside the company, for example: Gossage as president of HSCC (and subsequently of the reacquired Aqualon Group), and Richard Schwartz, previously president of Rockwell International's Rocketdyne Division, as head of HAC. Believing that technology will be a key source of competitive advantage for the company's advanced and specialized businesses, Hollingsworth announced plans to raise spending on R&D from under 3 percent to about 4 percent of sales by the mid-1990s. He also renewed the company's traditional emphases on safety, quality, and environmental issues, introducing formal programs across the company in all three areas.[100]

Hollingsworth was also concerned about organizational issues, including staffing and employment levels. These issues came to the fore when the company reported "disappointing financial results" in 1988—operating income declined by about one-third from the previous year (from $176.9 million to $129.7 million).[101] This performance in part reflected rising raw materials costs and investments not yet yielding returns, but it also underscored the need to lower costs and raise productivity. Indeed, even with the earnings power of HIMONT, Hercules faced a serious challenge to reduce its overhead. Accordingly, in the winter of 1988–1989, the company launched several cost-cutting initiatives, including an early retirement program, a six-month salary freeze, and a thorough study of the sources of indirect expenditures.

In September 1989, Hollingsworth announced a new corporate organization that resembles the business-unit structures common in other diversified companies. As designed by a study team headed by Crittenden, who moved back from Aqualon as senior vice president, the operating companies were liquidated in phases, with "business groups" reporting to the president and chief operating officer. The plan started with the breakup of the HSCC and HEPC units: new business groups, for example, included Absorbent and Textile Products (fibers and synpulp), Adhesive and Specialty Resins (resins and peroxides), Flavors and Food Ingredients (including food-grade CMC, formerly part of the Aqualon Group), Paper, and Packaging.

"The principles behind the moves," explained Hollingsworth,

> include structuring Hercules' business toward greater focus on customers; defining businesses according to strategic synergy; organizing for maximum effectiveness in global operations; adopting a teamwork operating structure; minimizing organizational layers; and keeping decision making at the lowest appropriate levels.

By decentralization, Hollingsworth meant "giving the business manager the resources and freedom to generate the best business results and to give them accountability." In practical terms, this entailed providing staff and support services such as human resources, R&D, finance and control, purchasing, and logistics within each business, and correspondingly less centralized support from the corporate level. Hercules Europe was also decentralized, and each business there directed to report to global business groups.[102]

In the fall of 1989, Hollingsworth signaled a marked departure from the style and culture of management under Giacco when he announced the formation of several corporate teams to implement the new organization, including a permanent corporate strategy council and a revived management committee. These steps were taken in the belief that "team decisions are usually superior to individual top-down decisions," and that teams are "best employed in a flatter, less hierarchical system with fewer layers between the top and the front-line managers."[103]

With the new structure and team-oriented management in place, Hollingsworth believes that Hercules is poised to realize its vision and strategy of growth for the 1990s: to build in advanced materials, grow in food and flavor ingredients, and maintain leading positions in the company's core businesses. Time will tell whether he is correct.

Reflections on Seventy-five Years: The Past as Prologue

Viewed from this end of Hercules' history, Hollingsworth's emphasis on growth marks—or will mark, if he succeeds—the start of a fifth era of strategic change.

During the first era (1913–1928—see Chapters 3–6), Hercules struggled to come to terms with its origins. A creation of public policy rather than of entrepreneurial effort or business opportunity, the company began as a going concern, but also as a mere collection of assets and as an organization without a strategy. During its first fifteen years, therefore, the company labored to define its place, first in the U.S. explosives industry, then in the larger U.S. chemical industry. Occupying a momentarily secure position as the number-two competitor (by a wide margin) in explosives, Hercules used its resources and built on its experiences in World War I to transform itself into a chemical producer. Sensing limits to growth in its base business, Hercules diversified after the war into new

Changing of the guard: At the company's annual meeting in March 1987, Al Giacco (right) stepped down as chairman and CEO. After a decade of restructuring, Hercules looked forward to a period of renewed growth under the leadership of Dave Hollingsworth.

Carrying on the tradition: In January 1989—as happened in 1919, 1946, and 1962—Hercules' management gathered offsite to review and endorse strategy. Present at the meetings held near Princeton, New Jersey, were, from left: (first row) Henry A. Schowengerdt, James E. Knox, Buckner, Hollingsworth, Engebretsen, Thomas L. Gossage, and Edward J. Sheehy; (second row) Theodore M. Bednarski, Ross O. Watson, Norman F. LeBlanc, John C. Allibashi, Jon B. DeVault, James R. Rapp, C. Doyle Miller, Edouard de Croutte, James A. Hunter, and Louis F. Erb; (third row) Jack Garwood, Gary C. Dunn, Vincent J. Corbo, D. James MacArthur, Dominick W. DiDonna, George MacKenzie, Eric G. Bruce, Bent Jakobsen, Peter A. Bukowick, and Terence W. Rave; (fourth row) Klaus Petersen, S. Maynard Turk, David T. Blake, Harry J. Tucci, Gary R. Muir, Frank H. Wetzel, G.C. (Buzz) Sutton, Harry N. van der Vlugt, Robert E. Gallant, and Lawrence C. Cessna; (fifth row) William E. Hosker, Alexander L. Searl, Walter W. Schultz, Thomas V. McCarthy, Donal W. Stitz, E.J. (Woody) Rice, James D. Beach, Donald R. Kirtley, Keith T. Smith, Norman F. Whiteley, W. Wells Hood, Robert J.A. Fraser, David A. Needham, George A. Ward, Harold N. Kenton, and Francesco Bonmartini.

In April 1990, the air-launched Pegasus rocket built by Orbital Sciences Corp. and Hercules roared into earth orbit carrying a small satellite, inaugurating a new era in the commercial development of space.

areas: the related chemistry of commercial nitrocellulose and the unrelated field of naval stores. In this period, the company manifested beliefs and behaviors that would characterize it for many years: financial conservatism and concern with profitability as opposed to growth; engineering resourcefulness in improving existing technology (smokeless powder, low-viscosity nitrocellulose) and developing creative ways to make scarce chemicals (acetone); an emphasis, carried over from its explosives experience, on technical sales and service; and a willingness to tackle big assignments, such as management of wartime ordnance operations.

In the next era (1928 to the mid-1950s—see Chapters 7–9), Hercules built on these characteristics to transform itself into a specialty chemicals company. The need to maintain growth and defend its position in commercial nitrocellulose prompted the company to branch out into other cellulose derivatives, often by licensing and improving upon the discoveries of other (especially European) chemical producers. At the same time, Hercules brought modern methods of research, production, and marketing to the naval stores industry, developing an impressive array of synthetic chemicals based on rosin, turpentine, and pine oil. PMC, acquired in 1931, eventually became a strong addition to the fold and an abiding source of profitable specialty chemicals. Hercules' most important new characteristics in this era were a heightened emphasis on R&D, starting with the opening of the Experiment Station in 1931 and culminating with the scientific research policy of research director Emil Ott; increasing ties to European chemical companies as sources of new technology and products; the continuing transfer of sales techniques and strategies from explosives to other industries, including protective coatings, plastics, adhesives, paper, and plastics; and, in World War II, an emerging expertise in specialized military technologies such as solid-fuel rocket motors.

The third era of Hercules' history dawned in the mid-1950s and concluded with the first energy crisis in 1973 (see Chapters 10–11). In this period of aggressive expansion, the company diversified into petrochemicals, developing rapidly growing businesses in natural-gas derivatives, terephthalates, and polyolefins. Hercules also grew vigorously in chemical propulsion, reconfiguring the Bacchus Works and using the 1958 acquisition of Young Development Laboratories to increase its involvement in defense rocketry, especially in the Navy's fleet ballistic missile programs. The company extended its reach overseas by establishing a growing network of foreign investments and joint ventures. Finally, Hercules accelerated its growth via acquisition and, under the ill-fated New Enterprise Department, ventured into nonchemical areas such as modular housing, airport security equipment, and credit card validation systems. For the entire American economy, this was an era of unparalleled optimism and growth. Hercules characteristics developed during this period included a willingness to assume bigger risks, to invest in capital and asset-intensive businesses, and to assume debt to finance growth; a reliance on portfolio analysis and other modern management techniques to administer a complex, diversified, multinational corporation; and a rising sensitivity to the concerns of external stakeholders such as governments and consumer and environmental groups.

The fourth era (1974–1987—this chapter) witnessed profound changes in the company as it restructured in a hostile economic environment. New leadership, two major reorganizations, the divestment or joint-venturing of more than twenty businesses, and the sorting of its remaining operations into three general areas—aerospace, engineered polymers, and specialty chemicals—left the company vastly altered. More new characteristics were added to the company's way of doing business: a shifting and uneasy balance between centralization and decentralization of operations; a still greater reliance on modern techniques of portfolio analysis and an intensive use of information technology to manage the corporation; a broader understanding of risk to include not only finances but social and environmental issues; and a greater emphasis on the *D* in *R&D*.

It is much too early to say whether Hollingsworth is correct in depicting Hercules as standing on the threshhold of a new strategic era. For the company to succeed in the future, however, it clearly must deal with pressing strategic issues in each of its major businesses (see Chapter 1). Indeed, Hercules' fate in the 1990s and beyond will depend on how well it copes with matters such as the apparently fundamental change in the business of defense—and hence, of military aerospace—as East-West tensions diminish, the geopolitical map of Europe is redrawn, and the U.S. government struggles to control mounting deficits; its proficiency in defending its value-added position in engineered polymers between oil companies and polymer producers on the one hand, and customers for fibers, film, and other applications on the other; and its ability to capitalize on a global scale and find new growth opportunities in specialty chemicals. And, of course, the problem of growth for the corporation as a whole, says Hollingsworth, "has got to be the big challenge" ahead.[104]

In confronting the future, Hercules can draw encouragement from the experience gained during its first seventy-five years. Although its overall pattern of growth mirrors that of most large, diversified corporations in capital-intensive industries, its story nonetheless features distinctive twists and turns, as well as lessons learned.

To begin with, although past performance never guarantees future achievement, Hercules has employed successfully all of the weapons in the modern corporation's strategic arsenal: growth through horizontal extension and vertical integration, continuing investment in R&D and new products, expansion overseas, acquisitions and joint ventures, and diversification. Its experience with these techniques and strategies, moreover, is long and broad because of the unusual circumstances of its origins and early years.

From its historical roots in commercial explosives, for example, Hercules branched out horizontally into enduring businesses such as nitrocellulose (and its later extensions into cellulose products and water-soluble polymers) and chemical propulsion. Participation in one business, moreover, often yielded unexpected opportunities in others. To ensure inexpensive and steady sources of key ingredients for making smokeless powder, Hercules integrated backward to produce nitric acid and chemical cotton. These operations, in turn, contributed to growth in new areas such as nitrogen chemicals, natural-gas derivatives, and new cellulose products. The process technology involved in making nitrogen chemicals helped Hercules to develop a better understanding of organic chemistry and to upgrade the engineering skills that proved useful when the company later entered petrochemicals. Close work with customers of chemical cotton gained Hercules its first exposure to the new technologies of plastics and synthetic fibers. From its base in explosives, Hercules also used acquisitions to expand its business or extend its capabilities. Its successes with Virginia Cellulose and Young Development, however, are counterbalanced by its difficult experiences with Aetna Explosives and, perhaps, Simmonds.

Hercules also succeeded in unrelated diversification at an early age, although the strategy was not then—or subsequently—simple to execute, nor quick to pay returns. When the company entered naval stores, for example, the task of bringing order to a fragmented and backward industry proved far more complex than originally anticipated, requiring nearly two decades before Hercules' strategy could properly be called a success. In polyolefins, the company achieved rapid growth but earned significant returns only with the sale of its stake in HIMONT—thirty years after it had started making linear, crystalline polyethylene. As Hercules has discovered, particularly during the New Enterprise era, the risks of unrelated diversification are considerable.

Although Hercules learned obvious lessons from these experiences and gained increasing sophistication in the techniques of screening and analysis, as well as greater understanding of the kinds of businesses in which it could make its mark, the salient fact about the company—and the source of many of its constraints and opportunities—is its relative size and position in the economy. As a middle-sized competitor in the chemical industry, and an industrial producer occupying a middle place in the value-added chain running from the extraction of raw materials to the delivery of finished goods to end-users, Hercules enjoyed its greatest successes in specific kinds of industries and markets.

The constraints of Hercules' position are many. In general, its size and available resources militated against (and fed its reluctance about) competing in asset-intensive businesses or placing huge bets, such as entering major petrochemical markets. For this reason, after World War II, Hercules refused to buy its way into polystyrene and other petrochemicals or to capitalize on its breakthrough process advantage to compete in the commodity petrochemical, phenol. Its initial forays into polyester intermediates and polyolefins represented more modest investments—guaranteed in the case of DMT—but the company was eventually unable to sustain them in the face of competition from bigger oil and chemical companies. Even in polypropylene, a business in which Hercules was the first and largest producer worldwide, it ultimately bowed to rivals that were better able to finance development of new process technology.

By the same token, its size hindered Hercules in its attempts to pursue another option open to its larger rivals: financing of many growth prospects at once. The failures of the late 1960s and early 1970s resulted at least partly from inadequate resources—placing too many bets that were each too small. Although much larger companies such as Du Pont also retreated from aggressive diversification into nonchemical areas, the greater wealth of larger companies affords more possibilities and minimizes the costs of failure. Apart from the ventures of the New Enterprise era—and taking this lesson from them—Hercules has limited its options lest it spread itself too thin.

Finally, Hercules' medium size constrained the possibilities of vertical integration. Although it integrated backward in naval stores and cellulose products to control some of its raw materials, and forward to perform some activities carried out by customers, these were relatively small businesses by chemical industry standards. Such a strategy was never realistic in petrochemicals: the company simply could not afford to buy an oil company. (Nor would such a move have made sense in any event, given the tiny fraction of oil company operations devoted to propylene.) At the other end of the chain, Hercules lacked sufficient scale to compete successfully in downstream applications of major product lines such as protective coatings or paper. Even in polypropylene, in which the company was partly integrated backward and well positioned in downstream applications such as fibers and film, these arrangements proved ultimately too expensive to sustain—as the formation of HIMONT and its eventual sale apparently signal.

Hercules' relative size and position, however, provided opportunities as well as constraints. In particular, the hallmarks of middle-sized organizations—tight communications loops, an ability to make decisions quickly, and a disciplined approach to new opportunities—worked often to Hercules' advantage. Similarly, in specialty chemicals and in niche markets, which do not require huge capital investments or economies of scale for long-term success, Hercules has excelled. From the beginning, when Hercules salesmen shot dynamite for their customers, charging not only for the product but also for the service and support, the company has thrived in businesses in which it is more technically advanced or specialized than its customers. In subsequent years, Hercules continued this approach in supplying industries such as paper, protective coatings, plastics, and, to some extent, foods and flavorings. The technical nature of Hercules' products and its ongoing investments in R&D are formidable sources of competitive advantage, as well as barriers to backward integration by its customers.

It follows that Hercules fares best when it focuses on applications engineering—a strategy, incidentally, much admired in contemporary Japanese corporations. Although the company occasionally discovered and developed original products—toxaphene and Kymene® are the most outstanding examples—it prospered more often by advancing on the discoveries or processes of others. This ability was manifested in its infancy, when Hercules improved the British cordite process and found innovative ways to make acetone from unlikely raw materials. Since then, Hercules has achieved similar breakthroughs many times, besting larger or more experienced competitors in developing economical processes for making low-viscosity nitrocellulose, refined wood rosin, cellulose acetate, CMC, oxychemicals, and many other products. In the 1950s, Hercules transformed Ziegler's polyolefin process from a concept in a laboratory notebook into a full-scale commercial operation within four years. More recently, the company quickly advanced beyond its partners and competitors in graphite fiber and composite technology because of its assiduous and resourceful approach to development.

Should the Hercules of the 1990s find opportunities where it can take advantage of its historical strengths and proven capabilities, the climate for renewal appears promising. The key moments of transition in Hercules' history came when senior managers were concerned about prospects for growth, and when they had at their command ample financial reserves and human resources to support change. In the 1920s, such circumstances provided the backdrop for turning an explosives company into a chemical company; during the post–World War II decade, similar circumstances lay behind Hercules' transformation into a diversified, multinational corporation and a significant competitor in petrochemicals. Now, with a new leader eager to see the company grow, a balance sheet remade after fifteen years of restructuring, and experienced and skilled personnel, Hercules seems ready to perform its next labor.

APPENDIX I

Oral History Interviews

I. TAPED INTERVIEWS

Interviewee	*Date*
Abbott, Marvin	September 9, 1987
Babcock, Leon W.	November 6, 1986
Beasley, Warren F. "Sam"	March 1, 1989
Bednarski, Theodore M.	February 10, 1988
Bewley, William W., Jr.	October 10, 1986 February 27, 1989
Bossardet, George	January 10, 1988 August 24, 1988
Brown, Werner C.	October 8, 1986 March 1, 1989
Buckner, Fred L.	March 10, 1988
Butler, Clifford T. "Pat"	July 22, 1987
Cain, Gilbert E.	April 7, 1988
Chaddock, Richard E.	February 22, 1989
Chris, Marjorie D.	July 21, 1987
Crittenden, Eugene D., Jr.	December 2, 1987 May 3, 1989
Cruise, Robert H.	March 23, 1988
Crum, Edward G.	October 8, 1986
DeVault, Jon B.	April 6, 1988
Douglas, Richard B.	November 13, 1987
Dunn, Charles	April 7, 1988
Dunn, Gary C., Maris DeLanger, Jack Garwood, Frank Giles, and Mylene Vanderburg	May 18, 1988
Ely, Albert R. "Slick"	November 12, 1987
Engebretsen, Arden B.	May 4, 1988
Eurenius, Carl	November 12, 1987 June 20, 1988
Frawley, John P.	March 16, 1989
Gardner, Charles C.	November 11, 1988
Giacco, Alexander F.	December 15, 1987

Glasebrook, Arthur L.	October 12, 1988 October 13, 1988
Gossage, Thomas L.	September 22, 1988
Grant, Edward L.	May 11, 1988
Greer, John E., Jr.	May 3, 1989
Hawkesworth, M. William, Jr.	February 28, 1989
Hendricks, R. Michael	March 17, 1989
Hollingsworth, David S.	October 8, 1986 September 1, 1988
Ivett, Reginald	December 2, 1987
Jenkins, Michael E.	November 13, 1987
Jennings, Earp F.	April 6, 1988
Johnson, John R.L., Jr	January 28, 1987
Keim, Gerald I.	May 4, 1989
Knox, James E.	September 22, 1988
Leahy, Robert J.	April 27, 1988
Little, Duard H. "Tex"	July 22, 1987
Martin, John M. "Jack"	October 9, 1986 September 21, 1988
MacArthur, D. James	May 2, 1989
McCauley, Fred D.	September 8, 1987
Miller, C. Doyle and Rodman B. Teeple	March 16, 1989
Reeves, Henry	January 30, 1989
Ryan, John R.	October 10, 1986
Schowengerdt, Henry A.	April 26, 1988
Sheehy, Edward J.	March 23, 1988
Sheffield, Donald H.	June 17, 1987
Skolnik, Herman	June 17, 1987
Smith, Keith T.	February 27, 1989
Spurlin, Harold M.	October 9, 1986
Steinberger, Rudolph	June 17, 1988
Taufen, Harvey J.	September 21, 1988 September 26, 1988 December 19, 1988
Turk, S. Maynard	October 10, 1986
Wetzel, Frank H., Jr.	April 27, 1988
Winer, Richard	May 18, 1988
Wohl, Georgiana and Betty Eurenius	June 30, 1988

II. INTERVIEWS AT HERCULES FACILITIES

Bacchus, Utah	June 20–23, 1988

L.E. Atkins
Bill Bernard
Bruce Biehler
Roger Blodgett
Perry Bruno
R.W. Buhr
Dail Butler

Ernie Cassler
Larry Claspill
Ken Dolph
Bob Folsom
Bill Gardner
Paul Kemp
Wayne Larabee
Debbie Lux
Marion "Mack" McGowan
Norm McLeod
Jim Mack
Jim Martin
Ernie Mettenet
Loren Morey
Gary Muir
Ruth Novak
Ralph Pacheco
Dewey Peterson
Steve Russell
Dick Sailer
Sam Starnes
Dave Thompson
Karen Watson
E.L. White
William Wilson

Brunswick, Georgia — September 9, 1987

Marian Cotney
Larry Deeg
Lamar and Quincy Floyd
Alfred, Alex, William, and James Grant, Sr.
Lillian Meyer and Anncille Stringfellow
Marcus Thompson and Clarke Yancey
Jim Turner

Hercules Research Center — March 4–5, 1987

A.Z. Conner
Herb Espy
Mary Hilton
Horace Hood
Dan Monagle
Ben Repka
Keith T. Smith
George Ward

Hopewell, Virginia — November 9, 1988

Leon Bertram
Jim Doud
Jim Willcox

Kenvil, New Jersey — August 26, 1987

Dick Best
Mary Piero
Oakley Utter

Lake Charles, Louisiana April 24–25, 1989

L.C.Anderson
Harry Beasley, Clive Broadbent, Roy Farrior, Mike Guillory, and Bonnie Rasberry
Charlie Burton, Gerald Hansen, Veronica Joubert, Vernon McGuire, Ronnie Oertling, and Cecil Stogsdill
Bill Calgreen, Coupie Paulk, and Jim Szydlo
Denzil Cooper, Laral LaFleur, Milton Self, and Eddie A. Welch
Carol Landreneau
Jeff LeJeune and L.G. Thomas
Jesse Motes, Charlie Pratt, and Theo Richard

Oxford, Georgia April 26–27, 1989

Wanda Adams, Debbie McClain, and Elsie Morris
Steve Austin, W.L. Scroggs, and Charlie Ubanks
Dick Baltz and Don Chapman
Sylvester Belcher and Marvin Knight
Shirley Benton, Harrison Day, and Ernest Gilstrap
Joe Bonkowski and Bill Henderson
Martin Burton, Bob Henderson and Ray McFadden
Wayne Campbell, Ray Fambrough, and James Middlebrooks
Robert Chaney, Sonny Heath, and Bill Lunsford
Jack Clark and Bobby Myhand
Helen Dawkins and Ed Hixon
Dave Delmonte and Jim Modrak
W.J. Fisher and J.G. Smith
Elsie Haigwood, Wayne Hunt, and Louise Murphy
Mel Harter and Roger Jensen
Rose Hatcher
Jimmy Little, Richard Ridling, and Nat Williams
Stan Nechvatal and Blair Burch
Bob Outlaw
Roger Tingler and Jack Stirzaker
H.A. Wise

Parlin, New Jersey August 27, 1987

Jim Cannon
Bill Farrell
Rich Grossmeier
Phil Grossmeier
Bill Nagel
Joseph Nebus
Walter Ososki
Ziggy Szatkowski

Radford, Virginia November 10–11, 1988

James Dickenson
Charles C. Gardner
John Horvath
Robert Jenrette
John Pierce
Les Pugh
Cannonball Stauffer
Paul Steele

Savannah, Georgia	September 8, 1987
Will Bates	
Jack Flaherty	
Norman McCarthy	
Fred McCauley	
Kevin "Bo" Russom	
Curt Stroup	
Terre Haute, Indiana	February 20–21, 1989
Mel Brinson	
Steve Doyle	
Tom Grimes	
Ralph Hiatt	
Mike McCullough	
Jim Oaks	
Don Simonds	

III. INTERVIEWS NOT TAPED

Giacco, Alexander F.	August 30, 1988
Ivett, Reginald	November 8, 1989
Kelly, Maurice J., Robert I. Mason, Charles A. Twigg	April 5, 1988
Don Pedersen	September 22, 1988

IV. INTERVIEWS CONDUCTED PRIOR TO THE BOOK PROJECT

Eurenius, Carl	November 14, 1979
Hinner, Elmer	May 13, 1980
Johnson, John R.L., Jr	June 30, 1980
Mayfield, Paul	June 1983

V. VIDEOTAPED INTERVIEWS CONDUCTED PRIOR TO THE BOOK PROJECT

Greer, John E., Jr.	March 13, 1987
Mettenet, Ernest M.	March 17, 1987
Muir, Gary	April 1, 1987
Novak, Ruth L.	April 1, 1987
Sailer, Richard	April 13, 1987
Schowengerdt, Henry A.	March 12, 1987
Thompson, David E.	March 18, 1987

APPENDIX II

Directors of Hercules, 1912–1990

Russell H. Dunham	1912–1956
Thomas W. Bacchus	1912–1941
George H. Markell	1912–1926
Clifford D. Prickett	1912–1941
James T. Skelly	1912–1936
Frederic W. Stark	1912–1929
Gould G. Rheuby	1918–1943
Norman P. Rood	1921–1935
Leavitt N. Bent	1928–1952
Charles A. Bigelow	1928–1944
Charles A. Higgins	1928–1954
Charles C. Hoopes	1928–1945
George M. Norman	1928–1945
William J. Lawrence	1932–1937
Anson B. Nixon	1932–1957
Philip B. Stull	1932–1959
Mahlon G. Milliken	1936–1955
Lloyd Kitchel	1937–1950
Petrus W. Meyeringh	1937–1951
William R. Ellis	1937–1953
Edward B. Morrow	1937–1964
Luke H. Sperry	1941–1946
Albert E. Forster	1941–1966
John B. Johnson	1945–1957
Ralph B. McKinney	1945–1946
Wyly M. Billing	1945–1961
John J.B. Fulenwider	1945–1966
Francis J. Kennerley	1947–1953
Reginald Rockwell	1947–1956
Elmer F. Hinner	1952–1970
Paul Mayfield	1952–1967
John M. Martin	1954–1978
John R.L. Johnson, Jr.	1954–1971
Leon W. Babcock	1954–1960
Ernest S. Wilson	1954–1959
John E. Goodman	1954–1970

Edward G. Crum	1958–1970
G. Fred Hogg	1958–1962
John H. Long	1958–1971
Henry A. Thouron	1958–1970
J.H. Tyler McConnell	1959–1966
Robert W. Cairns	1960–1971
Arthur F. Brown	1960–1961
Werner C. Brown	1963–1982
Jack D. Hayes	1963–1975
Donald H. Sheffield	1963–1970
John H. Lux	1964–1966
Alfred E. Van Wirt	1964–1966
Jack G. Copeland, Jr.	1966–1974
E. Langford Jones	1966–1974
Harvey J. Taufen	1966–1982
Paul L. Johnstone	1967–1974
John R. Ryan	1967–1982
Myron W. Krueger	1968–1975
William H. Kendall	1969–1975
Joseph A. Thomas	1968–1973
Stephen R. Clarke	1971–1983
Alexander F. Giacco	1971–1987
Spencer H. Hellekson, Jr.	1971–1979
Charles S. Maddock	1971–1976
Thomas M. Macioce	1971–1981
John R. Petty	1973–1989
W. Coleman Edgar	1974–1976
Robert J. Leahy	1974–1985
Arthur E. Larkin, Jr.	1975–1981
Arthur C. Nielsen, Jr.	1977–1990
Guy T. McBride, Jr.	1981–1990
Marcus V. Pratini de Moraes	1981–1983
Eugene D. Crittenden	1982–
David S. Hollingsworth	1982–
Stuart E. Eizenstat	1983–
Arden B. Engebretsen	1983–
Geoffrey R. Simmonds	1983–1987
Jerome Jacobson	1985–
Robert G. Jahn	1985–
Joan E. Spero	1985–
Fred L. Buckner	1986–
Henry A. Schowengerdt	1986–1989
S. Maynard Turk	1987–
Gaynor N. Kelley	1989–
Thomas L. Gossage	1989–
Richard Schwartz	1989–
Ralph L. MacDonald, Jr.	1989–
Manfred Caspari	1990–

GLOSSARY OF TECHNICAL TERMS

Compiled by
John A. Abbitt, Ashalata S. Hirwe, and Reginald Ivett
Hercules Incorporated

Abietic acid
: the principal component of wood rosin.

Acetic anhydride
: reactive reagent derived from acetic acid and used in the manufacturing of acetate derivatives, e.g., cellulose acetate.

Acetone
: dimethyl ketone; used as commercial solvent and reagent; produced with phenol from cumene hydroperoxide cleavage.

Agent Orange
: chlorophenoxyacetic acid esters; used as jungle defoliant during the Vietnam War and packaged in orange containers.

Aluminum alkyl
: highly reactive compound containing aluminum and alkyl radicals, e.g., triethylaluminum; used as initiator for catalysts in polymerizations, e.g., in polymerization of propylene to polypropylene.

Ammonia
: colorless gas with a strong odor; used in fertilizers and as reagent in chemical reactions.

ANFO
: explosive mixture of ammonium nitrate (AN) and fuel oil (FO) (94 percent AN, 6 percent FO).

Aquapel® internal sizing agent
: a synthetic paper sizing agent advantageous because resulting paper has excellent aging properties.

Carrageenan
: water-soluble gum extracted from carragheen or Irish moss; used as emulsifier, stabilizer, thickener, or gelling agent in foods, pharmaceuticals, and cosmetics.

Casein
: protein derived from milk; used in paper sizing agents.

Cellulose
: carbohydrate polymer that is found in cell walls of trees, cotton plants, and other vegetation; used as raw material in production of various cellulose derivatives.

Cellulose acetate
: ester of cellulose and acetic acid; used in fibers, packaging, lacquers, and plastics.

Cellulose ester
: compound derived from reaction of cellulose and an acid; includes cellulose acetate and nitrocellulose.

Cellulose ether
: compound derived from reaction of cellulose and organic halide or cyclic organic ether; includes sodium carboxymethylcellulose, ethylcellulose, hydroxyethylcellulose, and hydroxypropylcellulose.

Chemical cotton
: highly purified cotton linters produced in various grades for conversion to cellulose derivatives.

Chlorinated camphene (Toxaphene)
: a residual-type insecticide effective against a wide range of insects.

Chlorinated rubber
: natural or synthetic rubber reacted with chlorine to give film-forming polymer; used in coatings, including maintenance paints for metal, wood, and concrete as well as traffic and marine paints; sold by Hercules under the Parlon® trademark.

CMC
: sodium carboxymethylcellulose; product of reaction involving chloroacetic acid, alkali (e.g., sodium hydroxide) and cellulose; water-soluble polymer used as thickener, binder, stabilizer, suspending agent, and film former in industries such as cosmetics, pharmaceuticals, paper, paint, petroleum, adhesives, textiles, and detergents.

Composite
: material made from distinct parts, generally fibers reinforcing a solid; used in aircraft or rocket components made from glass, graphite, or carbon fibers wound and impregnated with expoxy resins.

Cotton linters
: fuzz of short fibers adhering to seeds after processing to obtain cotton fiber; used in preparation of cellulose derivatives.

Cumene
: isopropylbenzene; reactant in process yielding phenol and acetone.

DDT
: dichlorodiphenyltrichloroethane; one of the first widely used residual-type of insecticides; produced in aerosol form by Hercules in the 1940s.

Defoamer
: additive used in papermaking process to control foaming of wood pulp.

Delnav® insecticide
: agent effective against mites and other animal pests.

Denier
: measure of fiber fineness; unit weight of fiber expressed as weight in grams per 9,000 meters.

Destructive distillation
: thermal decomposition of organic compounds in absence of air with distillation of products, as in production of wood tar.

Di-Cup® dicumyl peroxide
: Hercules trademark; organic compound derived from isopropylbenzene (cumene); used as vulcanization and polymerization agent in the manufacture of synthetic rubber.

Dimethyl terephthalate
: ester of terephthalic acid and methanol; used in manufacture of polyester fibers and films.

Disproportionation
: catalytic treatment of rosin to increase oxygen stability.

Distillation
: removal of volatile components from a mixture followed by their recovery by cooling.

Dope
: preparation for giving a desired quality to a substance or surface; absorbent or adsorbent material used in manufacturing processes, as in the making of dynamite.

Double-based powder
: propellant that incorporates two active explosives in the mixture, e.g., nitrocellulose and nitroglycerin.

Dresinate® rosin soap
: sodium or potassium salts of rosin; used to impart tack to synthetic rubber.

Dynamite
: a high explosive usually consisting of nitroglycerin and ammonium nitrate mixed with absorbent materials.

Emulsifier
: surface active agent promoting the formation and stabilization of a heterogeneous mixture of immiscible liquids.

Epoxy resin
: thermosetting resin containing epoxide groups; used with carbon or graphite fibers in composites.

Ester
: reaction production of an acid and an alcohol.

Ester gum
: reaction product of glycerin and rosin.

Esterification
: thermal treatment to effect reaction of an acid and an alcohol.

Ether
: organic compound characterized by oxygen atom attached to two carbon atoms.

Ethylcellulose
: cellulose ether made by heating ethyl chloride with alkali cellulose; used in coatings, pharmaceuticals, and plastics.

Extraction
: treatment of a solid or liquid with a liquid to remove components soluble in it.

Extrusion
: process in which a molten polymer is forced through a die or aperture; used in many processes, including preparation of smokeless powder grains and polypropylene fibers and film.

Fatty acid
: straight-chain acids, normally containing twelve to eighteen carbon atoms; found in wood and many seeds.

Filament-winding
: wrapping of fiber or strand around or into a form; process used in preparing composite structures from graphite or carbon fiber with the fiber impregnated with expoxy resin and subsequently cured.

Fractionation or fractional distillation
: distillation in specialized equipment to separate volatile materials differing in boiling point.

Glycerin
: glycerol, trihydropropane; a colorless, syrupy liquid obtained by reacting fat with alkali; used as a plasticizer; starting material for nitroglycerin, a powerful explosive used in dynamite.

Graphite fiber
: also called carbon fiber, Hercules trademark for graphite fiber is Magnamite®; high-strength, lightweight, almost pure carbon fibers; used in aircraft and sports equipment such as fishing rods, tennis rackets, and golf clubs.

Guar gum
: gum derived from the seeds of an Indian plant resembling soya; used as a thickener and stabilizer in foods.

Herculon polypropylene fiber
: Former Hercules trademark for fiber made by spinning or drawing polypropylene into a filament; used in upholstery and carpets.

Hi-Fax® polyethylene
: Registered trademark of HIMONT Incorporated (transferred from Hercules) for high molecular–weight thermoplastic made from ethylene.

Hydrogenation
: catalytic reaction of hydrogen to a material to remove unsaturation and improve stability.

Injection-molding
: method for preparing plastic parts by melting the solid polymer, forcing it under pressure into a mold, and cooling to resolidify.

Isomerization
: rearranging the atoms of a molecule into a different structure (configuration); at Hercules, isomerization is important in rosin manufacture and in terpenes rearrangement.

Klucel®
: Hercules trademark for hydroxypropylcellulose; a nonionic water-soluble thermoplastic polymer made from cellulose and propylene oxide; used as an emulsion stabilizer, for molding or extrusion operations, and as a thickener.

Kymene® resins
: formulations developed and trademarked by Hercules to provide wet strength to paper.

Lacquer
: organic coating that dries to form a film following evaporation of the solvent; lacquers are one of the many applications for nitrocellulose.

Linear polymer
: polymer that consists of straight chains of monomers; examples at Hercules include polypropylene, polyethylene, and polyethylene terephthalate.

Liquid-molding
: method for transforming liquid monomers into solid parts; two reactive liquids are mixed in a mold where they undergo a chemical reaction resulting in a solid part; e.g., liquid-molding is used to produce automotive and recreational vehicle parts from METTON.®

Magazine
: building or container for storing explosives, usually built to rigid specifications.

METTON® liquid-molding resin
: Hercules trademark for poly (dicyclopentadiene); used to prepare high-strength, lightweight, tough molded parts for automobiles and recreational vehicles such as snowmobiles.

Natrasol®
: Hercules trademark for hydroxyethylcellulose; a nonionic water-soluble polymer made from cellulose (either chemical cotton or wood pulp); used as a thickener, protective colloid, binder, stabilizer, and suspending agent in a wide variety of industrial uses.

Natural gum
: colloidal polysaccharides or polymers derived from natural sources such as plants, e.g., guar gum.

Naval stores
: term applied by custom to all terpene and rosin products obtained from living pines, pine wood extraction, and paper production.

Nitration
: chemical reaction that forms a nitrate (a salt or ester of nitric acid containing the NO_2 or NO_3 groups).

Nitrocellulose
: nitric acid ester of cellulose; nitrate content affects solubility and stability.

Nonwoven fabric
: fibers held together by interlocking or bonding rather than by being woven or knitted; Hercules' synthetic pulp, Pulpex,® is used in nonwoven fabrics for applications such as absorbent pads, feminine hygiene products, and diapers.

Olefin
: an unsaturated, open-chain hydrocarbon containing at least one double bond; e.g., ethylene or propylene.

Oleoresin
: exudate from a living pine tree.

Orientation
: as applied to Hercules' film products, drawing (stretching) heating film, thus arranging the linear polymer chains so that they are aligned in the same direction; imparts greater strength to the film.

Oxidation
: reaction of oxygen with an organic compound, sometimes controlled and often occurring on prolonged exposure to air.

Oxychemical
: chemical that contains oxygen; examples at Hercules include phenol and cresol.

Paper size
: additive used in papermaking to prevent spread of ink on the finished paper.

Paraffin
: waxy, mostly straight-chain hydrocarbon mixture; used in coating and sealing, candles, rubber compounding, and pharmaceuticals and cosmetics; reacted with chlorine to make Clorafin® chlorinated wax.

Para-xylene
: dimethylbenzene; oxidized to terephthalic acid for forming DMT (dimethyl terephthalate).

Parlon® chlorinated rubber
: Hercules trade name for chlorinated rubber.

Pentaerythritol
: branched hydrocarbon with four hydroxyl groups; used to manufacture rosin esters and also in pentaerythritol tetranitrite (PETN), a powerful explosive.

Pentalyn® resin
: ester resin prepared from rosin and pentaerythritol; used in protective coatings and lacquers.

Petrex® resin
: Hercules trademark for an alcohol solution of a thermal-curing alkyd derived from resinous polybasics of terpene origin; used in protective coatings and lacquers.

Petrochemical
: any chemical product derived from petroleum; examples at Hercules include ethylene, propylene, and hydrocarbon resins.

Pexite® rosin
: rosin of light color obtained by solvent-refining of the crude resin produced by solvent extraction of pine wood.

Phenol
: hydroxybenzene; compounds containing one or more hydroxyl groups attached to an aromatic ring; used in many industrial applications, including the manufacture of plastics.

Photopolymer
: plastic material that cross-links and becomes insoluble when exposed to light; used in the manufacture of printing plates; an example at Hercules is Merigraph® resin used in newspaper printing plates.

Pigment
: insoluble coloring agent used for paints, inks, and plastics.

Pinene
: major component of turpentine.

Pine oil
: mixture of terpene alcohol found in pine wood extract or produced synthetically from terpenes.

Plasticizer
: chemical additive used to give flexibility, stretchability, or workability to resins or rubbers.

Polyester
: polymer obtained by condensation polymerization of dicarboxylic acids with alcohols; used in a wide variety of applications, especially textiles.

Polymer
: high molecular–weight material existing naturally or obtained by joining together many simple molecules (monomers), linked end to end or cross-linked.

Polymerization
: formation of very large molecules from small molecules, normally using a catalyst; the resulting material is a polymer.

Polyolefin
: resin or polymer produced by polymerization of an olefin such as ethylene, propylene, or butylene.

Polypropylene
: thermoplastic polymer of propylene.

Prill
: solid form of ammonium nitrate; small, spherical, free-flowing granules.

Pro-Fax® polymer
: registered trademark of HIMONT Incorporated (transferred from Hercules) for homopolymers or copolymers of propylene.

Protective coating
: continuous film formed over a surface (or substrate such as wood or metal) to protect against chemicals, corrosion, or other exposure.

Resin
: thermoplastic material, generally melting at a low temperature, frequently obtained from natural sources (rosin, shellac, amber) or chemical modification of such natural products.

Reten® retention aid
: registered trademark of Hercules for high molecular–weight, water-soluble acrylamide copolymers; used for improved retention of fillers and fibers in paper and paperboard manufacture.

Rosin
: mixture of polycyclic acids; a major component of pine oleoresin, pine wood extract, and tall oil.

Rosin size
: sodium or potassium salt of rosin used in the papermaking process to retard the spread of ink in paper.

Scale up
increase in the size of an operation, such as going from laboratory to pilot plant, or from pilot plant to industrial or commerical-size plant.

Smokeless powder
variety of gunpowders containing nitrocellulose and other agents, sometimes including nitroglycerin and nitroguanadine.

Solid-rocket fuel
explosive mixture that is cast and cured in a solid state and burns at a controlled rate.

Staple fiber
short synthetic fiber used in preparing nonwoven fabrics and spun yarns.

Staybelite® resin
Hercules brand name for hydrogenated wood rosin; used in adhesives, laminants, paper coatings and size, rubber, and certain synthetics.

Synpulp®
Hercules trademark for synthetic polyolefin pulp; used in many plastics applications.

Synthetic fiber
fiber from synthetic polymers such as nylon, polyester, or polypropylene.

Tackifier
additive to resins or rubber to make them sticky.

Tall oil
mixture of resin and fatty acids obtained as a by-product of paper manufacture.

Tenter
process/equipment used to make thin-gauge film by sequentially stretching a thick sheet in two perpendicular directions.

Terephthalate
ester of terephthalic acid, such as dimethyl terephthalate; used as an intermediate in production of fiber or film.

Terephthalic acid
a, 4-dicarboxybenzene; an aromatic dicarboxylic acid (i.e., two acid groups) that is esterified to produce terephthalates; alternative to DMT as an intermediate in the the manufacture of polyester.

Terpene
cyclic hydrocarbon found in essential oils and oleoresins.

Terpineol
terpene alcohol, which is a major component of pine oil.

Thanite®
Former Hercules brand-name insecticide for terpene derivative used in fly sprays.

Thermoplastic polymer
polymer that can be repeatedly softened by heating, as distinguished from thermoset polymer.

Thermoset polymer
polymer that cross-links on heating and cannot be reformed through melting.

TNT
: Trinitrotoluene: a solid crystalline substance formed from the nitration of toluene; explosive material used primarily by the military inside bombs and shells.

Tow
: untwisted strand of synthetic fibers.

Tubular
: process for making thin-gauge film—blowing a bubble from warm polymer, increasing the size of the bubble, collapsing the bubble through a nip, and winding the film on a roll.

Turpentine
: mixture of bicyclic terpenes found in pine oleoresin and pine wood extracts; used as a paint solvent and in chemical syntheses.

Urea formaldehyde resin
: thermosetting resin prepared by the reaction of urea with formaldehyde; widely used in plastics applications.

Vinsol®
: Hercules trademark for a dark-colored resin obtained as a by-product in Pexite® refining of wood rosin; many uses include stabilization of roadbeds.

Viscose rayon
: first commercially successful synthetic polymer fiber produced by treating cellulose with alkali and carbon disulfide; contrast with acetate rayon, made by treating cellulose with acetic acid and acetic anhydride.

Water-soluble polymer
: polymer soluble or swellable in water; Hercules' examples include cellulose ethers.

Wet-strength resin
: additive to paper that increases substantially the strength of the paper when wet.

Wood pulp
: cellulose fibers produced by mechanical or chemical treatment of wood.

ACKNOWLEDGMENTS

Writing a book is an inherently isolating experience. No shortcut eliminates masses of documents that historians must read, nor, despite the marvels of word processing, can a computer generate original text. That said, our task was made far easier and our labors much more enjoyable and productive by the abundant support and encouragement of many people.

As we explained in the preface, this is a commissioned history, and we again acknowledge the wholehearted support of Hercules Incorporated and commend its willingness to have its past probed by outsiders. To Alexander Giacco, CEO when the project began, and current CEO David Hollingsworth, we are grateful for steady cooperation and encouragement. We are particularly indebted to Dick Douglas, formerly director of public relations, the champion of this particular project. Even after his retirement from Hercules in February 1988, Dick retained a lively interest in the project and was a wise and sharp critic of the emerging manuscript. His successor, Don Kirtley, similarly maintained a high degree of enthusiasm for our work. Carolyn R. Miller, senior public relations representative, served well beyond the call of duty, putting in many long hours on our behalf as our day-to-day contact and as guide and host during our many visits to Wilmington and other Hercules locations. Barbara Adamese and Barbara Temple of the PR Department coordinated administrative matters cheerfully and with thorough competence.

Our periodic meetings with the Hercules History Advisory Committee—Ted Bednarski, Werner Brown, Eugene Crittenden, Dick Douglas, Carl Eurenius, Don Kirtley, John Lee Johnson, Glenn Porter, and Maynard Turk—were extremely helpful in shaping the outline and general approach of the book. The committee members also responded sagely and patiently to chapter drafts, improving them at every stage. Robert Hessler and Joyce Carlson of the PR Department read the entire manuscript for clarity and accuracy, along the way making many valuable suggestions. Reg Ivett also read the entire manuscript, spotting many passages where our technical expertise was deficient or our language unclear. Our account of the explosives industry, its products, and evolution was improved by John Abbitt's helpful criticisms. Art Glasebrook proved a particularly astute critic of our treatment of Hercules' entry into petrochemicals, described in Chapters 9 and 10.

Many other people at Hercules generously supported our work by volunteering to be interviewed, furnishing historical materials in their possession, and providing leads to additional sources. Herculites interviewed for the project are listed in Appendix I. To each of them, we are beholden for their time and support. Charles W. Gamble, corporate secretary, and Helen Calhoun from his staff provided access to the board minutes and other official governance records. Bill Bewley, Art Glasebrook, Reg Ivett, Bob Leahy, Loren Morey, and Keith Smith lent us (or provided copies of) documents and materials in their possession. Wendy Sapp and Stan Kersey proved to be valuable guides through the company's records management system, helping to locate key documents and arranging for us to use materials housed in the Hall of Records. Asha Hirwe coordinated our visits to the Research Center and helped steer us through the abundant sources there. Joanne Henderson and Tourica Fullman of the corporate library staff responded brightly and promptly to many requests and helped photocopy key documents. Maureen Tomczak and Carolyn Samluk introduced us to the formidable array of information technology at Hercules and helped us translate interview transcriptions from one computer standard to another. Jimmy Carter of the PR Department provided clear reproductions of many of the illustrations seen in the book.

Our stays in Wilmington were made comfortable and pleasant by Ray and Ange Rossetti and Scotty Chisholm of the Hercules Guest House, and we enjoyed many informal hours of conversation about the company with Dick Douglas, Don Kirtley, Carolyn Miller, George Frederick, and Jimmy Carter.

The manuscript was very much improved as the result of trenchant comments by several outside readers, including Louis Galambos, Peter H. Spitz, Jeffrey L. Sturchio, and one anonymous reviewer. We wish also to thank Eliza Collins, Carol Franco, and Natalie Greenberg, our editors at the Harvard Business School Press, for helping to shepherd the manuscript through the editorial and production processes.

Our colleagues at the Winthrop Group also contributed to the project in many ways. David Grayson Allen commented pointedly on chapter drafts, and we received many helpful suggestions about particular points from George Smith, Meg Graham, and Bettye Pruitt. Dan Jacoby, Abby Jungreis, and Paul Barnhill assisted with research and helped prepare many of the exhibits. Susan McWade, Linda Tillman, and Pam Bracken proved to be, as always, terrific transcribers.

Finally, we wish to thank Janice McCormick and Leila (Kakatsakis) Sicilia, who put up with frequent trips and long absences and endured countless hours of solitude even when we were ostensibly at home. Coincidentally, independently, happily, and remarkably, Janice and Leila agreed to become our wives during the course of the project.

NOTE ON SOURCES

This book is based primarily on sources owned and maintained by Hercules Incorporated. As mentioned in the preface, we relied heavily on periodic reports of the company's operating and auxiliary departments, which are preserved in Hercules' Hall of Records in Wilmington. These reports, which are virtually complete between 1913 and 1978, proved an invaluable and nearly inexhaustible source of information about the company's activities and operations and helped flesh out the bones of official records such as board minutes. Although the departmental records became over time more formal and leaner in terms of commentary, we were able to supplement them from other sources, including company publications (such as the *Hercules Mixer,* the *Explosives Engineer,* the *Hercules Chemist, Hercules Horizons,* and many others in libraries at the Home Office or the Research Center); books and industry journals (*Chemical Week, Chemical and Engineering News,* and many others); and the oral history interviews listed in Appendix I.

Although the vast majority of our research involved sources proprietary to Hercules, we found extensive information about the company and the chemical industry in the public record, as well as in several research libraries. First, the Hagley Museum and Library near Wilmington contains copious materials relating to the early explosives industry, the Du Pont company, the origin of Hercules, and its first five years of existence. In addition, the Hagley Library houses abundant secondary sources about Du Pont, Hercules, and the explosives and chemical industries. The Beckman Center for the History of Chemistry at the University of Pennsylvania in Philadelphia is another tremendous resource for primary and secondary information about the chemical industry, especially about the development of polymer chemistry. In the Boston area, we drew heavily on the seemingly inexhaustible resources of the libraries at Harvard University and the members of the Boston Library Consortium, especially Massachusetts Institute of Technology. To the librarians and staff of all these institutions, we are extremely grateful.

The notes provide full citations to original sources. In addition, at the conclusion of the project, the authors' notes and files were returned to Hercules, where they are part of a historical archive. The authors devoutly wish that these materials and other historical records will be made accessible to business and technology historians in the future.

NOTES

Preface

1. At the time of our visit to Hopewell, the Aqualon Group was jointly owned by Hercules and the German consumer products company, Henkel KGaA.

2. Alfred D. Chandler, Jr., *Strategy and Structure: Chapters in the History of the American Industrial Enterprise* (Cambridge, Mass., 1962).

3. Arnold Thackray, Jeffrey L. Sturchio, P. Thomas Carroll, and Robert Bud, *Chemistry in America, 1876–1976: Historical Indicators* (Boston, 1985); Joseph L. Bower, *When Markets Quake: The Management Challenge of Restructuring Industry* (Boston, 1986); Sheldon Hochheiser, *Rohm and Haas: History of a Chemical Company* (Philadelphia, 1986); Peter H. Spitz, *Petrochemicals: The Rise of an Industry* (New York, 1988); Robert Stobaugh, *Innovation and Competition: The Global Management of Petrochemical Products* (Boston, 1988); David A. Hounshell and John Kenly Smith, Jr., *Science and Corporate Strategy: Du Pont R&D, 1902–1980* (New York, 1988).

4. R.H. Dunham, "Twenty-five Years," *Hercules Mixer,* 20 (January 1938), 3; Hercules Powder Company, Annual Report for 1962; Hercules Incorporated, Annual Report for 1987.

5. Edith T. Penrose, *The Theory of the Growth of the Firm* (New York, 1959), and Penrose, "The Growth of the Firm: A Case Study: The Hercules Powder Company," *Business History Review* (1960), 1–23.

Chapter 1

1. "It's Tough Up There," *Forbes* (July 13, 1987), 145.

2. Hercules Incorporated, Annual Report for 1987.

3. *Fortune* (April 25, 1988).

4. *BusinessWeek* (April 15, 1988).

5. This account is based on a tour of Bacchus West by the authors in June 1988, as well as on "Bacchus West: A Hercules Commitment to Large Booster Production," undated typescript available at Bacchus Works, Magna, Utah.

6. Edward J. Sheehy interview, March 23, 1988.

7. David S. Hollingsworth interview, September 1, 1988.

8. This section is based on information gathered in visits to Brunswick by the authors in October 1987 and September 1988.

9. Hollingsworth interview, September 1, 1988; Thomas L. Gossage interview, September 22, 1988.

10. This section is based on information gathered on a visit to the Terre Haute plant in March 1989.

11. C. Doyle Miller and Rodman B. Teeple interview, March 16, 1989.

12. Alexander F. Giacco interview, December 15, 1987; Miller and Teeple interview, March 16, 1989.

13. James E. Knox interview, September 22, 1988.

14. Fred L. Buckner interview, March 10, 1988.

15. Hollingsworth interview, September 1, 1988.

16. This account is based on many visits to Hercules Plaza and on interviews with Alexander F. Giacco, December 15, 1987, and John E. Greer, Jr., May 3, 1989. See also Thomas Walton, *Architecture and the Corporation: The Creative Intersection* (New York, 1988), chap. 4; and *Hercules Mixer* 65, no. 1 (1984).

17. The architects conceived of the building not as a sentinel but as part of a "gateway" complex that opens from the Brandywine River to greet visitors from the north into downtown Wilmington. A corresponding gateway is developing at the southern end of the city along the Christina River.

18. Walton, *Architecture and the Corporation,* 77–89; Greer interview, May 3, 1989.

19. Hercules Incorporated, *The Hercules Integrated Telecommunications Network,* corporate brochure (n.d.).

20. Hollingsworth interview, September 1, 1988.

21. For general descriptions of the economics of the industry, see Jules Backman, *The Economics of the Chemical Industry* (Washington, D.C., 1970); Joseph L. Bower and Hassell H. McClellan, "Note on the U.S. Chemical Industry" (Harvard Business School Case Services 375–371); Arnold Thackray, Jeffrey L. Sturchio, P. Thomas Carroll, and Robert Bud, *Chemistry in America, 1876–1976: Historical Indicators* (Boston, 1985), chap. 4; and Robert Stobaugh, *Innovation and Competition: The Global Management of Petrochemical Products* (Boston, 1988), app. A.

22. U.S. Department of Commerce, International Trade Administration, *1988 U.S. Industrial Outlook,* 12–1; U.S. Department of Commerce, Bureau of the Census, *Statistical Abstract of the United States,* 108th ed. (Washington, D.C., 1988), 700.

23. This has been true since the 1920s. See Williams Haynes, *American Chemical Industry,* vol. 5, *Decade of New Products* (New York, 1954), chap. 2, esp. 23–26.

24. Alfred D. Chandler, Jr., "Scale, Scope, and Organizational Capabilities," in Thomas K. McCraw, ed., *The Essential Alfred Chandler: Essays toward a Historical Theory of Big Business* (Boston, 1988), 475.

25. Alfred D. Chandler, Jr., *The Visible Hand: The Managerial Revolution in American Business* (Cambridge, Mass., 1977), 256.

26. Chandler, "Scale, Scope, and Organizational Capabilities," 475.

27. *1988 U.S. Industrial Outlook,* 12–1.; Thackray, et al., *Chemistry in America,* 97; Bower and McClellan, "Note on the U.S. Chemical Industry." Note that in this comparison, the chemical industry includes pharmaceuticals manufacturers, which typically spend as much as 10 percent or more of sales on R&D.

28. Bower and McClellan, "Note on the U.S. Chemical Industry," 5.

29. Alfred D. Chandler, Jr., *Scale and Scope: The Rise of Industrial Capitalism* (Cambridge, Mass., 1990), chap. 5.

30. Thackray, et al., *Chemistry in America,* 93–94.

31. Booz, Allen & Hamilton, *The Worldwide Chemical Industry: Challenges for Future Growth* (New York, 1988), 8.

Chapter 2

1. Minutes of first meeting of incorporators of Hercules Powder Company, October 29, 1912, bound in a volume labeled "HPCo. Stockholders 1912," vol. 1 of the series in the secretary's vault at Hercules Incorporated headquarters (hereafter cited as HPC Stockholders, with volume number and date).

2. Certificate of Incorporation of Hercules Powder Company, October 17, 1912, in ibid., 7.

3. Minutes of first meeting of incorporators, in ibid.

4. *In the District Court of the United States for the District of Delaware. In Equity, no. 280. The United States of America, Petitioner,* v. *E.I. du Pont de Nemours & Company and Others, Defendants. Opinion of Court and Final Decree* (June 13, 1912), passim (hereafter cited as *Final Decree*).

5. *In the District Court of the United States for the District of Delaware. In Equity, no. 280. The United States of America, Petitioner, v. E.I. du Pont de Nemours & Company and Others, Defendants. Opinion of the Court and Interlocutory Decree* (June 11, 1971) (hereafter cited as *Interlocutory Decree*).

6. George E. Mowry, "Election of 1912," in Arthur M. Schlesinger, Jr., and Fred L. Israel, eds., *History of American Presidential Elections, 1789–1968,* vol. 3 (New York, 1971), 2145.

7. Donald F. Anderson, *William Howard Taft: A Conservative's Conception of the Presidency* (Ithaca, N.Y., 1973), 78–82.

8. U.S. Department of Commerce, Bureau of the Census, *Historical Statistics of the United States: Colonial Times to 1970,* vol. 1 (Washington, D.C., 1976), 8, 11, 224.

9. T.W. Bacchus, *Dynamite—The New Aladdin's Lamp* (n.p., n.d. [Wilmington, 1922?], 14. (This pamphlet originally appeared as an article in *Hercules Mixer* 4 [April and May 1922]), 91–94, 127–130.

10. U.S. Department of Commerce, Bureau of the Census, *Thirteenth Census of the United States, Taken in the Year 1910,* vol. 10, *Manufactures, 1909. Reports for Principal Industries* (Washington, D.C., 1913), 567.

11. Bureau of the Census, *Historical Statistics of the United States,* 685; Williams Haynes, *American Chemical Industry,* vol. 1, *Background and Beginnings* (New York, 1954), 401–402; Solomon Fabricant, *The Output of Manufacturing Industries, 1899–1937* (New York, 1940), 232. We have added the totals for ammunition and fireworks to those for explosives in Fabricant's tables.

12. *Encyclopedia of Chemical Technology,* 3d ed., vol. 9 (New York, 1980), 561, 620; see also Tenney L. Davis, *The Chemistry of Powder and Explosives,* 2 vols. in 1 (New York, 1943), 2–3.

13. Arthur Pine Van Gelder and Hugo Schlatter, *History of the Explosives Industry in America* (New York, 1927), 321–322.

14. For an excellent brief description of powdermaking, see Norman B. Wilkinson, *Explosives in History: The Story of Black Powder* (Wilmington, 1966), 21–27.

15. Norman B. Wilkinson, *Lammot du Pont and the American Explosives Industry, 1850–1884* (Charlottesville, Va., 1984), 38–40.

16. Williams Haynes, *Cellulose: The Chemical That Grows* (Garden City, N.Y., 1953), 57.

17. Van Gelder and Schlatter, *Explosives Industry,* 340–348.

18. E.M. Symmes, *The Manufacture of Dynamite and Gelatin* (n.p., n.d. [Wilmington, 1919?]), 4–16.

19. E.A.W. Everitt, *Smokeless Powder. Presented before the Hercules Club of Wilmington, April 19, 1917* (n.p., n.d. [Wilmington, 1917?]), 3–4; "The Development of Sporting Powders," *Hercules Mixer* 14 (November 1932), 200.

20. Arthur Marshall, *A Short Account of Explosives* (Philadelphia, 1917), 12–13.

21. Wilkinson, *Explosives in History,* 31, 52; Van Gelder and Schlatter, *Explosives Industry,* 959.

22. *In the District Court of the United States for the District of Delaware. In Equity, No. 280. The United States of America, Petitioner, v. E.I. du Pont de Nemours & Company and Others, Defendants. Defendants' Additional Proofs,* Exhibit A (March 2, 1912), 112–116 (vol. 6 of the trial documents at the Hagley Museum and Library) (hereafter cited as *Defendants' Additional Proofs*); Wilkinson, *Explosives in History,* 52–55.

23. *Defendants' Additional Proofs,* Exhibit A, 112–116.

24. Bureau of the Census, *Historical Statistics of the United States,* 732; Van Gelder and Schlatter, *Explosives Industry,* 1006–1008, 1020, 1044, 1050.

25. Van Gelder and Schlatter, *Explosives Industry,* 1071; Williams Haynes, *American Chemical Industry,* vol. 3, *The World War I Period, 1912–1922* (New York, 1945), 463.

26. Clifford T. "Pat" Butler interview, July 22, 1987.

27. Van Gelder and Schlatter, *Explosives Industry,* 123–125, 767–797.

28. Marshall, *Short Account of Explosives,* chap. 5; Van Gelder and Schlatter, *Explosives Industry,* 929–956.

29. *Interlocutory Decree,* 30; William S. Stevens, "The Powder Trust, 1872–1912," *Quarterly Journal of Economics* 26 (1912), 478; E.I. du Pont de Nemours & Co., Annual Report for 1907.

30. The account that follows is based on Van Gelder and Schlatter, *Explosives Industry,* 126–158, 402–430; Wilkinson, *Lammot du Pont,* chaps. 9 and 11; and Alfred D. Chandler, Jr., and Stephen S. Salsbury, *Pierre S. du Pont and the Making of the Modern Corporation* (New York, 1971), 57–62.

31. Van Gelder and Schlatter, *Explosives Industry,* 126–128, 287; Wilkinson, *Explosives in History,* 52; Wilkinson, *Lammot du Pont,* 188.

32. Van Gelder and Schlatter, *Explosives Industry,* 219–252. See also Wilkinson, *Lammot du Pont,* 228, for a list of Du Pont's acquisitions of other powder companies between 1870 and 1882.

33. Chandler and Salsbury, *Pierre S. du Pont,* 57.

34. Van Gelder and Schlatter, *Explosives Industry,* 230–231.

35. Ellis W. Hawley, "Antitrust," in Glenn Porter, ed., *Encyclopedia of American Economic History: Studies of the Principal Movements and Ideas* (New York, 1980), 773–775; William Letwin, *Law and Economic Policy in America: The Evolution of the Sherman Antitrust Act* (New York, 1965), chaps. 5–6.

36. Van Gelder and Schlatter, *Explosives Industry,* 499.

37. Ibid., 420–422.

38. Wilkinson, *Lammot du Pont,* 258–259, 264–265.

39. Chandler and Salsbury, *Pierre S. du Pont,* 61.

40. Alfred D. Chandler, Jr., has written most extensively about the reorganization of Du Pont under the three cousins in *Strategy and Structure: Chapters in the History of the American Industrial Enterprise* (Cambridge, Mass., 1962), chap. 2; (with Salsbury) in *Pierre S. du Pont,* esp. chaps. 3–10; and in *The Visible Hand: The Managerial Revolution in American Business* (Cambridge, Mass., 1977), 438–450.

41. Chandler, *Visible Hand,* 439; cf. Chandler and Salsbury, *Pierre S. du Pont,* 56, 118–119.

42. The debates summarized here are described at length in Chandler and Salsbury, *Pierre S. du Pont,* 104, 110–116.

43. The most complete account of Du Pont's antitrust troubles is in Chandler and Salsbury, *Pierre S. du Pont,* chap. 10.

44. *In the District Court of the United States for the District of Delaware. In Equity, No. 280. The United States of America, Petitioner, v. E.I. du Pont de Nemours & Company and Others, Defendants. Brief for the United States (August 5, 1907)* 1:2.

45. Ibid., 3–5.

46. *Interlocutory Decree,* 44–45, 2–3.

47. The following summary is based on *Final Decree,* 3–8.

48. The correspondence can be followed in three sources at the Hagley Museum and Library: Coleman du Pont's presidential papers, papers of E.I. du Pont de Nemours & Company (hereafter abbreviated at EIDPDN, with series, part, box, and accession numbers), series II, part 3, boxes 131 and 132, accession 472; the correspondence between Pierre du Pont and his cousin William, in papers of Pierre S. du Pont, accession 616, Hagley Museum and Library (hereafter cited as accession 616); and Hamilton M. Barksdale's papers, EIDPDN, series II, part 2, boxes 1006 and 1007, accession 518. See also, Chandler and Salsbury, *Pierre S. du Pont,* 277–290, for an account of the negotiations from Pierre's point of view.

49. Pierre S. du Pont to William du Pont, July 20, 1911, and October 25, 1911, accession 616; Chandler and Salsbury, *Pierre S. du Pont,* 277–279, 280–281, 283–284.

50. Chandler and Salsbury, *Pierre S. du Pont,* 278–281. Coleman's efforts may have been responsible for Attorney General Wickersham's "very violent language" and extreme tactics on March 4, 1912, as the dialogue approached its conclusion.

51. *Interlocutory Decree,* 3–4; James A. Wickersham to Coleman du Pont, October 20, 1911, EIDPDN, series II, part 3, box 132, accession 472: Pierre S. du Pont to executive committee, October 18, 1911, EIDPDN, series II, part 3, box 132, accession 472; Chandler and Salsbury, *Pierre S. du Pont,* 285–286.

52. Chandler and Salsbury, *Pierre S. du Pont,* 276, 279.

53. Pierre S. du Pont to Hamilton M. Barksdale, August 15, 1911, EIDPDN, series II, part 2, box 1006, accession 518.

54. Pierre S. du Pont to Hamilton M. Barksdale, September 28, 1911, EIDPDN, series II, part 2, box 1006, accession 518. At this time, Pierre was not prepared to sacrifice L&R's smokeless powder facility at Haskell, N.J., which Du Pont had upgraded considerably after 1902.

55. Pierre S. du Pont to executive committee, October 18, 1911, EIDPDN, series II, part 3, box 132, accession 472.

56. See four letters of Pierre S. du Pont dated October 18, 1911, in EIDPDN, series II, part 3, box 132, accession 472. Two of the letters are addressed to Du Pont's executive committee, and two to its lawyers. See also Pierre S. du Pont to James M. Townsend, October 26, 1911, EIDPDN, series II, part 3, box 132, accession 472. The dynamite capacity Du Pont was willing to relinquish could be packaged in one of two ways: through three plants (Forcite, Kenvil, and Emporium) totaling thirty-eight million pounds, or through one (Hercules, Calif.) totaling forty million pounds. The blasting powder plants were in Rosendale, N.Y., Columbus, Kans., Pittsburg, Kans., Pleasant Prairie, Wis., Youngstown, Ohio, and Chattanooga, Tenn. The black sporting powder plant was in Schaghticoke, N.Y. For a history of these plants, see Exhibits 2.1–2.3. For the Emporium plant, see Van Gelder and Schlatter, *Explosives Industry,* 686–692.

57. Pierre du Pont to executive committee, November 10, 1911, EIDPDN, series II, part 3, box 132, accession 472. The dynamite plants to be spun off now included Hercules, Forcite, and Emporium. The black sporting powder plants were Schaghticoke (an old L&R facility) and Wayne (originally a Du Pont plant).

58. Galley proofs of Du Pont's proposal, with the penciled date "December 15, 1911," are in EIDPDN, series II, part 2, box 1006, accession 518. The three dynamite plants were the Forcite plant in Hopatcong, N.J., the Emporium plant in western Pennsylvania, and the Hercules, Calif. plant. For histories of these plants, see Exhibits 2.1–2.3. See also Pierre S. du Pont to William du Pont, letters of October 25, 1912, January 23, 1912, and April 3, 1912, accession 616; Chandler and Salsbury, *Pierre S. du Pont,* 285, 287–288; Pierre S. du Pont to H.M. Barksdale, November 30, 1911, EIDPDN, series II, part 2, box 1006, accesssion 518.

59. William S. Hilles to President's Office, E.I. du Pont de Nemours Powder Company (a subsidiary of Du Pont), December 18, 1911, EIDPDN, series II, part 3, box 132, accession 472.

60. EMR [?] to Cong. William H. Heald, February 14, 1912, and S.A. Yorks to T.C. du Pont, May 14, 1912, EIDPDN, series II, part 3, box 132, accession 472.

61. *In the District Court of the United States for the District of Delaware. In Equity, No. 280. The United States of America, Petitioner,* v. *E.I. du Pont de Nemours & Company and Others, Defendants. Petition of E.I. du Pont de Nemours & Company and Others* (n.d. [March 4, 1912?]) (vol. 7 of the printed trial records at the Hagley Museum and Library [microfilm M60.9, reel 2]), 22–23. The dynamite plants were Hercules, Emporium, Forcite, and Vigorite; the blasting powder mills were in Columbus and Pittsburg, Kans., Pleasant Prairie, Wis., Rosendale, N.Y., Shenandoah and Riker, Pa., and Santa Cruz, Calif.

62. Pierre S. du Pont to William du Pont, April 3, 1912, accession 616; see also *In the District Court of the United States for the District of Delaware. In Equity, No. 280. The United States of America, Petitioner,* v. *E.I. du Pont de Nemours & Company and Others, Defendants. Appendix to Brief Abstract of Proceedings* (n.d. [March 4, 1912?]) (vol. 4 of the printed trial documents at the Hagley Museum and Library [microfilm M60.9, reel 2]), 24.

63. Pierre S. du Pont to William du Pont, December 11, 1911, and January 23, 1912, accession 616.

64. Chandler and Salsbury, *Pierre S. du Pont,* 280–282.

65. William S. Hilles to President's Office, December 18, 1911, EIDPDN, series II, part 3, box 132, accession 472.

66. Richard B. Tennant, *The American Cigarette Industry: A Study in Economic Analysis* *and Public Policy* (New Haven, 1950), 60–61. Unfortunately, the criteria used by the court in dissolving the tobacco trust are not clear from Tennant's account: "output" may refer either to capacity or to production levels in the tobacco industry.

67. Pierre S. du Pont to executive committee, March 18, 1912, EIDPDN, Series II, part 2, box 1006, accession 518.

68. Ibid.

69. Pierre du Pont to executive committee, April 19, 1912, EIDPDN, series II, part 2, box 1006, accession 518. This document includes a copy of Glasgow's letter to Du Pont's lawyers, dated April 17. See also Hamilton M. Barksdale to Pierre S. du Pont, April 19, 1912, and Pierre S. du Pont to executive committee, April 20, 1912, EIDPDN, series II, part 2, box 1006, accession 518.

70. Pierre S. du Pont to executive committee, April 20, 1912, EIDPDN, series II, part 2, box 1006, accession 518: see also James Townsend to Pierre du Pont, February 7, 1912, EIDPDN, series II, part 2, box 1006, accession 518. Pierre S. du Pont to executive committee, May 7, 1912, EIDPDN, series II, part 2, box 1006, accession 518.

71. See T. Coleman du Pont to David Watson, October 6, 1911: "Mr. Glasgow wants to divide us up into so many parts that it will not be possible for us to accept the proposition" (EIDPDN, series II, part 3, box 132, accession 472).

72. Henry F. Pringle, *The Life and Times of William Howard Taft,* vol. 2 (New York, 1939), 667; Tennant, *American Cigarette Industry,* 60–62, 65; Chandler and Salsbury, *Pierre S. du Pont,* 289.

73. W.S. Simpson to H.M. Barksdale, June 28, 1916, 2–3, EIDPDN, series II, part 2, box 1023, accession 518. Simpson was manager of the statistical division in the Du Pont company's sales office. The government appears to have decided that the black sporting and smokeless sporting powder businesses were too small to split up in this fashion.

74. Chandler and Salsbury, *Pierre S. du Pont,* 289–290.

75. Van Gelder and Schlatter, *Explosives Industry,* 657–658, 646.

76. L. du Pont to F.L. Connable, October 28, 1912, EIDPDN, series II, part 2, box 1007, accession 518.

77. Van Gelder and Schlatter, *Explosives Industry,* 307, 309.

78. William A. Glasgow to Messrs. Townsend and Button, April 17, 1912, Pierre du Pont to executive committee, April 19, 1912, and Hamilton M. Barksdale to Pierre du Pont, April 19, 1912, EIDPDN, series II, part 2, box 1006, accession 518.

79. See Exhibits 2.1–2.3 and 2.5 and the sources there cited.

80. Chandler and Salsbury, *Pierre S. du Pont,* 289; Pierre S. du Pont to William du Pont, April 3, 1912, accession 616. In fact, Pierre was right to be concerned. At the time he wrote to his cousin, the government was willing to allow Du Pont to retain all of the income bonds ($10 million) in its treasury. On April 29, however, the government insisted that Du Pont distribute half of the income bonds to its stockholders, thereby reducing Du Pont's surplus by another $5 million. Pierre's view of this offer was that, "while objectionable to our side, [it] will not work any serious injury" (unknown [Pierre du Pont?] to F.M. Andrews, April 29, 1912, EIDPDN, series II, part 3, box 132, accession 472; and Pierre du Pont to William du Pont, May 8, 1912, accession 616).

81. Pierre S. du Pont to finance committee, May 25, 1912, 1, 8, EIDPDN, series II, part 3, box 131, accession 472.

82. Ibid; *Final Decree,* 11.

83. Chandler and Salsbury, *Pierre S. du Pont,* 298.

84. Pierre S. du Pont to finance committee, May 25, 1912, 7, and unpaginated page attached at the back of the document, EIDPDN, series II, part 3, box 131, accession 472. It appears from the last sheet of this document that Hamilton Barksdale also voted (for Dunham as first choice), but perhaps after the du Ponts had already chosen him.

85. Frank Turner succeeded Dunham as Du Pont's comptroller on September 18. J.A. Haskell to Frank Turner, September 18, 1912, EIDPDN, series II, part 2, box 1007, accession 518. For the appointment of the other officers, see Pierre S. du Pont to

William du Pont, October 2, 1912, accession 616; Pierre S. du Pont to Hamilton M. Barksdale and Charles L. Patterson, October 4, 1912, EIDPDN, series II, part 2, box 1007, accession 518; *Hercules Mixer* 35 (January–February 1954), 19.

86. R.H. Dunham, "Object of Accounting," paper delivered to the Du Pont High Explosives Operating Department superintendents' meeting no. 33, at New York, N.Y., April 25, 1911, EIDPDN, series II, part 2, box 577, accession 641, 15, 18; Chandler and Salsbury, *Pierre S. du Pont,* 71–72; H. Thomas Johnson, "Management Accounting in an Early Integrated Industrial: E.I. du Pont de Nemours Powder Company, 1903–1912," *Business History Review* 49, no. 2 (Summer 1975), 184–204; Pierre S. du Pont to William du Pont, April 7, 1911, and July 16, 1912, accession 616; *Hercules Mixer* 1 (March 1919), 17; *Hercules Mixer* 39 (February 1958), 10–11, 27; Hercules Powder Co., *Past Directors* (Wilmington, 1960); Hercules Incorporated, Public Relations Department, Directors' Biographies, PR 3021.

87. Van Gelder and Schlatter, *Explosives Industry,* 522–523; *Hercules Mixer* 1 (April 1919), 43; Hercules Powder Company, *Past Directors,* PR 3021.

88. "About Herculights," *Hercules Mixer* 1 (May 1919), 77; Hercules Powder Company, *Past Directors,* PR 3021.

89. "About Herculights," *Hercules Mixer* 1 (June 1919), 111; Hercules Powder Company, *Past Directors,* PR 3021.

90. "About Herculights," *Hercules Mixer* 1 (August 1919), 179; Hercules Powder Company, *Past Directors,* PR 3021. Stark testified in the antitrust trial on June 10, 1909. See *In the District Court of the United States for the District of Delaware. In Equity, No. 280. The United States of America, Petitioner,* v. *E.I. du Pont de Nemours & Company and Others, Defendants. Defendants' Record. Testimony,* 1:199–206.

91. *Hercules Mixer* 1 (July 1919), 145; Hercules Powder Company, *Past Directors,* PR 3021.

Chapter 3

1. Alfred D. Chandler, Jr., *Strategy and Structure: Chapters in the History of the American Industrial Enterprise* (Cambridge, Mass., 1962), Introduction and passim.

2. Minutes of meetings of board of directors, 1912–1914, in the secretary's vault at Hercules Incorporated headquarters (hereafter cited as Board Minutes, with date).

3. These discussions can be followed in Hamilton M. Barksdale's papers, EIDPDN (see chap. 2, note 48), series II, part 2, boxes 1007, 1021, and 1023, accession 518.

4. Treasurer's Report, December 1919.

5. J.T. Skelly to the board, September 15, 1919, lists only twenty-five sales personnel in Wilmington in 1913, for example. Special Reports to the Directors, 1917–1920, Hall of Records, Hercules Incorporated.

6. Hercules may also have worked with a list of available managers like the one Pierre du Pont had prepared in choosing the top men for each company (see chap. 2). For an example of Du Pont offering a man that Hercules initially refused, see below, n. 49.

7. Pierre S. du Pont to William du Pont, October 2, 1912, in papers of Pierre S. du Pont, accession 616, Hagley Museum and Library (hereafter cited as accession 616); see also correspondence between F.L. Connable and H.M. Barksdale as to the assignment of trade, EIDPDN, series II, part 2, box 1007, accession 518.

8. *In the District Court of the United States for the District of Delaware. In Equity, No. 280. The United States of America, Petitioner,* v. *E.I. du Pont de Nemours & Company and others, Defendants. Defendants' Report on Compliance with Final Decree of June 13, 1912 (January 15, 1913)*, 5–6, 77–78 (hereafter cited as *Defendants' Report on Compliance*). See also *In the District Court of the United States for the District of Delaware. In Equity, No. 280. The United States of America, Petitioner,* v. *E.I. du Pont de Nemours & Company and Others, Defendants. Additional and Supplemental Report of Compliance with the Final Decree of June 13, 1912* (February, 1913), 18–24 (hereafter cited as *Additional and Supplemental Report on Compliance*).

9. Bacchus also supervised production of smokeless powder, which began in 1914.
10. See 71–74 for a discussion of acquisition of the Midwest dynamite plant.
11. *In the District Court of the United States for the District of Delaware. In Equity, No. 280. The United States of America, Petitioner,* v. *E.I. du Pont de Nemours & Company and Others, Defendants. Opinion of the Court and Final Decree* (June 13, 1912), 8 (hereafter cited as *Final Decree*).
12. Board Minutes, December 28, 1912.
13. Ibid.
14. Memorandums of the board of directors attached to the Sales Department Report, July 1913.
15. Some of these reports are preserved in Special Reports, 1917–1920 (see note 5). Others are included along with the monthly reports of various departments. Still other reports, including one on the proposed changes in capital structure, have apparently been lost.
16. Treasurer's Report, January 1913.
17. *Final Decree,* 4–7. The court specified only the manufacturing facilities to be transferred to Hercules, not the sales offices. The court also referred to the plants by different names. The names used here are those adopted by Hercules.
18. Arthur Pine Van Gelder and Hugo Schlatter, *History of the Explosives Industry in America* (New York, 1927), 655–658 (Marquette), 232–233 (Rosendale). Hercules received "a trifle over $5000 from the sale of scrap and buildings from the Rosendale powder line" by July. Operating Department Report, July 1913.
19. Division Superintendent (Lammot du Pont) to F.L. Connable, October 28, 1912, EIDPDN, series II, part 2, box 1007, accession 518.
20. Operating Department Reports, December 1913 and December 1914. See also, C.D. Prickett to the board, May 1, 1914, included with Operating Department Report, March 1914.
21. Van Gelder and Schlatter, *Explosives Industry,* 283–291. The Santa Cruz mill was idle on September 23 and 24, 1913, while the entire work force was fighting a forest fire. Operating Department Report, September 1913.
22. Operating Department Reports for 1913 and 1914. The report for December 1914 includes a 10-year summary of output at the plant; J.T. Skelly to the board, August 21, 1913, included with Sales Department Report, July 1913. See also C.D. Prickett and G.H. Markell to the board, September 6, 1913, filed with the Operating Department Report, September 1913.
23. C.D. Prickett to G.H. Markell, January 22, 1913, with accompanying report on the explosion addressed to the board of directors, January 29, 1913, included with Operating Department Reports for 1913. By May, the work force at Hazardville had been let go; the plant remained under the supervision of the superintendent and a watchman. Operating Department Report, May 1913.
24. Operating Department Report, February 1914; Board Minutes, March 22, 1913.
25. Treasurer's Report, December 1913.
26. Data gleaned from Sales Department Reports for 1913.
27. T.W. Bacchus to George M. Norman, February 15, 1913, in papers of G.M. Norman, accession 1150, Hagley Museum and Library, letter included among papers placed in the back of an unmarked buckram noteback. See also, G.H. Markell to the board, March 17, 1913, letter filed with the Treasurer's Reports for 1913. Over the next two decades, the board periodically considered building a dynamite plant in Arizona; see Board Minutes, May 5, 1914, for the first serious study of the issues.
28. C.D. Prickett and G.H. Markell to the board, September 6, 1913, filed with Operating Department Reports for 1913; *Hercules Mixer* 1 (October 1919), 215–220, 238; Van Gelder and Schlatter, *Explosives Industry,* 526–530. According to Bacchus, the plant was ready to operate by the end of November 1914. The plant did not formally open until nearly six months later, on April 1, 1915, because of lack of business. T.W.

Bacchus to board, November 27, 1914, included with Operating Department Report, October 1914; Board Minutes, October 7 and December 3, 1913.

29. Sales Department Report, January–February 1913.

30. For example, Sales Department Reports, January–February, March, and April 1913.

31. Board Minutes, June 3 and August 6, 1913.

32. Board Minutes, August 6, 1913.

33. Van Gelder and Schlatter sketch the history and operations of these companies in *Explosives Industry,* 529–532, 541–560.

34. Board Minutes, November 5, 1913. See also Board Minutes, August 6 and 7, 1913.

35. Board Minutes, October 7, 1913, and February 24, 1914. According to the latter entry, Rood actually approached Hercules with a proposition to sell.

36. Capacity figures taken from Operating Department Report, December 1914.

37. Board Minutes, February 24, 1914. Bacchus's letter to Dunham is dated January 15, 1914.

38. Ibid. Independent had valued its buildings, machinery, and real estate at $252,938; Bacchus believed these assets were worth $324,670, a difference of $71,732.

39. R.H. Dunham to board, February 24, 1914, in ibid.

40. Ibid.

41. Board Minutes, June 2 and September 1, 1914.

42. Operating Department Report, December 1914.

43. On Rood and Bent, see chap. 5; on Talley, see chap. 4, and Van Gelder and Schlatter, *Explosives Industry,* 530–531.

44. *Additional and Supplemental Report of Compliance,* 37–38.

45. Sales Department Report, June 1913.

46. See above, 33.

47. "Certificate of Chief Engineer" of Du Pont, William G. Ramsay, February 10, 1913, included as Exhibit E in *Additional and Supplemental Report of Compliance,* 47–48.

48. H.M. Barksdale to H.F. Brown, July 23, 1912, EIDPDN, series II, part 2, box 1007, accession 518. Dunham's final acceptance of the terms is noted in a letter to Barksdale dated November 26, 1912, EIDPDN, series II, part 2, box 1007, accession 518.

49. Van Gelder and Schlatter, *Explosives Industry,* 893–894. Bacchus estimated that the semismokeless capacity could be added for about $30,000. He deferred the question of whether the facility should include a nitrocotton or guncotton plant until later. See Sales Department Report, April 1913, 10; Board Minutes, June 3 and August 5, 1913; T.W. Bacchus to the board, August 29, 1913, included with Operating Department Report, July 1913. As for Troxler, Bacchus had rejected an offer to employ him late in 1912 and sought a different candidate. By the summer of 1913, however, Bacchus had evidently changed his mind. T.W. Bacchus to H.M. Barksdale, November 29, 1912, EIDPDN, series II, part 2, box 1007, accession 518.

50. T.W. Bacchus to the board, April 2, 1914, included with Operating Department Report, February 1914.

51. Board Minutes, November 4, 1914.

52. Operating Department Report, December 1914.

53. Board Minutes, December 26, 1912; *Final Decree,* 7–8.

54. Pierre S. du Pont to William du Pont, July 22, 1913, 1–2, accession 616.

55. G.H. Markell to the board, March 17, 1913, letter filed with Treasurer's Reports for 1913.

56. G.H. Markell to the board, March 28, 1913, letter accompanying Treasurer's Report, January–February 1913.

57. Board Minutes, June 3, 1913.

58. Board Minutes, October 7 and December 3, 1913.

59. E.I. du Pont de Nemours & Co., Annual Reports, 1907–1912.
60. G.H. Markell to the board, March 28, 1913, letter accompanying Treasurer's Report, January–February 1913.
61. Board Minutes, December 2, 1913; minutes of special meeting of stockholders, January 22, 1914, HPC Stockholders, vol. I, 1912 (see chap. 2, note 1).
62. R.H. Dunham, "Object of Accounting," paper delivered to the Du Pont High Explosives Operating Department superintendents' meeting no. 33, at New York, N.Y., April 25, 1911, 17, EIDPDN, series II, part 2, box 577, accession 641; H. Thomas Johnson, "Management Accounting in an Early Integrated Industrial: E.I. du Pont de Nemours Powder Company, 1903–1912," *Business History Review* 49, no. 2 (Summer 1975), 187–190. For a discussion of Du Pont's similar policies, which Dunham helped devise, see Alfred D. Chandler, Jr., and Stephen S. Salsbury, *Pierre S. du Pont and the Making of the Modern Corporation* (New York, 1971), 210, 251–253.
63. R.H. Dunham to H.M. Barksdale, November 26, 1912, EIDPDN, series II, part 2, box 1007, accession 518. Cf. Johnson, "Management Accounting in an Early Integrated Industrial," 203, for Du Pont's similar practice before 1912.
64. *Defendants' Report in Compliance,* 79–81; and *Additional and Supplemental Report of Compliance,* 3. How these arrangements worked out can be followed in EIDPDN, series II, part 2, boxes 1007 and 1021, accession 518.
65. *Defendants' Report in Compliance,* 84–87; Chandler and Salsbury, *Pierre S. du Pont,* 298–299.
66. The reports of the Du Pont Chemical Department furnished to Hercules, dating from 1909 to 1916, are at the Hall of Records. For Hercules' use of this department, see EIDPDN, series II, part 2, box 1021, accession 518.
67. Van Gelder and Schlatter, *Explosives Industry,* 527.
68. R.H. Dunham to H.M. Barksdale, March 3, 1913, EIDPDN, series II, part 2, box 1021, accession 518. Dunham also wrote, "It seems to us that for this amount of money . . . we might possibly be able to organize a department on our own which would be as satisfactory and probably as profitable to us as it would be . . . to continue the proposed arrangements for an extended period."
69. Operating Department Report, December 1913. Cf. U.S. Department of Commerce, Bureau of the Census, *Thirteenth Census of the United States Taken in the Year 1910,* vol. 10, *Manufactures, 1909. Reports for Principal Industries* (Washington, D.C., 1913), 569–570.
70. These concerns surfaced as early as the spring of 1913; see Board Minutes, May 6, 1913. A year later, on June 2, 1914, the board formally requested the Operating Department henceforth to report detailed information on its purchased inputs. Board Minutes, June 2, 1914. See also, EIDPDN, series II, part 2, boxes 1007 and 1021, accession 518.
71. Board Minutes, September 2, 1913.
72. Sales Department Report, March 1913.
73. Sales Department Report, August 1914.
74. Sales Department Report, January–February 1913.
75. J.T. Skelly to the board, December 13, 1913, 1–3, filed with Sales Department Report, October 1913.
76. Bureau of the Census, *Thirteenth Census of the United States,* 570; Williams Haynes, *American Chemical Industry,* vol. 3, *The World War I Period, 1912–1922* (New York, 1945), app. 17, 463.
77. Edward Banks to J.T. Skelly, February 28, 1913, passim, included with Sales Department Report, January–February 1913.
78. *In the District Court of the United States for the District of Delaware. In Equity, No. 280. The United States of America, Petitioner,* v. *E.I. du Pont de Nemours & Company and Others, Defendants. Defendants' Record. Testimony,* vol. II, part 1, 201.
79. Sales Department Reports, March and April 1913.
80. Sales Department Reports, July and August 1913.

81. Sales Department Report, January 1914.
82. Sales Department Report, September 1914.
83. Sales Department Report, December 1914; cf. Operating Department Report, December 1914.
84. W.S. Simpson to H.M. Barksdale, June 14, 1916, EIDPDN, series II, part 2, box 1023, accession 518.
85. J.T. Skelly to the board, August 21, 1913, included with Sales Department Report, July 1913.
86. Sales Department Report, June 1914.
87. Sales Department Report, October 1914.
88. Hercules Powder Company, Annual Reports for 1913 and 1914.
89. Board Minutes, September 2, 1913; Treasurer's Report, September 1913.
90. Board Minutes, December 1, 1914; Treasurer's Report, November 1914.

Chapter 4

1. For general overviews of technological innovation during the war, see Guy Hartcup, *The War of Invention: Scientific Developments, 1914–1918* (London, 1988), and John Terraine, *White Heat: The New Warfare 1914–18* (London, 1982).
2. John Keegan, *The Face of Battle* (London, 1976), 235, 238–240.
3. Ibid., 238–239, 260, 285.
4. Calculation based on fig. 9 in Benedict Crowell, *America's Munitions, 1917–1918* (Washington, D.C., 1919), 104; see also Gerd Hardach, *The First World War, 1914–1918* (Berkeley, 1977), chap. 4. For the general context of the chemical industry during the war, see Williams Haynes, *American Chemical Industry,* vol. 2, *The World War I Period: 1912–1922* (New York, 1945), chaps. 1–15; Ludwig F. Haber, *The Chemical Industry, 1900–1930* (Oxford, 1971), chap. 7.
5. Hercules Powder Company, Atlas Powder Company, and E.I. du Pont de Nemours & Co. Annual Reports for 1917.
6. Vincent P. Carosso, *Investment Banking in America: A History* (Cambridge, Mass., 1970), 216.
7. In the 1930s, Du Pont, Bethlehem Steel, J.P. Morgan & Co., and other prominent companies were subject to a congressional investigation into wartime profiteering. A popular book of the era, H.C. Englebrecht and F.C. Hanighen, *Merchants of Death: A Study of the International Armament Industry* (New York, 1934), also provoked scrutiny. Although none of the companies under scrutiny was sanctioned in any way, the controversy affected their approach to dealing with the government and has intrigued historians ever since. See chap. 8.
8. Haynes, *American Chemical Industry,* 2:86–87.
9. Ibid., 2: chap. 14; for Hercules' kelp operations, see 98–100.
10. Williams Haynes, *The Chemical Front* (New York, 1943), 41. The remaining ingredients among the light oils are benzene (ten pounds), naphthalene (six pounds), and anthracene (ten ounces).
11. Haynes, *American Chemical Industry,* vol. 3, *The World War I Period: 1912–1922* (New York, 1945), 129.
12. Haynes, *Chemical Front,* 46–48; and Haynes, *American Chemical Industry,* 2: chap. 12.
13. Haynes, *American Chemical Industry,* 2:277 (app. 1).
14. T.W. Bacchus to the board, letter accompanying Operating Department Report, December 1914; Board Minutes, December 1, 1914, and January 5, 1915.
15. Paul Guinn, *British Strategy and Politics, 1914 to 1918* (Oxford, 1965), 76–80.
16. Report of the Sales Department covering special business for the month of October 1917. This report, which alters the format of reporting on wartime contracts, mentions J.P. Morgan & Co. as the customer for the initial cordite contracts. On the U.S.

industrial banks' financing of Allied munitions, see Carosso, *Investment Banking in America,* chap. 10, esp. 201–216.

Lack of munitions capacity also worried the French government. In the fall of 1914, the French arranged to buy eight million pounds of pyro cannon powder and 1.25 million pounds of guncotton from Du Pont. Alfred D. Chandler, Jr., and Stephen S. Salsbury, *Pierre S. du Pont and the Shaping of the Modern Corporation* [New York, 1971], 368.

17. Board Minutes, February 10 and 23, 1915; J.T. Skelly to the board, "Report on the Special War Business," June 5, 1916, letter included with Sales Department Report, May 1916. In the vertical files in the library at Hercules Incorporated are copies of the official British specifications for cordite; however, they make no mention of any financial arrangements or delivery schedules. See also Chandler and Salsbury, *Pierre S. du Pont,* 368–369, for similar contracts between Du Pont and the Allies.

18. Arthur Marshall, *Explosives,* 2d ed., vol. 1, *History and Manufacture* (Philadelphia, 1917), 304–308. The original formula for cordite (cordite Mk.I) is 37 percent guncotton, 58 percent nitroglycerin, and 5 percent mineral jelly; for cordite M.D., the formula is 65 percent guncotton, 30 percent nitroglycerin, and 5 percent mineral jelly.

19. Arthur Pine Van Gelder and Hugo Schlatter, *History of the Explosives Industry in America* (New York, 1927), 895; Operating Department Reports for 1915; Board Minutes, March 2, March 24, and April 6, 1915.

20. Van Gelder and Schlatter, *Explosives Industry,* 895–897. At the same time, Hercules developed an ingenious way to blend the strings of cordite, whose length posed unusual problems. Engineers designed a kind of conveyor bucket system into which operators stacked and sorted the strings so that they could be packed in the fourteen separate blending houses at Kenvil. See also William B. Williams, *History of the Manufacture of Explosives for the Great War, 1917–1918* (n.p., 1919), 27–21; William B. Williams, "Hercules Powder Co. in the World War," *Munitions Manufacture in the Philadelphia Ordnance District* (Philadelphia, 1921), 437–438.

21. Operating Department Reports, February–May 1915; "Hercules' Achievements during the Period of the War," Part I, *Hercules Mixer* 1 (April 1919), 20–21.

22. Skelly, "Report on the Special War Business."

23. The following account is based on the minute books of the Union Powder Corporation found at the Hall of Records, Hercules Incorporated; and on Van Gelder and Schlatter, *Explosives Industry,* 899–900, 912–915. Hugo Schlatter was the plant manager of the Parlin plant discussed below.

24. Minute Book of the Union Powder Corporation, January 4, 1916; "Union Now Manufacturing Many Pyroxylin Solutions," *Hercules Mixer* 1 (April 1919), 29, 39; Van Gelder and Schlatter, *Explosives Industry,* 899. Parlin's original capacity was 25,000 pounds of guncotton per day; by the end of the war, Hercules had upgraded capacity to 70,000 pounds per day.

25. Skelly, "Report on the Special War Business."

26. "Hercules' Achievements during the Period of the War," Part II, *Hercules Mixer* 1 (May 1919), 57.

27. R.H. Dunham to H.G. Haskell, October 26, 1914, EIDPDN (see chap. 2, note 48), series II, part 2, box 1021, accession 518; George H. Markell to the board, December 28, 1914, included with Treasurer's Report, November 1914.

28. John Albert Marshall, *The Manufacture and Testing of Military Explosives* (London, 1919), 127, 136–139; Arthur Marshall, *Explosives,* 1: 261–267.

29. "Hercules' Achievements," Part II, 56–57.

30. H.M. Barksdale to H.G. Haskell and C.L. Reese, May 13, 1915, EIDPDN, series II, part 2, box 1021, accession 518.

31. "Hercules' Achievements," Part II, 57–58; B.B. Tufts, "Hercules, Our Largest Plant—Its History and War Record," *Hercules Mixer* 1 (December 1919), 284–285.

32. Skelly, "Report on the Special War Business."

33. "Hercules' Achievements during the Period of the War," Part IV, *Hercules Mixer* 1 (July 1919), 126–127.
34. Skelly, "Report on the Special War Business."
35. "Hercules' Achievements," Part I, 21.
36. Skelly, "Report on the Special War Business." There were actually three separate contracts for the acetone. The first called for Hercules to supply 4.8 million pounds, for 21.5¢ per pound, during 1916 by shipping 300,000 pounds per month from January through June, and 500,000 pounds per month for the balance of the year. The second contract obliged Hercules to deliver 2.4 milion pounds, for 34¢ per pound, at the steady rate of 200,00 pounds per month. Finally, Hercules contracted to borrow 900,000 pounds of acetone from the British government at a price of 23¢ per pound. See also "Hercules' Achievements," Part I, 21; Van Gelder and Schlatter, *Explosives Industry,* 897–898.
37. Marshall, *Explosives,* 1: 340.
38. These efforts can be followed in "Historical Materials Relating to the Manufacture of Acetone from Kelp Plants at the San Diego Plant during the Years 1915 to 1920s," Hall of Records, Hercules Incorporated (hereafter cited as San Diego Kelp Plant Files).
39. George H. Markell to Hamilton M. Barksdale, April 6, 1916; EIDPDN, series II, part 2, box 1021, accession 518.
40. "Hercules' Achievements," Part I, 23; cf. Van Gelder and Schlatter, *Explosives Industry,* 897–898; Marshall, *Explosives,* 1: 340–349.
41. Van Gelder and Schlatter, *Explosives Industry,* 895–896; *Hercules Mixer* 1 (April 1919), 23–24; Marshall, *Explosives,* 3: 102–103.
42. Van Gelder and Schlatter, *Explosives Industry,* 897–898; "Hercules' Achievements," Part I, 23–24.
43. George H. Markell to Hamilton M. Barksdale, April 6, 1916, EIDPDN, series II, part 2, box 1021, accession 518.
44. Board Minutes, May 3, 1916.
45. H.H. Eastman to the board, October 26, 1915, included with Treasurer's Report, September 1915; George H. Markell to George W. Simmons, January 12, 1916, Hall of Records, Hercules Incorporated; see also, Williams Haynes, *American Chemical Industry,* vol. 6, *The Chemical Companies* (New York, 1954), 440–448, for a general history of U.S. Industrial Chemicals (successor to U.S. Industrial Alcohol).
46. "Hercules' Achievements," Part I, 23.
47. Ibid., 24.
48. Van Gelder and Schlatter, *Explosives Industry,* 898.
49. Board Minutes, May 3, 1916; Skelly, "Report on the Special War Business."
50. J.T. Skelly to the board, January 11, 1917, included with Sales Department Report, December 1916; J.T. Skelly to the board, February 1, 1917, included with Sales Department Report, January 1917.
51. See the correspondence between Hercules and the Du Pont Chemical Department from July 1915–January 1916 in the San Diego Kelp Plant Files. See also, Peter Neushul, "Seaweed for War: California's World War I Kelp Industry," *Technology and Culture,* vol. 30, no. 3 (July 1989), 561–583.
52. "About Herculights," *Hercules Mixer* 1 (July 1919), 145.
53. C.A. Higgins, "Recovery of Potash from Kelp," *Journal of Industrial and Engineering Chemistry* 10, no. 10 (October 1918), 832–833; C.A. Higgins, "Solvents from Kelp," ibid., 858–859. According to Neushul, who consulted USDA sources in the National Archives, Hercules learned in the summer of 1915 of a small-scale English experiment to make acetate and butyrate of lime from fermented seaweed. Neushul, "Seaweed for War," 572.
54. Neushul, "Seaweed for War," 564–569; Haynes, *American Chemical Industry,* 2, chaps. 13–15.
55. T.W. Bacchus to the board, October 5, 1915, filed with Operating Department

Report, August 1915; G.H. Markell to T.W. Bacchus, October 16, 1915, San Diego Kelp Plant Files.

56. Messrs. Pringle, Wright & Small (patent attorneys) to Hercules Powder Company, December 6, 1915, San Diego Kelp Plant Files. This letter mentions that Hercules inquired as to marine harvester patents on September 22.

57. George H. Markell to Herbert Talley, January 14, 1916, and George H. Markell to George W. Simmons, February 5, 1916, San Diego Kelp Plant Files.

58. Cecil W. Weaver, "Outline Specification of Self Propelled Kelp Sea Weed Dredge," January 17, 1916, and George H. Markell to Herbert Talley, January 19, 1916, San Diego Kelp Plant Files; Neushul, "Seaweed for War," 576–578.

59. "The California Kelp Operations of the Hercules Powder Company," *Metallurgical and Chemical Engineering* 18, no. 2 (June 1, 1918); W.S. Brubaker to George H. Markell, January 11, 1916, San Diego Kelp Plant Files; Neushul, "Seaweed for War," 572–573.

60. Higgins, "Solvents from Kelp," 858.

61. "Hercules' Achievements," Part I, 25.

62. J.T. Skelly to the board, June 6, 1917, included with Sales Department Report, May 1917.

63. George H. Markell to the board, February 13, 1915, filed with Treasurer's Report, December 1914.

64. Information gleaned from Hercules Powder Company, Annual Reports, Treasurer's Reports, and Board Minutes for 1915 and 1916.

65. Hartcup, *War of Invention,* 51; Arthur Marshall, *Explosives,* 2d ed., vol. 3 (Philadelphia, 1932), 89.

66. Hartcup, *War of Invention,* 52–54; see also, Leonard Stein, *The Balfour Declaration* (London, 1961), passim. Weizmann's discovery helped launch a significant political career. An ardent Zionist, he immersed himself during the war in the struggle for a Jewish homeland. Although there is no direct evidence linking his wartime activities to the Balfour Declaration, the 1927 commitment by the British government "to view with favour the establishment in Palestine of a national home for the Jewish people," his contributions surely helped. Weizmann became the first president of Israel in 1948; Neushul, "Seaweed for War," 578–579.

67. J.T. Skelly to the board, January 11, 1917, included with Sales Department Report, December 1916; J.T. Skelly to the board, March 5, 1917, included with Sales Department Report, February 1917.

68. See the correspondence in EIDPDN, series II, part 2, box 1021, accession 518, file labeled "Hercules General, 10/23/14 to 12/20/17."

69. "Hercules' Achievements during the Period of the War," Part III, *Hercules Mixer* 1 (June 1919), 59.

70. Hercules Powder Company, Annual Reports for 1916 and 1917; see also *In Re Title III of H.R. 16763, entitled "A Bill to Increase the Revenue and for Other Purposes," said Title III being designated as "Munition Manufacturer's Tax." Memorandum of the Hercules Powder Company, a Manufacturer of Gunpowder and Other Explosives* (Wilmington, n.d. [1917?]), passim.

71. J.T. Skelly to the board, November 6, 1916, included with Sales Department Report, October 1916; J.T. Skelly to the board, April 19, 1917, included with Sales Department Report, March 1917; J.T. Skelly to the board, June 6, 1917, included with Sales Department Report, May 1917.

72. J.T. Skelly to the board, March 5, 1917, included with Sales Department Report, February 1917; "Hercules' Achievements," Part II, 58.

73. J.T. Skelly to the board, June 6, 1917, included with Sales Department Report, May 1917; J.T. Skelly to the board, July 26, 1917, included with Sales Department Report, June 1917; J.T. Skelly to the board, August 31, 1917, included with Sales Department Report, July 1917; reports of the Sales Department covering special business for the months of December 1917, January 1918, and March 1918, included with Sales Department Reports for December 1917, January 1918, and March 1918, respectively.

74. "Hercules' Achievements," Part II, 58.

75. U.S. Department of Labor, Bureau of Labor Statistics, Employment and Occupational Outlook Branch, *History and Disposition of a Powder Plant Project, Nitro, West Virginia, 1917–1942,* Historical Study no. 78 (Washington, D.C., May 1945), 2–5.

76. Ibid., 5; "Hercules' Achievements," Part III, 90. Ground was broken on January 11, one week before the construction contract was finalized. The following account is based on these sources and U.S. War Department, U.S. Government Explosives Plants, Daniel C. Jackling, Director, *A Report of the Construction of the United States Government Explosives Plant "C" at Nitro, West Virgina,* vol. 1 (Nitro, W. Va., 1919), passim.

77. Department of Labor, *History and Disposition of a Powder Plant Project,* 5–8; "Hercules' Achievements," Part III, 90; Williams, *Munitions Manufacture,* 440; Van Gelder and Schlatter, *Explosives Industry,* 902–903. The estimate for the government advance is based on Treasurer's Reports, December 1918 and December 1919.

78. "Hercules' Achievements," Part III, 90.

79. Department of Labor, *History and Disposition of a Powder Plant Project,* 6–7; "Hercules' Achievements," Part III, 91–92.

80. "Hercules' Achievements," Part III, 90–91.

81. Ibid., 92; Department of Labor, *History and Disposition of a Powder Plant Project,* 5, 7–8; Van Gelder and Schlatter, *Explosives Industry,* 840–841.

82. On McKinney, see *Hercules Mixer* 2 (December 1920), 423. An incomplete set of monthly reports of the Purchasing Department from 1914 to 1921 is kept at the Hall of Records.

83. P.S. du Pont and H.M. Barksdale to (Du Pont) executive committee, December 9, 1915, EIDPDN, series II, part 2, box 1021, accession 518.

84. "The Treasurer's, Accounting, and Purchasing Departments," *Hercules Mixer* 1 (July 1919), 117, 142; R.B. McKinney, "Purchasing," and "Discussion," in *Minutes of Meeting of the Superintendents of the Hercules Powder Co. No. 1. Held in Hotel Traymore, Atlantic City, New Jersey, February 18th to 21st, 1919* (n.p., privately printed, 1919) (see chap. 5, note 17).

85. P.S. du Pont and H.M. Barksdale to (Du Pont) executive committee, December 9, 1915, EIDPDN, series II, part 2, box 1021, accession 518; on Du Pont's research on acetone, see 96–97.

86. "About Herculights," *Hercules Mixer* 2 (October 1920), 339.

87. "Our Engineering Department's Organization and Work," *Hercules Mixer* 2 (August 1920), 239–242, 260.

88. "The Organization and Work of the Chemical Department," *Hercules Mixer* 2 (October 1920), 307–310.

89. We are grateful to Barbara B. Chapman of the Technical Information Division of the Research Center at Hercules Incorporated for this information. See Chapman's letter to C.R. Miller, May 6, 1988. For Jackman's role, see the profile of A.B. Nixon in *Hercules Mixer* 7 (May 1925), 129.

90. George Norman, "TGM—Relocation of Experimental Station," a report to the board, June 13, 1929, 4. We are grateful to Barbara Chapman for providing a copy of this document.

91. At this distance in time, however, it is impossible to say how "tight" this coincidence actually was. Hercules continued to use Du Pont's research support for its commercial explosives business. The company's payments to Du Pont for research services were as follows: 1913, $80,685.12; 1914, $59,898.63; 1915, $61,016.48; 1916, $77,583.68; 1917, $46,182.89. Russell H. Dunham to E.I. du Pont de Nemours & Co., Chemical Department, December 7, 1917, EIDPDN, series II, part II, accession 1662.

92. "The Equipment and Work of Our Experimental Station," *Hercules Mixer* 1 (August 1919), 158–161; Board Minutes, August 12, 1915, and March 7, 1916; "About Herculights" (Profile of Bierbauer), *Hercules Mixer* 5 (April 1923), 103; Board Minutes, August 7, 1918.

93. "The Equipment and Work of Our Experimental Station," 158–159; "The Organization and Work of the Chemical Department," 308–310. See also chap. 5.

94. Hercules Powder Company, Annual Reports for 1914 and 1918.

95. Final report of Sales Department covering special business after completion or cancellation of all contracts, June 10, 1919, filed with Sales Department Reports for 1919.

Chapter 5

1. R.H. Dunham, "Opening Address," in *Managers Meeting—Operating,* 20 (see note 17 below).

2. Final report of Sales Department covering special business, June 10, 1919, 2–4, filed with Sales Department Report, April 1919.

3. Treasurer's Reports, December 1918 and December 1919.

4. Board Minutes, January 9, 1917.

5. Board Minutes, December 18, 1918.

6. C.A. Higgins, "The Development of New Products," in *Managers Meeting—Operating,* 443 (see note 17 below).

7. Treasurer's Report, December 1918.

8. Board Minutes, February 13, 1917.

9. Board Minutes, November 7 and December 4, 1918; *Hercules Mixer* 1 (October 1919), 247.

10. *Hercules Mixer* 1 (July 1919), 117, 142; R.B. McKinney, "Purchasing," and "Discussion," in *Managers Meeting—Operating,* 472, 485 (see note 17 below).

11. T.W. Bacchus to the board, January 30, 1919, filed with Special Reports to the Board of Directors, 1917–1921, Hall of Records, Hercules Incorporated.

12. Ibid.

13. Ibid. According to the Treasurer's Report, December 1919, Hercules' total payroll was 2,027 people.

14. Treasurer's Report, December 1919; see also H.H. Eastman to the board, September 13, 1919, and J.T. Skelly to the board, September 15, 1919 (Special Reports, 1917–1921) for explanations of the growth of the Accounting and Sales departments.

15. Bacchus to the board, January 30, 1919.

16. Albert R. "Slick" Ely interview, November 12, 1987.

17. Proceedings of these meetings are printed in two volumes labeled *Minutes of Meeting of the Superintendents of the Hercules Powder Co. No. 1. Held in Hotel Traymore, Atlantic City, New Jersey, February 18th to 21st, 1919*. They were privately printed in 1919 and are available in the library at the Hercules Home Office in Wilmington. The volumes bear similar titles, but one contains a record of joint sessions plus sessions of particular interest to the Operating Department, while the other contains the same record of joint sessions plus sessions of interest to the Sales Department. They are cited herein as *Managers Meeting—Operating* and *Managers Meeting—Sales*.

18. Dunham, "Opening Address," 20.

19. C.C. Gerow, "Business Conditions and Outlook," in *Managers Meeting—Sales,* 26.

20. Williams Haynes, *American Chemical Industry,* vol. 3, *The World War I Period: 1912–1922* (New York, 1945), app. 42, 463.

21. Operating Department Report, December 1919; Sales Department Report, December 1919.

22. Monthly Treasurer's Reports for 1919.

23. Dunham, "Opening Address," 19–20; Gerow, "Business Conditions," 26; cf. discussion of Gerow's paper in *Managers Meeting—Operating,* 30, 31.

24. J.T. Skelly to board, February 28, 1913, included with Sales Department Report, January–February 1913; Sales Department Reports, March to December 1913, May 1915, March 1916, January 1917, and February 1917.

25. Sales Department Report, January 1919.

26. Sales Department Reports, February–March, May, August, and December 1919, and January–February 1920, Special Report, March 4, 1920.

27. C.F. Bierbauer, "The Equipment and Work of Our Experimental Station," *Hercules Mixer* 1 (August 1919), 161; Statement of Experimental Costs from January 1 to December 31, 1918, filed with Operating Department Report, December 1918.

28. G.C. O'Brien, "Chemicals from Kelp," in *Managers Meeting—Sales,* 98–104; "Union Now Manufacturing Many Pyroxylin Solutions," *Hercules Mixer* 1 (April 1919), 27–29, 39; J. Boisseau Wiesel, "Soluble Cotton, Solvents, and Uses of Pyroxylin Solutions," *Hercules Mixer* 1 (May 1919), 52–54. For an excellent discussion of the development of celluloid plastics, see Robert Friedel, *Pioneer Plastic: The Making and Selling of Celluloid* (Madison, 1983), esp. chap. 1.

29. Treasurer's Report, December 1918; cf. Sales Department Report, December 1918, for a listing of all chemical products sold between October 1, 1917 and January 1, 1919.

30. "Chemical Sales Division's Organization Enlarged," *Hercules Mixer* 1 (April 1919), 41–42.

31. Sales Department Reports for December 1918, January 1919, and February–March 1919.

32. Sales Department Reports, January 1919, and February–March 1919.

33. Sales Department Report, April 1919.

34. G.H. Markell to the board, May 14, 1919, letter accompanying Treasurer's Report, March 1919.

35. Board Minutes, June 3, 9, and 10, 1919.

36. Board Minutes, August 12, 1919.

37. Higgins, "Development of New Products," 441; cf. Wiesel, "Soluble Cotton," 53–54.

38. Although few experts realized it, the market for celluloid plastics was about to peak in any event. See Friedel, *Pioneer Plastic,* chap. 5.

39. Wiesel, "Soluble Cotton," 52–54; "Union Now Manufacturing Many Pyroxylin Solutions," 39. See also Hugo Schlatter, "How Our Union Plant Manufactures Nitrated Cotton," *Hercules Mixer* 2 (April 1920), 103–106; C.A. Lambert, "The Manufacture of Pyroxylin Solutions and Paste at Union," ibid., 107–108; G.C. O'Brien, "The Many Ultimate Uses of Hercules Pyroxylin Solutions and Plastics," ibid., 109–111; and Williams Haynes, *Cellulose: The Chemical That Grows* (Garden City, N.Y., 1953), 207–211.

40. Atlas Powder Co., Annual Reports for 1918 and 1920.

41. Alfred D. Chandler, Jr., *Strategy and Structure: Chapters in the History of the American Industrial Enterprise* (Cambridge, Mass., 1962), 83–91; Alfred D. Chandler, Jr., and Stephen S. Salsbury, *Pierre S. du Pont and the Making of the Modern Corporation* (New York, 1971), 381–386.

42. Higgins, "Development of New Products," 441.

43. Ibid., 443.

44. Treasurer's Report, December 1919.

45. Norman Rood, "Industrial Research," *Managers Meeting—Operating,* 27, draws heavily on this source and another: "New Industrial Research Department Launched," *Hercules Mixer* 1 (March 1919), 15–16; and *Hercules Mixer* 5 (January 1923), 25. The complete IRD reports are kept in the library at Hercules Plaza, Wilmington.

46. Rood, "Industrial Research," 27, 36.

47. Higgins, "Development of New Products," 444.

48. Board Minutes, March 4, 1918.

49. The account of this meeting, including quotations, is drawn from IRD Report, February 15, 1919.

50. IRD Reports, February 22 and February 24, 1919. Although a writer for the *Hercules Mixer* stated that the IRD considered "hundreds of possibilities," the ideas listed in Exhibit 5.2 are the only alternatives to naval stores recorded in the IRD Reports ("Naval Stores Selected as New Industry in Which Our Company Will Engage," *Hercules Mixer* 2 [March 1920], 70).

51. "Naval Stores Selected as New Industry," 69–70; Williams Haynes, *This Chemical Age: The Miracle of Man-Made Materials* (Garden City, N.Y., 1942), 156.

52. Thomas Gamble, "Naval Stores Monopolies Created by the Pilgrims and Puritans," in Thomas Gamble, ed., *Naval Stores: History, Production, Distribution, and Consumption* (Savannah, Ga., 1921), 51–53. See also Lewis C. Gray, *History of Agriculture in the Southern United States to 1860,* vol. 1 (Washington, D.C., 1933).

53. J. Merriam Peterson, "History of the Naval Stores Industry in America," Part I, *Journal of Chemical Education* (May 1939), 206; Thomas Gamble, "The Production of Naval Stores in the United States," in Gamble, *Naval Stores,* 78.

54. U.S. Department of Agriculture, Report of the Division of Forestry (1892), 332–333.

55. Peterson, "History of the Naval Stores Industry," Part I, 210–211; J. Merriam Peterson, "History of the Naval Stores Industry in America," Part II, *Journal of Chemical Education* (July 1939), 320; IRD Report, August 15, 1919.

56. Peterson, "History of the Naval Stores Industry," Part I, 209–210; Peterson, "History of the Naval Stores Industry," Part II, 317; J.E. Lockwood, "Creating Greater Demand for Naval Stores Is the Most Serious Problem Facing the Entire Industry," in *Gamble's Naval Stores Year Book for 1929–1930,* 2; Haynes, *This Chemical Age,* 154.

57. Quoted in Haynes, *This Chemical Age,* 154; Peterson, "History of the Naval Stores Industry," Part I, 210–211; Lockwood, "Creating Greater Demand," 2; Gamble, "Production of Naval Stores," 78; Peterson, "History of the Naval Stores Industry," Part II, 320; USDA, Report of the Division of Forestry (1892), 332–335.

58. John T. Schlebecker, *Whereby We Thrive: A History of American Farming, 1607–1972* (Ames, Iowa, 1975), 147.

59. USDA, Report of the Division of Forestry (1892), 293–358.

60. Most notable was the work of Charles H. Herty of the U.S. Bureau of Forestry, who conducted successful trials between 1901 and 1905 with the "cup and gutter" method of tapping trees. This method eventually gained widespread acceptance in the South. Secretary of Agriculture Edwin T. Meredith report to U.S. Senate, "The Life of the Naval Stores Industry as at Present Carried on in the South" (1920), in Gamble, *Naval Stores,* 89; Peterson, "History of the Naval Stores Industry," Part I, 211; Eldon Van Romaine, "Naval Stores, 1919–1939," *Chemical Industries* 45 (October 1939), 404.

61. Albrecht H. Reu, "A Brief History of the Brunswick, Georgia, Plant of Hercules Incorporated" (1966), typescript in Hercules library.

62. Authors' conversation with Charles Penniman (Franklin Institute), January 19, 1988 (see Franklin Institute case file no. 1350); "Homer T. Yaryan: An Autobiography," *Hercules Mixer* 10 (1928), 140.

63. In 1905, a rival inventor, George Walker succeeded in extracting volatile oils and rosin from light woods and later built a 200-barrel-per-week plant in Conway, South Carolina, but it is not clear if Yaryan was aware of his work. Van Romaine, "Naval Stores, 1919–1939," 406; N.P. Rood, "Hercules Operates the Yaryan Plants," *Hercules Mixer* 2 (June 1920), 172.

64. Peterson, "History of the Naval Stores Industry," Part II, 318; Haynes, *American Chemical Industry,* vol. 2, *The World War I Period: 1912–1922* (New York, 1945), 260.

65. "Homer T. Yaryan: An Autobiography," 139–142; Haynes, *American Chemical Industry,* 2:260; Peterson, "History of the Naval Stores Industry," Part II, 318.

66. Peterson, "History of the Naval Stores Industry," Part II, 318; Haynes, *This Chemical Age,* 155.

67. IRD Report, July 27, 1919.
68. Ibid.; IRD Report, January 26, 1920; Haynes, *American Chemical Industry,* 2:261.
69. Haynes, *American Chemical Industry,* 2:261; IRD Report, July 28, 1919.
70. Even after a fire destroyed the Brunswick plant on March 17, 1916, Yaryan rebuilt and reopened the works within months. Haynes, *American Chemical Industry,* 2:256–257, 260; Rood, "Hercules Operates the Yaryan Plants," 171–172.
71. Haynes, *This Chemical Age,* 153.
72. U.S. Department of Commerce, Bureau of the Census, *Twelfth Census of the United States, Taken in the Year 1900,* vol. 9 (Washington, D.C., 1902), pt. 3, 1003.
73. Many factors went bankrupt when bank loans were called and when the Food and Fuel Administration set maximum prices for foodstuffs, which the factors normally sold at inflated rates to their sharecroppers. Haynes, *American Chemical Industry,* 2:257–259; IRD Report, August 18, 1919.
74. IRD Report, August 5, 1919; Gamble, "The Production of Naval Stores in the United States," 79; Haynes, *This Chemical Age,* 155.
75. Sales Department Reports, January and February 1918.
76. Ibid.; "Land Clearing and Wood Utilization by Distillation," *Hercules Mixer* 2 (February 1920), 42.
77. IRD Report, February 15, 1919, and "Land Clearing and Wood Utilization by Distillation," 42–45, 61.
78. "Land Clearing and Wood Utilization by Distillation," 43; Sales Department Report, August 1918.
79. "Land Clearing and Wood Utilization by Distillation," 42–45, 61.
80. Sales Department Report, November 1920. The figure as of November was 3,281,440 pounds. (The December report is missing at Hercules Incorporated.)
81. Ibid.
82. Gladys L. Baker et al., *Century of Service: The First 100 Years of the Department of Agriculture* (Washington, D.C., 1963), 41; Gladys L. Baker, *The County Agent* (Chicago, 1939), 26, 31–32, 45.
83. The secretary of agriculture established the States Relations Committee on June 15, 1914. On July 1, 1915, it was replaced by the States Relations Service, which comprised the Office of Experiment Stations, the Office of Extension Work in the South, the Office of Extension Work in the North and West, and the Office of Home Economics. Baker et al., *Century of Service,* 80–82; Baker, *County Agent,* 8, 24, 36, 207.
84. Mississippi authorized county departments of agriculture in 1908. Baker et al., *Century of Service,* 82; Dunbar Rowland, ed., *Publications of the Mississippi Historical Society,* centenary series, vol. 3 (Jackson, 1919), 250, 260.
85. Nollie W. Hickman, "Mississippi Forests," in Richard A. McLemore, ed., *A History of Mississippi,* vol. 2 (Jackson, 1973), 223.
86. IRD Reports, March 30, April 2, and April 4, 1920, and "Land Clearing and Wood Utilization by Distillation," 44–45.
87. IRD Reports, February 15, 1919, and February 22, 1919.
88. Baker et al., *Century of Service,* 110. See also, Homer E. Socolofsky, *Arthur Capper: Publisher, Politician, and Philanthropist* (Lawrence, Kans., 1962).
89. USDA, Forest Service, "Timber Depletion, Lumber Prices, Lumber Exports, and Concentration of Timber Ownership" (the Copper Report) (Washington, D.C., 1920), 3.
90. Ibid., 29–30.
91. Ibid., 29.
92. IRD Report, July 28, 1919.
93. Ibid.; IRD Report, August 12, 1919. According to Haynes, one (unnamed) government authority even warned Hercules of the difficulty in establishing the steam-solvent distillation business. Haynes, *This Chemical Age,* 158.
94. Quoted in Haynes, *This Chemical Age,* 156. Contrary to Bent's apparent expecta-

tion that prices would remain high, the *Hercules Mixer* observed just after the Yaryan acquisition that "in entering this field, our Company is not basing its calculations on the abnormal profits prevailing at the present time." "Naval Stores Selected as New Industry," 70.

95. Newport was not for sale, although its owner told Rood that he would welcome the entrance of a powerful firm such as Hercules to the industry and would "co-operate" to stabilize the market, perhaps even consider a merger at a later date. IRD Report, July 27, 1919.

96. The destructive distillation companies were American Tar & Turpentine, Florida Wood Products, Pensacola Tar & Turpentine, Chatham Manufacturing, Georgia Pine Turpentine, and Atlantic Turpentine & Pine Tar. IRD Report, January 21, 1920. The author of this document is not sure if Pensacola Tar & Turpentine Co. was acquired by Glidden. IRD Report, January 22, 1920.

97. Quoted in IRD Report, January 26, 1920.

98. Ibid.; IRD Reports, February 15, August 15, and September 15, 1919, and February 4, 1920; E.J. Kahn, Jr., *The Problem Solvers: A History of Arthur D. Little, Inc.* (Boston, 1986), 20, 29, 37, 41, 49, 58.

99. IRD Reports, January 26 and February 18, 1920.

100. IRD Reports, August 6, 1919, January 26, January 30, February 18, and April 30, 1920.

101. IRD Reports, February 27, March 16, March 19, and April 23, 1920. In the fall of 1919, Hercules showed an interest in cooperative research with the Florida Wood Products Co. because of that company's work in the large-scale production of camphor, an ingredient used in the making of Herculoid, which was chronically in short supply. IRD Report, October 10, 1919.

102. IRD Reports, February 6, 1919, and March 1, 1920. Williamson also revealed to Rood that while he was not actively promoting the sale of the plant, he had communicated with representatives of the Du Pont company about the matter. Du Pont's interest, however, seems to have been related to its experiments at Hopewell with craft paper manufacture; indeed, the chemical company may have had hopes of selling its papermaking apparatus to the FIC. IRD Report, March 1, 1920.

103. IRD Reports, April 10 and May 7, 1920.

104. Board Minutes, March 2, 1920.

105. IRD Report, March 30, 1920.

106. IRD Reports, April 2 and April 30, 1920.

107. DeCoriolis believed that rail connections were more important than water access because three-quarters of all naval stores transportation costs were for shipping raw materials. IRD Reports, January 26, March 16, and April 13, 1920.

108. IRD Report, February 23, 1920; "Hattiesburg, Miss., for Naval Stores Plant," *Hercules Mixer* 2 (May 1920), 146, 160.

109. "Hattiesburg, Miss., for Naval Stores Plant," 146, 160.; IRD Report, April 13, 1920; "Latest Hattiesburg Construction News," *Hercules Mixer* 2 (September 1920), 285.

110. Reu, "Brief History," 25; Rood, "Hercules Operates the Yaryan Plants," 171.

111. Dunham, "Discussion of Industrial Research," and "Opening Address," in *Managers Meeting—Operating*, 30, 20.

Chapter 6

1. Alfred D. Chandler, Jr., and Stephen S. Salsbury, *Pierre S. du Pont and the Making of the Modern Corporation* (New York, 1971), 381–386 and chaps. 16–17.

2. Williams Haynes, *American Chemical Industry,* vol. 3, *The World War I Period: 1912–1922* (New York, 1945), 166–167, 426, and vol. 4, *The Merger Era: 1923–1929* (New York, 1948), chap. 3, esp. 33–43, 46; Graham D. Taylor and Patricia E. Sudnik,

Du Pont and the International Chemical Industry (New York, 1984), 80–81; Alfred D. Chandler, Jr., *Strategy and Structure: Chapters in the History of the American Industrial Enterprise* (Cambridge, Mass., 1962), 374–378; W.J. Reader, *Imperial Chemical Industries, A History,* vol. 1 (London, 1970), 469–470.

3. Ludwig F. Haber, *The Chemical Industry, 1900–1930* (Oxford, England, 1970), 279–309; Taylor and Sudnik, *Du Pont and the International Chemical Industry,* chap. 7.

4. Arthur Pine Van Gelder and Hugo Schlatter, *History of the Explosives Industry in America* (New York, 1927), 541–542; Chandler and Salsbury, *Pierre S. du Pont,* 340.

5. Van Gelder and Schlatter, *Explosives Industry,* 542–543.

6. Ibid., 450–451, 543; "The Aetna Company," *Hercules Mixer* 3 (June 1921), 153–156. Market share data from "Statement Showing High Explosives Sales by Competitors Including I.M.E. Reports and Estimates for Those Companies Who Do Not Report to I.M.E.," a mimeographed document at Hall of Records, Hercules Incorporated.

7. Van Gelder and Schlatter, *Explosives Industry,* 544.

8. Board Minutes, November 2 and 18, 1920.

9. "The Aetna Transaction," *Hercules Mixer* 3 (June 1921), 151–152.

10. "Statement Showing High Explosives Sales by Competitors." Hercules' share of the black powder market had declined during World War I, and the addition of Aetna's mills brought it up to roughly 15 percent.

11. *In the District Court of the United States for the District of Delaware. In Equity, No. 280. The United States of America, Petitioner,* v. *E.I. du Pont de Nemours & Company and Others, Defendants. Petition of the Hercules Powder Company* (May 4, 1921), *273 Federal Reporter,* 871.

12. Ibid., 871, 875; "Aetna Transaction," 155–156, 178.

13. Board Minutes, June 3 and September 14, 1921; Hercules Powder Company, Annual Report for 1921. In 1922, Hercules increased its authorized issue of preferred stock, which it exchanged for the Aetna bonds over the next several years. In 1927, when the last bonds were retired, the company dissolved the Hercules Explosives Corporation. Board Minutes, February 22, 1927.

14. "Readjustments Follow Aetna Purchase," *Hercules Mixer* 3 (July–August 1921), 182; "A.P. Van Gelder Retires," *Hercules Mixer* 4 (February 1922), 52.

15. Sales Department, Annual Report for 1923.

16. G.H. Markell, "Operating Department—Adjustment of Organization," April 18, 1922, document filed with Special Reports, 1922–1926, Hall of Records, Hercules Incorporated.

17. "Aetna Company," 154.

18. Leon W. Babcock interview, November 6, 1986.

19. Albert R. "Slick" Ely interview, November 12, 1987.

20. Operating Department (C.A. Bigelow), Annual Report for 1924.

21. Operating Department, Annual Reports for 1923 and 1924; S.G. Prickett, "Jefferson Plant," *Hercules Mixer* 4 (February 1922), 31–36, 58; "The Plant at McAdory," *Hercules Mixer* 6 (October 1924), 209–212, 232; Board Minutes, January 15, 1924.

22. "Plant at McAdory," 211; Operating Department, Annual Report for 1925.

23. Sales Department, Annual Report for 1924; Operating Department, Annual Report for 1928.

24. G.H. Markell, "The Status of Our Industry," *Hercules Mixer* 5 (November 1923), 240.

25. U.S. Department of Commerce, Bureau of the Census, *Historical Statistics of the United States: Colonial Times to 1970* (Washington, D.C., 1976), 589, 592, 598, 599, 602, 603, 606.

26. Hercules Powder Company, Annual Report for 1923.

27. Operating Department, Annual Report for 1927; Sales Department, Annual Report for 1924.

28. Sales Department, Annual Report for 1925; Operating Department, Annual Reports for 1925 and 1928. In 1928, total net profit on smokeless powder was $132,500, or 3 percent of the company's total net profit for the year.

29. Markell, "Status of Our Industry," 242, 240.

30. "*The Explosives Engineer*'s Trophies Awarded," *Hercules Mixer* 8 (June 1926), 141.

31. T.W. Bacchus, "A New Method of Blasting," *The Explosives Engineer* 1 (April 1923), 33–36; J. Barab, "Hercoblasting," *The Explosives Engineer* 2 (January 1924), 41–43; "Hercoblasting," *Hercules Mixer* 8 (January 1926), 5–6, 23.

32. Sales Department Report, December 1922; Operating Department, Annual Report for 1928.

33. See R.H. Dunham's remarks in Hercules Powder Company, Annual Report for 1924.

34. Operating Department, Annual Report for 1927; Ernest Symmes, "Nitric Acid by Ammonia Oxidation," *Hercules Mixer* 10 (July–August 1928), 170.

35. See R.H. Dunham's letter in Hercules Powder Company, Annual Report for 1926.

36. Williams Haynes, *Cellulose: The Chemical That Grows* (Garden City, N.Y., 1953), 67.

37. Operating Department, Annual Report for 1923.

38. Haynes, *Cellulose,* 58; Everett P. Partridge, "Developments in Nitrocellulose Production," *Industrial and Engineering Chemistry* 21, no. 11 (November 1929), 1044–1045.

39. Partridge, "Developments in Nitrocellulose Production," 1045.

40. Operating Department, Annual Report for 1922.

41. Operating Department, Annual Report for 1923; Leavitt N. Bent to the board, June 13, 1925, Special Reports, 1922–1926.

42. Sales Department Report, December 1922.

43. Treasurer's Report, December 1921. The exact numbers are $445,933.63 in sales, with a net loss of $75,861.96.

44. Sales Department Report, December 1923.

45. Alfred D. Chandler, Jr., ed., *Giant Enterprise: Ford, General Motors, and the Automobile Industry* (New York, 1964), 145–175.

46. Stuart W. Leslie, *Boss Kettering: Wizard of General Motors* (New York, 1983), 191.

47. Haynes, *Cellulose,* 210–211; Leslie, *Boss Kettering,* 191–194.

48. Haynes, *Cellulose,* 211.

49. Partridge, "Developments in Nitrocellulose Production," 1046–1047; A.B. Nixon, "The Manufacture of Nitrocellulose," *Hercules Mixer* 8 (March 1926), 56–57; J.B. Wiesel, "Nitocellulose: Recent Developments in Manufacture and Use," *Hercules Mixer* 10 (October 1928), 222–223; Haynes, *Cellulose,* 211.

50. Operating Department, Annual Report for 1924.

51. Operating Department, Annual Report for 1925.

52. A.B. Nixon to the board, June 13, 1925, Special Reports, 1922–1926.

53. Operating Department, Annual Report for 1926; J.H. Rile, Jr., "Estimated Future Sales of Nitrocotton as Affecting Plant Extension," January 14, 1926, Special Reports, 1922–1926, Operating Department, Annual Report for 1927; Partridge, "Developments in Nitrocellulose Production," 1047; Haynes, *Cellulose,* 211.

54. Operating Department, Annual Report for 1923.

55. Charles A. Higgins to Leavitt N. Bent, June 3, 1926, "The Virginia Cellulose Company," 5–6, letter contained in a notebook labeled "The Virginia Cellulose Company," box 166–23, Hall of Records, Hercules Incorporated (hereafter coded as VCC Notebook).

56. The February letter is described on p. 2 of P.B. Stull's letter to Hercules of April 10, 1926, VCC Notebook.

57. Haynes, *Cellulose,* 68–69.

58. Ibid., 71–72; Higgins to Bent, June 3, 1926, 3–4, VCC Notebook; "Our New Industry in Virginia," *Hercules Mixer* 8 (September 1926), 183–184.

59. P.B. Stull to Hercules Powder Company, April 10, 1926, VCC Notebook; Higgins to Bent, June 2, 1926, 11, VCC Notebook; "Our New Industry in Virginia," 184.
60. Higgins to Bent, June 3, 1926, 6–8, VCC Notebook; Board Minutes, July 9, 1926.
61. Virginia Cellulose Company, Annual Report for 1929.
62. Albrecht H. Reu, "A Brief History of the Brunswick, Georgia, Plant of Hercules, Incorporated" (1966), typescript in Hercules library, 1–2, 18–19, 24–28; N.P. Rood, "Hercules Operates the Yaryan Plants," *Hercules Mixer* 2 (June 1920), 171–172; Williams Haynes, *This Chemical Age: The Miracle of Man-made Materials* (Garden City, N.Y., 1942), 160.
63. Reu, "Brief History," 24.
64. Ibid., 27A.
65. Rood, "Hercules Operates the Yaryan Plants," 171–172.
66. C.M. Sherwood, "The Manufacture of Naval Stores from the Dead Wood of Southern Pines," *Chemical and Metallurgical Engineering* 25 (November 30, 1921), 994; Operating Department, Annual Report for 1924.
67. Sherwood, "Manufacture of Naval Stores from Dead Wood," 994; V.R. Croswell, "The Manufacture of Naval Stores," *Hercules Mixer* 3 (April 1921), 91–94, 117. Unlike the destructive distillation of wood (see chap. 5), steam treatment did not alter the chemical composition of the wood.
68. Croswell, "Manufacture of Naval Stores," 94, 117. This refining and finishing process was continuous at Brunswick and Gulfport under the Yaryan method but was performed in batches at Hattiesburg.
69. Hercules Powder Company, Annual Report for 1921; *Savannah Weekly Naval Stores Review and Journal of Trade* (hereafter cited as *Naval Stores Review*) (September 3, 1921); Reu, "Brief History," 27–28.
70. *Gamble's Naval Stores Year Book for 1928–1929* (Savannah, Ga., 1928), 32–33; Operating Department, Annual Report for 1924. Gulfport was operated during part of 1923.
71. Hercules Powder Company, Annual Report for 1923.
72. Operating Department Reports, January–May, 1922, and Annual Report for 1923.
73. Operating Department, Annual Report for 1923.
74. George H. Markell, untitled report to the board, July 9, 1920, filed with Special Reports, 1917–1921, Hall of Records, Hercules Incorporated; and *Naval Stores Review* (July 3, 1920) (cited by Markell in his report).
75. Williams Haynes, *American Chemical Industry,* vol. 4, *The Merger Era: 1923–1929* (New York, 1948), 169–186, 348–361.
76. Ibid., 169.
77. *Naval Stores Review* (August 20, 1921, and June 10, 1922).
78. *Naval Stores Review* (August 12, August 19, September 16, and November 18, 1922).
79. *Naval Stores Review* (November 3, 1923).
80. *Naval Stores Review* (March 12, 1927, and March 3, 1928).
81. Thomas D. Clark, *The Greening of the South: The Recovery of Land and Forest* (Lexington, Ky., 1984), 48–49, 67, 110; Willard Range, *A Century of Georgia Agriculture, 1850–1950* (Athens, 1954), 208–210; Thomas R. Cox et al., *This Well-Wooded Land: Americans and Their Forests from Colonial Times to the Present* (Lincoln, Nebr., 1985), 205, 210–211; Richard A. McLemore, ed., *A History of Mississippi,* vol. 2 (Jackson, 1973), 224.
82. Between 1920 and 1928, for example, the percentage of treated railroad ties increased from 41 to 76 percent.
83. Sherry H. Olson, *The Depletion Myth: A History of the Railroad Use of Timber* (Cambridge, Mass., 1971), 121–132, 182–185.
84. Ibid., 142, 144, 178–180.

85. *Naval Stores Review* (June 22, 1929).
86. Leavitt N. Bent, "The Value of Research to Naval Stores Industry," *Naval Stores Review* (August 31, 1929).
87. Low naval stores prices during the early 1920s discouraged production, causing demand for rosin and turpentine to exceed supply by the 1925–1926 growing season. *Gamble's Naval Stores Year Book for 1928–1929,* 32–33.
88. Nathan Rosenberg, "Technological Interdependence in the American Economy," *Technology and Culture* 20 (January 1979), 32.
89. Jacob Schmookler, *Invention and Economic Growth* (Cambridge, Mass., 1966), 202.
90. Reu, "Brief History," 30; Operating Department Reports for 1922.
91. Operating Department, Annual Report for 1925.
92. Operating Department, Annual Report for 1925; Board Minutes, 1925 and 1926, passim.
93. Operating Department, Annual Report for 1925.
94. Haynes, *This Chemical Age,* 158; Reu, "Brief History," 37.
95. Operating Department, Annual Report for 1925; Reu, "Brief History," 37; V.R. Croswell and R. Rockwell, "Some Engineering Features of the Naval Stores Industry," paper presented at the December 8–10, 1930, meeting of the American Institute of Chemical Engineers, 3–7.
96. The innovation, in turn, prompted modifications at other points in the wood-handling system. To handle the large, intact stumps, a "cut-up plant" was constructed at Brunswick and the openings to the hogs were eventually widened. Haynes, *This Chemical Age,* 159; Reu, "Brief History," 37; Croswell and Rockwell, "Some Engineering Features," 3–7; Board Minutes, November 29, 1927, April 10, July 10, and October 3, 1928.
97. Operating Department, Annual Report for 1926; Croswell and Rockwell, "Some Engineering Features," 3–5.
98. *Naval Stores Review* (September 3, 1921). Similarly, in an internal memorandum to J.T. Skelly, Lockwood advocated the resumption of production "to maintain our reputation as a dependable source of supply or lose the valuable good will built up" (J.E. Lockwood to J.T. Skelly, "Sales Policy," Sales Department Report, August 1921).
99. *Naval Stores Review* (February 4, 1922).
100. *Naval Stores Review* (September 10, 1921).
101. Except for special citations, the following discussion of turpentine marketing strategies is based on N.S. Greensfelder and J.G. Pollard, Jr., "A Method of Marketing Turpentine That Begins with the Customer's Point of View," National Industrial Advertisers Association (New York, 1929), reprinted as "How Hercules Turpentine Is Marketed," in *Hercules Mixer* 11 (January 1929), 3–8; and N.S. Greensfelder, "How We Merchandise Turpentine," *Hercules Mixer* 8 (January 1926), 7–9, 24.
102. *Naval Stores Review* (December 17, 1921, August 26 and December 23, 1922).
103. *Naval Stores Review* (April 2, 1927, April 18 and September 19, 1925, March 20, 1926, and December 21, 1929).
104. "With Turp and Tine in Japan," *Hercules Mixer* 8 (September 1926), 186.
105. "Hercules Steam-Distilled Pine Oil," *Hercules Mixer* 10 (September 1928), 195–198; George C. O'Brien, "How and Where We Sell Our Naval Stores," *Hercules Mixer* 8 (May 1926), 105–107.
106. Haynes, *American Chemical Industry,* 4:356.
107. "Hercules Steam-Distilled Pine Oil," 195–198.
108. Sales Department, Annual Report for 1925. Ad headlines are from various issues of the *Naval Stores Review* in the 1920s.
109. "Advertising Appropriation," November 24, 1928, Special Reports, 1926–1928; *Naval Stores Review* (July 3 and 17, 1926, and January 12 and 26, 1929).
110. Irvin W. Humphrey, "Our Naval Stores Experimental Work," *Hercules Mixer* 8 (March 1926), 53.

111. Reu, "Brief History," 27, 31.
112. Haynes, *This Chemical Age,* 163; Operating Department, Annual Report for 1924.
113. Haynes, *This Chemical Age,* 163; Operating Department, Annual Report for 1923.
114. Reu, "Brief History," 31–32; Humphrey, "Our Naval Stores Experimental Work," 53–54, 76.
115. Operating Department, Annual Report for 1926.
116. These were patents Nos. 1,715,083–1,715,088. Hercules also applied for British patents for these processes in late 1926. *Naval Stores Review* (November 13, 1926).
117. U.S. patent No. 1,715,083.
118. *Naval Stores Review* (November 13, 1926); Haynes, *This Chemical Age,* 160.
119. Research Department, Annual Report for 1927; Board Minutes, April 15, 1927; Reu, "Brief History," 38–39; *Naval Stores Review* (March 17, 1928); Arthur Langmeier, "Naval Stores," *Hercules Mixer* 10 (August 1928), 7.
120. *Naval Stores Review* (March 17, 1928); Hercules Powder Company, Annual Report for 1928.
121. *Gamble's Naval Stores Year Book for 1928–1929,* 48–49.
122. U.S. patent No. 1,715,084.
123. *Naval Stores Review* (January 1, 1924); *Gamble's Naval Stores Year Book for 1928–1929,* 92–93.
124. Reu, "Brief History," 41.
125. Sales Department, Annual Report for 1922; Naval Stores Department, Annual Report for 1928.
126. *Gamble's Naval Stores Year Book for 1928–1929,* 92–93; Haynes, *American Chemical Industry,* 4:356; *Naval Stores Review* (February 3, 1923, and December 1, 1928).
127. Haynes, *American Chemical Industry,* vol. 6, *The Chemical Companies* (New York, 1954), 303, and 4:356.
128. Naval Stores Department, Annual Report for 1928; *Gamble's Naval Stores Year Book for 1928–1929,* 30, 32–33; *Naval Stores Review* (December 10, 1927).
129. Naval Stores Department, Annual Report for 1928.
130. Experimental Station, Annual Report for 1928; Safety and Service Division, Annual Report for 1928.
131. Development Department Reports for 1927.
132. In 1923, Hercules created a holding company, the North Star Explosives Company, to take over and operate the Prescott plant. The facility, valued at about $20,000 at that time, continued to operate until 1937, when it was shut down permanently. Board Minutes, July 24, 1923; *Hercules Mixer* 20 (January 1938), 7.
133. Van Gelder and Schlatter, *Explosives Industry,* 714–715; CME Reports for December 1926 and January 1929.
134. Anon., *N.V. Hercules Powder Company: The First Twenty-Five Years* (n.p. [Rotterdam?], 1950), 5.
135. *Hercules Mixer* 32 (July–August 1950) (special issue commemorating the twenty-fifth anniversary of N.V. Hercules Powder Company), 9, 11.
136. Hercules Powder Company, Annual Report for 1928.
137. "George H. Markell," *Hercules Mixer* 8 (April 1926), 79; Board Minutes, March 16 and 24, 1926. Leon Babcock, who knew Markell, recalls him as "a great guy. He was the Lee Iacocca of Hercules. Really great." Babcock interview, November 6, 1986.
138. In 1928, monthly departmental reports were generally discussed in the following order: Treasurer; Black Powder Operating; Purchasing; Traffic; Legal; Operating (everything but Black Powder); Experimental Station; Sales; Development; CME; VCC; and N.V. Hercules. See Board Minutes, August 7–21, 1928.
139. Robert J. Coolahan, "Soluble Cotton and Its Importance Today," *Hercules Mixer* 7 (March 1925), 53.
140. Board Minutes, September 4, 1928.
141. Anon., "Changes in Organization," *Hercules Mixer* 10 (October 1928), 250.

142. Chandler, *Strategy and Structure,* 376.
143. Hercules Powder Company, Annual Report for 1927.

Chapter 7

1. Thomas D. Darlington, "The New Era of Synthesis," *Wilmington* (July 1930), 28.

2. Williams Haynes, *American Chemical Industry,* vol. 3, *The World War I Period: 1912–1922* (New York, 1945), 353–370, esp. 354–355. We are grateful to Professor John Kenly Smith, Jr., for drawing our attention to the phenomenon of the chemicalization of industry.

3. Alfred D. Chandler, Jr., *Strategy and Structure: Chapters in the History of the American Industrial Enterprise* (Cambridge, Mass., 1962), 374–378.

4. Requests above this limit required the approval of the executive committee in concert with the finance committee, or of the board of directors as a whole.

5. Board Minutes, May 29 and June 26, 1929; C.C. Hoopes to the finance committee, July 19, 1929, with Finance Committee Minutes, July 1929 (hereafter cited as Finance Committee Minutes, with date). (Minutes of the Finance and Executive Committees, created in 1928, are bound with the Board Minutes; these volumes are housed in the Secretary's vault at Hercules Incorporated.)

6. C.C. Hoopes, "Indirect Expenses," March 20, 1930, filed with Special Reports, 1930; Board Minutes, February 26, 1930.

7. C.C. Hoopes and E.B. Morrow, "Auxiliary Department Reports to the Executive Committee," March 6, 1930, Special Reports, 1930.

8. David A. Hounshell and John Kenly Smith, Jr., *Science and Corporate Strategy: R&D at Du Pont, 1902–1980* (New York, 1988), introduction, esp. 5–7.

9. Ibid., 5–6.

10. Ibid., 136, and chap. 5 generally.

11. Harry Kaiser to G.M. Norman, October 13, 1928, and R.E. Zink to G.M. Norman, October 13, 1928, Special Reports, 1928–1929; cf. Helen C. Kelly, "The History of Our Experiment Station," *Hercules Mixer* 32 (January–February 1950), 4–5.

12. L.N. Bent, C.A. Bigelow, A.B. Nixon, and G.M. Norman [to the executive committee?] (n.d.), included with G.E. Ramer, "Project 2985: Study of Experimental Station Relocation," April 2, 1929. The date of the committee report was probably December 12, 1928, according to a letter from G.E. Ramer to G.M. Norman, March 13, 1930, Special Reports, 1930.

13. Messrs. Bent, Bigelow, Nixon, and Norman [December 12, 1928?]; G.M. Norman, "Why Relocate the Experimental Station," *Hercules Mixer* 12 (May 1930), 108–109.

14. Board Minutes, July 31, 1929; George M. Norman papers, Hagley Museum and Library, accession 1073; G.E. Ramer to G.M. Norman, March 13, 1930, Special Reports, 1930.

15. Dunham's speech at the dedication is printed in "Hercules New Experimental Station," *Hercules Mixer* 12 (July–August 1929), 157.

16. The word *experimental* in the unit's name was changed to *experiment* when the new station opened in 1931. According to Marjorie Chris, the change was made to avoid the suggestion that the station itself was a corporate experiment. Marjorie D. Chris interview, July 21, 1987.

17. "Hercules Experiment Station and Research Laboratories, Hercules, Del.," *Hercules Mixer* 14 (February 1932), 44–45; "Fortifying a Technical Institution at Its Foundation," *Chemical and Metallurgical Engineering* 38, no. 7 (July 1931), 384–385.

18. Technical Department, Annual Report for 1929. Cf. the almost identical language in H.E. Kaiser to G.M. Norman, March 12, 1930, Special Reports, 1930.

19. [H.E. Kaiser?] to G.M. Norman, February 17, 1928, "1927 Annual Summary," filed with records of the technical development committee, Hall of Records, Hercules Incorporated.

20. John M. "Jack" Martin interview, September 2, 1988.
21. RI reports from the Kenvil Experimental Station (1915–1932) are housed at the Hall of Records, Hercules Incorporated.
22. G.M. Norman to executive committee, July 7, 1930, Special Reports, 1930 (duplicate in Special Reports, 1931).
23. Technical Department Report (covering experimental work), Fourth Quarter 1931.
24. The following is based on O.A. Pickett, "Hercules Research—A Critical Review," April 4, 1933, copy in George M. Norman papers, Hagley Museum and Library, accession 1150.
25. Cf. Malcolm Ross, ed., *Profitable Practice in Industrial Research* (New York, 1932), passim.
26. H.E. Kaiser to the technical development committee, June 26, 1933, 3ff, records of the technical development committee, Hall of Records, Hercules Incorporated.
27. "Kaiser Resigns Research Post," *Hercules Mixer* 15 (July–August 1933), 131; "Experiment Station Changes," *Hercules Mixer* 16 (March 1934), 52.
28. Finance Committee Minutes, October 29, 1928; Board Minutes, October 31, 1928; Hercules Powder Company, Annual Report for 1929.
29. Stanley Lebergott, *The Americans: An Economic Record* (New York, 1984), 444; Alfred D. Chandler, Jr., and Richard S. Tedlow, *The Coming of Managerial Capitalism: A Casebook on the History of American Economic Institutions* (Homewood, Ill., 1985), 582–584.
30. Executive Committee Minutes, May 12 and December 1, 1931, and May 25, 1932 (hereafter cited as Executive Committee Minutes, with date). (See above, note 5.)
31. Hercules Powder Company, Annual Reports for 1930–1932.
32. Executive Committee Minutes, April 15, 1930; Board Minutes, May 27, 1931.
33. Most of Hercules' lacquer sales were made to Du Pont, which in turn relied heavily on General Motors for sales; see Cellulose Products Department, Annual Reports for 1930–1932.
34. Executive Committee Minutes, October 7, 1930.
35. Cellulose Products Department, Annual Report for 1932.
36. "History Resume," in Naval Stores Department, Annual Report for 1932.
37. These figures do not include operating expenses, including the exorbitant amount spent on "abnormal repairs" for the plants.
38. "History Resume," in Naval Stores Department, Annual Report for 1932; Executive Committee Minutes, May 21, 1929; Development Department Report, November 1926.
39. Executive Committee Minutes, September 10, 1930; Development Department Reports, November 1926, March and August 1927, August and October 1928, January, February, and August 1929, January–February 1930; Hercules Powder Company, Annual Report for 1928.
40. Another reason cited was that Du Pont, a large consumer of camphor, had stayed out of the synthetic camphor business. Executive Committee Minutes, October 6, 1931.
41. Executive Committee Minutes, April 2 and September 10, 1930.
42. Development Department Report, January–February 1930.
43. Albrecht H. Reu, "A Brief History of the Brunswick, Georgia, Plant of Hercules, Incorporated" (1966), 48, typescript in Hercules library, Hercules Incorporated. Reu is uncharacteristically imprecise about the timing of this episode, and we have not located corroborating accounts. Although he writes that it occurred after the company had been in the naval stores business for thirteen years (thus in 1933), he also states more specifically that the scaling back of the Brunswick operations that followed the meeting with Kloss and Lambert began on October 22, 1931.
44. Ibid., 48–49.
45. Executive Committee Minutes, September 15, 1931. The three men also owned

19,997 shares of the Superior Pine Products Co., a Georgia timber company, but these assets were not included in the original offer.

46. Ibid.; Hercules Powder Company, Annual Report for 1931.

47. *Hercules Mixer* 17 (March 1935), 52; *Naval Stores Review* (March 5, 1932 [see chap. 6, note 69]).

48. *Naval Stores Review* (March 5, 1932); "Paper Makers Chemical Corporation [PMC] Consolidated with Hercules Powder Company," *Hercules Mixer* 13 (November 1931), 188.

49. "[PMC] Consolidated with Hercules"; *Naval Stores Review* (February 2, 1924).

50. "PMC in Canada," *Hercules Mixer* 18 (June 1936), 136.

51. "[PMC] Consolidated with Hercules," 188–189; *Naval Stores Review* (June 23, 1928, and March 5, 1932).

52. Paper Makers Chemical Corporation Associates, *Superior Facts,* 1 (June 1928), 11.

53. David C. Smith, *History of Papermaking in the United States (1691–1969)* (New York, 1970), 372. Because gum rosin, uncharacteristically, was selling for less than wood rosin, PMC's consumption of wood rosin declined three times faster than its use of gum rosin during 1931. PMC Department, Annual Report for 1931.

54. Naval Stores Department, Annual Report for 1929.

55. Hercules Powder Company, Annual Report for 1931.

56. U.S. Department of Commerce, Bureau of the Census, *Historical Statistics of the United States: Colonial Times to 1970,* vol. 2 (Washington, D.C., 1975), 668, 673; Louis Tillotson Stevenson, *The Background and Economics of American Papermaking* (New York, 1940), 79–80; Smith, *History of Papermaking,* 372, 407–408.

57. See Williams Haynes, *American Chemical Industry,* vol. 5, *Decade of New Products* (New York, 1954), chap. 2.

58. Typed transcripts of meetings with Edie from April 1933 to November 1937 are bound in a volume entitled "Discussions with Dr. Edie," in the Hercules library.

59. Hercules Powder Company, Annual Report for 1933.

60. Darlington, "The New Era of Synthesis," 28.

61. Williams Haynes, *Cellulose: The Chemical That Grows* (Garden City, N.Y., 1953), 346–347, 351.

62. Ibid., 118; D.C. Coleman, *Courtaulds: An Economic and Social History,* vol. 2, *Rayon* (Oxford, 1969), 14–17.

63. Haynes, *Cellulose,* 142; Coleman, *Courtaulds,* 2:178–184.

64. Virginia Cellulose Department, Annual Report for 1936.

65. Executive Committee Minutes, July 20, 1937.

66. L.N. Bent, "Development of Cellulose Acetate," February 14, 1936, Special Reports, 1936.

67. Development Department Reports, March–May 1936; Executive Committee Minutes, May 22 and June 16, 1936; James A. Lee, "Cellulose Acetate by Hercules," *Chemical and Metallurgical Engineering* 45, no. 8 (August 1938), 404–408; "Cellulose Acetate: A Superior Product with a Brilliant Future," *Hercules Mixer* 20 (October 1938), 228–231, 250.

68. Board Minutes, July 27, 1937, July 27, 1938, and January 25, 1939; Lee, "Cellulose Acetate by Hercules," 405; Cellulose Products Department, Annual Report for 1939.

69. Hercules Powder Company, *A Survey of the Paint, Varnish, and Lacquer Industry* (Wilmington, 1945), 7. The following discussion draws heavily on this document.

70. M.G. Milliken, "1936 Greatest Year for Cellulose Derivatives," *Hercules Chemist* 1 (1937), unpaginated.

71. Cellulose Products Department, *LacQuotes,* 1954. (*LacQuotes* was a Hercules periodical.)

72. G.M. Norman to the board, January 14, 1937, filed with Development Department Reports for 1936.

73. Development Department Report, Second Quarter (March–May) 1932; R.A.

Coolahan, "Tornesit: Its Properties and Applications," *Chemistry and Industry* (July 20, 1934), unpaginated; Cellulose Products Department, Annual Reports for 1935 and 1936.

74. Naval Stores Department, Annual Report for 1939.

75. Between 1929 and 1939, Hercules' share of the wood turpentine market climbed 7 percent, while Newport's plummeted 15 percent. Naval Stores Department, Annual Reports for 1929 and 1940.

76. J.E. Lockwood, "The Outlook in Naval Stores," *Hercules Mixer* 14 (January 1932), 9; Naval Stores Department, Annual Reports for 1929, 1934, and 1935; P.W. Meyeringh to the board, filed with Special Reports to the Directors, 1930, Hall of Records.

77. Naval Stores Department, Annual Reports for 1932 and 1939; Meyeringh to the board (see above, note 76).

78. *Gamble's International Naval Stores Year Book for 1939–1940* (Savannah, Ga., 1939), 4, 48, 54; *Gamble's International Naval Stores Year Book for 1940–1941* (Savannah, Ga., 1940), 10, 12, 32; Naval Stores Department, Annual Report for 1929.

79. *Gamble's International Naval Stores Year Book for 1939–1940,* 46, 61–64. The gum distillers had already begun to follow the lead of the wood naval stores producers—with some success—through the "gum turpentine advertising campaign" and the use of modern packaging during the 1936–1937 season. *Naval Stores Review* (May 11, 1940).

80. Williams Haynes, *This Chemical Age: The Miracle of Man-made Materials* (New York, 1942), 160. For a description of Newport's fuller's earth process, see R.C. Palmer, "Turpentine and Rosin from Wood Wastes by Steam and Solvent Process," *Industrial and Engineering Chemistry* 26 (July 1934), 703–706.

81. Munn was utilized primarily in soap making, Tenex in paints and varnishes. Other Newport rosins developed by 1934 included Solros (FF) (Dosol, No. 45, Dover) Newport FF, and Triangle FF. *Naval Stores Review* (June 22, 1929, March 5, 1932, September 17, 1932, December 3, 1932, June 3, 1933, and January 6, 1934); Haynes, *This Chemical Age,* 160.

82. The suit concerned patent No. 1,715,085 issued May 28, 1929. *Naval Stores Review* (February 5, 1938).

Some indication of the uncertain deterrent effect of patents can be seen in a 1929 report by Leavitt Bent:

> We have been hesitant to duplicate Solros [Newport's soluble, noncrystallizing, non-precipitating FF wood rosin] on account of the patent situation and have attempted to meet this competition by #20 and limed rosin. We are not yet sure whether it would be better for us to duplicate Solros or to continue in our efforts to develop other products suitable for this class of trade and which will not be an infringement of the Newport patent.

Naval Stores Department, Annual Report for 1929.

83. The original members of the export association were Dixie Pine Products Co., Phoenix Naval Stores Co., and Mackie Pine Products Co. Continental and Newport did not join. The new organization's office was in the Delaware Trust Building, and A.B. Nixon was its secretary. Naval Stores Department, Annual Reports for 1929 and 1935; *Naval Stores Review* (September 28, 1935).

84. The naval stores products were Abalyn® (liquid resin), Brisgo (thermoplastic compound), Cabinol (terpene chemicals), Dianex (pine oil), Dri-Tex (pine oil base compound), Hercolyn® (white liquid resin). Hybrex (pine oil compounds), Petrex® (synthetic resin base), Solvenol (terpene derivative), and Vinsol® (black insulating resin); the paper chemicals were dry size and Paracol (a colloidal wax emulsion). "New Hercules Chemicals," *Hercules Mixer* 16 (January 1934), 10.

85. The turning point came in early 1936, when the Experiment Station discontinued semiplant production of sulphuric acid polymerized rosin and Hyex rosin (made by hydrogen exchange reaction) and expanded its hydrogenation work. Technical Department Reports, First and Fourth Quarters 1935, and First Quarter 1936.

86. *Naval Stores Review* (October 22, 1938); "Hydrogenated Rosin Now Available Commercially," *Hercules Chemist* 5 (1938); "Step along with Staybelite," *Hercules Mixer* 22 (July-August 1940), 213.

87. Vinsol® was nontacky, nonoxidizing, and insoluble in water, petroleum, and other hydrocarbon solvents but soluble in alcohol, acetone, esters, and related solvents. Naval Stores Department, Annual Report for 1934; Reu, "Brief History," 55–56; Hercules Powder Company, Annual Report for 1950; George Bossardet interview, September 9, 1987; "New Hercules Chemicals," *Hercules Mixer* 16 (January 1934), 10; "Vinsol," *Hercules Chemist* 0 (1936); "Vinsol," *Hercules Chemist* 1 (First Quarter 1937); "Vinsol," *Hercules Chemist* 2 (Second Quarter 1937); "Vinsol," *Hercules Chemist* 3 (1939).

88. Reu, "Brief History," 55–56; Bossardet interview, September 9, 1987; Technical Department Reports, Fourth Quarter 1934, Second Quarter 1935, and Fourth Quarter 1937.

89. In 1939, Hercules sold 50,900 standard units of Vinsol® to eight major markets. *Naval Stores Review* (February 22, 1936); "Modern Conveniences Depend on Electricity," *Hercules Mixer* 21 (June 1939), 156; Technical Deparment Report, First Quarter 1938; Naval Stores Department, Annual Report for 1939.

90. *Chemical Markets* 32 (April 1933), 352; Naval Stores Department, Annual Reports for 1929 to 1936.

91. Naval Stores Department, Annual Report for 1929.

92. Hercules Powder Company, Annual Report for 1950.

93. *Hercules Mixer* 18 (February 1936), 54; "Synthetics Department Holds Three-Day Session," *Hercules Mixer* 22 (January 1940), 10.

94. *Naval Stores Review* (April 9, 1932); "Abalyn and Hercolyn," *Hercules Chemist* 0 (1936).

95. "Petrex," *Hercules Chemist* 2 (Second Quarter 1937).

96. Synthetics Department, Annual Report for 1937; "Petrex 22," *Hercules Chemist* 1 (First Quarter 1937).

97. Its tendency to oxidize made Petrex® a poor surface coating when exposed to the elements.

98. Paint and varnish makers could also reutilize kettles that had been idled when old varnishes were replaced by alkyds instead of investing in the new, closed kettles that would be needed for synthetic resins.

99. "Synthetics Department Holds Three-Day Session," 10.

100. John M. "Jack" Martin interview, October 9, 1986.

101. *Hercules Mixer* 14 (March 1932), 65.

102. Ibid.

103. "[PMC] Consolidated with Hercules," 190.

104. Executive Committee Minutes, April 24, 1934, September 17, 1935, and January 14, 1936.

105. PMC Department, Annual Report for 1936.

106. Harry N. Scheiber, Harold G. Vatter, and Harold U. Faulkner, *American Economic History* (New York, 1976), 387–388; PMC Department, Annual Report for 1937.

107. John R.L. Johnson, Jr., interview, January 28, 1987. After taking his leave, Lawrence set up his own office at Kalamazoo to continue working the naval stores and paper industries and devoted more of his attention to several businesses in the South and Midwest in which he held interests. In the fall of 1938, he became president and general manager of the Bryant Paper Co. of Kalamazoo. *Naval Stores Review* (May 5 and 22, 1937, and September 10, 1938).

108. PMC Department, Annual Report for 1937.

109. Ibid.

110. PMC Department, Annual Report for 1938.

111. "The Legend of Jimmy Foxgrover," *Encore* (October 1982), 5–7; Maurice J. Kelly, Robert I. Mason, and Charles A. Twigg interview, April 5, 1988; Robert J. Leahy interview, April 27, 1988.
112. Explosives Department, Annual Reports for 1928 and 1939. Cf. E.M. Symmes, "Processes, Products, and Personnel Link Explosives Manufacture to Other Chemical Engineering Industries," *Chemical and Metallurgical Engineering* (April 1928), 234–235.
113. Hercules Powder Company, Historical Report of Missouri Ordnance Works [government-owned, Hercules Powder Company–operated], Louisiana, Missouri, August 18, 1941 through 1945, 13–14 [hereafter cited as Historical Report of Missouri Ordnance Works].
114. For an account of ammonia synthesis from natural gas in the 1920s and 1930s, see Kendall Beaton, *Enterprise in Oil: A History of Shell in the United States* (New York, 1957), 522–523; Peter H. Spitz, *Petrochemicals: The Rise of an Industry* (New York, 1988), 84–85.
115. Spitz, *Petrochemicals,* 14–16. For the Shapleigh furnacing process, see "A Work-Horse in the Chemical Industry," *Hercules Chemist* 27 (June 1956), 14–16; Engineering Department, Reports for the Fourth Quarter 1936, and Third Quarter 1938.
116. Historical Report of Missouri Ordnance Works, 15; Executive Committee Minutes, January 12 and September 14, 1937, February 22 and August 16, 1938, and January 24 and February 21, 1939.
117. Norman to Executive Committee, July 7, 1930, filed with Special Reports for 1930; Explosives Department, Annual Reports for 1934 and 1939.
118. "Hercules Patent Department and How It Grew," *Hercules Mixer* 21 (October 1939), 292–297; Patent Department, Annual Report for 1937.
119. "The Hercules Powder Company," *Fortune* (September 1935), 57.
120. Board Minutes, May 29, 1929, March 23, 1932, and November 25, 1936.

Chapter 8

1. Hercules Powder Company, Historical Report of Radford Ordnance Works (government-owned, Hercules Powder Company–operated), Radford, Virginia, August 16, 1940, through 1945, 1:12–14, box 179–44, Hall of Records, Hercules Incorporated (hereafter cited as Historical Report of Radford Ordnance Works); Constance McLaughlin Green, Harry C. Thomson, and Peter Roots, *The Ordnance Department: Planning Munitions for War* (Washington, D.C., 1955), 42–48.
2. H.C. Englebrecht and F.C. Hanighen, *Merchants of Death: A Study of the International Armament Industry* (New York, 1934), passim, but esp. chap. XIII.
3. Other powder companies shared this feeling. See Harry C. Thomson and Lida Mayo, *The Ordnance Department: Procurement and Supply* (Washington, D.C., 1960), 32; Henry L. Stimson and McGeorge Bundy, *On Active Service in Peace and War* (New York, 1948), 353.
4. Green, Thomson, and Roots, *Ordnance Department: Planning,* 53; John R.L. Johnson comment, Hercules History Advisory Committee Minutes, June 1988; John E. Wiltz, *In Search of Peace: The Senate Munitions Inquiry* (Baton Rouge, La., 1963).
5. Historical Report of Radford Ordnance Works, 1:14–18.
6. Explosives Department, Annual Report for 1939; Executive Committee Minutes, August 30, 1938, January 31, February 14, and March 7, 1939, and February 20, 1940; "Herculite R.L. Stern's Distinguished Service Cited by Army Award," *Hercules Mixer* 25 (November 1943), 341.
7. Executive Committee Minutes, September 5 and 19, October 10, and November 14, 1939; Explosives Department, Annual Report for 1939.
8. Board Minutes, March 22, 1940 (a copy of the contract is inserted).
9. Executive Committee Minutes, May 21, June 11, and June 25, 1940.

10. "Hercules Concern Expands Facilities," *New York Times,* September 13, 1940, 17; Board Minutes, August 8, 1940; Executive Committee Minutes, August 15, 1950.

11. Unless otherwise noted, the following account is based on Hercules Powder Company, Safety and Service Department, Major Works Accident and Fire Report No. 713, Smokeless Powder Solvent Recoveries Explosion and Fire [at Kenvil Plant], September 12, 1940, a report by Henry N. Marsh to W.R. Ellis, February 10, 1941 (hereafter cited as Major Works Accident and Fire Report); and *New York Times,* September 13, 1940, 1, 16–17.

12. Quoted in *Wilmington Morning News,* September 13, 1940, 6.

13. Major Works Accident and Fire Report, 13–14, and blueprint No. 1.

14. Historical Report of Radford Ordnance Works, 1:66–71.

15. Monetary totals from Major Works Accident and Fire Report, and letter of H.I. Sturtevant to L.W. Babcock, March 5, 1953 (inserted between pp. 24 and 25 of copy No. 8 of the report); lost production statistics from Explosives Department, Annual Report for 1940.

16. U.S. Army Armament Munitions Chemical Command and Hercules Incorporated, *Radford Army Ammunition Plant* (n.p., 1987), 2; Thomson and Mayo, *Ordnance Department: Procurement,* 32.

17. Donald M. Nelson, *Arsenal of Democracy: The Story of American War Production* (New York, 1947), 92; William E. Leuchtenburg, *Franklin D. Roosevelt and the New Deal, 1932–1940* (New York, 1963), 300; Green, Thomson, and Roots, *Ordnance Department: Planning,* 41, 67; Lt. Gen. Levin H. Campbell, Jr., *The Industry-Ordnance Team* (New York, 1946), 11–12.

18. U.S. Army and Hercules Incorporated, *Radford Plant,* 3; Thomson and Mayo, *Ordnance Department: Procurement,* 105. General Campbell, who served as chief of ordnance for the U.S. Army after 1943, puts the number of GOCO plants built during World War II at seventy-three; see Campbell, *Industry-Ordnance Team,* 102.

19. Historical Report of Radford Ordnance Works, 1:viii; Thomson and Mayo, *Ordnance Department: Procurement,* 32.

20. Thomson and Mayo, *Ordnance Department: Procurement,* 113; R. Elberton Smith, *The Army and Economic Mobilization* (Washington, D.C., 1959), 282, 297–298.

21. Historical Report of Radford Ordnance Works, 1:100.

22. "The Army Salutes Radford," *Hercules Mixer* 23 (April 1941), 115.

23. Historical Report of Radford Ordnance Works, 2:708.

24. Official histories of these plants are found in boxes 179–44 to 179–47, Hall of Records, Hercules Incorporated. Brief histories of each operation are found in the "homecoming" issue of the *Hercules Mixer* (October 1945). Hercules also made TNT at Virginia Ordnance Works, a satellite plant to Radford, in 1941 and 1942.

25. Percentage figures obtained by comparing Hercules' total production (Hercules Powder Company, Historical Report of Missouri Ordnance Works [government-owned, Hercules Powder Company–operated], Louisiana, Missouri, August 18, 1941 through 1945 [hereafter cited as Historical Report of Missouri Ordnance Works], 6) with figures for total national production in Campbell, *Industry-Ordnance Team,* 253–254; Smith, *Army and Economic Mobilization,* 501.

26. Historical Report of Missouri Ordnance Works, 5–6, and Table 8.x.

27. Hercules Powder Company, Historical Report of Sunflower Ordnance Works (government-owned, Hercules Powder Company–operated), Lawrence, Kansas, May 11, 1942 through 1945, 649–650 (hereafter cited as Historical Report of Sunflower Ordnance Works).

28. Historical Report of Radford Ordnance Works, 115–118, 154.

29. Ibid., 419–428.

30. "Hercules Army-Navy 'E' Award Winners," *Hercules Mixer,* 27 (October 1945), 261.

31. Duard H. "Tex" Little interview, July 22, 1987; Paul Steele interview, November 10, 1988. For similar stories, see Werner C. Brown interview, October 8, 1986; John M.

"Jack" Martin interview, October 9, 1986; John Ryan interview, October 10, 1986; Alexander F. Giacco interview, August 30, 1988, and *Hercules Mixer,* volumes 23 to 27.

32. Martin interview, October 9, 1986.

33. Green, Thomson, and Roots, *Ordnance Department: Planning,* 351–352; Martin interview, October 9, 1986.

34. Green, Thomson, and Roots, *Ordnance Department: Planning,* 356–357.

35. Henry N. Marsh, "The Development and Production of Rocket Propellants in World War II," *Chemical Industries* (July 1945); Green, Thomson, and Roots, *Ordnance Department: Planning,* 353–354.

36. Historical Report of Sunflower Ordnance Works, 44, 698–700, 870–873; "Important Invasion Weapon Made by Hercules: Rocket Powder a 'Secret Weapon' of Allies," *Hercules Mixer* 26 (June 1944), 184, 213; Lillian R. Fisher, "Sunflower—A Review," *Hercules Mixer* 27 (December 1945), 350. Hercules also made rocket powder at Badger Ordnance Works. See also Exhibit 8.1.

37. Williams Haynes, *The Chemical Front* (New York, 1943), 5; Maj. Gen. Gladeon M. Barnes, *Weapons of World War II* (New York, 1947), 73–74; Green, Thomson, and Roots, *Ordnance Department: Planning,* 367; Historical Report of Radford Ordnance Works, vol. 2, 1334–1336.

38. The literature on the VT fuze is substantial. The following discussion is based on Ralph B. Baldwin, *The Deadly Fuze: Secret Weapon of World War II* (San Rafael, Calif., 1980), passim; Green, Thomson, and Roots, *Ordnance Department: Planning,* 361–366; and two articles in *Hercules Mixer* 27: "Disclosing a Secret Weapon" (November 1945), 288–289, and "This Secret Weapon Panicked the Enemy—Hercules Made Millions of Detonators" (December 1945), 337–338.

39. "This Secret Weapon Panicked the Enemy," 338.

40. Baldwin, *Deadly Fuze,* 279–280.

41. "This Secret Weapon Panicked the Enemy," 337.

42. Daniel J. Kevles, *The Physicists: The History of a Scientific Community in Modern America* (1977; reprint, Cambridge, Mass., 1987), chap. XX, esp. 307–308.

43. Harold G. Vatter, *The U.S. Economy in World War II* (New York, 1985), 21. Hercules' experience mirrored that of the industry as a whole. See John Kenly Smith, Jr. , "World War II and the Transformation of the American Chemical Industry," in E. Mendelsohn, M.R. Smith, and P. Weingart, eds., *Science, Technology and the Military,* Sociology of the Sciences Yearbook, 12 (Boston, 1988), 307–322.

44. Cellulose Products Department, Annual Reports for 1940–1942; "Defying the Elements: Hercules Clorafins Help Protect Against Fire, Water, Earth, Air," *Hercules Chemist* 14 (March 1945), 7–8.

45. Cellulose Products Department, Annual Reports for 1941, 1943, and 1944; "Ethyl Cellulose Plastics," *Hercules Chemist* 12 (March 1943), 14–16; Board Minutes, June 7, 1944.

46. Cellulose Products Department, Annual Reports for 1944 and 1945; Williams Haynes, *Cellulose: The Chemical That Grows* (Garden City, N.Y., 1953), 250.

47. Executive Committee Minutes, March 3 and June 30, 1942, and November 2, 1943; Cellulose Products Department, Annual Report for 1943.

48. Executive Committee Minutes, November 7, 1944; Board Minutes, November 29, 1944; Executive Committee Minutes, May 29 and October 30, 1945; Board Minutes, May 31 and October 31, 1945. With the doubling of its capacity for cellulose acetate, Hercules also increased its annual capacity of acetic anhydride to sixty-four million pounds in 1945 (Board Minutes, August 29, 1945).

49. Virginia Cellulose Department, Annual Report for 1941.

50. Virginia Cellulose Department, Annual Reports for 1940 and 1941.

51. Virginia Cellulose Department, Annual Report for 1942.

52. Ibid.; David A. Hounshell and John Kenly Smith, Jr., *Science and Corporate Strategy: R&D at Du Pont, 1902–1980* (New York, 1988), 168.

53. Virginia Cellulose Department, Annual Reports for 1940 and 1945.

54. Naval Stores Department, Annual Reports for 1940 and 1943.
55. Naval Stores Department, Annual Report for 1941.
56. Naval Stores Department, Annual Report for 1942.
57. Synthetics Department, Annual Report for 1943; "Recent Purchase Rounds out Hercules' Synthetic Resin Line," *Hercules Mixer* 23 (July–August 1941), 224–225. These assets included trade names, inventories, and small plants in Mansfield, Massachusetts, and Brunswick, Georgia. Board Minutes, June 9, 1941.
58. Naval Stores Department, Annual Reports for 1942 to 1944; Albrecht H. Reu, "A Brief History of the Brunswick, Georgia, Plant of Hercules, Incorporated" (1966), 48, typescript in Hercules library. Forster's prediction came true in February 1945, when the War Production Board began to control the distribution of rosin.
59. Naval Stores Department, Annual Reports for 1940 and 1942; "This War May Never End," *Hercules Mixer* 23 (July 1941), 288–291.
60. Naval Stores Department, Annual Reports for 1942 to 1944. See also Emil Ott, "The Team Approach to Research and Development," *Chemical and Engineering News,* vol. 28, no. 4 (June 12, 1950), 1944–1946; see also chap. 9.
61. Naval Stores Department, Annual Reports for 1943 and 1945; Board Minutes, March 21, 1944; Executive Committee Minutes, May 2, 1945.
62. "The Dresinates Are Useful and Versatile Rosin Derivatives," *Hercules Chemist* 13 (January 1944), 13–15.
63. PMC Department, Annual Report for 1942.
64. Vernon Herbert and Attilio Bisio, *Synthetic Rubber: A Project That Had to Succeed* (Westport, Conn., 1985), 126–127; Peter J.T. Morris, *The American Synthetic Rubber Research Program* (Philadelphia, 1989), 31–32.
65. Herbert and Bisio, *Synthetic Rubber,* 33–34.
66. Donald H. Sheffield interview, June 17, 1987; Reginald Ivett interview, December 2, 1987; Harold M. Spurlin interview, October 9, 1986; Ivett comment, November 8, 1989; Morris, *American Synthetic Rubber Research Program,* 31–32.
67. PMC Department, Annual Reports for 1942, 1943, and 1945. Dresinates® were used by the synthetic rubber industry not only as an emulsifier, but also, in its losing battle against rayon tire cord, as a strengthening agent for cotton tire cord.
68. James C. Cox, Jr., "Emil Ott, 1902–1963," in Miles D. Wyndham, ed., *American Chemists and Chemical Engineers* (Washington, D.C., 1976), 378–379; "Dr. Emil Ott, Head, Research Department, Experiment Station," *Hercules Mixer* 21 (April 1939), 86.
69. Ivett interview, December 2, 1987; Spurlin interview, October 9, 1986.
70. The publication is Ott, ed., *Cellulose and Cellulose Derivatives* (New York, 1943). Among the contributors from Hercules were Ott, Dr. J. Barsha, Walter Gloor, Elmer Hinner, Dr. R.H. Osborn, G.H. Pfeiffer, and Spurlin. A second edition in three parts edited by Ott, Spurlin, and Grafflin was published in 1954 and 1955. In the 1970 volume that supplemented the original study, Spurlin sketches Ott's role in the publication in an introductory appreciation, "Emil Ott and Cellulose Chemistry," xi–xiii. Ott outlined his own philosophy of research in his article, "The Team Approach to Research and Development": see note 60 above.
71. Research Department, Annual Reports for 1940 and 1943.
72. Research Department, Annual Report for 1940; Ivett interview, December 2, 1987.
73. Research Department, Annual Report for 1941.
74. Ivett comment, November 8, 1989. On the conservatism of senior management, see chap. 9.
75. Ivett interview, December 2, 1987.
76. Research Department, Annual Report for 1940; Spurlin interview, December 2, 1987; Bruce Fader, "Phenols by Oxidation," *Chemical Processing* (October 1955), 11, 247; Ivett comment, November 8, 1989.
77. Research Department, Annual Report for 1943.
78. Hercules Powder Company, Annual Report for 1944.

Chapter 9

We are grateful to Dr. Arthur L. Glasebrook, a retired Hercules researcher, for a close and critical reading of the second part of this chapter and for many suggestions and improvements.

1. Charles A. Higgins, "Why I Work for Hercules," *Hercules Mixer* 28 (April 1946), 115; Charles A. Higgins, "To Project for Our Company a Greater Future . . . A Brighter Future for All Our Employees," ibid., 112–113.

2. Albert E. Forster, quoted in Donald P. Burke, "Planning for the Next Half Century," *Chemical Week* (October 6, 1962), 47.

3. John Kenneth Galbraith, *American Capitalism* (Boston, 1952), 63–83; Werner C. Brown interview, October 8, 1986. The behavior of Ed Morrow, chairman of the finance committee, provides a vivid illustration of the depression psychosis at work at Hercules. As recalled by Carl Eurenius, the source of Morrow's fear was "the postwar recession that happened in 1920 and 1921. He always remembered that. He assumed that after World War II we were bound to have a recession or big depression. Year after year after that, he still believed it was coming. A financial analyst from New York would come and talk to us and bring out the fact that things looked great and that next year would be better, but Ed always said, 'The hell with that. You're wrong.' " Hercules History Advisory Committee Minutes, October 11, 1988.

4. Harvey J. Taufen interview, September 21, 1988.

5. James W. Dickenson interview, November 11, 1988; Duard H. "Tex" Little interview, July 22, 1987.

6. Hercules Incorporated, *Resumé of Facility Capabilities at Allegany Ballistics Laboratory* (n.p., 1971), unpaginated; Loren E. Morey, history of Hercules' aerospace operations, (undated) typescript in Morey's possession; Richard Winer interview, May 18, 1988.

7. Morey, history of Hercules' aerospace operations; Winer interview, May 18, 1988.

8. Winer interview, May 18, 1988; Rudolph Steinberger interview, June 17, 1988.

9. "Radford and Sunflower: Back on the Job," *Hercules Mixer* 34 (May–June 1952), 6–9, 36.

10. Explosives Department, Annual Report for 1952.

11. *Chemical Week* 74 (April 17, 1954), 20.

12. Explosives Department, Annual Report for 1954.

13. John M. "Jack" Martin interview, September 21, 1988.

14. Martin interview, September 21, 1988; Dan J. Forrestal, *Faith, Hope and $5,000: The Story of Monsanto* (New York, 1977), 109–117; Melvin A. Cook, *The Science of High Explosives* (New York, 1958), 13; Robert B. Hopler, "The History of Explosives," *Explosives Engineer,* supplement 2 (January 1980).

15. The following section is based on Cellulose Products Department, Annual Reports for 1946–1954.

16. Cellulose Products Department, Annual Reports for 1948 and 1949.

17. "A Service That Builds Markets," *BusinessWeek* (October 7, 1950), 88, 91.

18. Leland H. Burt, "Carboxymethylcellulose," *Drug Standards* 19 (May–June 1951), 106; Williams Haynes, *Cellulose: The Chemical That Grows* (Garden City, N.Y., 1953), 314; Eugene D. Klug, "Cellulose Derivatives," in *Encyclopedia of Chemical Technology* (New York, 1949), 385.

19. Klug, "Cellulose Derivatives," 385; Herman Skolnik interview, June 17, 1987.

20. Skolnik interview, June 17, 1987; Haynes, *Cellulose,* 314.

21. Klug, "Cellulose Derivatives," 386–387; Burt, "Carboxymethylcellulose," 106.

22. Klug, "Cellulose Derivatives," 387; T.S. Morse, "Sodium Carboxymethylcellulose," typescript, Hercules library; "What Do You See in CMC?" *Hercules Mixer* 28 (October 1946), 280; Cellulose Products Department, Annual Report for 1946.

23. Leland Burt, a Yale graduate with a background in water-soluble resins, was put in charge of sales and market development for CMC ("What Do You See in CMC?" 280; Cellulose Products Department, Annual Reports for 1946 and 1947; Carl Eurenius interview, November 12, 1987).

24. Cellulose Products Department, Annual Reports for 1948 and 1949.

25. Cellulose Products Department, Annual Reports for 1949–1951.

26. The following section is based on Naval Stores Department, Annual Reports for 1946–1954.

27. Hercules Powder Company, Annual Reports for 1946 and 1953; Naval Stores Department, Annual Reports for 1949 and 1953.

28. Naval Stores Department, Annual Reports for 1951–1954.

29. Naval Stores Department, Annual Reports for 1946, 1947, and 1951.

30. Naval Stores Department, Annual Reports for 1946 and 1947.

31. Naval Stores Department, Annual Report for 1953.

32. Albrecht H. Reu, "A Brief History of the Brunswick, Georgia, Plant of Hercules, Incorporated" (1966), 80, typescript in Hercules library; Hercules Powder Company, Annual Report for 1949; Donald H. Sheffield interview, June 17, 1987.

33. Reu, "Brief History," 80; Hercules Powder Company, Annual Report for 1951; *New York Times,* March 29, 1950, 43; Hercules Powder Company, Board Minutes, November 29, 1950; George Bossardet interview, August 24, 1988; Sheffield interview, June 17, 1987.

34. Bossardet interview, August 24, 1988; Executive Committee Minutes, December 8, 1953.

35. Harry Schieber, Harold Vatter, and Harold Faulkner, *American Economic History* (New York, 1976), 453–454.

36. Samuel P. Epstein, *The Politics of Cancer* (San Francisco, 1978), 31–32.

37. Charles Dunn interview, April 7, 1988; Reu, "Brief History," 72; "Testing Thanite," *Hercules Mixer* 28 (April 1946), 118.

38. J.W. Dolson, "Toxaphene," *Hercules Mixer* 31 (January–February 1949), 5–6; Dunn interview, April 7, 1988; Hercules Powder Company, "Toxaphene Manual: Toxicology" (July 1955) (copy in library at Hercules Incorporated), Naval Stores Department, Annual Report for 1946.

39. Naval Stores Department, Annual Reports for 1947 and 1948; Reu, "Brief History," 72, 75; Executive Committee Minutes, August 29, 1947, and May 24, 1948.

40. Naval Stores Department, Annual Report for 1947; Dolson, "Toxaphene," 5–6.

41. "A Review of Toxaphene in [1949]," *Hercules Mixer* 32 (January–February 1950), 10–11, 40; E.N. Woodbury, "The Effectiveness of Toxaphene," *Agricultural Chemical* (March 1949), 31–32; Lawrence P. Killilea, "A True Chemurgic Insecticide," *Chemurgic Digest* (April 1948), 30–31.

42. Dolson, "Toxaphene," 5–6.

43. Naval Stores Department, Annual Report for 1947; Executive Committee Minutes, December 21, 1948; "A Review of Toxaphene in [1949]," 10–11, 40.

44. Naval Stores Department, Annual Report for 1951; Richard B. Douglas interview, November 13, 1987.

45. Dunn interview, April 7, 1988.

46. Naval Stores Department, Annual Reports for 1948, 1949, and 1951; "A Review of Toxaphene in [1949]," 10–11, 40; Dolson, "Toxaphene," 5–6.

47. The broad licensing provision was rescinded in 1957. Executive Committee Minutes, June 13, 1950, August 1, 1950, and August 29, 1950; "Toxaphene Made in England," *Manufacturing Chemist* 22 (January 1951), 28.

48. Board Minutes, June 28, 1950; *New York Times,* July 5, 1950, 47.

49. Hercules was so desperate to secure these commodities that it considered loaning Algonquin $1 million for ten years at 4.5 percent for rights to purchase 20,000 tons of chlorine and 22,000 tons of caustic soda a year at market prices for two years; Hercules ultimately agreed to lower quantities, without a loan. The Dow carbon tetrachloride agreement was approved the same day Hercules voted to license its phenol process to Dow (see below), although it is not clear if the two agreements were negotiated together (Board Minutes, October 31, 1951; Executive Committee Minutes, December 4, 1951, and March 25, 1952; Naval Stores Department, Annual Report for 1950).

50. About $1.37 million had been spent on the two projects. Reu, "Brief History," 82–83; Executive Committee Minutes, August 19 and December 23, 1952.

51. Naval Stores Department, Annual Report for 1953.

52. Paul Mayfield, "A Sound Future for Agricultural Chemicals," *Agricultural and Food Chemistry* 2 (February 17, 1954), 172–176.

53. That year, the company opened the extensive new Agricultural Chemicals Laboratory at the Experiment Station. Hercules Powder Company, Annual Report for 1954; Naval Stores Department, Annual Report for 1954; "New Agricultural Chemicals Laboratory," *Hercules Mixer* 36 (January–February 1954), 12–13.

54. PMC Department, Annual Reports for 1947 and 1953; Hercules Powder Company, Annual Report for 1954.

55. Gerald I. Keim interview, May 4, 1989.

56. Ibid.

57. The following discussion of the paper chemicals business is based on interviews with Robert J. Leahy, April 27, 1988, and with Maurice J. Kelly, Robert I. Mason, and Charles A. Twigg, April 5, 1988.

58. PMC Department, Annual Report for 1947.

59. PMC Department, Annual Reports for 1946 and 1951.

60. Martin recalls the cost of the plant as in the range of $100 million, but that figure is certainly too large. Shell Chemical acquired the facility (including the butadiene plant) for $30 million in 1953 (John M. "Jack" Martin interview, September 21, 1988). See Kendall Beaton, *Enterprise in Oil: A History of Shell in the United States* (New York, 1957), 680; see also Brown interview, October 8, 1986.

61. Eugene D. Crittenden, Jr., interview, December 2, 1987.

62. Martin interview, September 2, 1988.

63. Brown interview, October 8, 1986; Taufen interview, September 21, 1988; Alexander F. Giacco interview, November 15, 1987.

64. Eugene D. Crittenden comment, Hercules History Advisory Committee Minutes, October 11, 1988.

65. Information in the following paragraphs is drawn from Peter H. Spitz, *Petrochemicals: The Rise of an Industry* (New York, 1988), 253–268, 305–310.

66. Spitz, *Petrochemicals,* 302–310.

67. Frank M. McMillan, *The Chain Straighteners: Fruitful Innovation; The Discovery of Linear and Stereoregular Synthetic Polymers* (London, 1979), 75; Spitz, *Petrochemicals,* 338–339, 342.

68. Harold M. Spurlin interview, October 9, 1986; Sheffield interview, June 17, 1987; Crittenden interview, December 2, 1987; "New Jersey Site Chosen for Hercules Hydrocarbon Chemicals Plant," *Hercules Mixer* 34 (May–June 1952), 3.

69. "New Jersey Site Chosen," 3; Bruce Fader, "Phenols by Oxidation," *Chemical Processing* (October 1955), 9–11 ff.; John M. "Jack" Martin interview, October 9, 1986.

70. "Hercules—From Gunpowder to Phenol," *BusinessWeek* (January 19, 1952), 66–67; "New Jersey Site Chosen," 3.

71. "Hercules—From Gunpowder to Phenol," 67; Executive Committee Minutes, September 19, 1950.

72. Territorial rights for HPC Ltd. were later extended to include Luxembourg, the Belgian colonies, and Austria. Executive Committee Minutes, April 10 and October 2, 1951, April 22, 1952, and December 8, 1953; "Hercules—From Gunpowder to Phenol," 67.

73. Executive Committee Minutes, November 30, 1951; Board Minutes, December 10, 1951.

74. Crittenden interview, December 2, 1987.

75. Executive Committee Minutes, November 30, 1951; Crittenden interview, December 2, 1987; Sheffield interview, June 17, 1987.

76. Hercules contracted for a supply of propylene from the Socony-Vacuum Oil Co. in early 1953. Du Pont's Repauno plant was nearby as well. "New Jersey Site Chosen," 3; Executive Committee Minutes, February 17, 1953.

77. Executive Committee Minutes, November 4, 1952; Spitz, *Petrochemicals,* 318.

78. Spitz, *Petrochemicals,* 292.

79. W.J. Reader, *Imperial Chemical Industries: A History,* vol. 2, *The First Quarter Century 1926–1952* (London, 1975), 381–388; Spitz, *Petrochemicals,* 274–279, 282–290; David A. Hounshell and John Kenly Smith, Jr., *Science and Corporate Strategy: R&D at Du Pont, 1902–1980* (New York, 1988), chaps. 12–13, 18.

80. Spitz, *Petrochemicals,* 286–287.

81. Construction overruns may have eroded the confidence of the department by driving the cost of the new plant to nearly $3 million by the end of 1947. Taufen interview, September 21, 1988; Board Minutes, January 1, 1946, March 31, and July 30, 1947.

82. Crittenden interview, December 2, 1987; Taufen interview, September 21, 1988.

83. Taufen interview, September 21, 1988.

84. Ibid.; Executive Committee Minutes, July 10, August 25, September 18, and November 27, 1951.

85. Taufen interview, September 21, 1988; Executive Committee Minutes, June 26, 1952. According to Arthur L. Glasebrook, Hercules later discovered that Standard Oil of California held patents on a similar xylene oxidation process. Thereafter, Hercules paid royalties to Standard Oil on production in the United States, and to Imhausen on production in the rest of the world. Arthur L. Glasebrook, comment to authors (undated).

86. Harvey J. Taufen interview, September 26, 1988.

87. Synthetics Department Reports, April, July, and September 1952; Executive Committee Minutes, June 20, 1952.

88. Executive Committee Minutes, April 21, 1953; Synthetics Department Report, March 1953.

89. Synthetics Department Reports, November 1952, and June–July 1953.

90. In 1952 and 1953, Hercules considered licensing two processes to augment its xylene oxidation work: one from Richfield Oil Co. for its process for converting crude xylene from a hydroforming operation into pure toluic acid, to supplement the Imhausen process; and the second from Scientific Design Co. for a process for making TPA from *para*-diisopropyl benzene by air-oxidation. The first licensing proposal was rejected because Hercules was able to lower the cost of xylenes sufficiently without it; the second never came through because changing market prices made *para*-diisopropyl benzene prices uncompetitive with *para*-xylene. Scientific Design was a petrochemical technology R&D firm that developed its own process of air-oxidizing *para*-xylene in 1956 and licensed it to Amoco Chemical Co. Synthetics Department Reports, July, September, October, and November 1952, and February, May, November, and December 1953; Executive Committee Minutes, December 22, 1953; Board Minutes, December 30, 1953; Spitz, *Petrochemicals,* 38, 326–331.

91. Crittenden interview, December 2, 1987.

92. Skolnik interview, June 17, 1987.

93. Martin interview, September 2, 1988; Brown interview, October 8, 1986.

94. Information in the following paragraphs is drawn from McMillan, *Chain Straighteners,* 3–45.

95. Peter J.T. Morris, *Polymer Pioneers* (Philadelphia, 1986), 79.

96. Fortunately for PCL, this contract was one of the few exclusive agreements Ziegler ever signed; in addition, it was worded broadly enough to encompass subsequent developments in polymer chemistry. McMillan, *Chain Straighteners,* 38–43.

97. Arthur L. Glasebrook, handwritten narrative in Glasebrook's possession. For slightly different versions of the following account, see McMillan, *Chain Straighteners,* 44; and "Hercules Makes up Its Mind," *BusinessWeek* (October 15, 1955), 122.
98. Glasebrook narrative; Arthur L. Glasebrook interview, October 12, 1988; McMillan, *Chain Straighteners,* 44–45; "Hercules Makes up Its Mind," 124. The German coal companies that sponsored the Max Planck Institute, of course, had free rights to the fruits of its research.
99. McMillan, *Chain Straighteners,* 53–54, 77.
100. Arthur L. Glasebrook to David Wiggam, "Ziegler Processes," report No. 4, May 31, 1953, file in Glasebrook's possession.
101. Glasebrook, comment to authors (undated). Note that Glasebrook's version of this key event differs slightly from that given by McMillan, *Chain Straighteners,* 56–59.
102. Arthur L. Glasebrook to J.J.B. Fulenwider, September 1, 1953, file in Glasebrook's possession.
103. Glasebrook interview, October 13, 1988.
104. McMillan, *Chain Straighteners,* 62–63.
105. Ibid., 10–13; Spitz, *Petrochemicals,* 332.
106. Glasebrook, comment to authors (undated); Glasebrook interview, October 13, 1988; McMillan, *Chain Straighteners,* 76–77.
107. Glasebrook narrative.
108. Glasebrook, comment to authors (undated); McMillan, *Chain Straighteners,* 77–78; "Hercules Makes up Its Mind," 125–126.
109. Executive Committee Minutes, September 13, 1954; McMillan, *Chain Straighteners,* 77–78; "Hercules Makes up Its Mind," 124.
110. Donald Whitehead, *The Dow Story: A History of Dow Chemical Company* (New York, 1968), 236.
111. Edward B. Morrow, "Methods and Means of Financing Business Expansion," and "Formula for Calculating Required Return on the Total Hercules Powder Company Plant Investment," reprints of financial seminars presented to Hercules executives in 1952 (copy in library at Hercules Incorporated).
112. W.K. Gutman, *Money Magic in Chemicals* (New York, 1953), 177–180. For another analyst's characterization of Hercules as "conservative," see "Wall Street of Chemistry," *Chemical and Engineering News* 28 (December 18, 1950), 4500.

Chapter 10

1. Quoted in "Metamorphosis at Hercules," *Forbes* (June 1, 1961), 13.
2. "The New Force at Hercules," *Chemical Week* (August 27, 1960), 56.
3. On the competitive dynamics of the petrochemicals industry, see Peter H. Spitz, *Petrochemicals: The Rise of an Industry* (New York, 1988), esp. 537–541; and Robert Stobaugh, *Innovation and Competition: The Global Management of Petrochemical Products* (Boston, 1988), part I and apps. A-C.
4. The startup date was sometime during the week of January 13, according to the *Hercules Mixer* ("Higgins Plant Begins Operations," *Hercules Mixer* 37 [February 1955], 9), or March 1, according to the Naval Stores Department, Annual Report for 1955. See also, Board Minutes, June 30, and September 29, 1954; "Bright and Beautiful Higgins Plant," *Hercules Mixer* 37 (November 1954), 1; "602 Neighbors Came to See Us at Gibbstown," *Hercules Mixer* 37 (June 1955), 6–10.
5. Naval Stores Department, Annual Report for 1958. The terms of the agreement (the legal fine points of which we have not detailed) seem to have evolved in the course of negotiating with Shell, which initially rejected several offers, including a straight 3 percent royalty. The executive committee minutes for this period also note a contract with the Organic Chemical Industry of Yugoslavia for $750,000 plus royalties. Executive Committee Minutes, February 18 and August 5, 1957, May 26, 1958, November 2 and 16, 1959, and September 7, 1960.

6. Naval Stores Department, Annual Reports for 1955 and 1956; Executive Committee Minutes, June 25, 1956; "Gibbstown to Double Para-Cresol Capacity," *Hercules Mixer* 38 (September 1956), 25.

7. Its high-temperature stability also made Di-Cup® attractive to makers of polyester premixes and companies that polymerized polyesters and styrene.

8. At various times, some of these products were also made at the company's other naval stores plants. Naval Stores Department, Annual Reports for 1955–1960.

9. "Gibbstown to Double Para-Cresol Capacity," 25; Naval Stores Department, Annual Reports for 1955–1960.

10. J.W. Kneisley, "Faith, Air, and Xylene," *Hercules Chemist* 42 (June 1961), 15.

11. Among Hercules' customers were Toyo Rayon Co. and Teikoku Rayon Co. (Japan), Société Rhodiaceta (France), Algemene Kunstzijde Unie (Holland), Vereinigte Glanzstoff-Fabriken and Hoechst (Germany), and Montecatini (Italy); Kneisley, "Faith, Air, and Xylene," 14–17.

12. Synthetics Department, Annual Report for 1957; Executive Committee Minutes, November 7, 1960; Board Minutes, November 23, 1960; Kneisley, "Faith, Air, and Xylene," 14–17; Synthetics Department, Annual Report for 1960.

13. Marjorie D. Chris interview, July 21, 1987; Synthetics Department Report, December 1957.

14. Hercules apparently played a role in Amoco's entrance into DMT production by unwittingly helping it gain both a large customer and a manufacturing process. According to Harvey Taufen, Goodyear, long a major consumer of Hercules' rubber chemicals, wanted to negotiate a five-year contract to buy approximately sixteen million pounds of DMT annually from Hercules to make polyester tire cord. When Goodyear called for a price protection clause (stipulating that if a competitor offered it DMT of equivalent quantity and quality, Hercules would meet the competitor's price or risk losing the account), Philip Stull and perhaps other directors balked. Goodyear then called on Amoco, requesting that it become a DMT supplier. Harvey J. Taufen interviews, September 21 and December 19, 1988. Estimate of Goodyear's annual requirements of DMT is from Executive Committee Minutes, November 11, 1957.

15. John M. "Jack" Martin interview, September 21, 1988; Taufen interview, December 19, 1988; Spitz, *Petrochemicals,* 326–327.

16. David A. Hounshell and John Kenly Smith, Jr., *Science and Corporate Strategy: R&D at Du Pont, 1902–1980* (New York, 1988), 419; Kneisley, "Faith, Air, and Xylene," 14–17.

17. Synthetics Department, Annual Report for 1961; Hercules Powder Company, Annual Report for 1961.

18. Well aware of the situation at Mülheim, Glasebrook advised against sending a team, believing it was not worth the effort. Arthur L. Glasebrook interview, October 13, 1988; "Hercules' Ziegler Polyethylene Plant Goes on Stream," *Chemical Processing* (August 1957), 4; Werner C. Brown, "Developing through Internal Diversification," in John Glover and Gerald A. Simon, eds., *The Chief Executive's Handbook* (Homewood, Ill., 1976), 356; Elmer Hinner interview, May 13, 1980.

19. Hercules Powder Company, *The Achievement of Crystalline Polypropylene: An Outstanding New Plastic* (Wilmington, n.d.), 9–17.

20. Karl Winnacker, trans. David Goodman, *Challenging Years: My Life in Chemistry* (London, 1972), 124, 144–145, 152, 165–166, 182, 185–191, 282.

21. Winnaker, *Challenging Years,* 207; Glasebrook interview, October 13, 1988. According to "Hercules Makes up Its Mind," *BusinessWeek* (October 25, 1955), 126, Hinner learned of the progress of Hoechst from Wiggam's reports.

22. Glasebrook interview, October 13, 1988; Winnaker, *Challenging Years,* 193.

23. Glasebrook interview, October 13, 1988; Cellulose Products Department, Annual Report for 1955; "Hercules Makes up Its Mind," 129.

24. Board Minutes, February 9, 1955; Glasebrook interview, October 13, 1988; Hinner interview, May 13, 1980; Winnaker, *Challenging Years,* 218–219. The *BusinessWeek* article (see note 21) errs in stating that Walter Gloor, not Glasebrook, accompanied the executives to Frankfurt.

25. Winnaker, who developed friendships with Forster and other Herculites, donated a set of fine china to the Hercules Guest House that can still be found there. Glasebrook interview, October 13, 1988.

26. Hercules Powder Company, *The Achievement of Crystalline Polypropylene,* 8.

27. "The Hydrogen Effect on Ziegler Catalysis," *Hercules Chemist* 46 (February 1963), 6–11. The circumstances of Vandenberg's discovery demonstrate that the arrangement between Hercules and Hoechst was a true technical "exchange," even though payments flowed in one direction (from Hercules to Hoechst). A Hoechst researcher named Herbert Bestian was the first to recognize the importance of Vandenberg's discovery, which was incorporated into the workings of the Hoechst pilot plant and finally, thanks to Hoechst, the Hercules pilot plant. Glasebrook interview, October 13, 1988; Frank M. McMillan, *The Chain Straighteners: Fruitful Innovation; The Discovery of Linear and Stereoregular Synthetic Polymers* (London, 1979), 137.

28. Executive Committee Minutes, September 12, 1955; William Gardner interview, June 20, 1988; Earp F. Jennings interview, April 6, 1988; Cellulose Products Department, Annual Report for 1955.

29. Hinner later estimated that Hercules' collaboration with Hoechst saved "from 2 to perhaps 5 years in getting to the market with a saleable product," "E.F. Hinner Acceptance Speech," CCDA Award Dinner, March 18, 1969, *Commercial Development Journal* (November 1969), 7. On another occasion, Hinner estimated that Hercules had saved as much as ten years in the development process. "CCDA Honors Elmer Hinner," *Chemical and Engineering News* (March 3, 1969), unpaginated reprint.

The cost of constructing the plant was higher than originally planned, reaching nearly $11 million. Hinner interview, May 13, 1980; Board Minutes, June 27, 1956.

30. Hinner interview, May 13, 1980; Forster quotation from "Metamorphosis at Hercules," 14; "We Enter a Vast New Field of Chemistry with Bright Promise for Hercules' Growth as Parlin's Multimillion Dollar Hi-Fax Plant Begins Operation," *Hercules Mixer* 39 (July–August 1957), 1–8; McMillan, *Chain Straighteners,* 83.

31. "Hercules' Ziegler Polyethylene Plant Goes on Stream," 3.

32. J. Paul Hogan and Robert L. Banks, "History of Crystalline Polypropylene," in Raymond B. Seymour and Tai Cheng, eds., *History of Polyolefins* (Dordrecht, Holland, 1986), 110; McMillan, *Chain Straighteners,* 116.

33. Cellulose Products Department, Annual Report for 1956; McMillan, *Chain Straighteners,* 116.

34. Glasebrook interview, October 13, 1988; Hogan and Banks, "History of Crystalline Polypropylene," passim; McMillan, *Chain Straighteners,* chaps. 7 and 9; David B. Sicilia, "A Most Invented Invention," *American Heritage of Innovation and Technology* 6 (Spring–Summer, 1990).

35. McMillan, *Chain Straighteners,* 116; Cellulose Products Department, Annual Report for 1956.

36. Cellulose Products Department, Annual Report for 1956.

37. "E.F. Hinner Acceptance Speech," 7; McMillan, *Chain Straighteners,* 118; "Metamorphosis at Hercules," 13.

38. McMillan, *Chain Straighteners,* 118; "E.F. Hinner Acceptance Speech," 7.

39. Gardner interview, June 20, 1988; "E.F. Hinner Acceptance Speech"; Hinner interview, May 13, 1980; Hercules, *Achievement of Crystalline Polypropylene,* 11.

40. Cellulose Products Department, Annual Reports for 1957 and 1958; "Metamorphosis at Hercules," 13.

41. Perrin Stryker, "Chemicals: The Ball Is Over," *Fortune* (October 1961), 215; Polymers Department, Annual Report for 1961.

42. Cellulose Products Department, Annual Reports for 1957–1960. In 1960 and 1961 (but never again after that), output of Hi-Fax® again exceeded output of Pro-Fax.®

43. Executive Committee Minutes, July 26, 1959; Board Minutes, July 28, 1959; Spitz, *Petrochemicals,* passim.

44. Polymers Department, Annual Report for 1961; Stryker, "Chemicals: The Ball Is Over," 210, 215; "Metamorphosis at Hercules," 15.

45. Stryker, "Chemicals: The Ball Is Over," 215.

46. Cellulose Products Department, Annual Reports for 1955–1959.

47. Hinner interview, May 13, 1980.

48. Ibid.; Board Minutes, October 10, 1960; Executive Committee Minutes, March 20, 1961; C. Doyle Miller and Rodman B. Teeple interview, March 16, 1989.

49. "Fiber Development Department Formed," *Hercules Chemist* 41 (February 1961), 18–19; Gilman S. Hooper, "A Career in Fibers," *Hercules Chemist* 42 (June 1961), 4–5.

50. Donald W. Simonds interview, February 20, 1989; Hinner interview, May 13, 1980; "Oriented Thermoplastics," *Hercules Chemist* 44 (February 1962), 10–11.

51. Fred L. Buckner interview, March 10, 1988; "Oriented Thermoplastics," 10–11; Herman Skolnik interview, June 17, 1987; Executive Committee Minutes, June 22, 1959; Board Minutes, June 25, 1959; Cellulose Products Department, Annual Reports for 1959 and 1960.

52. George Taylor, "Profits in a Shrinking Market," *Hercules Chemist* 44 (February 1962), 12–15.

53. Hinner interview, May 13, 1980; Hercules Powder Company, Annual Report for 1960; "Oriented Thermoplastics," 11.

54. Cellulose Products Department, Annual Report for 1960; Polymers Department, Annual Report for 1961; Reginald Ivett interview, December 2, 1987; Buckner interview, March 10, 1988.

55. Forster quoted in "Metamorphosis at Hercules," 14; Eugene D. Crittenden comment, Hercules History Advisory Committee Minutes, December 6, 1988; Hercules Incorporated, Missouri Chemical Works, typescript history, June 1958 (copy in authors' possession); "Hydrogen," *Hercules Chemist* 27 (June 1956), 14–15; "Nitrogen: The Powerful Element," *Hercules Chemist* 30 (June 1957), 5–9.

56. "Hercules Chemistry on the Mississippi," *Hercules Mixer* 39 (October 1957), 1–9; *Chemical Week* 78 (May 12, 1956), 104, 106; Explosives Department, Annual Report for 1956; Executive Committee Minutes, January 23, 1961.

57. "Hercules Pentaerythritol," *Hercules Chemist* 27 (June 1956), 20–21; "Hercules Chemistry on the Mississippi," 1–9; "More Pentaerythritol," *Hercules Chemist* 30 (June 1957), 14–15.

58. *Chemical Week* 74 (April 17, 1954), 20; "More Pentaerythritol," 14–15; *Chemical Week* 77 (December 17, 1955), 94, 96, 98. In early 1956, Hercules announced "improved technical" or "zero ash" pentaerythritol—composed of 88 percent mono-PE, 12 percent di-PE, and almost no residuals—aimed at makers of special alkyd resins, synthetic drying oils, core oils, and certain polyesters. "Hercules Pentaerythritol," 20–21; *Chemical Week* 78 (January 7, 1956), 14; Synthetics Department, Annual Reports for 1949 and 1955.

59. "UN-32," *Hercules Chemist* 33 (June 1958), 15–17.

60. The common ratio in Southern pine was 46 percent resin acids, 40 percent fatty acids, and 14 percent unsaponifiables. L.G. Zachary, H.W. Bajak, and F.W. Eveline, eds., *Tall Oil and Its Uses* (New York, 1965), 3; John Drew and Marshall Propst, *Tall Oil* (New York, 1981), 1; "PAMAK Tall Oil Fatty Acids: A New Hercules Pine Tree Product," *Hercules Chemist* 28 (October 1955), 6–8.

61. Drew and Propst, *Tall Oil,* 87–88; Ralph H. McKee and Helmer L. Blengsli, "Historical Development of the Tall Oil Industry," *Paper Trade Journal,* technical association section (September 17, 1936), 34; *Naval Stores Review,* November 22, 1930, and February 13, 1937 (see chap. 6, note 69).

62. Technical Department Report, First Quarter 1937. See also, McKee and Blengsli, "Historical Development of the Tall Oil Industry"; and Ralph H. McKee and Helmer L.

Blengsli, "A New Method of Refining Tall Oil," *Paper Trade Journal,* technical association section (October 1, 1936), 33–34.

63. *Naval Stores Review* (March 9, 1940); Drew and Propst, *Tall Oil,* 88–89.

64. Thomas D. Clark, *The Greening of the South: The Recovery of Land and Forest* (Lexington, Ky., 1984), passim; Zachary, Bajak, and Eveline, *Tall Oil and Its Uses,* 3–5; Carl Eurenius interview, November 14, 1979.

65. Executive Committee Minutes, May 16 and July 18, 1955; Eurenius interview, November 14, 1979; Eugene D. Crittenden, Jr., interview, December 2, 1987.

66. Board Minutes, August 31, 1955; Crittenden interview, December 2, 1987; "PAMAK Tall Oil Fatty Acids," 8; Reginald Ivett comment to authors, November 8, 1989.

67. Fred D. McCauley interview, September 8, 1987; Crittenden interview, December 2, 1987.

68. "PAMAK Tall Oil Fatty Acids," 8.

69. Pine and Paper Chemicals Department, Annual Report for 1961; Zachary, Bajak, and Eveline, *Tall Oil and Its Uses,* 3.

70. Naval Stores Department, Annual Reports for 1955–1960.

71. Hercules Powder Company, Annual Report for 1960; Pine and Paper Chemicals Department, Annual Report for 1961; Albrecht H. Reu, "A Brief History of the Brunswick, Georgia, Plant of Hercules, Incorporated" (1966), typescript in Hercules library, 91–93, 96–98; "Corbu Industrial, S.A.," *Hercules Chemist* 29 (February 1957), 12; "New International Outposts," *Hercules Chemist* 43 (October 1941), 24; Taufen interview, December 19, 1988; "Mexican Wood Naval Stores Plant . . ." *Hercules Mixer* 40 (April 1958), 19–21.

72. Between 1936 and 1941, the company acquired several small operations to augment its business in paper chemicals and synthetics. These deals represented expenditures of $100,000 or less and were made as simple extensions of existing product lines; see chaps. 7 and 8.

73. Quoted in "The New Force at Hercules," 52.

74. Executive Committee Minutes, September 8, 1958; Hercules Powder Company, Annual Report for 1960; Pine and Paper Chemicals Department, Annual Report for 1961.

75. Sheldon Hochheiser, *Rohm and Haas: History of a Chemical Company* (Philadelphia, 1986), 55–73, passim; Hounshell and Smith, *Science and Corporate Strategy,* 204, 476, 480, 513.

76. Executive Committee Minutes, March 26, April 9, and June 18, 1956; *Hercules Mixer* 38 (July–August 1956), 30; *Hercules Mixer* 39 (January 1957), 21.

77. Board Minutes, November 27, 1957.

78. Martin interview, September 21, 1988; Graham D. Taylor and Patricia E. Sudnik, *Du Pont and the International Chemical Industry* (Boston, 1984), 166–168; *Hercules Mixer* 38 (July–August 1956), 30.

79. Executive Committee Minutes, November 9, 1959; *Hercules Mixer* 42 (February 1960), 19.

80. Donald H. Sheffield interview, June 17, 1987. As part of the purchase, Hercules also agreed to endorse notes payable in ten years or less for about $2 million. Executive Committee Minutes, August 21, 1961; Hercules Powder Company, Annual Report for 1961; Gale E. Peterson, "The Discovery of 2,4-D," *Agricultural History* 41 (1967), 243–253.

81. "The New Force at Hercules," 62; "This Is Harbor Beach, Michigan," *Hercules Mixer* 40 (October 1958), 1–11.

82. Leland H. Burt, "Now Wheat Has Been Added," *Hercules Chemist* 29 (February 1957).

83. "Case of the Multimillion-Dollar Gray Bag," *Hercules Mixer* 39 (January 1957), 1–7.

84. "This Is Harbor Beach, Michigan," 1–11; Executive Committee Minutes, November 19 and December 31, 1956; "Case of the Multimillion-Dollar Gray Bag," 1–7.
85. Virginia Cellulose Department, Annual Reports for 1958–1961.
86. Board Minutes, December 30, 1959.
87. The following account is based on Imperial Color Chemical & Paper Corp., "How Imperial Serves Its Many Customers" (n.p., n.d.), and "This Is Imperial," a company brochure based on an article that appeared in the *American Ink Maker* (undated).
88. Crittenden interview, December 2, 1987.
89. Richard B. Douglas interview, November 13, 1987.
90. The following quotations are drawn from Imperial Color and Chemical Department, Annual Report for 1961.
91. Note that *Paper* was dropped from the name in 1961, the year before that part of the business was sold.
92. Martin interview, September 21, 1988; Explosives Department, Annual Reports for 1955–1960.
93. Explosives Department, Annual Reports for 1956 and 1957.
94. Hercules halted publication of the *Explosives Engineer* at the end of 1961. "Farewell to 'The Black' at Hercules, California," *Hercules Mixer* 38 (March 1956), 18; Explosives Department, Annual Reports for 1955–1961; *Hercules Mixer* 37 (July–August 1955), 25.
95. Walter LeFeber, *America, Russia, and the Cold War, 1945–1980* (New York, 1980), 199.
96. Walter L. McDougall, *The Heavens and the Earth: A Political History of the Space Age* (New York, 1985), 128.
97. Henry A. Schowengerdt interview, April 26, 1988; Richard Winer interview, May 18, 1988.
98. David E. Thompson interview, June 22, 1988; Richard E. Young, "History and Potential of Filament Winding," *Reprint Book of the Thirteenth Annual Technical and Management Conference, Reinforced Plastics Division, Society of the Plastics Industry* (Chicago, February 4, 1958), 1.
99. Young, "History and Potential of Filament Winding," 1.
100. "Newest Member of the Hercules Family," *Hercules Chemist* 35 (February 1959), 23–24.
101. "In the Race to Outer Space," *Hercules Mixer* 40 (October 1958), 14–24; Duard H. "Tex" Little interview, July 22, 1987; Winer interview, May 18, 1988.
102. "For Future Space Travel?" *Hercules Mixer* 40 (April 1958), 8–11.
103. Explosives Department, Annual Report for 1958; John F. Kushnerick, "How Hercules Hurdled Its Competitors," *Aerospace Management* (March 1963), unpaginated reprint. Brunswick was encouraged to enter the field by ABL when the government demanded a second source of filament-wound structures. Thompson interview, June 22, 1988; Winer interview, May 18, 1988; Schowengerdt interview, April 26, 1988.
104. *Hercules Mixer* 41 (January 1959), 0–1.
105. Ibid.; *Hercules Mixer* 42 (February 1960), 2–8.
106. Except where other sources are noted, the following account of the Minuteman program is based on Roy Neal, *Ace in the Hole: The Story of the Minuteman Missile* (New York, 1962).
107. These needs were (1) a lightweight vectoring nozzle (heavy metal nozzles were cooled by liquids in liquid-fuel rockets); (2) new manufacturing and curing processes for the solid propellants to handle motors as large as 100,000 pounds (the largest motors to date were 7,000 pounds); (3) higher performance propellant (from 230 to at least 245 pounds per second); (4) improvement of the mass fraction by using stronger and lighter cases and better loading; and (5) finding a way to stop the burn.
108. Little interview, July 22, 1987.

109. Robert L. Perry, "The Atlas, Thor, Titan, and Minuteman," in Eugene M. Emme, ed., *The History of Rocket Technology: Essays on Research, Development, and Utility* (Detroit, 1964), 158–159.
110. Neal, *Ace in the Hole,* 136.
111. *Hercules Mixer* 41 (June 1959), 1–3.
112. Neal, *Ace in the Hole,* 8–9.
113. This account follows Wyndham D. Miles, "The Polaris," in Emme, *History of Rocket Technology,* 162–175.
114. Little interview, July 22, 1987; Ernest A. Mettenet interview, June 22, 1988; Ruth L. Novak interview, June 21, 1988; *Hercules Mixer* 42 (November 1960), 15.
115. McDougall, *The Heavens and the Earth,* 129.
116. "How Hercules Helped Put the Paddlewheel Satellite into Orbit," *Hercules Mixer* 41 (September 1959), 12; Richard B. Douglas, "New Faces, New Buildings at Rocky Hill Mark First Year of Tremendous Growth," *Hercules Mixer* 42 (January 1960), 5–11, 29.
117. Kushnerick, "How Hercules Hurdled Its Competitors"; Hercules Incorporated, *Hercules' Experience in Chemical Propulsion* (n.p., n.d.); "In the Race to Outer Space," 14–24.
118. Quoted in "Metamorphosis at Hercules," 16.
119. Harvey J. Taufen interview, December 19, 1988.
120. Taylor and Sudnik, *Du Pont and the International Chemical Industry,* 161–168; Spitz, *Petrochemicals,* 197–225, 313–318.
121. Mira Wilkins, *The Maturing of Multinational Enterprise: American Business Abroad from 1941 to 1970* (Cambridge, Mass., 1974), 342–343.
122. Taufen interview, December 19, 1988.
123. "Looking Back on Hercules' Fifty Years of Growth," *Hercules Mixer* 45 (October 1963), 25.
124. Hercules was not alone in its response to the European communities. Also "encouraged by the formation of the European Economic Community," Du Pont organized its own International Department the same year as Hercules, and Monsanto set up a European headquarters at Brussels in 1960. "How Hercules Is Growing Overseas," *Hercules Mixer* 43 (March 1961), 27; Taufen interview, December 22, 1988; "The Changing Pattern of World Trade," *Hercules Mixer* 43 (September 1961), 5; Hounshell and Smith, *Science and Corporate Strategy,* 511; Dan J. Forrestal, *Faith, Hope, and $5,000: The Story of Monsanto* (New York, 1977), 181.
125. "Hercules Enters New Era," *Hercules Mixer* 42 (January 1960), 2–3; "Chemistry Is International," *Hercules Chemist* 39 (June 1960), 14; "New International Outposts," 24.
126. "Hercules Enters New Era," 2–3; Taufen interview, December 22, 1988.
127. "Chemistry Is International," 14; "New International Outposts," 21; "Swedish Plant Begins Operation," *Hercules Mixer* 43 (September 1961), 28; "Start-up of New Dutch Plant," *Hercules Chemist* 43 (September 1961), 23.
128. "Chemistry Is International," 14; "New International Outposts," 23.
129. "The Changing Pattern of World Trade," 5, 10–12, 15, 18.
130. Taufen interview, December 22, 1988.
131. Hercules Powder Company, *Operating Departments: Organization, Personnel, Products* (Wilmington, 1961), unpaginated.
132. Ibid.; "The New Force at Hercules," 55. The most conspicuous company governed by an internal board was Standard Oil of New Jersey.
133. "Metamorphosis at Hercules," 13–16.
134. "Hercules Powder Accents Plastics, Petrochemicals," *Investor's Future* (November 1958), 6, 24; "The New Force at Hercules," 49ff.; Donald P. Burke, "Planning for the Next Half Century," *Chemical Week* (October 6, 1962), 53.
135. Hercules spent $1.25 million on improvements for its new Home Office. "Her-

cules Tower," *Hercules Mixer* 42 (September 1960), 0–1; Board Minutes, November 25, 1959.

136. Quoted in "Metamorphosis at Hercules," 15.

137. Forster told *Chemical Week,* "We've never turned a project down for lack of funds. Oh, I suppose we could have grown a lot faster if we had done things differently. But would it have been as sound as the way we've worked it?" "The New Force at Hercules," 64.

138. Quoted in "Metamorphosis at Hercules," 14.

Chapter 11

1. The company was governed by eight operating and eight auxiliary departments, seven vice presidents, a thirteen-member board and eight-member executive committee chaired by Forster, and a five-member executive committee chaired by Edward Morrow. Operating departments are discussed at the end of chap. 10; the auxiliary departments were Advertising, Engineering, Legal, Medical, Personnel, Purchasing, Research, and Traffic. Hercules Incorporated, Annual Report for 1962; *Fortune* 68 (July 1963), 180.

2. *Wilmington Evening Journal* (July 17, 1962), 31.

3. Advertising Department Reports, First, Second, and Third Quarters 1962, Hall of Records, Hercules Incorporated.

4. "Our Golden Opportunity," *Hercules Mixer* 44 (October 1962), 18–22.

5. *Hercules Mixer* 45 (January 1962), 0–2.

6. Perrin Stryker, "Chemicals: The Ball Is Over," *Fortune* (October 1961), 125 ff.

7. William W. Bewley, Jr., interview, January 26, 1989.

8. William W. Bewley, Jr., interview, February 27, 1989.

9. David A. Hounshell and John Kenly Smith, Jr., *Science and Corporate Strategy: Du Pont R&D, 1902–1980* (New York, 1988), 509.

10. See Robert Sobel, *The Rise and Fall of the Conglomerate Kings* (New York, 1984); and Alfred D. Chandler, Jr., and Richard S. Tedlow, eds., *The Coming of Managerial Capitalism: A Casebook on the History of American Economic Institutions* (Homewood, Ill., 1985), 737–775.

11. Werner C. Brown interview, March 1, 1989; *Hercules Mixer* 46 (September 1964), 20–23.

12. Eugene D. Crittenden, Jr., interview, December 2, 1987; Bewley interview, February 27, 1989; Brown interview, March 1, 1989.

13. Crittenden interview, December 2, 1987; Board Minutes, April 1, 1964.

14. *Hercules Mixer* 46 (September 1964), 20–22; Board Minutes, August 3, 1964.

15. Haveg Industries, Annual Reports to the Hercules Executive Committee for 1964–1970.

16. For example, Hercules paid about $2.2 million for Glascote but sold it for approximately $560,000. Haveg Industries, Annual Reports to the Hercules Executive Committee for 1967–1970; S. Maynard Turk interview, October 10, 1986; and Executive Committee Minutes, August 8, 1966, and December 27, 1971.

17. Brown interview, March 1, 1989.

18. Bewley interview, February 27, 1989.

19. C&PP, E&CP, P&PC, Imperial Color and Chemical, and Synthetics departments, Annual Reports for 1962–1970.

20. U.S. Department of Commerce, Bureau of the Census, *Historical Statistics of the United States: Colonial Times to 1970,* vol. 1 (Washington, D.C., 1975), 210, 224.

21. Peter H. Spitz, *Petrochemicals: The Rise of an Industry* (New York, 1988), 375–376; M.Y. Yoshino, "Japan as Host to the International Corporation," in Charles P. Kindleberger, ed., *The International Corporation* (Cambridge, Mass., 1970), 345–347.

22. Spitz, *Petrochemicals,* 377.

23. Except where additional sources are cited, the following discussion of Teijin-Hercules is based on interviews with Dr. Harvey J. Taufen on September 21 and December 19, 1988.
24. Spitz, *Petrochemicals,* 378–381.
25. *Hercules Mixer* 44 (April 1962), 12.
26. "New DMT Plant in Japan," *Hercules Chemist* 47 (September 1963), 23; International Department, Annual Report for 1964.
27. International Department, Annual Reports for 1964–1973.
28. Ibid.
29. *Hercules Mixer* 47 (April 1965), 24.
30. Under this agreement, the World Bank supplied about $35 million for the project, Hercules $12 million, the Dawoods (a wealthy Pakistani family) $12 million, and the International Finance Corporation (the investing arm of the World Bank) $3 million; there was a small reserve of Agency for International Development (AID) money, held by the Pakistani government at no interest, that was to be made available to the venture if the plant operated satisfactorily. Taufen interview, December 19, 1988.
31. Taufen interviews, September 21 and December 19, 1988.
32. Warren F. "Sam" Beasley interview, March 1, 1989.
33. Ibid.
34. Henry Reeves interview, January 30, 1989.
35. "Hercules Bares Key to Growth Abroad," *Wilmington Morning News* (May 30, 1963); Joseph O. Bradford, "Hercules Circles the Globe," *Hercules Mixer* 55 (September 1967), 7–12.
36. See Mira Wilkins, *The Maturing of Multinational Enterprise: American Business Abroad from 1941 to 1970* (Cambridge, Mass., 1974), 350, 380–381.
37. Beasley interview, March 1, 1989.
38. Taufen interview, September 21, 1988.
39. Taufen interview, December 19, 1988.
40. *Hercules Mixer* 46 (July–August 1964), 3–5; and *Hercules Mixer* 54 (1972), 29–30.
41. *Hercules Mixer* 54 (1972), 29–30; Wilkins, *Maturing of Multinational Enterprise,* 334–336; Bradford, "Hercules Circles the Globe," 7–12.
42. Wilkins, *Maturing of Multinational Enterprise,* 334–401.
43. Hercules Incorporated, Annual Report for 1973; International Department, Annual Reports for 1962–1973; Wilkins, *Maturing of Multinational Enterprise,* 375.
44. Board Minutes, February 24, 1965; *Hercules Mixer* 47 (May 1965), 9–11.
45. Synthetics Department, Annual Report for 1966; Executive Committee Minutes, July 5, 1966, and March 23, 1970; Hercules Incorporated, Annual Reports for 1968, 1972, and 1973.
46. Hercules Incorporated, Annual Reports for 1971–1973; Executive Committee Minutes, November 8, 1971.
47. Hercules Incorporated, Annual Reports for 1969–1973.
48. *Hercules Mixer* 48 (July–August 1966), 2–9.
49. Hercules Mixer 50 (May–June 1968), 2–5; Karl Winnaker, trans. David Goodman, *Challenging Years: My Life in Chemistry* (London, 1972), 209–210, 369.
50. Winnaker, *Challenging Years,* 369–372; *Hercules Mixer* 48 (July–August 1966), 2–9.
51. *Hercules Mixer* 50 (May–June 1968), 2–5; Winnaker, *Challenging Years,* 372–373.
52. Brown interview, March 1, 1989.
53. Board Minutes, February 25, 1970. Payments were made on a deferred basis and included some interest on the loans to Hystron.
54. Brown interview, March 1, 1989.
55. *Hercules Mixer* 50 (May–June 1968), 2–5.

56. Synthetics Department, Annual Report for 1964; "Can DMT Duo Turn the Tables on TPA," *Chemical Week* (June 30, 1976), unpaginated clipping; Executive Committee Minutes, December 22, 1972.

57. Brown interview, March 1, 1989.

58. Hercules Incorporated, Annual Report for 1972; Executive Committee Minutes, February 2, 1972.

59. Brown interview, March 1, 1989.

60. Richard B. Douglas comment, Hercules History Advisory Committee Minutes, February 7, 1989.

61. Group interviews at Oxford plant, April 26–27, 1989.

62. C. Doyle Miller and Rodman B. Teeple interview, March 16, 1989.

63. Ibid.

64. Ibid.; Fibers Development Department, Annual Report for 1963; Fibers and Film Department, Annual Reports for 1965 and 1966; *Hercules Mixer* 48 (February 1966), 3.

65. Miller and Teeple interview, March 16, 1989; Hounshell and Smith, *Science and Corporate Strategy,* 268–274.

66. Richard B. Douglas interview, November 13, 1987; Advertising Department Reports, First Quarter 1965 and First Quarter 1966.

67. Miller and Teeple interview, March 16, 1989.

68. Advertising Department, Annual Reports for 1969 and 1971; Miller and Teeple interview, March 16, 1989.

69. James E. Knox interview, September 22, 1988.

70. Miller and Teeple interview, March 16, 1989; *Hercules Mixer* 51 (1969), 28.

71. Miller and Teeple interview, March 16, 1989; *Hercules Mixer* 51 (1969), 28; Hounshell and Smith, *Science and Corporate Strategy,* 441.

72. Polymers Department, Annual Report for 1962.

73. Fred L. Buckner interview, March 10, 1988; Donald W. Simonds interview, February 20, 1989; Polymers Department, Annual Report for 1964.

74. Polymers Department, Annual Reports for 1963 and 1964; Simonds interview, February 20, 1989; Buckner interview, March 10, 1988.

75. Buckner interview, March 10, 1988; and Fibers and Film Department, Annual Report for 1967.

76. Simonds interview, February 20, 1989.

77. Buckner interview, March 10, 1988.

78. Polymers Department, Annual Report for 1972.

79. Buckner interview, March 10, 1988.

80. Board Minutes, May 27, 1970; Brown interview, March 1, 1989; *Hercules Mixer,* 57 (1975), 28.

81. Werner C. Brown interview, October 8, 1986.

82. Brown interview, March 1, 1989.

83. "Brown Steering Hercules to Billion Mark," *Chemical Week* (July 15, 1970), 38; *New York Times,* November 1, 1970, 5.

84. *Wall Street Transcript* (July 17, 1972), 29248.

85. Ibid.

86. Brown interview, March 1, 1989.

87. The following discussion of BCG's approach is based on Richard G. Hamermesh, *Making Strategy Work* (New York, 1986), 9–16; and Arnoldo C. Hax and Nicholas S. Majluf, "The Use of the Growth-Share Matrix in Strategic Planning," in Arnoldo C. Hax, ed., *Readings on Strategic Management* (Cambridge, Mass., 1984), 61–75.

88. William W. Bewley, Jr., interview, May 23, 1989.

89. Lucien G. Maury became head of the NED in 1972, and Ross O. Watson in 1973.

90. NED, Annual Report for 1968.

91. NED, Annual Report for 1970.

92. Ibid.

93. NED Reports, December 1969 and December 1970.
94. NED, Annual Reports for 1970 and 1971; NED Report, Fourth Quarter 1972.
95. NED, Annual Report for 1970; University of Delaware, *University News* (Fall 1970), 6–8.
96. Douglas interview, November 13, 1987; Arden B. Engebretsen interview, May 4, 1988; NED, Annual Reports for 1972 and 1973.
97. NED, Annual Reports for 1970–1973.
98. Ibid.; "Data Source Corporation: A New Hercules Enterprise," *Hercules Mixer* 52 (June Supplement 1970), 20–23; "Short-Circuiting the Card Sharp," *Hercules Mixer* 55 (1973), 23–26; Frank H. Wetzel, Jr., interview, April 27, 1988.
99. NED, Annual Reports for 1970 and 1971.
100. NED, Annual Reports for 1971–1973; Engebretsen interview, May 4, 1988.
101. NED, Annual Report for 1973.
102. NED, Annual Report for 1970.
103. NED, Annual Reports for 1970–1973; *Hercules Mixer* 52 (June Supplement 1970); *Hercules Mixer* 53 (1971), 28–29.
104. NED, Annual Report for 1969; "A New Labor for Hercules—Building Modular Housing," *Hercules Chemist* 62 (July 1971), 30–32.
105. "New Labor for Hercules," 31.
106. Ibid.; *Hercules Mixer* 53 (1971), 11.
107. "New Labor for Hercules," 31–32.
108. Douglas interview, November 13, 1987; Hercoform Department Reports, September and December 1970; "New Labor for Hercules," 32.
109. Hercoform Department Reports, September and December 1970.
110. Ibid.; Hercoform Department, Annual Report for 1971.
111. Hercoform Department Reports, October–December 1972 and January–March 1973; *Hercules Mixer* 53 (1971), 11.
112. Engebretsen interview, May 4, 1988; Hercoform Department Reports, September 1970 and December 1971.
113. The integration of mass-production and mass-distribution in American history is described in Alfred D. Chandler, Jr., *The Visible Hand: The Managerial Revolution in American Business* (Cambridge, Mass., 1977), 285–376.
114. NED, Annual Report for 1973; *Hercules Mixer* 55 (1973), 32.
115. The plant was leased in 1976 and sold in 1979. R.G. Guenter, "Snowden Lightweight Aggregate Spurs Concrete Construction," *Hercules Chemist* 63 (December 1971), 18–24; Industrial Systems Department, Annual Reports for 1970–1973.
116. Polymers Department, Annual Report for 1973; Wetzel interview, April 27, 1988.
117. David S. Hollingsworth interview, October 8, 1986; Engebretsen interview, May 4, 1988; Brown interview, October 8, 1986.
118. Advertisement for Data Source, *Hercules Mixer* 54 (1972), back cover.
119. Because Monsanto's documented history lacks much financial data or critical analysis, the fate of that company's New Enterprise ventures is deduced from their minimal presence in later discussions. Dan J. Forrestal, *Faith, Hope and $5,000: The Story of Monsanto* (New York, 1977), 187–191, 210.
120. Hounshell and Smith, *Science and Corporate Strategy,* 509–540.
121. For example, conglomerates were examined by the Legal Department in 1966, and discussed by Henry Thouron in 1969. Legal Department Report, First Quarter 1966; *Hercules Mixer* (Fiftieth anniversary issue, 1969), 6.
122. Brown interview, October 8, 1986.
123. NED, Annual Report for 1971; Brown interview, October 8, 1986.
124. Brown interview, October 8, 1986.
125. Eugene D. Crittenden interview, May 2, 1989.
126. See Louis Galambos and Joseph Pratt, *The Rise of the Corporate Commonwealth:*

U.S. Business and Public Policy in the Twentieth Century (New York, 1988). The concept of countervailing power is developed in John Kenneth Galbraith, *The Affluent Society* (New York, 1958).

127. Martin V. Melosi, *Coping with Abundance: Energy and Environment in Industrial America* (New York, 1985), 295–298; Walter A. Rosenbaum, *The Politics of Environmental Concern,* 2d ed. (New York, 1977), passim.

128. Quoted in Frank Graham, Jr., *Since Silent Spring* (Greenwich, Conn., 1970), 14.

129. Rachel Carson, *Silent Spring* (New York, 1962), 128.

130. Ibid., 131.

131. Ibid., 131–132.

132. Medical Department, Annual Report for 1962; Legal Department Reports, Second and Third Quarters 1963.

133. John P. Frawley, "Process of Discovering and Developing a Marketable Pesticide" (1960), typescript in Hercules library.

134. Medical Department, Annual Reports for 1962–1964; Synthetics Department, Annual Report for 1964.

135. Medical Department, Annual Report for 1964; Charles Dunn interview, April 7, 1988.

136. Dunn interview, April 7, 1988.

137. Hercules Incorporated, Annual Report for 1970; Advertising Department, Annual Reports for 1962 and 1970; Advertising and Public Relations Department, Annual Report for 1972; Synthetics Department, Annual Report for 1969.

138. Medical Department, Annual Report for 1964.

139. The other members were Emil E. Christofano, an industrial hygienist, and John P. Frawley, the company's first toxicologist (Medical), C.D. Ender (Administrative), J.D. Floyd (C&PP), Thomas R. Hunt (Legal), W.T. Laffey (Engineering), T. Stegura and M. William Hawkesworth (Advertising and Public Relations), and Ed Crum (Executive). *Hercules Mixer* 52 (1970), 11–12; Richard E. Chaddock interview, February 22, 1989.

140. Chaddock interview, February 22, 1989.

141. Ibid.; M. William Hawkesworth, Jr., interview, February 28, 1989; Sports Foundation Awards (1973) booklet in Hercules library vertical files.

142. *Hercules Mixer* 52 (1970), 13; *Hercules Mixer* 54 (1972), 11; *Hercules Mixer* 55 (1973), 21; Hawkesworth interview, February 28, 1989.

143. Bruce W. Dickerson, "No Short Cut to Waste Disposal," *Hercules Chemist* 52 (April 1966), 16; *Hercules Mixer* 54 (1972), 9; Hercules Incorporated, Annual Report for 1973; Sports Foundation Awards booklet.

144. *Hercules Mixer* 51 (1969), 34; Edward J. Sheehy interview, March 23, 1988.

145. *Hercules Mixer* 51 (1969), 33–34; *Hercules Mixer* 49 (September–October 1967), 1–3; *Hercules Mixer* 52 (1970), 5.

146. Hercules owned 80 percent of the company and limited its investment to a maximum of $1.2 million. *Hercules Mixer* 52 (1970), 5; Chaddock interview, February 22, 1989; Executive Committee Minutes, March 29, 1971.

147. "Here's a Totally New Method of Sewage Treatment," reprint from *House and Home* (1972) in Hercules library vertical files; "About Hercules," *Hercules Chemist* 63 (December 1971); William H. Gardner interview, June 20, 1988.

148. Gardner interview, June 20, 1988; *Wilmington Morning News* (May 19, 1971); *Hercules Mixer* 52 (1970), 8.

149. *Hercules Mixer* 52 (1970), 8.

150. Gardner interview, June 20, 1988; *Hercules Mixer* 53 (1971), 10–12.

151. Medical Department, Annual Report for 1965.

152. Medical Department, Annual Reports for 1964 and 1966; Medical Department Reports, July–December 1972; John P. Frawley interview, June 20, 1988.

153. Chaddock interview, February 22, 1989.

154. Medical Department, Annual Reports for 1965 and 1969; Medical Department Reports, July–December 1970, January–June 1972, and January–June 1973; John P. Frawley interview, March 16, 1989.
155. Forrestal, *Faith, Hope and $5,000,* 195; Ralph Nader and William Taylor, *The Big Boys: Power and Position in American Business* (New York, 1986), 143–195.
156. Sheldon Hochheiser, *Rohm and Haas: History of a Chemical Company* (Philadelphia, 1986), 167–177; Hounshell and Smith, *Science and Corporate Strategy,* 570.
157. Medical Department Reports, July–December 1972, and January–June 1973.
158. Medical Department Reports, July–December 1972.
159. Medical Department Reports, January–June 1970.
160. Polymers Department, Annual Report for 1964.
161. Michael Krueger, address before Boston Security Analysts Society, February 7, 1972, quoted in *Wall Street Transcript* (March 6, 1972), 27504.
162. Quoted in *Wall Street Transcript* (July 17, 1972), 29249.
163. Brown interview, March 1, 1989.

Chapter 12

1. Hercules Incorporated, Annual Report for 1974.
2. Keith T. Smith interview, February 27, 1989; Robert J. Leahy interview, April 27, 1988; Organics Department Reports, Fourth Quarters of 1974 and 1975.
3. Arden B. Engebretsen interview, May 4, 1988.
4. Hercules Incorporated, Annual Report for 1975.
5. David S. Hollingsworth interview, October 8, 1986; see also Frank H. Wetzel, Jr., interview, April 27, 1988.
6. Werner C. Brown to Carolyn R. Miller, July 20, 1989 (copy in authors' possession.); Werner C. Brown interview, March 1, 1989; Brown comment, Hercules History Advisory Committee Minutes, July 27, 1989.
7. Alexander F. Giacco interview, December 15, 1987.
8. Brown to Miller, July 20, 1989; Brown interview, March 1, 1989; Brown comment, Hercules History Advisory Committee Minutes, July 27, 1989; Harvey G. Taufen interviews, September 21 and December 19, 1988.
9. "Can DMT Duo Turn the Table on TPA," *Chemical Week* (June 30, 1976), 19–20. SRI International, *Chemical Economics Handbook* (August 1983), 543.1000A.
10. Alexander F. Giacco interview, August 30, 1988; Richard B. Douglas interview, November 13, 1987; Brown interview, March 1, 1989; Brown to Miller, July 20, 1989; Brown comment, Hercules History Advisory Committee Minutes, July 27, 1989.
11. Giacco interview, December 15, 1987; "Hercules, Petrofina Sign DMT/TPA Agreement," *Chemical Age* (June 18, 1976), unpaginated clipping. Hercules' apparent risk in occupying the middle ground between the oil companies and the fiber producers was noticed as early as 1954, when it was preparing to produce its first commercial DMT. "One to Make Ready . . . ," *Chemical Week* (January 30, 1954), 64.
12. A.F. Giacco, speech to annual stockholders' meeting, March 20, 1979.
13. Fred L. Buckner interview, March 10, 1988.
14. James E. Knox interview, September 22, 1988.
15. Hercules Incorporated, Annual Report for 1977.
16. Hercules Incorporated, Annual Report for 1976.
17. The following paragraphs draw heavily on the interviews with Giacco on December 15, 1987, and August 30, 1988.
18. Giacco had first proposed a matrixlike structure to Brown in 1974, but the plan was rejected then as too radical. Giacco interview, December 15, 1987.
19. Giacco interviews, December 15, 1987, and August 30, 1988; Theodore M. Bednarski interview, February 10, 1988. Bob Leahy described the confusion in this way: "In 1977, when we reorganized into the matrix, that was an interesting event. I think I took over . . . in October; we had the board meeting in Sea Island, when he [Giacco] was

given the job. I went out the following month to the West Coast and had dinner with George Weyerhaeuser. George had a thing about reorganizing. Every time I called at Weyerhaeuser I had to figure out what their new organization was. One time the manager was making the decision of what they are going to be doing; the next time the decisions would be made centrally. So I told him, 'George, you reorganize more than anybody I know. Hercules has never reorganized before in the true sense. Now we are going to the matrix system.' He said, 'Oh my God. You are going to need a consultant.' I said, 'Now George, we don't want a consultant. Consultants are guys that can't find work. We really don't want anybody coming in messing up our place and walking away. We've got to do this ourselves.' He said, 'I don't mean a management consultant. I agree with you on that. You're going to need an industrial psychologist.' " Leahy interview, April 27, 1988.

20. For example, Arden Engebretsen speculates that a decision such as the one to build the Wateree plant might never have been made if Hercules had operated under the matrix structure at the time. Engebretsen interview, May 4, 1988; see also Bednarski interview, February 10, 1988; Edward J. Sheehy interview, March 23, 1988; Wetzel interview, April 27, 1988.

21. For an alternative view in defense of the departmental structure, see Brown interview, March 1, 1989.

22. Leahy interview, April 27, 1988.

23. Wetzel interview, April 27, 1988.

24. Eugene D. Crittenden, Jr., interview, May 3, 1989.

25. Reginald Ivett interview, December 2, 1987; Giacco interview, December 15, 1987.

26. For example, see David A. Hounshell and John Kenly Smith, Jr., *Science and Corporate Strategy: Du Pont R&D, 1902–1980* (New York, 1988), chap. 25.

27. Bednarski interview, February 10, 1988.

28. "Hercules to End Old 'Bureaucracies,' " *Wilmington News Journal* (October 9, 1977), 18; "Hercules: Excising the Losers in a Campaign to Consolidate," *BusinessWeek* (April 3, 1978), 95. This trend was accelerated by Hercules' simultaneous decision to make extensive use of computers and telecommunications networks to reduce order processing and other paperwork and help control overhead costs. See, "Hercules' Worldwide Data Processing Network Assures Cost-effective Transport Management," *Traffic Management International* (Spring 1978), 21–25.

29. Wetzel Interview, April 27, 1988.

30. The system never lived up to its billing, however, and became something of an in-house joke among top management insiders. See D. James MacArthur interview, May 2, 1989.

31. A.F. Giacco, speech to European Financial Community, May 1983. Copy in possession of William W. Bewley.

32. Brown comment, Hercules History Advisory Committee Minutes, July 27, 1989; Joseph L. Bower, *When Markets Quake: The Management Challenge of Restructuring Industry* (Boston, 1986), 38–39, 42–43; Giacco, speech to European Financial Community, May 1983; Giacco, speech to annual stockholders' meeting, March 26, 1985; William Bewley interview, February 27, 1989.

33. Giacco, speech to European Financial Community, May 1983.

34. Giacco, speech to annual stockholders' meeting, March 20, 1979; R. Michael Hendricks interview, March 17, 1989.

35. Crittenden interview, May 3, 1989; Douglas interview, November 13, 1987; Giacco, speech to European Financial Community, May 1983. According to Douglas, three of the fourteen casualties in the Advertising Department would have left the company within six months of the decision. See also, E. D. Crittenden, Jr., letter to the editor, *Wilmington Evening Journal* (April 1, 1980); Harold Seneker, "Mr. Nice Guy He Wasn't," *Forbes* (March 31, 1980), 116.

36. R.G. Eccles and J. Linder, "Hercules Incorporated: Anatomy of a Vision" (Boston: Harvard Business School Case Services, 186–305), 22.

37. Giacco, speech to annual stockholders' meeting, March 20, 1979; "Opening Remarks," "Introduction," and "Investment Profile—1984," in Hercules Incorporated, Financial Analysts Seminar, Wilmington, November 29, 1979; Hercules Incorporated, Annual Report for 1979.

38. Brown and Giacco often published their views on value-adding: see Werner C. Brown and Alexander F. Giacco, "Hercules Incorporated: A Study in Creative Chemistry" (New York: Newcomen Society in North America, 1977); Giacco, "Hercules Value-Added Concept," speech to Financial Analysts of Wilmington, November 17, 1976; Hercules Incorporated, Annual Report for 1979; see also, Giacco interviews, December 15, 1987, and August 30, 1988.

39. Henry A. Schowengerdt interview, April 26, 1988.

40. Ruth L. Novak interview, June 21, 1988.

41. Peter H. Schuck, *Agent Orange on Trial: Mass Toxic Disasters in the Courts* (Cambridge, Mass., 1987), 45.

42. Douglas interview, November 13, 1987.

43. Schuck, *Agent Orange on Trial,* 33; S. Maynard Turk interview, October 10, 1986; Turk comment, Hercules History Advisory Committee Minutes, February 7, 1989.

44. Synthetics Department, Annual Report for 1968, and for 1961–1971, generally. In 1971, the plant was leased to Vertac Chemical Corporation, which acquired it several years later.

45. Schuck, *Agent Orange on Trial,* passim, esp. chap. 12.

46. Ibid., 16–20.

47. John P. Frawley interview, March 16, 1989; Turk interview, October 10, 1986; Schuck, *Agent Orange on Trial,* 86, 100.

48. Schuck, *Agent Orange on Trial,* chaps. 7 and 8, esp. 114–115, 166.

49. Turk interview, October 10, 1986; Schuck, *Agent Orange on Trial,* 156, 159; "Hercules Settles Agent Orange," *Wilmington News Journal* (May 8, 1984), 1, 10.

50. Douglas interview, November 13, 1987.

51. Giacco, "Roles and Responsibilities in Our Technological Society," *Chemtech* (November 1986), 650–653; Werner C. Brown interview, October 8, 1986; Turk interview, October 10, 1986.

52. Sheldon Hochheiser, *Rohm and Haas: History of a Chemical Company* (Philadelphia, 1986), 162–164; Peter H. Spitz, *Petrochemicals: The Rise of an Industry* (New York, 1988), chaps. 12 and 13; Bower, *When Markets Quake,* passim.

53. Hounshell and Smith, *Science and Corporate Strategy,* chap. 25; Graham D. Taylor and Patricia E. Sudnik, *Du Pont and the International Chemical Industry* (Boston, 1984), 205.

54. Bower, *When Markets Quake,* 36, and chap. 2, generally.

55. Giacco interview, December 15, 1987.

56. "Giacco Reshuffles 'Old Gang' at the Top," *Wilmington Evening Journal* (November 30, 1979), 18.

57. Giacco interviews, December 15, 1987 and August 30, 1988; Giacco, speech to European Financial Community, May 1983.

58. Buckner interview, March 10, 1988.

59. Novak interview, June 21, 1988.

60. Ernest A. Mettenet interview, June 22, 1988.

61. Jon B. DeVault interview, April 6, 1988.

62. Steven E. Russell interview, June 22, 1988.

63. DeVault interview, April 6, 1988.

64. Giacco interview; December 15, 1987; Schowengerdt interview, April 26, 1988; David E. Thompson interview, June 22, 1988. According to Hercules' Annual Report for 1983, another motive for the acquisition was the belief that Simmonds would help the paper chemicals group provide systems to the paper industry.

65. Giacco interview, December 15, 1987.

66. Ibid.

67. Buckner interview, March 10, 1988.

68. C. Doyle Miller and Rodman B. Teeple interview, March 16, 1989; W.J. Fisher and J.G. Smith interview, April 26, 1989.

69. Buckner interview, March 10, 1988; Miller and Teeple interview, March 16, 1989; Eccles and Linder, "Hercules Incorporated," 8.

70. Giacco interview, December 15, 1987.

71. David S. Hollingsworth interview, September 1, 1988.

72. G. DiDrusco, "The Spheripol Process," HIMONT Inc., Financial Analysts Presentation, Ferrara, Italy, September 19, 1984, 24–34; Hendricks interview, March 17, 1989; Bewley interview, February 27, 1989; F.J. Aguilar, J.L. Bower, and B. Gomes-Casseres, "Montedison, S.p.A. (A)," (Boston: Harvard Business School Case Services, 385–065), abridged in Bower, *When Markets Quake,* chap. 9.

73. Aguilar, Bower, and Gomes-Casseres, "Montedison, S.p.A. (A)," 18.

74. Ibid.; Bower, *When Markets Quake,* 160.

75. Aguilar, Bower, and Gomes-Casseres, "Montedison, S.p.A. (A)," 18; Bower, *When Markets Quake,* 160. Giacco's version of the story is not incompatible, although he emphasizes different details: "Maybe [relating some of the] conversation between me and Mario Schimberni will sort of edify this [situation]. . . . We were getting ready to try and put together HIMONT and we kind of decided that, yeah, we'd put the plastics business together, but Mario says to me, 'Why don't you put your fiber and film businesses together?' I said, 'Well, why don't you put the drug business in?' Because we only had a piece of it that we were going to get payment for. He says, 'Al, are you going to take my crown jewels, they'll laugh at me in Italy!' I said, 'Well, you know, we're interested in drugs.' He said, 'You are?' I said, 'Yeah.' He said, 'Would you be willing to spend 15 percent of your sales on research?' I said, 'You're crazy!' He said, 'You're not interested in drugs!' It's a completely different culture. Okay, so you gotta fish, and you gotta fish for quite a while. Now, the whole function of Adria was oncology, the cancer drugs, cancer therapies, and they have a hell of an R&D on this thing." Giacco interview, December 15, 1987; see also Giacco interview, August 30, 1988.

76. Hendricks interview, March 17, 1989; Bower, *When Markets Quake,* 161.

77. Hercules Incorporated, Annual Reports for 1983; Aguilar, Bower, and Gomes-Casseres, "Montedison, S.p.A. (A)," 18–20; Bower, *When Markets Quake,* 160–161.

78. Hollingsworth interview, September 1, 1988.

79. *Hercules News* (September 9, 1983), 1; Crittenden interview, May 3, 1989.

80. Hercules Incorporated, Annual Report for 1984; Eccles and Linder, "Hercules Incorporated," 3.

81. Mettenet interview, June 22, 1988.

82. Eccles and Linder, "Hercules Incorporated," 5.

83. Hercules Incorporated, Annual Report for 1985.

84. Knox interview, September 22, 1988.

85. Hercules Incorporated, Annual Report for 1983.

86. Thomas L. Gossage interview, September 22, 1988.

87. Hercules Incorporated, Annual Report for 1980; Robert G. Fajans, "Hercules Pinex Program," *Naval Stores Review* 90 (November–December 1980), 11–13 (see chap. 6, note 69); Smith interview, February 27, 1989.

88. MacArthur interview, May 2, 1989; Gossage interview, September 22, 1988.

89. MacArthur interview, May 2, 1989.

90. Giacco interview, December 15, 1987.

91. Wetzel interview, April 27, 1988.

92. Wetzel interview, April 27, 1988; See also Crittenden interview, May 3, 1989.

93. Crittenden interview, May 3, 1989; "Some New Players in Water-Soluble Polymers," *Chemical Week* (June 4, 1986).
94. Hollingsworth interview, September 1, 1988; Crittenden interview, May 3, 1989; Hendricks interview, March 17, 1989.
95. Hercules Incorporated, Annual Report for 1987.
96. D.S. Hollingsworth, Vision Statement, January 25, 1989, printed in Hercules Incorporated, Annual Report for 1988.
97. D.S. Hollingsworth, speech to annual stockholders' meeting, March 22, 1988.
98. D.S. Hollingsworth, Teleprompter Speech, January 25, 1989.
99. "Research and Development Provided Options for Vision," *Hercules Horizons* (March 3, 1989), 1–2.
100. The programs were the Hercules PROACT Safety Management System (1987) for line managers, supervisors, and employees across the company, and "We Care!" (1988), which focuses on concern for the environment as "a moral obligation." Maintaining that "excellence in quality is the single, most important strategic issue facing Hercules in the years ahead," Hollingsworth also formed a corporate quality council and promoted quality awareness programs for all employees. To direct these activities, in January 1989, Frank Wetzel was appointed to a new position, vice president of corporate quality, the first such corporate officer in Hercules' history. Hollingsworth interview, September 1, 1988; Hollingsworth, speech to annual stockholders' meeting, March 22, 1988; Hercules Incorporated, Annual Reports for 1987 and 1988.
101. Hercules Incorporated, Annual Report for 1988.
102. "Hercules Announces Corporate Restructuring," and "Hollingsworth Responds to Reorganization Questions," *Hercules Horizons* (September 8, 1989), 1–3.
103. "Hollingsworth Responds to Reorganization Questions"; "Reorganization Moves to Next Phase," *Hercules Horizons* (September 15, 1989), 1.
104. Hollingsworth interview, September 1, 1988.

Index

COLOPHON

The LABORS OF A MODERN HERCULES was composed in Galliard type by Achorn Graphic Services on a Linotron 202N. Galliard was designed by Matthew Carter in 1978 for exclusive use in phototypesetting. Offset preparation, including halftone camera work, was done by Jay's Publishers Services, Inc. The book was printed and bound by the Courier Corporation in their Kendallville, IN and Westford, MA plants respectfully, using Glacier Opaque paper manufactured by Cross Pointe Paper Corporation using Hercules Hercon 70 as the alkaline sizing agent. This acid free characteristic will give the book an effective life of at least three hundred years. The Roxite book cloth was supplied by the Holliston Mills.

HERCULES